# Employing Recent Technologies for Improved Digital Governance

Vasaki Ponnusamy
*Universiti Tunku Abdul Rahman, Malaysia*

Khalid Rafique
*Azad Jammu and Kashmir Information Technology Board, Pakistan*

Noor Zaman
*Taylor's University, Malaysia*

A volume in the Advances in Electronic Government, Digital Divide, and Regional Development (AEGDDRD) Book Series

IGI Global
DISSEMINATOR OF KNOWLEDGE

Published in the United States of America by
IGI Global
Information Science Reference (an imprint of IGI Global)
701 E. Chocolate Avenue
Hershey PA, USA 17033
Tel: 717-533-8845
Fax: 717-533-8661
E-mail: cust@igi-global.com
Web site: http://www.igi-global.com



Library of Congress Cataloging-in-Publication Data

Names: Ponnusamy, Vasaki, 1974- editor. | Rafique, Khalid, 1973- editor. | Zaman, Noor, 1972- editor.
Title: Employing recent technologies for improved digital governance / Vasaki Ponnusamy, Khalid Rafique, Noor Zaman, editors.
Description: Hershey, PA : Information Science Reference, [2020] | Includes bibliographical references and index. | Summary: "This book explores recent advances in the development, application, and impact of technologies in digital governance"-- Provided by publisher.
Identifiers: LCCN 2019034106 (print) | LCCN 2019034107 (ebook) | ISBN 9781799818519 (h/c) | ISBN 9781799818526 (s/c) | ISBN 9781799818533 (eISBN)
Subjects: LCSH: Public administration--Data processing. | Municipal services--Data processing. | Internet in public administration.
Classification: LCC JF1525.A8 E5857 2020 (print) | LCC JF1525.A8 (ebook) | DDC 352.3/80285--dc23
LC record available at https://lccn.loc.gov/2019034106
LC ebook record available at https://lccn.loc.gov/2019034107

This book is published in the IGI Global book series Advances in Electronic Government, Digital Divide, and Regional Development (AEGDDRD) (ISSN: 2326-9103; eISSN: 2326-9111)

British Cataloguing in Publication Data
A Cataloguing in Publication record for this book is available from the British Library.



For electronic access to this publication, please contact: eresources@igi-global.com.

# Advances in Electronic Government, Digital Divide, and Regional Development (AEGDDRD) Book Series

Zaigham Mahmood
University of Derby, UK & North West University, South Africa

ISSN:2326-9103
EISSN:2326-9111

## Mission

The successful use of digital technologies (including social media and mobile technologies) to provide public services and foster economic development has become an objective for governments around the world. The development towards electronic government (or e-government) not only affects the efficiency and effectiveness of public services, but also has the potential to transform the nature of government interactions with its citizens. Current research and practice on the adoption of electronic/digital government and the implementation in organizations around the world aims to emphasize the extensiveness of this growing field.

**The Advances in Electronic Government, Digital Divide & Regional Development (AEGDDRD)** book series aims to publish authored, edited and case books encompassing the current and innovative research and practice discussing all aspects of electronic government development, implementation and adoption as well the effective use of the emerging technologies (including social media and mobile technologies) for a more effective electronic governance (or e-governance).

## Coverage

- Knowledge Divide, Digital Divide
- Issues and Challenges in E-Government Adoption
- Social Media, Web 2.0, and Mobile Technologies in E-Government
- Citizens Participation and Adoption of E-Government Provision
- Adoption of Innovation with Respect to E-Government
- Case Studies and Practical Approaches to E-Government and E-Governance
- ICT Infrastructure and Adoption for E-Government Provision
- ICT within Government and Public Sectors
- Electronic Government, Digital Democracy, Digital Government
- Frameworks and Methodologies for E-Government Development

IGI Global is currently accepting manuscripts for publication within this series. To submit a proposal for a volume in this series, please contact our Acquisition Editors at Acquisitions@igi-global.com or visit: http://www.igi-global.com/publish/.

The Advances in Electronic Government, Digital Divide, and Regional Development (AEGDDRD) Book Series (ISSN 2326-9103) is published by IGI Global, 701 E. Chocolate Avenue, Hershey, PA 17033-1240, USA, www.igi-global.com. This series is composed of titles available for purchase individually; each title is edited to be contextually exclusive from any other title within the series. For pricing and ordering information please visit http://www.igi-global.com/book-series/advances-electronic-government-digital-divide/37153. Postmaster: Send all address changes to above address. 

## Titles in this Series

*For a list of additional titles in this series, please visit: www.igi-global.com/book-series*

***Political Decision-Making and Security Intelligence Recent Techniques and Technological Developments***
Luisa Dall'Acqua (University of Bologna & LS TCO, Italy) and Irene M. Gironacci (University of Swinburne, Australia)
Information Science Reference • ©2020 • 170pp • H/C (ISBN: 9781799815624) • US $195.00

***Tools, Strategies, and Practices for Modern and Accountable Public Sector Management***
Graça Azevedo (University of Aveiro, Portugal) Jonas Oliveira (Instituto Universitário de Lisboa (ISCTE-IUL), Portugal) Rui Pedro Marques (University of Aveiro, Portugal) and Augusta Ferreira (University of Aveiro, Portugal)
Information Science Reference • ©2020 • 372pp • H/C (ISBN: 9781799813859) • US $215.00

***Examining the Roles of IT and Social Media in Democratic Development and Social Change***
Vikas Kumar (Chaudhary Bansi Lal University, India) and Geetika Malhotra (Chaudhary Bansi Lal University, India)
Information Science Reference • ©2020 • 380pp • H/C (ISBN: 9781799817918) • US $195.00

***Leveraging Digital Innovation for Governance, Public Administration, and Citizen Services Emerging Research and Opportunities***
Nidhi Vij Mali (University of Mississippi, USA)
Information Science Reference • ©2020 • 194pp • H/C (ISBN: 9781522554127) • US $195.00

***Learning Cities, Town Planning, and the Creation of Livelihoods***
Idowu Biao (University of Botswana, Botswana)
Information Science Reference • ©2019 • 312pp • H/C (ISBN: 9781522581345) • US $215.00

***Global Campaigning Initiatives for Socio-Economic Development***
Luísa Cagica Carvalho (Instituto Politécnico de Setúbal, Portugal & Universidade de Évora, Portugal) and Maria José Madeira (University of Beira Interior, Portugal)
Information Science Reference • ©2019 • 325pp • H/C (ISBN: 9781522579373) • US $195.00

***Constitutional Knowledge and Its Impact on Citizenship Exercise in a Networked Society***
Ana Melro (University of Aveiro, Portugal) and Lídia Oliveira (University of Aveiro, Portugal)
Information Science Reference • ©2019 • 332pp • H/C (ISBN: 9781522583509) • US $195.00

***Private Sector Innovations and Technological Growth in the MENA Region***
Maryam Ebrahimi (Azad University, Iran)
Information Science Reference • ©2019 • 217pp • H/C (ISBN: 9781522570868) • US $195.00

IGI Global
DISSEMINATOR OF KNOWLEDGE

701 East Chocolate Avenue, Hershey, PA 17033, USA
Tel: 717-533-8845 x100 • Fax: 717-533-8661
E-Mail: cust@igi-global.com • www.igi-global.com

*Editors dedicated this book to their respective families.*

# Editorial Advisory Board

# Table of Contents

# Detailed Table of Contents

Information communication technology plays a vital role in countries' economic growth, and at the same time, it changes the form of traditional government to digital government. Digital governance is about having a competent and responsible government to deliver efficient, better, and 24/7 public services. It improves citizen participation, transparency, and overcomes administrative bottlenecks. Slackness of any government organisation in providing quality services will instantly become a topic of debate on print, electronic, and social media, forcing the political leaders to take corrective actions to save their own skin. However, for developing countries, this is going to be difficult to attain the same level of efficiency and flexibility. This study will examine the state of digital governance in developing countries particularly in the Pakistan government and will highlight the facts that through the implementation of digital governance in Pakistan (i.e., challenges and difficulties as well as opportunities that will help us to promote the application of digital governance).



ICT has been used as a model for education in many parts of the world. At the end of the conflict in 2009, the Government of Sri Lanka embarked on an e-Sri Lanka initiative aimed at introducing ICT across several sectors. The education sector received substantial investment in ICT, but only a section of this sector was targeted. This study explores the barriers of introducing and implementing ICT. The analysis of the case studies together with the variables identified from the literature review provided the basis for developing the conceptual framework for the study. The target population for this study were teachers and principals in the rural schools of Trincomalee representing the rural population in Sri Lanka. A combined sample of 70 teachers and principals were considered for this survey across 20 schools in the district. The results indicate that internet access, resources, personnel, and security are essential for introducing and implementing ICT education in the rural areas.



With the ubiquitous low-cost sensor devices and widespread use of IoT, the paradigm is shifted from urban areas towards a smart city. A smart city is an urban area that uses IoT technologies to collect data and manage resources efficiently. The vision is to improve the capabilities and to solve the citizens' problems (e.g., energy consumption, transportation, recycling, intelligent security, etc.) in an efficient way. A smart city is a multidimensional term including a smart economy, smart mobility, smart living, smart environment, smart people, and smart governance. Although the concept of a smart city is increasing and currently there exist many such cities in many developed countries, one of the key challenges faced by these cities is good governance. Smart cities need smart governance to run the city in a smarter way, and effective digital governance is a solution to this end. Digital governance refers to the use of digital technology in government practices.



Information and communication technologies (ICTs) know a steady evolution that leads to the development of digital technologies. These affect human lives and organizations' activities. In particular, they have become ubiquitous in the public sector almost throughout the world. The extensive use of ICTs in public administration (PA) contributes to the emergence of digital government. It is perceived as a means of public value creation and represents a part of a paradigm shift in public management. This chapter intends to review the milestones of PA reforms in a developing country, namely Tunisia, to highlight how it is using ICTs to create public value. A content analysis of data on public sector reforms and e-government strategies is performed. The results show progressive ICTs deployment by Tunisian PA over time. They illustrate a search to public value creation through the digital government.



The acronym IoT stands for internet of things. The IoT ecosystem can be envisioned as a set of physical electronic devices embedded with intelligence, connected through a network, enabling them to collect and exchange data, and allowing these devices to be sensed and handled remotely between the physical and cyber worlds. The devices connected through the internet has been influencing all walks of our life ranging from individual, societal, educational, industrial, entrepreneurial, and related to governance as well. As we are connected and surrounded with a plethora of connected smart devices, it seems there is a great risk of security and privacy in several aspects, such as device authentication, data theft, device manipulation, data falsification, etc., to name a few. Hence, the current chapter has been undertaken to explore and comprehend the security and privacy related implications, opportunities, future directions, and challenges involved in implementing digital governance with IoT.



Information and communication technology (ICT) has a great role in the development, growth, and economy of the country. ICT refers to any system, device, or product that can store, manipulate, and receive and transmit digital data effectively. ICT brings clear differences in the economy of countries wherever it is successfully implemented. It becomes a reason of timely, accurate, and efficient operations and becomes a source of user satisfaction by providing them fast, better, easy service around the clock. As timely access of information and communication plays a vital role in the success of any organization either public or private, many developed countries are already competing in digitalization. It is not only convenience, but it is the prime ingredient of good governance and an indispensable part of modern life. This chapter discusses the successful implementation of ICT infrastructures in the context of good governance and builds a vibrant society, the main agenda of Vision 2030 of KSA.



Transportation is an essential part of human activity that allows people to move from one place to other places. Modern transportation systems refer to intelligent transportation systems (ITS). It has been proven that information and communication technology (ICT) has a significant role to improve the quality and security transportation services. ITS can be described as an integration of computer, electronics, and communication technologies and management strategies that can increase the efficiency and safety of transportation. The application is a platform that manages the vehicles, drivers, passengers, road operators, and managers in an environment that is suitable for everyone to communicate and interact. This chapter presents the history of transportation systems and the role of ICT to bring ITS in different transportation modes. The main motivation of the chapter is to survey the various technology-enabled ITS systems that have been in place, analyzing their strengths and providing recommendations for such ITS systems in developing countries to improve the governance.



The adoption of social media by large businesses has influenced small medium enterprises (SMEs) to be part of this phenomenon. Evaluating the results of the online efforts is a prerequisite to the success of social media adoption. As more branding activities move online, marketers are faced with the realisation that brands and businesses are not always relevant in all the social media facets. The study is aimed at solving the question of how SMEs in Malaysia are evaluating their online marketing efforts. The quantitative survey was deployed across six states in Peninsular Malaysia to get feedback from 125 active firms from different size, sector, year of establishment, and period of use of social media. The pattern of all these participants conducting the evaluation phase on the social media platform is descriptively examined from quantitative findings. It is used to formulate new information into literature and body of knowledge. Research implications in the practical and social form are also discussed.

Raising awareness on social engineering is becoming more essential. When we are connected to the internet, we are at the risk of becoming victims of numerous attacks. It is very difficult to protect ourselves from all of these threats, but the least we could do is not jeopardize our own safety. In order to achieve this, the need to have substantial knowledge on social engineering techniques and how to safeguard one from being victimized are mandatory. This research presents the overview of social engineering attacks. This plays a crucial role in supporting victims of cyber assaults.



This chapter intends to provide a review of cooperation between public and private sectors towards cybersecurity governance. With the partnership, government can have confidence towards the safety and protection of their national critical digital infrastructure. The goal of this chapter is achieved by analyzing some of the cybersecurity frameworks adopted by the developed and developing nations. The analysis is further carried out by investigating the public-private policy initiatives in their national cybersecurity framework. The chapter also investigates the effectiveness of the National Institute of Standards and Technology Cybersecurity Framework (NIST) adopted by the US government.



In any democratic nation, electoral processes pave the way to democratic governance. In regards to this, a fully digitized online voting system should be introduced, and governments should invest efforts in studying the possibilities in adopting online voting system to curb the issues mentioned above. In that sense, any user should be allowed to cast his/her vote from any remote location, and the processes should be held in full confidentiality without any intervention or possible vulnerabilities. Therefore, this chapter investigates how governments can learn various initiatives and provide funding to further conduct research and implement such methodologies. The chapter begins by exploring some traditional and current practice of electoral system and further embarks into a review of secured online voting systems that has been proposed by the body of knowledge. The chapter then presents how some countries have adopted this secured online voting systems and giving an avenue to secured digital electoral governance.

The purpose of this chapter is to analyze the effect of FinTech and political incertitude on economic growth through a multiple regression. Thus, the authors employ the method of generalized least square (GLS) with panel data. The sample concerns 21 African countries during (2001-2014-2017). The authors use a wide range of measures from Global Findex Database 2017, the World Bank platform, the World Bank national accounts data, and the OECD National Accounts data files base in the context of Africa. Empirical results show that FinTech is a driver of economic growth unless it is actively used in a developed digital infrastructure. In fact, the authors prove that, when financial technologies are used in both transactions (receive and made digital payment), they significantly contribute to the economic cycle. Passive use like simple consumption actions are not a significant lever for the economy. The principal contribution is to highlight that the active use of financial innovations and not passive one and the developed digital infrastructure do promote economic growth in African countries.



The delivery of healthcare in remote rural areas is a global phenomenon. Using digital governance, a discretized cluster of node relaying time-sensitive, and reliable data, life-critical health services can be ensured. Here tactile internet-based internet of things (IoT) and its variant internet of medical things (IoMT), as well as tactile internet, is introduced. Its relationship with 5G is explored. A model for e-health implementation using unmanned autonomous vehicles (UAVs) in the backhaul is presented. The link is analyzed over fiber optics, satellite, and 5G network in terms of latency and reliability: the crucial metrics for robotic surgery. It is also shown that 5G implemented in Italy, Zimbabwe, and Cooks Island have supported mobile hospitals in rural regions. Hence, induction of 5G networks is a feasible approach. Finally, a critical review of expected future technologies including telematics is presented. Hence, tactile internet-based robotic surgery can reduce the digital divide.



In the context of digital governance, manmade activities will have the most impact on the environment. The concept of internet of things (IoT) is much more profound, supposedly having an impact on the way environmental sampling of data takes place. The cost benefits of unmanned aerial vehicles (UAVs) as alternate to satellite or high-altitude platform system (HAPS) is established as these are more flexible, mission specific, and versatile. UAV in sensor network context and communication relaying as well as data harvesting application is explored. Proposal to upgrade environmental laws and implementing network infrastructure security and cyber insurance for HetNets involving UAVs and environmental monitoring will be covered as well.

# Foreword

Digital-Governance or E-Governance is a term that refers to the interface between the public, ICT ecosystem and local government to provide fast and efficient services via online mediums. According to The World Bank (2002), e-governance is ''the use by government agencies of information technologies that have the ability to transform relationships with citizens, businesses, and other arms of government.'' Although this term might have existed for long and a lot of work has been carried out in the ICT Ecosystem and governance, the trends in the Internet is growing rapidly from cloud computing, FOG computing, Internet of Things (IoT), Big data analytics and more advanced techniques using Artificial Intelligence. Many smart concepts are either being implemented partially or in the line like smart cities, smart homes, smart devices, which is a new trend that combines the ideas from IoT, cloud computing, big data and many more. All these new trends should contribute to the public or community instead of just running as independent systems. This book encourages the integration of new trends in ICT into digital governance whereby local government should improve the service delivery through smart technology. I believe that this book has a dual objective such as, educating the academic and research community more on digital governance and encouraging more research in deploying recent trends in collaborative digital governance.

The ultimate objective of this volume is to provide the academic, industrial and local government communities with recent advances in development, application and impact of technologies for the initiative of E-Governance. Further this will assist the academics and industrial communities with up to date information related to the recent technologies for improved digital governance, mainly focusing on several areas of daily life and research. In addition, this book will encourage more research in this domain.

*Mamoona Humayun*
*Jouf University, Saudi Arabia*

## REFERENCES

World Bank. (2002). *E-Governments*. Retrieved from: http://documents.worldbank.org/curated/en/527061468769894044/266390WP0E1Gov1gentina1Final1Report.doc#:~:targetText=E%2Dgovernment%20refers%20to%20the,and%20other%20arms%20of%20government

# Preface

The unprecedented progressive development in digital technologies electronic and information processing capacities- in recent decades is unfolding an emerging ICT ecosystem shaping up the digital world the dividends of which are widespread in terms of knowledge sharing, effective and efficient information sharing, smart businesses and digital governance. The public at large has acknowledged the dividends of digital governance in terms of transparency, efficiency and effectiveness. Thus, the governments around the world are fast deploying ICT technologies to re-engineer their governance capacities to meet the challenges of this digital era. Thus, Digital Governance refers to the deployment of emerging ICT tools to re-engineer the working models of traditional administration, economy and public service delivery. Although digital governance or e-governance has evolved over the years but the recent improvement in high-speed internet connectivity, digital content creation, growing use of smart phone and allied hand-held gadgets, growing number of sensor technologies, social media use, exponential rise in data creation and mobile applications is enabling the governments to strengthen this digital governance phenomenon in public interest. Based on the developments discussed above, new trends of Cloud Computing, Meta databases, Data analytics, Smart cities, Internet of things, Artificial Intelligence etc. are dynamically interactive and contributing towards a smart digital world with manifold improved capacities of efficient resource utilization. Thus the improved ICT infrastructure, big databases and digital applications combine together to come up with new business and governance models serving the community in a much smarter way than was ever envisioned in the past. However, in order to extend fairly the dividends of all these evolving technologies and application to the world, digital governance architecture calls for a synergy between developed and developing countries to address the concept of Digital Divide. The digital divide implies a widening of the gap between the "haves" and "have-nots" with respect to ICT across the world (Ashraf et al,. 2009). As the developing countries are struggling to cope with the challenges faced due to insufficient ICT infrastructure development and capacity dearth- both in terms of poor governance and technology literacy – the growing digital divide between developed and developing countries is increasing with enormous social implications. The digital divide issue, unless taken seriously by the Governments in terms of a clear roadmap and framework to integrate technology into governance and development policies, would hamper the progress of the society in developing countries.

The ultimate objective of this volume is to provide the academic, industrial and local government communities with an insight to the recent advances in development, application, policies and impact of technologies for the digital governance initiatives. This editing book has the dual objective. The foremost is to share the experiences of deploying digital technologies for governance. The second objective is to encourage more research in this area and publish the best research being conducted in this domain today. The book has multitude of area from digital governance in school curriculum design, digital social media practices for governance, smart city with employing digital technologies for efficient resource utilization, ICT implementations, intelligent transportation systems, digital governance challenges and success stories. The book also discusses cyber security governance for secured ICT implementation and finally concludes in the area of IoT implementations and smart system implementations.

This book will be the first outcome on employing latest technologies for Collaborative Digital Governance.

## THE ORGANIZATION OF THE BOOK

### Chapter 1: Digital Curriculum Importance for New Era Education

Education is the key to success for any society, societies can grow well by achieving the current era education in the right way. The quality of education is directly linked with the curriculum it contains, and the mode of delivery, that how much it is effective, as well as the volume of dispersion among the society. The new era brought the concept of Digital Curriculum (DC) which changes the tradition of old teaching and bringing higher results in different domain. This chapter examines how (DC) has played a main role in Education in society at various levels and the importance of Digital Curriculum in education for teachers and students. The digital curriculum has been upgraded to enhance the status of education in a new era. Change provides a greater focus on building era of education. Moreover, the effect and the future of DC will be illustrated. This research will further bring several ways of DC implementation in a different domain, different benefits out of it, and its impact on economic growth as well as in digital governance.

### Chapter 2: Digital Era Governance and Social Media – The Case of Information Department Brunei

Digital technology is at the forefront of transforming how governments operate around the world. Using Brunei's Information Department (InfoDept) as a case study, this chapter looks at how the agency has evolved from its inception in the 1950s to 2019 in its embrace of both Old and New Media to pursue its mission and objectives as a government-run media agency. The results demonstrate how New Media, such as Social Media, can complement Old Media if done right. The case study on InfoDept contributes to the growing field of research related to the increased advancement, development, application, and impact of new technologies in bolstering the digital governance process. This chapter also provides strong evidence on how governments can improve its general governance process and unlock the digital dividend in the 21st century by incorporating New Media into its public policy architectonic. Salient lessons for policy-makers and practitioners on digital governance have also been presented in this chapter.

## Chapter 3: Digital Governance for Developing Countries Opportunities, Issues, and Challenges in Pakistan

Information Communication Technology plays a vital role in countries economic growth, and at the same time it changes the form of traditional government to digital government. Digital Governance is about having a competent and responsible government to deliver efficient, better and 24/7 public services. It improves citizen participation, transparency and overcomes administrative bottlenecks. Slackness of any government organization in providing quality services will instantly become topic of debate on print, electronic and social media, forcing the political leaders to take corrective actions to save their own skin. However, for developing countries this is going to be difficult to attain same level of efficiency and flexibility. This study will examine the state of digital governance, in developing countries particularly in Pakistan government and will highlight the facts that through the implementation of Digital Governance in Pakistan, i.e. challenges and difficulties as well as opportunities which will help us to promote the application of digital governance

## Chapter 4: Digital Governance in Rural Schools of Sri Lanka – Digital Governance in Rural Schools

ICT has been used as a model for education in many parts of the world. At the end of the conflict in 2009, the Government of Sri Lanka embarked on an e-Sri Lanka initiative aimed at introducing ICT across several sectors. The education sector received substantial investment in ICT however only a section of this sector was targeted. This study explores the barriers of introducing and implementing ICT The analysis of the case studies together with the variables identified from the literature review provided the basis for developing the conceptual framework for the study. The target population for this study was Teachers and Principals in the rural schools of Trincomalee representing the rural population in Sri Lanka. A combined sample of 70 teachers and principals were considered for this survey across 20 schools in the district. The results indicate that internet access, resources, personnel and security are essential for introducing and implementing ICT education in the rural areas.

## Chapter 5: Smart Cities and Digital Governance

With the ubiquitous low-cost sensor devices and widespread of IoT, the paradigm is shifted from urban areas towards a smart city. A smart city is an urban area that uses IoT technologies to collect data and manage resources efficiently. The vision is to improve the capabilities and to solve the citizens' problems e.g. energy consumption, transportation, recycling, intelligent security, etc. in an efficient way. A smart city is a multidimensional term including a smart economy, smart mobility, smart living, smart environment, smart people and smart governance. Although the concept of a smart city is tremendously increasing and currently there exist a lot of such cities in many developed countries, however; one of the key challenges faced by these cities is good governance. Smart cities need smart governance to run the city in a smarter way, and effective digital governance is a solution to this end. Digital governance refers to the use of digital technology in government practices.

## Chapter 6: From E-Government to Digital Government – The Public Value Quest in the Tunisian Public Administration

Information and Communication Technologies (ICTs) know a steady evolution that leads to the development of digital technologies. These affect human lives and organizations' activities. In particular, they have become ubiquitous in the public sector almost throughout the world. The extensive use of ICTs in public administration (PA) contributes to the emergence of digital government. It is perceived as a means of public value creation and represents a part of a paradigm shift in public management. This chapter intends to review the milestones of PA reforms in a developing country, namely Tunisia, to highlight how it is using ICTs to create public value. A content analysis of data on public sector reforms and e-government strategies is performed. The results show progressive ICTs deployment by Tunisian PA over time. They illustrate a search to public value creation through the digital government

## Chapter 7: The Role of IoT in Digital Governance

The acronym IoT stands for Internet of Things. The IoT ecosystem can be envisioned as a set of physical electronic devices embedded with intelligence, connected through a network, enabling them to collect and exchange data, and allowing these devices to be sensed and handled remotely between the physical and cyber worlds. The devices connected through the internet has been influencing all walks of our life ranging from individual, societal, educational, industrial, entrepreneurial and related to governance as well. As we are connected and surrounded with a plethora of connected smart devices, it seems there is a great risk of security and privacy in several aspects; such as device authentication, data theft, device manipulation, data falsification, etc., to name a few. Hence, the current chapter has been undertaken to explore and comprehend the security and privacy related implications, opportunities, future directions, and challenges involved in implementing digital governance with IoT.

## Chapter 8: Success Stories of ICT Implementation in Saudi Arabia

Information and Communication Technologies (ICT) has a great role in the development, growth, and economy of the country. ICT refers to any system, device or product that can store, manipulate, and capable of receiving and transmitting digital data effectively. ICT brings clear differences in the economy of countries wherever is successfully implemented. It becomes a reason of timely, accurate, and efficient operations as well as becomes a source of user satisfaction by providing them fast, better, easy service around the clock. As timely access of information and communication plays a vital role in the success of any organization either public or private many developed countries already competing in digitalization. It is not only convenience, but it is the prime ingredient of good governance and an indispensable part of modern life. This chapter discusses the successful implementation of ICT infrastructures in the context of good governance and builds a vibrant society, the main agenda of Vision 2030 of KSA.

## Chapter 9: Information and Communication Technology (ICT) for Intelligent Transportation Systems (ITS)

Transportation is essential part of human activity that allows people to move from one place to other places. In modern transportation systems that refers to intelligent transportation systems (ITS), it has been proven that information and communication technology (ICT) has significant role to improve the quality and security transportation services. ITS can be described as an integration of computer, electronics, and communication technologies and management strategies that can increase the efficiency and safety of transportation. The application is a platform that manages the vehicles, drivers, passengers, road operators, and managers in an environment that is suitable for everyone to communicate and interact. This chapter presents the history of transportation systems and the role of ICT to bring ITS in different transportation modes. The main motivation of the chapter is to survey on the various technology enables ITS systems that have been in place analyzing their strengths and providing recommendation for such ITS systems in developing countries to improve the governance.

## Chapter 10: The Evaluation of Social Media Practice in Malaysia by Small Medium Enterprises

The adoption of social media by large businesses has influenced Small Medium Enterprises (SMEs) to be part of this phenomenon. Evaluating the results of the online efforts is a prerequisite to the success of social media adoption. As more branding activities move online, marketers are faced with the realization that brands and businesses are not always relevant in all the social media facets. The study is aimed at solving the question of how SMEs in Malaysia are evaluating their online marketing efforts. The quantitative survey was deployed across six states in Peninsular Malaysia to get feedback from 125 active firms from different size, sector, year of establishment, and period of use of social media. The pattern of all these participants conducting the evaluation phase on the social media platform is descriptively examined from quantitative findings. It is used to formulate new information into literature and body of knowledge. Research implications in the practical and social form are also discussed.

## Chapter 11: Cybersecurity Governance on Social Engineering Awareness

Raising awareness of social engineering is becoming more essential nowadays. When we are connected to the internet, we are at the risk of becoming victims of numerous attacks. It is very difficult to protect ourselves from all of these threats, but the least we could do is not to jeopardize our own safety. In order to achieve this, the need to have substantial knowledge on social engineering techniques and how to safeguard one from being victims is mandatory. This research presents the overview of social engineering attacks, countermeasure against this attack and some. This plays a crucial role on countering attackers on cyber assaults

## Chapter 12: Fostering Public-Private Partnership – Between Governments and Technologists in Developing National Cybersecurity Framework

National cybersecurity governance in the era of Industry 4.0 becomes major critical factor A good cybersecurity governance in the private stakeholder site and cybersecurity defense at the national perimeter by the government is crucial. But this effort cannot be achieved without the collaboration between the government bodies and the privately-owned critical infrastructure by technologists. Therefore, this paper intends to provide a review of cooperation between public and private sectors towards cybersecurity governance. With the partnership, government can have confidence in the safety and protection of their national critical digital infrastructure. The goal of this paper is achieved by analyzing some of the cybersecurity frameworks adopted by the developed and developing nations. The analysis is further carried out by investigating the public-private policy initiatives in their national cybersecurity framework. The paper also investigates the effectiveness of the National Institute of Standards and Technology Cybersecurity Framework (NIST) adopted by the US government

## Chapter 13: Democratic Governance – A Review of Secured Digital Electoral Service Infrastructure

In any democratic nation, electoral processes pave the way to democratic governance. In regards to this, a fully digitized online voting system should be introduced and governments should invest efforts in studying the possibilities in adopting online voting system to curb the issues mentioned above. In that sense, any user should be allowed to cast his/her vote from any remote location and the processes should be held in full confidentiality without any intervention or possible vulnerabilities. Therefore, this paper investigates how governments can learn various such initiatives and provide funding to further conduct research and implement such methodologies. The paper begins by exploring some traditional and current practice of electoral system and further embarks into a review of secured online voting systems that has been proposed by the body of knowledge. The paper then presents how some countries have adopted this secured online voting systems and giving an avenue to secured digital electoral governance.

## Chapter 14: Economic African Development in the Context of FinTech

The purpose of this chapter is to analyze the effect of Fintech and political incertitude on economic growth through multiple regression. Thus, we employ the method of generalized least square (GLS) with panel data. Our sample concerns 21 African countries during (2001-2014-2017). We use a wide range of measures from Global Findex Database 2017, the World Bank platform, the World Bank national accounts data and the OECD National Accounts data files base in the context of Africa. Empirical results show that Fintech is a driver of economic growth unless it is actively used in a developed digital infrastructure. In fact, we prove that, when financial technologies are used in both transactions (receive and made the digital payment), they significantly contribute to the economic cycle. While, passive use like simple consumption actions are not significant lever for the economy. Our principal contribution is to highlight that the active use of financial innovations and not passive ones and the developed digital infrastructure do promote economic growth in African countries.

## Chapter 15: Tactile Internet and the Remote Surgeon

The delivery of health care in remote rural areas is a global phenomenon. Using digital governance, a discretized cluster ofs node relaying time-sensitive and reliable data, life-critical health services can be ensured. Here tactile internet-based Internet of Things (IoT) and its variant Internet of Medical Things (IoMT), as well as Tactile Internet, is introduced. Its relationship with 5G is explored. A model for e-health implementation using Unmanned Autonomous Vehicles (UAVs) in the backhaul is presented. The link is analyzed over fiber optics, satellite and 5G network in terms of latency and reliability: the crucial metrics for robotic surgery. It is also shown that 5G implemented in Italy, Zimbabwe, and Cooks Island have supported mobile hospitals in rural regions: hence induction of 5G networks is a feasible approach. Finally, a critical review of expected future technologies including telematics is presented. Hence, Tactile Internet-based robotic surgery can reduce the digital divide.

## Chapter 16: UAV-Based Smart Environmental Monitoring

In the context of digital governance man-made activities will have the most impact on the environment. The concept of Internet of Things (IoT) is much more profound supposedly having an impact on the way environmental sampling of data takes place. The cost benefits of Unmanned Aerial Vehicles (UAVs) as alternate to satellite or High-Altitude Platform System (HAPS) is established as these are more flexible, mission specific and versatile. UAV in sensor network context and communication relaying as well as data harvesting application is explored. Proposal to upgrade environmental laws and implementing Network Infrastructure Security and Cyber Insurance for HetNets involving UAVs and Environmental Monitoring will be covered as well.

*Khalid Rafique*
*Azad Jammu and Kashmir Information Technology Board, Pakistan*

*Vasaki Ponnusamy*
*Universiti Tunku Abdul Rahman, Malaysia*

*Noor Zaman*
*Taylor's University, Malaysia*

## REFERENCES

Ashraf, M., Hanisch, J., & Swatman, P. (2009). ICT intervention in the 'Chandanbari' Village of Bangladesh: Results from a field study. *Information Systems Frontiers*, *11*(2), 155–166. doi:10.100710796-008-9133-0

# Acknowledgment

We would like to acknowledge the help of all people and close reference research circle involved in this project, and more specifically, to the authors and reviewers that took part in the review process. Without their support, this book could not have become a reality.

First, the editors would like to thank each one of the authors for their contributions. Our sincere gratitude goes to the chapter's authors who contributed their time and expertise to this book.

Second, the editors wish to acknowledge the valuable contributions of the reviewers regarding the improvement of quality, coherence, and content presentation of chapters. Most of the authors also served as referees; we highly appreciate their double task.

Finally, we appreciate the cooperation, support and patience extended by our families for sacrificing family times to complete this project well in time.

# Chapter 1
# Digital Curriculum Importance for New Era Education

**Malak Z. Alamri**
https://orcid.org/0000-0002-3484-0994
*Jouf University, Saudi Arabia*

**N. Z. Jhanjhi**
https://orcid.org/0000-0001-8116-4733
*Taylor's University, Malaysia*

**Mamoona Humayun**
*Jouf University, Saudi Arabia*

## ABSTRACT

*Education is the key to success for any society. The quality of education is directly linked with the curriculum it contains, the mode of delivery, how effective it is, as well as the volume of dispersion among the society. Digital curriculum (DC) changes the tradition of old teaching and bring higher results in different domains. This chapter examines how DC has played a main role in education in society at various levels and the importance of digital curriculum in education for teachers and students. The digital curriculum has been upgraded to enhance the status of education in a new era. Change provides a greater focus on building the era education. Moreover, the effect and the future of the DC will be illustrated. This research will further bring several ways of implementing DC in different domains, different benefits of it, and its impact on the economic growth as well as in digital governance.*

## INTRODUCTION

As the current era is full of many cognitive and technological variables of rapid pace, which have implications for the personal, societal, political, artistic life and all aspects that are the manifestations of human life. Therefore, the educational curricula must have certain features that correspond to the requirements of living in this age and define the way To the learning that enables the individual to

DOI: 10.4018/978-1-7998-1851-9.ch001

acquire the characteristics of citizens of the twenty-first century, such as competition and the ability to innovate and the choice and flexibility and other qualities that educators believe that it should be one of the most important outputs of education, It was therefore necessary to identify these features so that the educational curriculum in turn to support the community to deal with and benefit from the data of digital and technological development.

Several people have confirmed that a country's capacity for contend in the global economy and allow its peoples to take all benefit of technological developments depend on improvement the quality of the schooling providing and confirming that which is educated is aligned with national primacies and international improvements.

In fact, people must now continue to improve and update their skills and knowledge in order to convoy in a row with the continuous innovations and novel improvements in the digital world. (Alexander & et al, 2017) Education is the key to future economic, political and social progress. The use of digital resources can improve student engagement and exam results.

## LITERATURE REVIEW

Today, the world is witnessing a technological and digital revolution in various fields, including education. The digital world has begun to increasingly spread in education and skills, as well as the use of technology and the Internet to provide education, knowledge and skills in novel and innovative methods. The usage of these digital technologies has helped transform learning and develop cognitive and scientific skills into a lifelong process. (Alexander & et al, 2017)

Today's society is seen as a digital society. Where people are using technology and digital media as an integral part of their daily lives and practices. Due to the process of examining and evaluating the educational process that is very important, therefore, it is necessary to work and contribute to educational reform by understanding the best digital technology and the life environment of students as they share new ideas and knowledge in learning. Extreme attention must also be given to the digital media process; In order for teachers to reach levels of progress in attention, ability and motivation to the attention of their students, In addition, emphasis should be placed on the kind of content, skills and knowledge that must be enhance by individuals, as well as on enhancing the capacity to adapt and use technology in education and society. Therefore, the process of learning and acquisition of knowledge depends on the educational curriculum and how to teach. (Chien, 2012)

### Digital Curriculum Pillars

The curriculum is the sum of the educational, cultural, social, sports, religious, environmental and artistic experiences that the educational institution prepares for its students inside and outside the institution in order to achieve their overall growth and modify their behavior.

It is the content and methods of teaching and classroom activities and extra-curricular methods and teaching methods and methods of appropriate assessment and keep abreast of the changes and current and future developments of society, and the individual output adapted to the requirements of his time, achieving his personal goals and the goals of his community. (Srivastava, 2014)

It is all classroom and extracurricular activities that aim to provide students with educational experiences and achieve the desired goals. Curriculum can be classified as mathematics, science, and other Curriculum, into five basic Pillars: (1) symbols, (2) patterns, (3) order, (4) relationships and / or categories. These groups are referred to as "pillars" because they can be considered individual and stand firmly without outside support, yet, when combined, they can maintain larger structures, in which case human beings learn. Shown the figure below. (Tokuhama-Espinosa, 2017)

## Digital Curriculum

In light of the rapid growth of knowledge and the information revolution, the need for new curricula and methods to cope with the demands of the age and its challenges has become apparent, such as the marked increase in demand for education, the lack of educational institutions, the information revolution, the increase in knowledge and the use of technology in education. Which helped to emerge one of the new types of education, e-learning that supported and helped the learner to learn at any time and place, through interactive content depends on multimedia and provides through electronic media such as: computer, the Internet and others.

The digital curriculum is a system of e-learning that includes a range of learning and scientific skills that are provided to the students by the huge potential offered via information and communication technology. It is a novel form of communication among the knowledge of the teacher and the student. This connection is digitally transmitted through the information network or through another digital medium such as CDs. (Justification, 2018)

*Figure 1. Overlap mapping of pillars and subject areas*
*Source: (Tokuhama-Espinosa, 2017)*

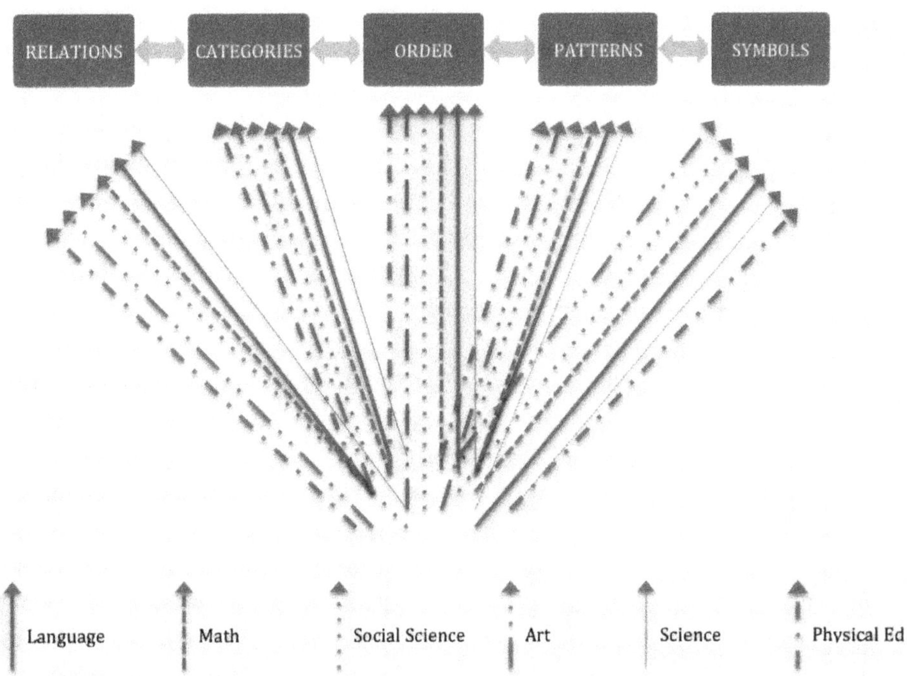

The digital curriculum is a collection of educational and technical knowledges delivered to the learner via the great potential offered by information and communication technology, which represents a new method of connect between the knowledge of the expert, the learner.

It is a system that contains a set of interrelated, functional and integrated elements and follows a comprehensive general plan, in which learners are provided with a set of technology-based learning opportunities in planning, implementation and evaluation in order to achieve the overall integrated growth of the learner. (ALCHEMY LEARNING, 2014)

## Critical Thinking and Evaluation

The learner today needs a method of teaching and learning that provides him with a broad and advanced educational horizons that help him enrich his knowledge and develop his different mental skills and train him to create and produce new and different by providing him with the appropriate sources and raising his interest through various programs and teaching strategies.

The key to exploiting and utilized the rule of information is the capacity for analyze info to ensure that it is relevant, reliable and up-to-date. Information evaluation is a basic and lifelong life skill for lifelong learning. The evaluation includes several components of the process that include objectives, personal judgment, knowledge improvement, negotiation and decision-making. Which is a complicated and difficult problem, and stresses the significance of being able to dangerous thinking.

Digital curriculum is one of the most important tools that contribute to the promotion of critical thinking, creativity and innovation. Critical thinking is an essential learning result for learners. Educational organizations have experimented with numerous different plans for assistance promote serious thinking as a means of developing information estimation and information awareness between learners; learners must therefore be encouraged to practice arguments when presenting evidence. Discussions and methodological presentations should also encourage analysis and evaluation of information.

## Cultural and Social Understanding

Teaching methods and practices should facilitate and enhance the student's ability, in an increasingly information-based society, to harness the power of information. Digital curriculum is the means of development and survival of nations is governed by social philosophies and aspects of life and cultural heritage left by previous generations and economic systems that prevail.

## Digital Literacy and Subject Areas

Literacy is no longer an aim for countries that seek to structure new and sophisticated information societies through providing their persons with the basic experiences to usage computer equipment's in their daily lives through which they can exploit and improve educational, financial, community or civilized opportunities for themselves, their families, or their communities in overall. (The Building Blocks of Digital Literacy, 2015)

Digital literacy is an emerging concept that mean the primary skill or capability to usage a computer definitely, security and efficiently, containing: the capability to usage software tools like word processors, email and management software, the capability for build and manage pictures, audio and video, and the capability for usage a web browser and search engines optimization (SEO) in Internet. (Furber, 2012)

*Figure 2. Framework for digital literacy*
*Source: (Belshaw, 2011)*

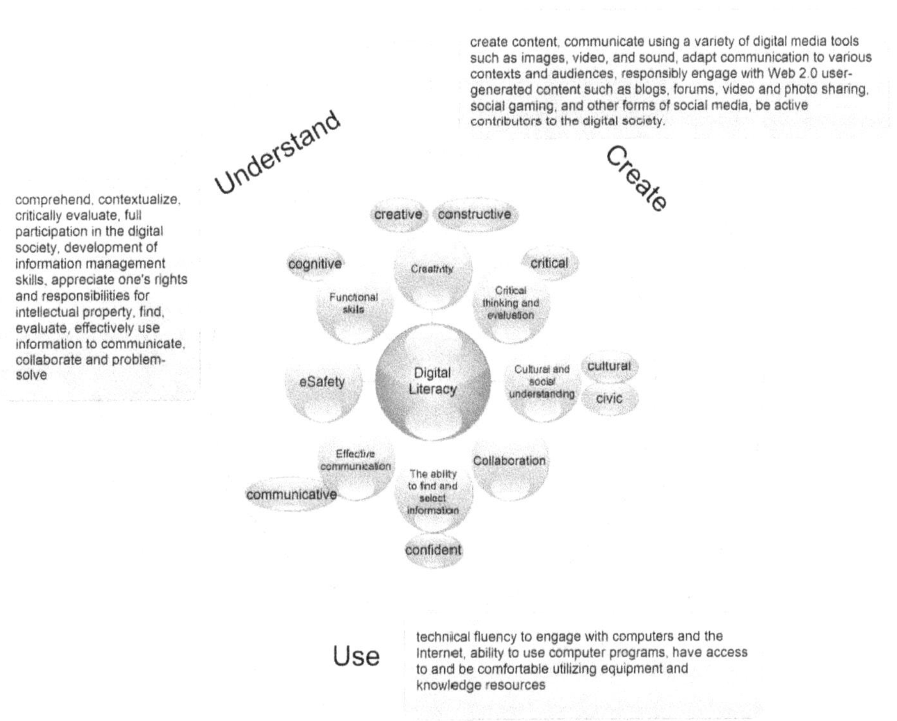


The figure above shows how digital literacy integrates many interrelated skills ranging from basic awareness to creative literacy and more complex outcomes. (Belshaw, 2011) The figure below illustrates the digital literacy model and a plan of action to attract a number of competencies. (Hobbs, 2010) (The Building Blocks of Digital Literacy, 2015)

*Figure 3. Model for digital literacy*
*Source: Adapted from Hobbs (2010)*

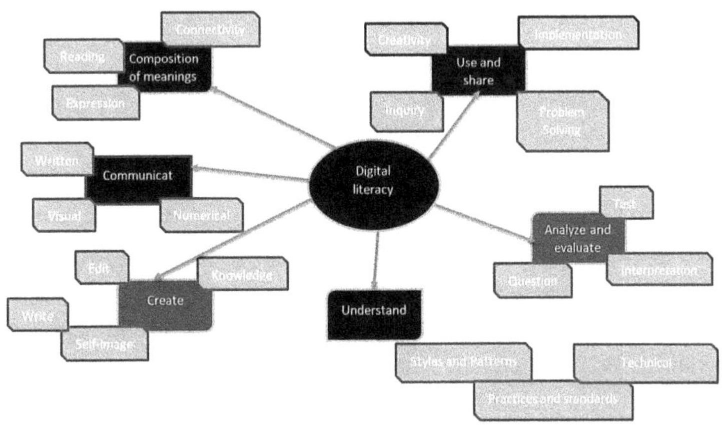

*Figure 4. Standards for digital skills*
*Source: dwq1Adapted from Vaizey and Boles (2016)*

The proposed model provides the most significant components and dimensions of digital literacy, and contains the numerous concepts and meanings attributed to digital literacy. As the proposed model provides a framework for the improvement of all elements of digital information. Which will enrich the curriculum and graduate programs as well as contribute to the improvement of team experiences and employment programs.

## Functional Skills and Beyond

Digital can be incorporated into both ICT and other educational topics. Completely as learners training writing across several school topics, thus must learners be improving digital literacy practices in several topics, including ICT. (Hague & Payton, 2011)

Promoting digital literacy means overcoming beyond functional skills and the capability to perform main online searches and PowerPoint presentations. It means providing learners the chance to usage a broad domain of technologies collaboratively, imaginatively and seriously. (Hague & Payton, 2011)

Basic digital skills enable us to work at the lower level within the community. Basic skills for basic functions, there is a growing consensus that basic digital performance corresponds to basic knowledge that occupies its niche alongside traditional written and arithmetic knowledge. Basic skills include equipment (such as keyboard usage and touch screen technology), software (such as word processing, file management in computers and privacy settings in mobile phones) and basic online processes (such as e-mail, search, or online form). The basic skills enrich our lives and enable us to interact with others and access to government, business and financial services.

Professional skills are the skills needed by ICT professionals such as computer programming and network management. These include artificial intelligence, large data, encryption, cybersecurity, object internet, mobile application development.

Online safety Skills is one of the most important skills in digital skills. It includes the cyber security, Age-appropriate content, Plagiarism and copyright, permissions.

## Accessibility

The digital curriculum can be considered as a new system of modern teaching that is depend on the delivery of learning content and the sending of experiences and models to the student on ICTs, the Internet and their multimedia in a method that permits the learner to interact actively with content and with the educator and colleagues concurrently or asynchronously at the time, location and speed that suit the conditions of the learner and his capacity, and the management of all scientific and educational events and requirements in automatic form through the automatic methods allocated to it. The digital curriculum is accessible at any time and from anywhere, and it is easy to store and navigate. It is digitally stored on laptops or smart phones. (Fletcher, 2014)

## Virtual Reality

In an era when technology has become an integral part of our lives, an era in which the evolution of applications and technologies is accelerating, an era in which easy contact to information is facilitated by intelligent devices such as laptop, iPads, smart phone and others. Our worlds are converted into a virtual world in which the individual lives, travels, also learns and studies. (Hu-Au & Lee, 2018)

Virtual reality (VR) indicates to a computer representation that generates a perception of the world that shows our senses like to the actual world. Through VR, knowledge and experiences can be transmitted to the mind in an attractive and interactive manner. (LuoChanggen & YuanXuemei, 2017)

VR technologies have come in numerous methods as technologies that can be harnessed to serve the science, learners and even teachers. VR is one of the most recently achieved and is in fact a means of transferring the education environment to a novel and advanced stage.

## Distance Live-Lab

Live Lab is one of the greatest distance learning; It is a direct interactive educational session held by teachers in their field to provide the benefit and knowledge to students and learners on the Internet, through the web browser. A live-lab contains tools like a computer, server, adapter, or router that the user can configure for free.

Live-Labs offer a real computer equipment, networked together and conveniently accessible over the internet using virtualization. The figure below shows triangulation between environment, approaches and results in live laboratories.

*Figure 5. The live lab triangle*
*Source: (Veeckman, Schuurman, Leminen, & Westerlund, 2013)*

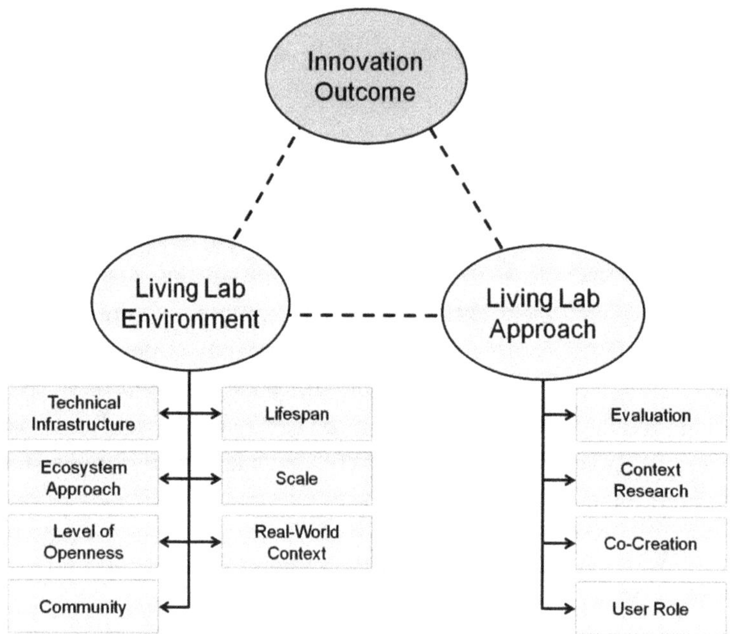


## Teachers

The teacher is the basis of the educational process; he/she builds minds, and educates the generations that will return to the benefit of their communities, which affects the learners, guide them to learning, and promotes in them the rest, tranquility. The teacher is able to know the weaknesses and strength of his student, which helps him to guide them properly reflect the principles of unity among them, and contributes to the refinement of their ethics, and provide them with the ability to think logically.

To achieve this in the light of the techniques and digital curriculum that have invaded education and the educational process, we must address some of the basic concepts that have become part of the educational process in the present age.

## Audio Enhancement

The Audio Enhancement process improves the teacher's voice - distributing it evenly everywhere the teaching location. Researches have shown that as up to 75% of the school day is expended in activities that inclosing listening. Audio Enhancement is identified as one of the main methods used in classroom today's. (Rossiter, Nortcliffe, Griffin, & Middleton, 2015)

## Live Chat/Instant Messaging

After the spread of smart phones and new technologies for the Internet, instant messaging programs became a basic means of communication and data transfer, which allowed for exploitation and use it in the educational process and the transfer of knowledge and experiences. Teachers can use live chat with learners and students through the Internet. This live chat system is also perfect for providing support to students, that teacher can chat live with them. (Simon, 2016)

## Live Video

Live video is one of the most prominent digital educational platforms that support digital curricula and contribute to providing information, knowledge and experiences to learners. Using a Live video platform, the teacher can communicate with students and provide knowledge and information to them remotely outside the classroom, the teacher can also gather his students with other students separated by large geographical areas.

## Training

Training is defined as a continuous systematic system effort pursued at performing or improving the teacher's knowledge, experiences and performance to achieve his work with a high degree of competence and validation. It is a systematic and continuous process in the life of the teacher aims to enhance the teacher's ability to achieve a high level of performance and professional growth, through the provision of information, skills and trends related to the field of work or specialization. (Boudersa, 2016)

## Virtual Conferencing

The virtual conference provides teachers and learners with the ability to attend events and meetings held digitally on a web site through their computers, regardless of the geographical distances that separate them. Virtual conference is presented fully online; teachers and learners have no need to meet in a conference area. By a conference website or video conferencing apparatuses designed especially for the virtual skill, members can easily join virtual meetings. (Gottfried, DeLancey, Watwood, & Hardin, 2015)

## Online Trainings

With the advent of the Internet, the interest in using digital technologies in the development of training programs has increased, and as a result, the concept of online training has emerged. It is very significant to allow teachers and learners to acquire novel knowledge, capabilities, and skills or to benefit from the experience gained at a time and place of their choice. At particular times in their careers, it may be more challenging for professionals to join live events due to time, movement, commercial or other limits. (Stern, 2004)

*Figure 6. Overview of ICT infrastructure in education*

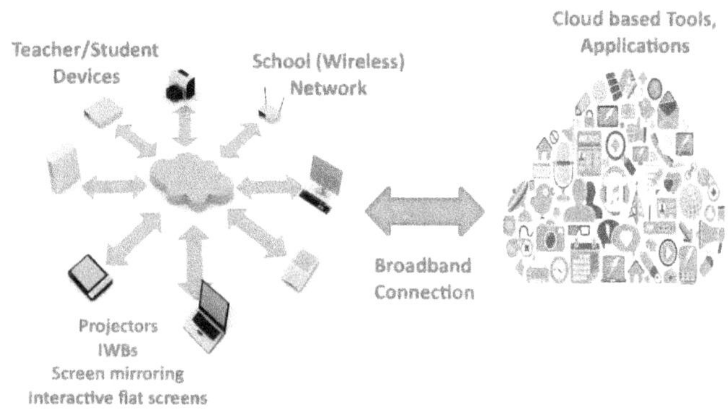


## ICT Technologies use

Educational institutions use a variety of ICT tools to generate, communicate, publish, collection, and manage information. (Blurton, 2000)When teachers are able to read and write digitally and are trained in the use of ICTs, these methods can enhance higher thinking skills, extend inspired and individualized choices for students and learners to rapid their understandings, develop and improve the willingness of students and learners to cope with technological change in culture and the workstation. (Goodwin, 2012) To ensure the use of ICT in digital education, an infrastructure must be available, as shown in the figure below. (Pdst)

ICT cases developers should reflect contain: in view of the over-all cost and advantage equation, providing and keeping of the required infrastructure, and confirming savings are coordinated with teacher sustenance and other rules goaled at usage of efficient ICT. (Enyedy, 2014) The use of ICT in digital education plays an active role in achieving the objectives of the educational process. See the following figure that shows the reasons for the use of ICT in education. (Vijayalakshmi, 2016)

## Students

Student is anyone who seeks to acquire the information, knowledge and new skills that he seeks to use in his scientific and practical life. The student is the main focus of the educational process, Student is the learner and what it has characteristics of psychological, mental and social, and the desire and motivation to learn and the basis in the educational process.

In the current era, it is possible for students to acquire knowledge, information and skills in modern and sophisticated ways through the use of digital technology and the Internet. In light of this, there are some digital concepts that relate to the student.

## Social Gaming

The usage of Social games in learning is one of the most significant means to catch the notice of students, so educational social games assistance to emphasis information and stability in the minds of students to the attention of learners throughout the usage.

Digital and social games provide students a range of benefits, the most important of which are the following:

- Digital educational games are not tied to a specific time, so the student can play at any time he wishes and for any length of time he wants.
- Digital learning games divide knowledge into small stages that require reaction and instant feedback, which emphases on the educational objective and drives the student to continue playing.
- By playing the student gets rid of the psychological pressure that falls on him from educational practices or socialization.

## Spell Checkers

The spelling checker is a feature of computer programs that are looking for spelling mistakes in text. Often features are in software programs, such as word processor, e-mail client, e-dictionary or search engine.

Digital technologies provide programs or feature in a computer programs designed to locate words that contain misspellings and to inform the student of the error. Depending on the spelling checker, the feature may either correct the word automatically or allow the student to choose from possible corrections to the misspelled word. (Rimbar, 2017)

## Animations

Animated images have become the means of communication on the Internet More than still images and even words, it has also been used in various fields including educational field and educational process. The student may resort to a set of animation that helps him understand the nature of the work of some things,

## Visual Advantages

Visual aids in the field of learning and education are all printed information vessels: books, pamphlets, manuals, newspapers, magazines and so on or all other graphic and graphic materials: maps, posters, educational boards, diagrams, models, Slides, Microscopes and others.

The advantages of Visual aids are often the combination of sound and image, a movement that attracts attention and attracts the viewer, interesting and exciting, unconventional in its programs, although traditional in its mind-based method of indoctrination without interaction. Dealing with all aspects of mental and related senses and providing what other sources cannot provide, explain, demonstrate, connect, comment and deal with issues that are practically important and do not occupy large areas in libraries. (Shabiralyani, Hasan, Hamad, & Iqbal, 2015)

## Self-Contained Learning System

Self-Contained Learning System is a deliberate procedural process in which the student tries to acquire the most knowledge, concepts, principles, attitudes, values and skills using or benefiting from technological and digital applications.

This kind of learning is one of the most significant education techniques that allow the usage of educational knowledge effectively, which contributes to the improvement of the student cognitive, behavioral and emotional, in addition to providing a weapon that can absorb the data in the future, the student decides when and where and starts, and any alternatives or means Which he chooses, he is responsible for teaching himself and about the decisions and results he takes.

Self-Contained Learning System is characterized by a set of features, the most important of which are the following: (Causton, Theoharis, & Orsati)

- The student is the basis of the educational process, supervising itself every step.
- Self-Contained Learning System is related to tendencies, something that a student loves.
- Achieve psychological comfort with the will and strong determination and desire to improve itself.
- Not feeling afraid of what he learns! Because he/she chose it.
- Gain experience from learning the mistakes facing the student.

## Digital Curriculum Recent Statics

The methods used in teaching and learning have changed. In the past, classes were limited to lectures, conversations, physical objects and printed curricula, but with the development of the world and the emergence of digital technologies, there was no longer a need to use traditional teaching methods. Educators and learners today have a digital toolbox – ranging from attractive devices to learning learners to online training and digital Curriculum. Digital teaching is the concept utilized for indicate to several online learning performs. Projections indicate that the digital education market international is expected to exceed US $ 243 billion by 2022. In the recent years, the digital education (self-paced) product market amounted to 46.67 billion US $, and in 2021is expected to decline to 33.5 billion US $. Showed a significant share of international faculty members in preparation for support fewer traditional and digital learning methods. Approximately 65 percent of faculty members supports the usage of open educational resources (OERs) in teaching, and 63 percent showed support for the competency-based teaching system. (Jaaskelainen)

## Digital Curriculum Growth

Throughout history, teaching aids have evolved, the three most important axes in the educational system, teacher, student, remained the method of education. There is no doubt that the technological technologies are considered in our current reality one of the most important and most powerful sources of learning because of its characteristics are difficult to limit. This is evidenced by the proliferation of educational sites and academies that are considered as an educational resource for many students or employees seeking to increase their knowledge.

With the revolution of computer technology, information and communications witnessed by this era, Many technological and informational innovations have emerged that have cast a shadow over global, regional and local educational systems and have had a profound impact on them, Which led to the creation of new educational methods, a learning system based on learning through a digital educational content different from what is presented in traditional textbooks, depends on multimedia, and provides through modern electronic media. (Al-Awidi & Aldhafeeri, 2017)

In present time, there has been a growing interest in the integration of ICT into the curriculum and all other elements of relevance to the teaching and learning process. They have paid interest to the case of improving the competencies of teachers to integrate technology into the curriculum and to enhance their capabilities to make technology-based education and learning environments. (Lim, Chai, & Churchill, 2011) Teachers must be prepared for transformational approaches, to transform traditional teaching materials into digital teaching and to create new resources. To achieve this transformation; teachers need to have several skills that will support them successfully embrace this change. These skills contain knowledge and experience in the usage of digital tools in several curriculum domains and creation students' learning spread beyond the teaching space. (Bonanno, 2011)

## Digital Curriculum Impact on Economic Growth

Many researchers and economists believe that modern education and digital curricula can contribute to improving the economic development of societies. Many studies confirmed that there is a relationship between digital education techniques and economic development, education can be used as a tool for economic development. (Williams, 2014)

Digital education has become an essential part of most development strategies. The process of technological expansion in education plays an active role in influencing the economic situation. (Hanushek & Wößmann, 2017)

## Digital Curriculum Role in Digital Governance

Digital governance is defined as a modern, digital-based system that aims to transform traditional governance into digital governance based on the use of computers and its applications. Digital governance also defines all functions and tasks defined by governance for employees by relying on digital means of communication, Electronic communication channels, and this type of modern governance plays an active role in the field of education, which leads to easy linking between the administrative body, teachers and students.

Digital Governance and digital education are important aspects of electronic communication of information. If digital government systems are dynamic to development of public services, before digital education is a boon to confirming qualitative education everywhere. (Panda & Swain, 2009)

Digital Governance and digital education are the developing terms of new ICT that are presented and applied at government and organized stages respectively to support effectual and active connection of digital information in order to bridge the gap of digital divide around the world. (Panda & Swain, 2009) To design an digital governance framework there are a set of requirements and priorities of legal and regulatory to consider. As well as risk and objectives like processes, policies, tools.

Digital Governance includes the usage of ICTs in dealing with the government actions. At the service stage, Digital governance abilities a full service existing 24 hours a day and seven days a week. At several stage of basic features like government responsibility and public acceptance of national organizations, Digital governance participate to the effective of democracy through providing government information on the Internet that is difficult to find or unavailable, and by online discussions and referendums. (Saxena, 2005)

*Figure 7. Digital Governance framework*
*Source: Adapted from Thomas (2014)*

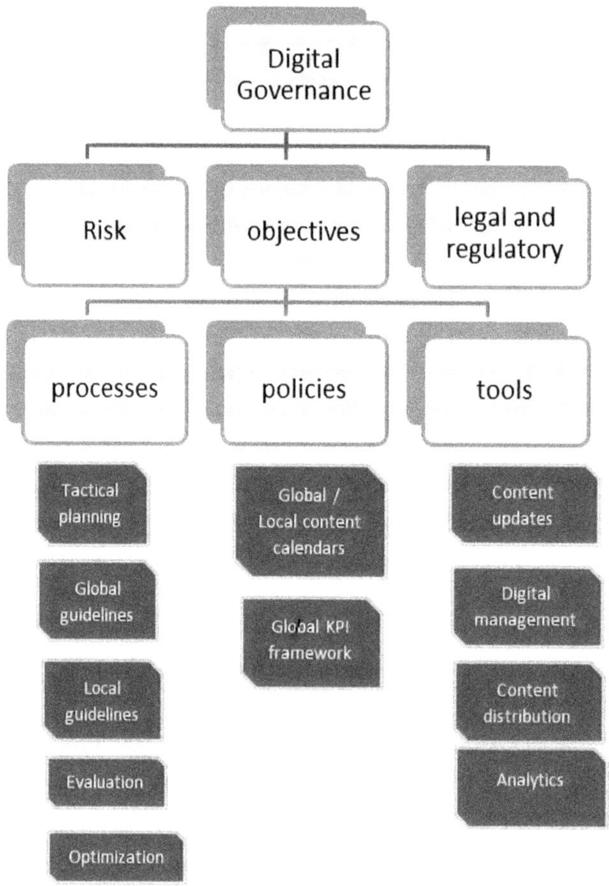


*Figure 8. The goals of digital governance*
*Source: (DavidZamora, CarlosBarahona, & IleanaPalaco, 2016)*

*Figure 9. Digital governance maturity curve*
*(WELCHMAN, 2017)*

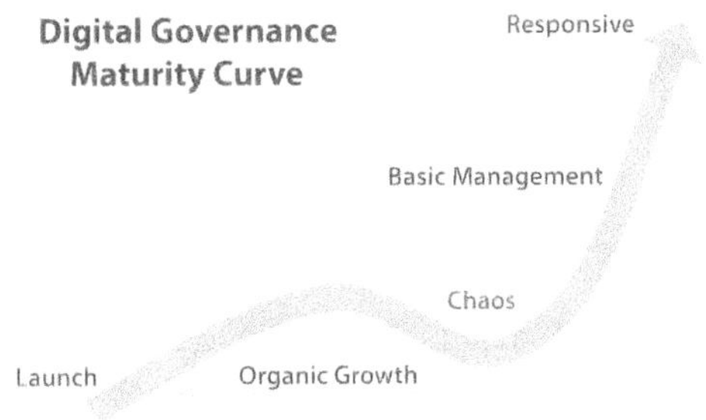


The following figure illustrates the maturity curve of digital governance that most educational and non-educational institutions transfer by when they presentation a digital channel (Web, mobile, or social). The progression of maturity starts with the resolution to promotion a novel channel, and it ends with the institution integrating the entire channel into the business to the range that governing dynamics and operative procedures are involuntary, departure the institution completely receptive to digital directions. (WELCHMAN, 2017)

## Future of Digital Curriculum

Some people believe that digital learning and digital literacy can help improve the learning environment that stimulates innovation. Curriculum development and teaching and evaluation methods, providing education services to all segments of society, Communicating around the clock ensures follow-up. Inter-stakeholder engagement ensures more efficient training and education, developing professional skills for teachers, raising the efficiency of education in the classroom, and enhancing the absorptive capacity of existing schools.

Digital curricula play an active role in improving the knowledge and experience of students and learners. Digital curricula are expected to become the primary source of information and knowledge in the future. Schools may also rely entirely on digital curricula in the classroom process.

# CONCLUSION

In light of the rapid growth of knowledge and the information revolution, the need for new curricula and methods to cope with the demands of the age and its challenges has become apparent, such as the marked increase in demand for education, the lack of educational institutions, the information revolution, the increase in knowledge and the use of technology in education. Which helped to the emergence of one of the new types of education is e-learning, which supported and helped the learner to learn at any time and place, through interactive content depends on multimedia and provides through electronic media such as: computer and the Internet and others.

This chapter examines how (DC) has played a main role in Education in society at various levels and the importance of Digital Curriculum in education for teachers and students. E-learning and digital curricula are an integral process that is not useful without a key role for the teacher, student, book or curriculum, which is the source of the first knowledge.

## REFERENCES

Al-Awidi, H., & Aldhafeeri, F. (2017). Teachers Readiness to implement Digital Curriculum In Kuwaite Schools. *Journal of Informationtechnology Education: Research, 16.*

Alchemy Learning. (2014). *The Evolution of Digital Curriculum and the Role of Cultural Institutions.* Author.

Alexander, C. (2017). *Digital learning education and skills in the digital age.* Corsham Institute and RAND Europe.

Belshaw, D. A. (2011). *What is 'digital literacy'? A Pragmatic investigation. Degree of Doctor of Education, Durham university.* Department of Education.

Blurton, C. (2000). *New Directions of ICT-Use in Education. United National Education Science and Culture Organization.* UNESCO.

Bonanno, P. (2011). Developing an instrument to assess teachers' readiness for technology-enhanced learning. *14th International Conference on Interactive Computer Aided Learning.* 10.1109/ICL.2011.6059622

Boudersa, N. (2016). *The Importance of Teachers' Training Programs and Professional Development in the Algerian Educational Context: Toward Informed and Effective Teaching Practices.* Ministere de Lenseignementt Superieur et de la Recherche Scientifique.

Causton, J., Theoharis, M. C., & Orsati, F. (n.d.). *Does self-contained special education deliver on its pronises? a critical inquiry in to research and practice.* Chaprnan University.

Changgen & Xuemei. (2017). The impact of virtual reality technology on digital media teaching. *3rd International Conference on Economics, Social Science, Arts, Education and Management Engineering*, 119.

Chien, J. (2012). *How digital media and Internet transforming education.* University of Denver.

Enyedy, N. (2014). *Personalized Instruction: New Interest, Old Rhetoric, Limited Results, and the Need for a New Direction for Computer-Mediated Learning.* Boulder, CO: National Education Policy Center.

Fletcher, G. (2014). *The accessibility of learning content for all students, including students with disabilities, must be addressed in the shift to digital instructional materials.* United States Leadership Technology Innovation Learning.

Furber, S. (2012). *Shut down or restart? The way forward for computing in UK schools.* The Royal Academy of Engeneering.

Goodwin, K. (2012). *Use of Tablet Technology in the Classroom*. Strathfield, Australia: NSW Curriculum and Learning Innovation Centre.

Gottfried, J., DeLancey, L., Watwood, C., & Hardin, A. (2015). Virtual conferencing and meeting systems: Resources for online connections. *College & Research Libraries News*, *76*(2), 98–101. doi:10.5860/crln.76.2.9265

Hague, C., & Payton, S. (2011). Digital literacy across the curriculum. *The Curriculum & Leadership Journal*, *9*(10).

Hanushek, E. A., & Wößmann, L. (2017). *The Role of Education Quality in Economic Growth*. Stanford University.

Hobbs, R. (2010). *DIGITAL and media literacy: A Plan of Action*. The Aspen Institute.

Hu-Au, E., & Lee, J. (2018). *Virtual reality in education: a tool for learning in the experience age*. Columbia University, Department of Communications.

Jaaskelainen, L. (n.d.). E-learning and digital education - Statistics & Facts. *Statista*.

Justification, S. (2018). The Effects of Digital Curriculum in the Classroom: Rigorous Evidence from Three Digital Math Programs. *Conference Symposium Justification*.

Lim, C. P., Chai, C. S., & Churchill, D. (2011). A framework for developing pre-service teachers' competencies in using technologies to enhance teaching and learning. *Educational Media International*, *48*(2), 69–83. doi:10.1080/09523987.2011.576512

Panda, B. P., & Swain, D. K. (2009). *Effective Communications through e-Governance and e-Learning*. IBS Bhubaneswar.

PDST. (n.d.). *Technology Overview in Schools - ICT Infrastructure*. PDST Technology in Education.

Rimbar, H. A. (2017). the influence of spell-checkers on students ability to generate repars of spelling errors. *Journal of Nusantara Studies*, *2*(1).

Rossiter, A., Nortcliffe, A., Griffin, A., & Middleton, A. (2015). Using student generated audio to enhance learning. *Journal of the Higher Education Academy*, *4*(2).

Saxena, K. B. (2005). Towards excellence in e-governance. *International Journal of Public Sector Management*, *18*(6), 498–513. doi:10.1108/09513550510616733

Shabiralyani, G., Hasan, K. S., Hamad, N., & Iqbal, N. (2015). Impact of Visual Aids in Enhancing the Learning Process Case Research: District Dera Ghazi Khan. *Journal of Education and Practice*, *6*(19).

Simon, S. (2016). Mobile instant messaging support for teaching and learning in higher education. *The internet and higher education*, 31.

Srivastava, A. (2014). Basics in Education - Textbook for B. Ed. Course. National Council of Educational Research and Training.

Stern, J. (2004). *Introduction to Online Teaching and Learning*. wlac.education.

The Building Blocks of Digital Literacy. (2015). *Digital Literacy 21st Century Competences for Our Age*. Department of eLearning.

Thomas, K. (2014). *10 steps to salvation: Creating digital governance that works*. Confab Europe Barcelona.

Tokuhama-Espinosa, T. (2017). *Elegant Complexity: The Theory of the Five Pillars ofNeuoconstructivism in the Brain*. Harvard University Extension School, Latin American Faculty for Social Science Research.

Vaizey, E., & Boles, N. (2016). *Review of publicly funded digital skills qualifications*. Skills Funding Agency.

Veeckman, C., Schuurman, D., Leminen, S., & Westerlund, M. (2013). *Linking Living Lab Characteristics and Their Outcomes: Towards a Conceptual Framework*. Technology Innovation Management Review.

Vijayalakshmi, M. (2016). *Information and Communication Technology In Education*. Academic Press.

Welchman. (2017). Understanding Digital Governance Maturity. Academic Press.

Williams, S. E. (2014). *Education and Economic Development: An Untapped Alliance*. The University of Southern Mississippi, The Aquila Digital Community.

Zamora, Barahona, & Palaco. (2016). Case: Digital Governance Office. *Journal of Business Research*, *69*(10).

# Chapter 2
# Digital Era Governance and Social Media: The Case of Information Department Brunei

**Abdul Malik Omar**
https://orcid.org/0000-0002-7623-8969
*University Brunei Darussalam, Brunei*

## ABSTRACT

*Digital technology is at the forefront of transforming how governments operate around the world. Using Brunei's Information Department (InfoDept) as a case study, this chapter looks at how the agency has evolved from its inception in the 1950s to 2019 in its embrace of both old and new media to pursue its mission and objectives as a government-run media agency. The results demonstrate how new media, such as social media, can complement old media if done right. The case study on InfoDept contributes to the growing field of research related to the increased advancement, development, application, and impact of new technologies in bolstering the digital governance process. This chapter also provides strong evidence on how governments can improve its general governance process and unlock the digital dividend in the 21st century by incorporating new media into its public policy architectonic. Salient lessons for policymakers and practitioners on digital governance have also been presented in this chapter.*

## INTRODUCTION

Governments around the world have taken steps to embrace and incorporate technology into their operative functions (Bertot, Jaeger, Munson, and Glaisyer, 2012). New Media such as social media and the internet are capitalized to boost the digital governance process (Milakovich, 2012). In embracing digital governance, governments can unlock digital dividends in the form of increased public service delivery, citizen engagement, and organizational efficacy (UNDP, 2016). Governments also apply new technology in response to the increased public participation of the public through New Media. These technological

DOI: 10.4018/978-1-7998-1851-9.ch002

changes are the natural consequence of creative destruction (Schumpeter, 1942) that necessarily test organizations to adapt to become better. Brunei's Information Department (henceforth called InfoDept) is no exception when it comes to tapping into the digital governance. While digital governance may not necessarily deliver automatically digital dividends (Peña-López, 2016) this work demonstrates that Brunei's InfoDept has attained success, with it being ranked by We Are Social Report (2018) as having the most active government Instagram account in the world.

There are three justifications as to why this study is done: The first reason is to contribute to the growing literature on digital governance through the deployment, application, and impact of social media as a governance tool (Mallard, 2003) to boost the government's ability to build "better, more efficient public services" (Blair, 2005) and to promote engagement and trust with citizens (Almunawar et al., 2012). This study adds to the growing literature of governments moving towards the digital governance trends, which entails the empowering of the public sector to treat citizens as collaborators and active participants in the governance process. Secondly, the case study selected is unique in such that the InfoDept operates in a small country with a low digital divide where there is a low gap between the "haves" and "have-nots" concerning ICT and internet access (Ashraf, Hanisch, and Swatman, 2009). There is a high degree of internet penetration and social media usage. We Are Social Report (2018) ranks Brunei as the 4th highest when it comes to social media penetration in the world, with 95% of its population of 410,000 having access (We Are Social, 2018). This chapter ultimately unpacks the digital governance architectonic using InfoDept as a case study in order to advance the growing global literature that intersects digital governance, social media, and public policy. The success of InfoDept and the high usage of social media in the nation provide strong evidence for policy-makers and practitioners of how governance can be reshaped to the realities in the digital century, as with salient lessons that can be learned.

## LITERATURE REVIEW

E-governance is "...the use by government agencies of information technologies that can transform relationships with citizens, businesses, and other arms of government" (Steins, 2002: 18). E-governance also entails the exploitation of digital storage and internet communications to transform citizen-government engagement and interactions (Dunleavy, Margetts, Bastow and Tickler, 2006; Ronaghan, 2002). Done right, the incorporation of digital technology can assist governments to reduce operational costs, enhance public service delivery, and deepen citizen-to-government engagement. Meanwhile, E-government (or E-gov), refers to the use of IT, ICT, and other web-based telecommunication technologies to augment inter-organizational relations (Ronaghan, 2002) and enhance the "efficiency and effectiveness of service delivery in the public sector" (Fang, 2002). Jeong and Hai (2007) described e-government as the interactions at the digital level between people and government (C2G), government and other government agencies (G2G), government and citizens (G2E), and government and enterprise (G2B). Through social media, the government can interact with all levels of government and businesses. The difference between the two concepts is how E-governance focuses on the application of ICT for governance purposes, whereby E-government aims at improving the means of ICT-based public service delivery. Both of the concepts essential works to enhance the "public views of, and interactions with, the government via electronic channels" (Gauld, Goldfinch, and Horsburgh, 2010: p177) in reducing digital gap (Noh and Yoo, 2008) in Brunei.

Digital Era Governance (DEG) is a relatively new concept that gained prominence in the academic literature over the past few years, "...founded on themes of disaggregation, competition, and incentivization." (Dunleavy, Margetts, Bastow and Tickler, 2006). It was a response to the advent of the digital era, which is "...now the most general, pervasive, and structurally distinctive influence on how governance arrangements are changing in advanced industrial states" (Dunleavy, Margetts, Bastow and Tickler, 2006). The literature further categorized DEG into three key modalities: Reintegration, Needs-based holism, and Digitization. Dunleavy argues that DEG is an evolutionary by-product of earlier bureaucratic models of governance such as the New Public Management (NPM). The NPM is the governance model that features "...formalism, vertical hierarchies, limited communication channels, and has been particularly resistant to the influence of the external world" (Wojciech, 2017). NPM seeks to adopt a market-driven, transaction-oriented approach to the management of public services — leaving little room for active citizen participation (Nam, 2011). The concept has origins in institutional economics and has elements of managerialism. It was developed in the United Kingdom and was later adopted by the USA, New Zealand and Australia, and other OECD countries (Gruening 2001). The increased deployment of new technologies and the emergence of the internet has challenged the NPM model, among which has led to far-reaching social, cultural, and technological alterations (Margetts and Dunleavy, 2013). Digital governance essentially means to govern in the cyberspace using new technology such as social media and the internet (Milakovich. 2012).

DEG addresses this as a model for transforming the outlook in the administrative and bureaucratic structures to be more streamlined, simplified, and flexible. The concept of DEG also provides the notional capacity to govern within "less government, more governance" approach through the emerging "centerless society" (Rhodes 2006). Wojciech (2017) argues that the transformation of the DEG model can be seen "as a question of time". In the context of InfoDept, they are now tapping into 'e-government,' which seeks to promote everything from more significant interaction with the government; the saving of 'taxpayer dollars' (Gauld, 2006) and the streamlining of citizen-government relations. Out of the greater integration with technology, there lies the emergence of a new social contract that empowers the public to participate in the running of their governments (Linders, 2012). Such a means can unlock for citizens "unimagined opportunities to do more for themselves" (Johnston & Hansen, 2011) and to "dramatically alter how the public and government interact, develop solutions, and deliver services" (Bertot, Jaeger, Munson, & Glaisyer, 2010). Further, there is the now evolution from e-Government (citizen as a customer) to We-Government (citizen as a partner), which causes the emergence of "a new kind of social contract" (Long, 2002) that may require governance be operated in a polycentric way (Kooiman, 2003) as exemplified by DEG literature.

## RESEARCH METHODOLOGY

The epistemological and ontological basis in carrying out the methodology falls within the parameters of pragmatism followed by an inductive approach to reasoning, or to present a premise followed by the furnishing of facts and data in support of the thesis based on international reports. Qualitative methods are utilized as a means of discovering, reorganizing, and extrapolating the data to furnish the case study, which in this case is Brunei's InfoDept. This chapter begins by giving a brief background of Brunei Darussalam, followed by the historical development of InfoDept from its inception in 1 April 1952 to 2019, the rise of the digital era and its effect to the Bruneian digital scene, the embrace of new technology by InfoDept, and the results reaped by the organization as a result. New Media in this chapter refers to social media, with a focus on Facebook, Instagram, and Twitter, whereas Old Media refers to print media and public outreach, such as awareness talks to villages and schools. The results of the incorporation of New Media and Old Media have been extrapolated based on observation of how InfoDept has evolved over the years. Secondary sources from high-ranking and reputed journals as well as books have been unpacked in regards to the concepts of Digital Governance, Social Media, and Public Policy. Data and reports from the BCW (Burson, Cohn, & Wolfe), United Nations, We Are Social, Borneo Bulletin, EGNC website, Ministry of Communications, and AITI White Paper have been synthesized and evaluated to see how efficacious the adoption of technology has been for InfoDept. Several limitations have been identified and presented before concluding.

## BRIEF BACKGROUND OF BRUNEI DARUSSALAM

Brunei Darussalam is a country with a population of less than half a million people located in Southeast Asia. It is rich in Oil and Gas and is ranked by the Global Finance Magazine (2019) the 5th richest nation in the world in 2019. However, recent economic setbacks have pressed the government to intensify its diversification efforts through its national vision, called Wawasan Brunei 2035. The vision aims to produce in Brunei a highly skilled and educated people, high quality of life, and a dynamic and sustainable economy (PMO, 2018). As a means of realizing these goals, digital governance has been tapped into. In the 15th July 2019, Sultan Haji Hassanal Bolkiah consented to the establishment of the Digital Economy Council, with economic masterplan under the works (Xinhua Net, 2019). Among the factors that have been crucial in the lead up of the council's establishment is how good the government has been when it comes to adopting technology as a means of governance. However, digital governance "will continue to evolve for the foreseeable future providing a dynamic environment for ongoing learning and action." (Dawes, 2008). This chapter explores how the InfoDept has followed global trends to capitalize on digital media to improve its governance tool (Bhatnagar, 2004; Carrizales, Holzer, Kim, and Kim, 2006) and trust to build up national unity (Almunawar, Low, Rahman, and Mohiddin, 2012) and development.

## HISTORY AND DEVELOPMENT OF INFODEPT

The Information Department (InfoDept), formerly known as Brunei Information Service, is a state-owned media agency established on the 1st April 1952. It was then named the Information Services Department Brunei. Its first office and reading desk was established in 1954 in the Belait District. Two years after, the Pelita Brunei, the nation's first official newspaper was published and printed using cyclostyle technology. In the 1st January 1956/58, Pengiran Mohd Yusof bin Pengiran Haji Abdul Rahim was appointed as the agency's first officer. More officers were established at Bandar Town, Temburong District, and Tutong District that following year. With the establishment of the Radio Services Brunei in 1957, the InfoDept provided subsidies for radio purchases. By 1958, the InfoDept sold 1048 radio sets. On 15 February 1959, the Pelita Brunei was printed using the printing press. A monthly English-language publication called "Brunei (The Abode of Peace)" was published and distributed in March 1960. In 1983, the Information Unit was upgraded to become the Information Department.

On 20 October 1986, it as placed under the purview of the Prime Minister's Office. It celebrated its first Jubilee on the 1st April 2002 (1952-2002). On the 9th July 2008, Pelita Brunei was published twice a week (every Wednesday and Saturday) while the online version, Pelita Brunei is updated every day and can be accessed through www.pelitabrunei.gov.bn. In the 18th July 2010, Pelita Brunei was published three times a week, with the Monday edition introduced and added to the Wednesday and Saturday editions. On the 1st April 2012, InfoDept celebrated its 60th anniversary. In 2014, they relocated to a new building located in Berakas. On the 8th August 2016, the E-paper version of Pelita Brunei was introduced and can be downloaded through smartphones and tablets through www.epaper.pelitabrunei.gov.bn. On the 9th May 2017, the InfoDept launched its App, 'InfoDeptBN.'

The mission of the organization is to "....nurturing an informed society" (Information.gov.bn, 2019). It supports the national vision of Wawasan Brunei 2035 through the dissemination of government-sanctioned information towards strengthening socio-political unity and well-being of the country. They have five objectives: "1) To clarify and explain the official information in order to deepen understanding, inspire confidence and reinforce citizen's and residents' support for the leadership by monarch and public service institutions, 2) To deliver and promote the official information on government policy and intention in order to enable citizens and residents understand, support and internalise effectively, 3) To link the citizens and residents with the government through departmental programmes and activities, 4) To deepen citizens' and residents' understanding of the Wawasan (Vision) 2035, and 5) To manage, coordinate and implement media relations activities."

The Old Media medium employed by the InfoDept for attaining their stated objectives is through the publication of the tri-weekly free newspaper, Pelita Brunei. The reach of the paper is comprehensive in its geographical scope. Pelita Brunei is distributed across government ministries, mosques, and are even posted to village heads and Penghulus across the four districts. Copies can also be secured from fuel stations too. The contents of the paper reflect the events are happening during the week. The Sultan's official visits and special occasions are prominently featured on the front page. Meanwhile, subsequent pages are filled with stories, reports, and articles atypical of what is covered by any media outlets but with a heavy emphasis on governmental and civic-based themes featured in the paper. The article is written in Malay, which is the nation's official language. Given that the government has increased the publication frequency from twice a week to thrice a week attests to the success of the medium in reaching out to the public, particularly the elderly generation.

The secondary medium used by InfoDept is through public talks. Periodic public talks that are given include the 'Informed Community' program (Borneo Bulletin, 2019a). It is aimed at "keeping the community well-informed of current developments and efforts of the Government to enhance the livelihoods of the Bruneian people" and to disseminate "official information on the policies and intentions of the government for the community to understand, support and embrace them in an effective way through briefings, dialogue sessions, and visits." InfoDept also delivers talks in schools on various topics and themes. Among the topics is the 'Get to Know ASEAN' theme, which is a talk given by the InfoDept to school students across the country on the importance of ASEAN for Brunei. The information necessarily reflects the interest of the Brunei government and aims to enhance awareness on critical issues or themes that would be beneficial for the government.

In keeping with the Whole-of-Nation approach espoused by the government of Brunei, the InfoDept serves a conduit for organisations such as the Wawasan Brunei 2035 Permanent Secretariat, Department of Agriculture and Agrifood, DARe (Darussalam Enterprise), Royal Brunei Police Force, and the Ministry of Development to reach out to the public either through talks or by organising the 'Muzakarah Session'. The 'Muzakarah Session' is a town hall meeting between government representatives from various governmental departments with the public. Two sessions were organized in the past several years, one in Muara in 2017 and another in Pangkalan Batu in 2018. The session was aimed at increasing exchanges of information and feedback between the government and the public. The Muzakarah Session tends to attract large audiences, many of whom who attend the events composed of pensioners and the elderlies. In the event, the audiences can relay their feedback or complaints directly to the government representatives. The Muzakarah or Town Hall, organized by InfoDept, helps to enhance trust and information exchange between citizens and the government sector.

Overall, the InfoDept has evolved over the last sixty years or so since its establishment, and yet its core mission remains the same, which is to nurture an informed society. The following section demonstrates the changing external forces that would eventually shape InfoDept's modus operandi in the 21st century.

## THE RISE OF THE DIGITAL ERA

The internet is shaping how the world operations. Since its mass introduction in the 1990s and its continued widespread use in the 21st century, the internet has "flattened the world" (Friedman, 2005) by connecting people and providing the power of information to the fingertips of the ordinary people. Never in history, there has been such democratization of information for the average person. The internet brings people closer together. Once a letter sent in the 19th century from London to Brunei would take months to arrive. Nowadays, it only takes a WhatsApp text to send a message to the person. Also, the cost of communication has gone dramatically down, further enabling people to be digitally closer to one another. As of 2019, there were more than 4.4 billion internet users (Internet World Stats, 2019). Universal internet access has been projected to be reached at least by 2050 (Guardian, 2019). The widespread diffusion and fast adoption of the technology bring about its unique challenges and opportunities, many of the likes that people have never been encountered before. So while the internet does change the world, it brings about its share of positive and negative changes, making it imperative for policy-makers to take steps to actively forecast and help make a decent change to their communities through digital technology.

One powerful byproduct of the internet is social media. Social media is the "interactive computer-mediated technologies that facilitate the creation and sharing of information, ideas, career interests and other forms of expression via virtual communities and networks" (Kietzmann, Hermkens, McCarthy, & Silvestre 2011). It is also "a group of Internet-based applications that build on the ideological and technological foundations of Web 2.0, and that allow the creation and exchange of User Generated Content." (Kaplan et al. 2010). One of the earliest social media used in Brunei was MSN Messenger, a chat software that became popular in the sultanate in the early 2000s. It was followed with the rise and subsequent descent of Myspace, Friendster, and Blogs. Subsequently, sites such as Facebook, Instagram, and Twitter arose into widespread prominence among Bruneians. As of January 2018, Brunei Darussalam ranks as the 4th highest penetration of social media in the world with 95% of its population of 410,000 having access to social media or 99 percent of its 'eligible' population (aged 13 and above) (We Are Social, 2018). The report ranks Brunei at fifth in the world with 22.9% when it comes to 'Facebook Organic Reach Rankings.' The Global Digital Report (Ibid) puts the total number of monthly active Facebook 'users' in the Sultanate to be 350,000 or 85 percent of the population. Eighty-one percent of the population accessed Facebook via mobile phones. In addition, Brunei has the highest Instagram penetration rate in the world, with 49% or with 210,000 users using Instagram (Ibid).

Amidst the rise of the internet and social media, the Old Media model is now being challenged. The new wave of technological "creative destruction" threatens the existence of news media outlets and newspapers (Schumpeter, 1942). Between 2004 to 2018, more than one in five newspapers or 1,800 local newspapers throughout the United States have close down (US News Deserts, 2018). Meanwhile, the easy access to torrenting and the broad wide range of entertainment options have made many old media channels to lose out in its appeal among the public. When people lose interest in print media, for example, newspaper sales revenue drop. Advertisers will also respond to the change by advertising in new media channels (such as Facebook and Instagram) instead of through newspapers. In 2016, The Brunei Times, an English newspaper agency, shut its doors down because of commercial losses arising due to the challenges from alternative media (The Borneo Post, 2016). The Brunei Times is not the only paper in Brunei that has experienced hardship. Even Borneo Bulletin, Brunei's only English newspaper, has increased its prices from BND$0.8 to BND$1.00 in August 2018 (Borneo Bulletin, 2018a). The revision of the newspaper sales price is perhaps indicative of the financial challenges posed by the internet. Indeed, New Media have a substantial effect on Old Media as evidenced by the effect it has onto Brunei's local newspapers.

Governing has also evolved in light of the digitization of a global society. The concept of e-governance was introduced and expanded upon in response governments' need to tap into technology to serve the public better (Saugata and Masud, 2007). The need to set up a governmental digital framework to both incorporate new technologies as well as to enhance citizen-state relations is paramount. The Brunei Government answered the call by amplifying its determination to realize the nation's aspiration to become a key player in the 4th Industrial Revolution, as per His Majesty the Sultan of Brunei's Speech in the Brunei Legislative Council 2019 (Borneo Bulletin, 2019b). Furthermore, the World Economic Forum Global Information Technology Report 2014, under the Network Readiness Index, ranks Brunei Darussalam 45$^{th}$ worldwide (World Economic Forum, 2014). There is also the National ICT White Paper (Ministry of Communications-AITI, 2016) commissioned by Ministry of Communications and the Authority for Info-communications Technology Industry of Brunei Darussalam (AITI) which aimed at growing the ICT industry by 2020. The White Paper states that 6% of ICT contributes to Brunei's GDP. The cost of broadband, a prerequisite in creating a digital economy, is 1.5% of average monthly income. It has 18,000 citizens trained in ICT literacy. There is even an innovation hub called iCentre (Omar, 2016). Finally, Brunei Darussalam is ranked 59th among 193 countries in the United Nations E-Government Survey 2018 (Borneo Bulletin, 2018b; United Nations, 2018). In light of these initiatives, the Brunei government has taken proactive steps in building the infrastructure and human capital for e-governance needs.

From another angle, the increasing use of social media by the public has necessitated the government to incorporate social media into its governance model. As of 2018, Brunei is ranked top 5 when it comes to social media and internet penetration. A Facebook 'Fanpage' called 'Brunei FM' dwarfs the government in its influence on social media. As of 1st August 2019, the fan page has over 260,684 'likes', which is close to 60% of the population figure of Brunei Darussalam of428,697. Smartphone usage is also substantial among the public. In a study done by Anshari and Alas (2015), fifty percent of 589 Bruneian respondents used e-news to get their information. 38% spent less than six hours per day using the Internet, 38% spent 6 to 12 h per day, and 25% indicated that they were always connected, which meant that nearly two-thirds of respondents accessed the Internet for more than six hours daily. Also, a survey was conducted, which concludes that 97% of the participants used smartphones for communication with the use of WhatsApp (Ibid). With Brunei being a tribal and collective society of less than half a million people, a viral message can reach the entirety of the country's population in less than

12 hours or even less. To illustrate the power of WhatsApp and internet connectivity in Brunei, when the New Zealand Shooting took place early in 2019, it only takes less than 4-5 hours for the dubious shotting video to reach the Bruneian virtual public. The democratization of information has its dubious side when the contents shared are explicit and violent, especially when the digital scene is heavily embedded into the society #.

The rise of fake news has also necessitated the government to introduce new laws and to clamp culprits down. As of 2019, it is a punishable offense for people to knowing share false news. The dissemination of false information is an offense under the Public Order Act, Chapter 148, which carries a penalty of three years' imprisonment, or a BND3,000 fine, or both (Borneo Bulletin, 2018c). As an agency, InfoDept reports directly to the Prime Minister Office, and whatever statement that is required to be put out to all government ministries would be channeled, among many others, through them. The letter or statement signed and approved by the government agencies through the InfoDept would then be circulated across social media. An interesting note is how the anti-fake news law was championed by a youth member of parliament, YB Khairunissa Ashari, who herself is internet savvy (Scoop, 2018a). The law was passed soon later that further saw the incorporation of new legislation in the e-governance process. The InfoDept played a role in the spread of awareness needed to clamp down the spread of fake news. People can also complain about fake news through a government hotline (Xinhua, 2018).

On the whole, the internet is an unstoppable force that has transformed how people communicate, operate, and connect in the world. Online medium such as social media became a powerful force that empowers the average individual to participate digitally. Whether their motivations and actions may be positive or negative is another question altogether, what matters is how government can respond to these changes by incorporating the same technologies the people use when it comes to improving its governance process. The case of the Bruneian population and its high rate of social media usage attests the need for the government to respond (and they did) when it comes to formulating policies that work and is coherent with the governance process. The case of the InfoDept and its incorporation of new technology have been expanded in the following section.

## INFODEPT EMBRACES DIGITIZATION

In light of the increased usage of the internet, the Brunei government, or more specifically, the InfoDept wisely embraced and adopted the technology in its public policy architectonic. It was a gradual change rather than an instantaneous one, as evidenced by the department's introduction of its website, e-newspapers, and Apps, as well as the phasing out of old technologies such as cyclostyle over the decades. These changes were done by stages until at last; it finally incorporates elements of social media to its core functions. These e-reforms were made in reflective of the appreciation and recognition of new technology to complement its old model of paper-based newspaper distribution. The was also the recognition of the dangers of becoming irrelevant, and out-of-touch they were with the general public lest they stick to the way how they do things in the past. However, time and time again, the InfoDept has been proactive in its embrace of new technology. From its very inception more than sixty years ago to 2019, they remain relevant to the Bruneian polity as a source of government-sanctioned information with the frequent injection of new investments and technologies as reflective of the development it has experienced over the years.

The InfoDept has been proactive when it comes to complementing new technology such as social media, to achieve its mission and objectives. If for the past fifty years or so, it manages to disseminate official information through newspaper medium, now it can set up and operate social media outlets in boosting its efficacy. By incorporating social media into its modus operandi, the government can reach the wider public, a majority of whom are digitally savvy and connected. By using social media, InfoDept is then able to produce critical updates and publish new information that would be useful for general society online, and hence strengthening citizen-government ties in the process. Its online newspapers uploaded can also serve as a conduit for local and international researchers to mine data on the history or events happening around Brunei, which means that people do not need to get the physical Pelita Brunei paper any longer or to go to the archival center to get updated with the latest happenings around the country. It only takes a google search to find out what is happening in the past week or so, for instance. Unlike other private sector media, the contents are provided for free. There is no paywall to block the public's access to this information, and hence, furthering InfoDept's mission and objectives through New Media.

The InfoDept planned out a systematic yet straightforward strategy when it comes to its social media operations. First, they have all the social media channels prepared. These channels include Facebook, Twitter, and Instagram. These three serve as the primary conduits upon where the InfoDept can reach out to the public. Next, they have their news desk agencies ready to publish new information or new articles through these mediums. In a given day, the InfoDept can publish up to 17 Instagram posts a day, as evidenced by the data provided by We Are Social (2018). The articles and accompanying photos published then serve as the "digital newspaper" to reach the public. Based on observation, they do get engagement in the form of comments, 'likes' and 'follows' to their posts. It is most likely that the social media operations are handled and centralized all the accounts within one central unit or body within the InfoDept. Within this central system and department are several individuals who are responsible for managing all the social media accounts. These social media accounts are being managed using a centralized social media management system. Centralizing the process also ensures that the department can easily monitor and control the social media process smoothly and to help overcome the organizational or institutional challenges that may be associated with the incorporation of digital technology (Criado, Sandoval-Almazan and Gil-Garcia, 2003).

There are three reasons identified as to why social media is tapped into as a new source channel to reach the public. The first reason is how social media is virtually free. No matter how much Facebook posts or Instagram uploads they have made, it cost them virtually zero dollars to do operate. As McLuhan and Fiore (1967) said, the medium has become the message. Given there are no charges to the system, it makes it appealing for the government officers to use it to complement its main operations. In light of the austerity measures put into place by the government in response to the oil downturn in 2014, the government has been continually instructed to be innovative and cost-conscious. It is thus no surprise that social media became a critical tool that was later adopted by the InfoDept. There are paid services offered by social media, such as Facebook's Paid Ad services, but there has been no sign that the InfoDept has paid for the service thus far. They remain conservative in its operations, depending heavily on its organic or natural reach to the general public.

The second reason why social media was used is because of factor endowment in the form of an internet-savvy public. Given the fact that more than 95% of its population have internet access can explain why social media has been successful as a public policy tool. On the note of government sector, more than 40% of the national workforce belong to the public sector (JPKE-ILO, 2017). Most also do already know what InfoDept is all about, so there is already a conscious awareness among the public of its existence as a public sector media arm. The high rate of social media literacy and low digital divide reduces any costs that may be associated with training the InfoDept officers on using social media, for they are heavy users of the technology. Similarly, the InfoDept being the conduit of government sanction information made it necessary to adopt social media as the public workforce would want to check the updates to learn essential announcements relating to their respective ministries. The streamlining of this process increases the government's efficacy and effectiveness to attain its mission to reach people throughout the people living in the four districts of the country.

Finally, it has got to do with national security. As mentioned before, cases such as Fake News can get viral and destabilize national security and harmony. To minimize the negative impact of fake news happening, the government through the InfoDept serves as that trusted and go-to resource to verify any stories that went viral or to dispel Fake News stories. The InfoDept then becomes one key frontline resource to combat any falsehoods that come in the way. The best part about social media is how the news published can be quickly and universally accessed by everyone instantly. Had the InfoDept solely rely on physical-based papers, a process that takes time, then it may have already been too late. The damage by fake news would have already been done. Social media augments the problem by providing almost instant publication of correct facts to the public online.

Whereas social media and the internet have made changes to how the InfoDept operates, it nevertheless retained its old model operations of publishing a newspaper and giving talks to the public. Rather than treating new media as a threat to old media, the officers wisely wedded the two together to further enhance the InfoDept's efficacy. An example is how Pelita Brunei has its e-paper version published online now. If a villager may have missed the physical version of Pelita Brunei, he or she can easily use the smartphone and download the e-paper version. Another example is how people can get daily updates regarding the news through posts made by InfoDept in social media. As long as they follow the InfoDept's social media accounts, they can get these updates and stories that may interest them. In keeping with its mission to create an informed society, having the organization positioned in the social media sphere enables them to digitally converge with the citizens. Meanwhile, the Old Media further supplements the process. All these changes unleash the synergistic effect between old and new media.

The InfoDept has also continued to polish its old media operations. Evidence of the improvements made includes the Pelita Brunei. Over the years, the Pelita Brunei has gain improvements to its design. The quality of phots and design print are more attractive and better than before. The changes made were a massive improvement since its early inception, where it was just black and white and published using 19$^{th}$-century cyclostyle technology. Now the newspaper is in full-color and printed using high-quality printing presses. It has even increased its publication from twice to three times a week. The circulation has also been increased over the years to attain a greater geographical reach. Nowadays, the council members, not just the Village Heads, are given free copies of the Pelita Brunei. The papers are sent strategically in such that it reaches out to its relevant audiences, namely village leaders and their committees. Public talks given to schools and other communities have also gained prominence over the years. Nowadays, a fresh batch of young Bruneian officers has been empowered to give these talks, thus adding to the color, relevance, and angle that further deepens societal ties and effective information dissemination to young people, especially. Whatever the changes made, the InfoDept continues to bolster its old media operations in light of the changes bought about by the internet and social media.

Auxiliary elements have necessarily played a role in the way in which the InfoDept operates in the digital world. The establishment of EGNC in 2014 saw the gradual standardization of ICT-related training of human capital in the government workforce. EGNC was entrusted with training InfoDept officers on areas relating digital technology as well as to enhance the InfoDept's team ICT literacy and competence. The training may have included social media management and consultancy that may have steered the InfoDept to take a new channel seriously. From a technical point of view, EGNC under the IT and E-Government unit (PMO) was also crucial in setting up the department's website. The second auxiliary element is the government's update to its legislation in keeping up with the technological changes in the society. These legislations include Chapter 153 Official Secrets Act An, Chapter 194 Computer Misuse Act, Chapter 196 Electronic Transactions Act, and the Chapter 180 Broadcasting Act - Internet Code of Practice. These legislations were introduced and reformed to diminish the chances of abuse or hacking, either from inside or outside the organization. Local ICT security agencies such as Brunei's Information and Cyber Security (ITPSS) also stand ever-ready to protect InfoDept from online threats and security incidents that may arise.

## RESULTS ACHIEVED BY INFODEPT

The consequence of InfoDept's systematic and consistent use of social media has produced exceptional results for the organization. For one, the department's Instagram post is ranked the top in 2018 worldwide, making it the most active account in the world. It has published a total of 6,252 posts and with over 17 posts made per day on average (We Are Social, 2018). Most of the posts published are high-quality posts with photos and good descriptions accompanying them, not just randomized stories. For a country with less than half a million people, the success earned by the agency is a remarkable and unique achievement to be celebrated. Given too that the InfoDept has worked its social media without any substantial investment in the software (social media is free to use) just further demonstrate the cost-effectiveness of the model in reaching out to the public as a public policy tool. The success is in keeping with the continual drive for digitization by the government. The achievement has also probably empowered the government to further focus on digital governance.

The following result is how there is a high level of awareness and outreach to the public through the digital space. Between October 1, 2017, and September 30, 2018, InfoDept recorded around 270,000 likes, 130,000 views, and around 800 news and videos on its social media platforms (We Are Social, 2018). Based on the data, InfoDept has secured significant to the Bruneian public, as compared to its Old Media. For example, there are hardly any young readers who read the physical newspaper Pelita Brunei nowadays, but a majority of them do have social media and this is where InfoDept reach out to them. Had the government solely focus on Old Media, they would have risked overlooking the youths, which makes up more than 70% of the national population. Local and international researchers can even google in the e-papers or online news articles of stories that are relevant to their needs. Brunei-related information can be hardly found in other news outlets or online publications, and thus, the stories produced by InfoDept adds to a unique angle that adds to the freshness and diversity of ideas relating to the country and its community. Overall, the InfoDept team's reach to a new audience through the digital platform has made the department more accessible and relevant to a broader readership base, be they youths or researchers.

Finally, it can be observed how social media has become a core model, not just a supplementary model that works to enhance the InfoDept's success. There is a strong team within InfoDept that handles the social media accounts to help the other teams that look into other non-online matters. Given the consistency of the posts and the high-quality output produced, it would be no surprise why they would not have a reliable and experienced team in place to carry the policy out. The quality control is also high and is consistent with how the department presents itself to the public audience. Social media is also used to deepen ties with the public. The public is encouraged to follow or like their social media accounts. The invitation to follow their social media has been extended to ads inserted in Pelita Brunei and even in public talks, which reflects the synergistic nature of Old and New media. Social media hence complements the agency's ability to strengthen and deepen digital ties or relationship with the citizens. From a strategic point of view, the InfoDept has successfully tapped into social media to strengthen its overall competence, both in the old and new medium, in enhancing general governance in the digital precinct.

## LIMITATIONS OF INFODEPT'S STRATEGY

Despite the accomplishments by InfoDept, it remains difficult to track actual engagement with the public. Indeed, there are chances that most of the followers or likes are akin to "bots," giving rise to the question of whether the InfoDept's efforts have been successful in deploying social media as technology or not. In using the platforms too, the agency is merely publishing content and can be seen to make little effort to engage with social media users. The consequence of this is how it can instead deepen one-way, top-down dimensions of governance, which can be counterproductive to the agency. Rather than permeating a form of genuine engagement, it would instead become superficial. Instead, it could actively – not periodically – engage with the public through the digital sphere. To InfoDept's credit, however, they do reply genuine commentators occasionally but to the effect that they ask to contact them through the proper channels, such as the agencies' office number and email. A possible reason why InfoDept may be less susceptible to reply to comments and to direct them to the aforementioned proper communication channels is to need to minimize the risk of potential controversies or fallout that may arise from replying to internet trolls.

The next limitation is the potentiality of InfoDept's social media system of being compromised by nefarious actors or foreign agents through cyber-hacking. The inherent danger associated with this includes it being hacked and misused by other irresponsible parties is a problem among stakeholders in Brunei, be they from the public and the private sector. In 2016, there was a total of 2,143 cybersecurity attacks recorded in Brunei (Scoop, 2018b). Unless the government takes preemptive action to defend its digital territory (Beslay and Hakala, 2007), hackers can seriously deal a lot of harm to websites, steal confidential information, and destroy valuable national data. Even worse is if the hackers would publish harmful materials that seek to permanently damage the reputation of the government. InfoDept mitigates such a situation by following strict protocols and guidelines that are outlined by the government. As stated earlier, national legislation has been put into place to diminish the chances of abuse or hacking, either from inside or outside the organization.

While online engagement has been used as a means to deepen ties with the citizenry, InfoDept does still depend on much of its agency's relevance to traditional forms of media – Pelita Brunei - and on-the-ground operations (delivering talks to school students and organizing Muzakarah events) to disseminate information across the country. The social media channels, therefore, gives us only one dimension to measure their real presence and accomplishments. A way to overcome the research limitation is how there ought to be a mixed research approach with qualitative and quantitative methods, including interviews with the officers from the InfoDept themselves to ascertain the facts and narrative of what is being presented in this chapter. Such a process would enhance further the perspectives of how the government works, and to the square in the factors that have been discussed in the previous section to whether the elements have significant effects when it comes to reorienting the government to embrace new technology to deliver public services. However, a case study would be sufficient as a book chapter given the apparent abundance of secondary data as well as theoretical insights and observations that were used to elaborate and justify the study on digital governance.

## RESULTS AND CONCLUSION

This study validates the success of the InfoDept as a state-media agency in capitalizing and deploying social media as digital technology to enhance its reach to the public. In the process, this chapter contributes to the growing fields of research related to the increased advancement, development, application, and impact of new technologies in bolstering the digital collaborative governance process. From its inception to 2019, InfoDept has proved resilient and innovative in its quest to rebrand and to make itself continually relevant to the Brunei market. Through the InfoDept, Brunei serves as an example of a nation that has successfully reaped what it sowed when it comes to the improvement, development, application, and impact of new technologies in strengthening the digital governance process. Finally, Old and New media does not need to be at loggerheads with one another. They do not need to be a dichotomy. Instead, they can be integrated together to produce a synergistic effect that can result in something substantial for the government, as with the example of InfoDept.

Policy lessons that can be derived from this chapter are that to govern in the digital era, a proper appreciation and application of new technologies, such as social media (Instagram, WhatsApp, and Facebook), have to be operationalized into the public policy architectonic of an organization. Whichever organization applies it consistently and systematically well can see its effectiveness and efficiency increase without expending too much on costs. Another important lesson from this chapter is that having the positive institutional attitude towards technology as a tool to transform how governments govern in the changing era is crucial to keep pace with the increased innovation that is being generated in the world. The advancements of technology would necessarily produce innovations that have to be taken into consideration in the public policy to further improve the general governance process, be it non-digital or digital.

## REFERENCES

Almunawar, M. N., Low, P. K. C., Rahman, M. H., & Mohiddin, F. (2012). From e-government to e-governance: winning people's trust. In *E-Governance and civic engagement: Factors and determinants of e-democracy* (pp. 613–636). IGI Global. doi:10.4018/978-1-61350-083-5.ch030

Anshari, M., & Alas, Y. (2015). Smartphones habits, necessities, and big data challenges. *The Journal of High Technology Management Research*, *26*(2), 177–185. doi:10.1016/j.hitech.2015.09.005

Ashraf, M., Hanisch, J., & Swatman, P. (2009). ICT intervention in the "Chandanbari" village of Bangladesh: Results from a field study. *Information Systems Frontiers*, *11*(2), 155–166. doi:10.100710796-008-9133-0

Bertot, J., Jaeger, P., Munson, S., & Glaisyer, T. (2010). Engaging the public in open government: Social media technology and policy for government transparency. *Computer*, *43*(11), 53–59. doi:10.1109/MC.2010.325

Bertot, J. C., Jaeger, P. T., & Hansen, D. (2012). The impact of polices on government social media usage: Issues, challenges, and recommendations. *Government Information Quarterly*, *29*(1), 30–40. doi:10.1016/j.giq.2011.04.004

Beslay, L., & Hakala, H. (2007). *Digital territory: bubbles. European visions for the knowledge age: A quest for new horizons in the information society*. Academic Press.

Bhatnagar, S. (2004). *E-government: from vision to implementation: a practical guide with case studies*. New Delhi: Sage Publications.

Blair, T. (2005). *Prime Minister's Mission for E-Government Unit*. London: Cabinet Office.

Borneo Bulletin. (2018a). *Revision in cover price for Borneo Bulletin and Media Permata*. Retrieved from https://borneobulletin.com.bn/revision-in-cover-price-for-borneo-bulletin-and-media-permata/

Borneo Bulletin. (2018b). *Brunei leapfrogs to 59th place in the United Nations E-Government Survey*. Retrieved from https://borneobulletin.com.bn/brunei-leapfrogs-to-59th-place-in-un-e-government-survey/

Borneo Bulletin. (2018c). *Fake news can destabilize the nation*. Retrieved from https://borneobulletin.com.bn/fake-news-can-destablise-the-nation/

Borneo Bulletin. (2019a). *'Informed Community' programme held at Information Department*. Retrieved from https://borneobulletin.com.bn/informed-community-programme-held-at-information-department/

Borneo Bulletin. (2019b). *Sultan says Brunei to see better growth in 2019*. Retrieved from https://borneobulletin.com.bn/sultan-says-brunei-to-see-better-growth-in-2019/

Carrizales, T., Holzer, M., Kim, S. T., & Kim, C. G. (2006). Digital governance worldwide: A longitudinal assessment of municipal web sites. *International Journal of Electronic Government Research*, *2*(4), 1–23. doi:10.4018/jegr.2006100101

Criado, J. I., Sandoval-Almazan, R., & Gil-Garcia, J. R. (2013). *Government innovation through social media*. Academic Press.

Dawes, S. S. (2008). The evolution and continuing challenges of e-governance. *Public Administration Review*, *68*, S86–S102. doi:10.1111/j.1540-6210.2008.00981.x

Dunleavy, P., Margetts, H., Bastow, S., & Tickler, J. (2006). New public administration is dead — Long live digital-era governance —. *Journal of Public Administration: Research and Theory*, *16*(3), 467–494. doi:10.1093/jopart/mui057

Fang, Z. (2002). E-government in the digital era: concept, practice, and development. *International Journal of the Computer, the Internet and Management, 10*(2), 1-22.

Gauld, R. (2006). E-government: What is it, and will it transform government? *Policy Quarterly, 2*(2).

Gauld, R., Goldfinch, S., & Horsburgh, S. (2010). Do they want it? Do they use it? The 'Demand-Side' of e-Government in Australia and New Zealand. *Government Information Quarterly*, *27*(2), 177–186. doi:10.1016/j.giq.2009.12.002

Global Finance Magazine. (2019). *Richest Countries in the World 2019*. Retrieved from https://www.gfmag.com/global-data/economic-data/richest-countries-in-the-world

Gruening, G. (2001). Origin and theoretical basis of New Public Management. *International Public Management Journal*, *4*(1), 1–25. doi:10.1016/S1096-7494(01)00041-1

Guardian. (2019, Jan. 8). Universal internet access unlikely until at least 2050, experts say. *The Guardian.*

Hai, C., & Jeong, I. (2007). *Fundamental of development administration*. Scholar.

Information.gov.bn. (2019). *Information Department's Mission and Objectives*. Author.

Internet World Stats. (2019). *Global Internet Users Statistics*. Author.

Johnston, E., & Hansen, D. (2011). Design lessons for smart governance infrastructures. In D. Ink, A. Balutis, & T. Buss (Eds.), *American Governance 3.0: Rebooting the Public Square?* National Academy of Public Administration.

JPKE-ILO. (2017). *Labour Force Survey 2017*. Department of Economic Planning and Development.

Kaplan, A. M., & Haenlein, M. (2010). Users of the world, unite! The challenges and opportunities of Social Media. *Business Horizons*, *53*(1), 59–68. doi:10.1016/j.bushor.2009.09.003

Kietzmann, J. H., Hermkens, K., McCarthy, I. P., & Silvestre, B. S. (2011). Social media? Get serious! Understanding the functional building blocks of social media. *Business Horizons*, *54*(3), 241–251. doi:10.1016/j.bushor.2011.01.005

Kooiman, J. (2003). *Governing as governance*. SAGE Publishing.

Linders, D. (2012). From e-government to we-government: Defining a typology for citizen coproduction in the age of social media. *Government Information Quarterly*, *29*(4), 446–454. doi:10.1016/j.giq.2012.06.003

Long, M. (2002). Beyond traditional boundaries: Government in the information age. *Australian Journal of Public Administration*, *61*(1), 3–12. doi:10.1111/1467-8500.00253

Mallard, T. (2003). *New Zealand E-Government Strategy - June 2003 - Update*. Wellington: Minister of State Services.

Margetts, H., & Dunleavy, P. (2013). The second wave of digital-era governance: a quasi-paradigm for Government on the Web. *Philosophical Transactions of the Royal Society A: Mathematical, Physical and Engineering Sciences, 371*(1987), 20120382.

McLuhan, M., & Fiore, Q. (1967) The medium is the message. New York: Academic Press.

Milakovich, M. (2012). *Digital governance: New technologies for improving public service and participation*. Routledge. doi:10.4324/9780203815991

Ministry of Communications-AITI. (2016). *National ICT White Paper for 2016-2020*. Author.

Nam, T. (2011). Suggesting frameworks of citizen-sourcing via Government 2.0. *Government Information Quarterly*, (29): 12–20.

Noh, Y. H., & Yoo, K. (2008). Internet, inequality, and growth. *Journal of Policy Modeling*, *30*(6), 1005–1016. doi:10.1016/j.jpolmod.2007.06.016

Omar, A. M. (2016). *Innovation for Economic Development*. Academic Press.

Peña-López, I. (2016). *World development report 2016: Digital dividends*. Academic Press.

PMO. (2018). *Prime Minister's Office. Vision, Mission, and Strategy*. Retrieved from http://www.pmo.gov.bn/SitePages/Vision,%20Mission%20and%20Strategy.aspx

Rhodes, R. A. W. (1996). The new governance: Governing without government. *Political Studies*, *44*(4), 652–667. doi:10.1111/j.1467-9248.1996.tb01747.x

Ronaghan, S. A. (2002). *Benchmarking e-government a global perspective: Assessing the progress of UN member states*. New York, NY: United Nations DPIDG.

Saugata, B., & Masud, R. R. (2007). Implementing E-Governance Using OECD Model(Modified) and Gartner Model (Modified) Upon Agriculture of Bangladesh. IEEE.

Schumpeter, J. (1942). Creative Destruction. *Capitalism, Socialism, and Democracy, 825*, 82-85.

Scoop. (2018a). *Darussalam Hotline now open for social media woes*. Retrieved from https://thescoop.co/2018/09/25/darussalam-hotline-now-open-for-social-media-woes/

Scoop Media. (2018b). *More than 2,000 cybersecurity attacks in Brunei last year*. Author.

Steins, C. (2002, Sept.). E-government: the top ten technologies. *Planning*.

The Borneo Post. (2016). *Brunei Times shuts down today after ten years*. Retrieved from https://www.theborneopost.com/2016/11/08/brunei-times-shuts-down-today-after-ten-years/

UNDP. (2016). *Citizen Engagement in Public Service Delivery The Critical Role of Public Officials*. Author.

United Nations. (2018). *The 2018 United Nations E-Government Survey: Gearing e-government to support transformation towards sustainable and resilient societies*. Author.

US News Deserts. (2018). *The Expanding News Desert*. Retrieved from https://www.usnewsdeserts.com/reports/expanding-news-desert/

We Are Social. (2018). *Global digital report 2018*. Retrieved from https://wearesocial. com/blog/2018/01/global-digital-report-2018

Wojciech, B. (2017). Digital Era Governance–a new chapter of public management theory and practice. *MAZOWSZE Studia Regionalne*, (22), 117-129.

World Economic Forum. (2014). *Network Readiness Index*. World Economic Forum.

Xinhua Net. (2018). *Brunei introduces hotline to address fake news on social media*. Retrieved from http://www.xinhuanet.com/english/2018-09/25/c_137492233.htm

Xinhua Net. (2019). *Brunei sets up Digital Economy Council to explore economic potential*. Retrieved from http://www.xinhuanet.com/english/2019-07/15/c_138229056.htm

# Chapter 3
# Digital Governance for Developing Countries Opportunities, Issues, and Challenges in Pakistan

**Bushra Hamid**
*PMAS-Arid Agriculture University Rawalpindi, Pakistan*

**N. Z. Jhanjhi**
https://orcid.org/0000-0001-8116-4733
*Taylor's University, Malaysia*

**Mamoona Humayun**
*College of Computer and Information Sciences, Jouf University, Saudi Arabia*

## ABSTRACT

*Information communication technology plays a vital role in countries' economic growth, and at the same time, it changes the form of traditional government to digital government. Digital governance is about having a competent and responsible government to deliver efficient, better, and 24/7 public services. It improves citizen participation, transparency, and overcomes administrative bottlenecks. Slackness of any government organisation in providing quality services will instantly become a topic of debate on print, electronic, and social media, forcing the political leaders to take corrective actions to save their own skin. However, for developing countries, this is going to be difficult to attain the same level of efficiency and flexibility. This study will examine the state of digital governance in developing countries particularly in the Pakistan government and will highlight the facts that through the implementation of digital governance in Pakistan (i.e., challenges and difficulties as well as opportunities that will help us to promote the application of digital governance).*

DOI: 10.4018/978-1-7998-1851-9.ch003

## INTRODUCTION

These days' internet is being used to provide the online services by the governments to its citizens (Layne & Lee, (2001). Due to ICT technologies and internet, governments can serve their citizens in better way (Misuraca & Viscusi, 2015). ICT technologies can be applied in different ways, for example, improved provision of public services to the population, improved cooperation with corporate organizations and industries, empowerment of citizens through access to information, or more efficient public administration (Khan & et al, 2010). Digital governance is a framework to establish roles, accountability and decision making authority for institute digital presence i.e. internal and external websites, mobile applications and social networks.

After the e-government model, public establishments all around the world are transforming from conventional to automated forms, as they realize the significance of creating services more efficient and reachable (Beardsley & et al, 2010). An uprising information communication technologies are not only altering the day-to-day life of public but also transforming processes of the collaboration among the governments and their people (Gil-García & et al, 2005). Now a days, digital governance is considered as standard tools of statecraft in developed country. E-governance is vigorous for development of economies and capacity-building for residents to contribute in ameliorating governance. The overarching goal of digital governance is provision of service and information to various stakeholders like citizen, business, government and employees, in an efficient and cost effective manner. These goals strengthen the Government by administrative reforms, employing technology, and new systems. The fundamental goals of digital governance based on e-Government, e-regulation and e-democracy. These fundamentals profoundly serves the object of delivering better services, overcome governmental bottlenecks, improve accountability, and enhance transparency and participation of citizen. Countrywide and native governments round the global have participated profoundly in introducing ICT based services since last two decades. However, till now the outcomes of these investments have not met the targets and give fruitful results as expected. This situation is particularly more observed in the public sector, where the take-up of ICT based public services has been comparatively less and conversion of the managements not as rapid and radical as it was anticipated (Misuraca & et al, 2010).

There are some good reasons to believe that transition towards an effective digital governance framework can be achieved by capacity building of government departments. The impact of transition is more evident when there is motivations and imperatives for adopting digital governance. To achieve success evaluation controls and strict monitoring system can also be employed in absence of capabilities and

*Figure 1. Digital governance*

motivation. Moreover, organizational structure also adds towards or against the objectives of digital governance. Organization lacking digital literacy, adaptability to change, e-readiness and inextricably processes proved to be decisive hindrances. Stakeholders often lead to confusion by lack of consistency in rules, regulations and policies of government.

Today, in this era of information economy, stakeholders needs a highly responsive and dynamic role of government and its leadership. To instill a trust affiliation between the stockholders, a firm commitment of government became unquestionable. Establishing a sound strategy of digital governance can uphold the expectations of stakeholders. The strategy should be capable of managing the conflicting goals, complex regulatory environment and Digital Supply Chain system.

To improve accountability, efficiency and productivity while providing better public services, government institutions often emphasis only on capacity building, whereas, the citizen feedback is not considered. For improving the digital governance, feedback serves as a catalyst by emphasizing on organizational requirements. Another major factor of digital governance framework is open data. This key factor enables the stakeholders in decision making, formulation of policies and alleviation of services. While considering the aforementioned factors, digital governance can bring Simplicity, Morality, Accountability, Responsiveness and Transparency in governance.

This chapter is structured into sections. In Section II a literature review of e-government has been presented. Section III discusses the status of e-government in neighboring countries. Section IV discusses the recent state of ICT in Pakistan, then in section V strengths and Challenges has been presented. Section VI conclude the chapter as well as provide suggestions and ideas for addressing the issues faced by Pakistan in implementation of e-Government.

## LITERATURE REVIEW

Moor & Norris (Moor & Norris (2005) described E-Government as "means of delivering government information and services". Another definition is "E-Government is the use of ICTs to promote more effective and efficient government that facilitates more accessible government services, improves public access to information, and makes government more accountable to citizens" (Alavi & et al, 2018) The survey conducted by United Nation for e-readiness shows a relative assessment of the 198 United Nation member countries. The key purpose for conducting this survey is to show each government stability and shortcomings within digital governance readiness capacity. Affiliated countries are at different levels of provision of digital government services. Few developed countries are in the introductory phase of the transfer of e-government to i-government. This transformation is the basis for the transition from a bureaucratic government to a public-centric government. Some countries are in the transactional phase of e-government, while others are in the preliminary phase of e-government, where very few services are available online (Ilyas, 2016).

ICT plays a vital role in the field of E-governance. Government use ICT to bring benefits both inside and outside; inside by providing benefits such as an improved staff interest, better political control, and an improved public image and outwardly by delivering economical improved services to those who be contingent on government (Heeks, 2001). E-government use ICT as an instrument to accomplish good governance. Good governance means providing public services in a convenient, customer-oriented, cost-effective and improved way (Song, 2004). Currently, most of developed as well as under developed countries have adopted ICT in public sector organization in order to provide good governance to their citizen which results in increasing trend of spending a good amount of GDP on adoption of ICT (Harfouche, A., & Robbin, A. (2015). From Figure 2 presents Worldwide ICT spending has grown by less than 5% every year since the financial crisis in 2009.

## Background of E-Government

The term e-government (e-Gov) first appeared in the late 1990s, but the history of computing in government organizations goes back to the early days of computer history. A literature on "computing in government" goes back at least to the 1970s. E-Government has been developed to use the information technologies comprehensively in government for the procurement of several public services and to improve the efficiency of managerial tasks, promoting democratic moral values and mechanisms as well. Information technology has the ability to convert the government structure and enhance the quality of government services. Information technology is providing two main benefits to government i.e. the operational efficiency is increased by minimizing cost and productivity is increased and also quality services are provided to government agencies (Ghayur, 2006).

# A GLANCE ON E-GOVERNMENT IN NEIGHBORING COUNTRIES (CHINA, INDIA, AND IRAN)

Pakistan shares its borders with four neighboring countries – People's Republic of China, Afghanistan, India, and Iran. The government of China has enthusiastically fortified and capitalized in the growth of the Internet to capture the vast marketable potential of technology, while put forth state control. China began to device programs that would facilitate e-government even before the country was fully connected to the global "network of networks" (Zhou, 2004). The government of China's eagerness in applying Internet technology to boost up the financial and commercial development is vibrantly proven by a series of important projects few of which are the Golden Bridge Project, the Golden Card Project

*Figure 2. Worldwide ICT spending*

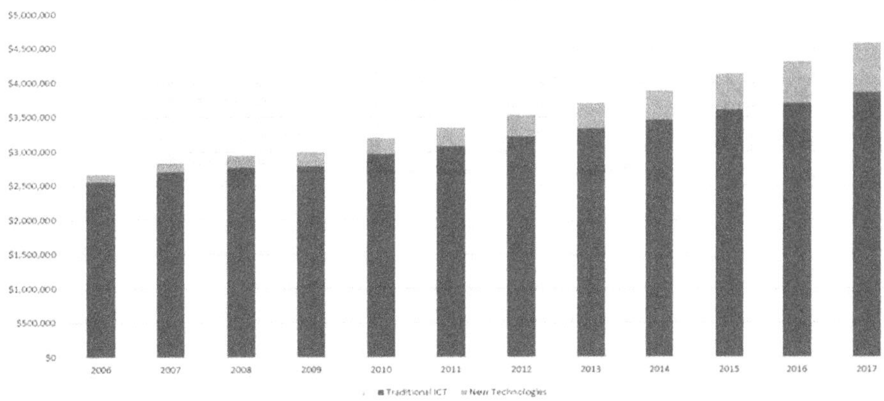

and the Golden Gate Project, all of these were initiated in the mid-1990s (Xu, 2010). China becomes a critical e-government case due to its sheer size. Despite the fact that progress is currently limited to surface change rather than fundamental reworking of the governance system, however the current efforts of china's government reflects that the coming years will witness remarkable e-government progress.

Indian government have undertaken a large number of e-governance initiatives at multiple over the years and have work hard to improve the delivery and quality of public services by simplifying the process to access e-governance services. In this regard different public organizations are working towards fixing specific, achievable objectives and deadlines for various processes within the government such as the Planning Commission, the Department of Administrative Reforms and Public Grievances, the Department of Information Technology and other Central Ministries at the Center and several state governments. The central government of India in 2014 has approved the Digital India program consist of a number of projects which includes projects to ensure availability of government services to citizens digitally and were of worth above 1 lakh crore INR (Approx. US$15 billion) (Kumar & et al, 2017). In India the national e-Governance plan on May 18, 2006 have been formulated by the department of electronics and information technology and department of administrative reforms and public grievances (DARPG) which takes a holistic view of e-Governance initiatives across the country (National Portal of India: https://india.gov.in/egovernance/national-e-governance-plan, 2016)

In Iran, establishment of e-governance is a promise that related organizations are now trying to fulfill as fast as possible. However the Iranian e-governance is in initial phase of its development, though the time spent on the process has been long, however, some aspects of the process have not been properly fulfilled. As already discussed earlier, establishing and advancing e-governance needs e-readiness which means Iranian authorities should think of sanctioning their electronic infrastructures (Shahghasemi & et al, 2013).

As china is a developed country, therefore they have well established e-governance systems. China is very advanced in use of technology and included in a list of countries who have high rank in online service index (OSI). Although due to large population China is still facing issues in providing digital services to their citizen and trying to improve its position in world e-government index. India is also struggling to provide good e-governance to its citizen and also secure good position among countries having high rank in OSI. Still India needs to initiate more projects in order to improve its rank in world e-government index. Iranian e-governance is in initial phase of its development, though the time spent on the process has been long, however, some aspects of the process have not been properly fulfilled. One of the major reasons is non-technological structure of Iran as well as many other aspects such as f nonproliferation culture among citizens and government domination of governmental organizations and their attitudes that causes hurdles in providing good e-governance to citizen of Iran (Tohidi, 2011).

## PAKISTAN AT A GLANCE

Pakistan is located in South Asia and neighboring Afghanistan, China, India and Iran. Pakistan is the seventh most populous country in the world. After Africa, the South and Central Asia region has the lowest numerical e-readiness index. In the last five e-readiness surveys, South Asia's regional average is extremely below the global average. In South Asia, most countries are in the development, enhancement or interactive phases of the digital presence. Recognizing the importance of e-government, Pakistan's Ministry of Information and Technology has incorporated e-Government as a priority area in its first National IT Policy and Action Plan, developed in 2000 (Azam & et al, 2013). Recognizing that e-government plays a catalytic role in building good governance by improving the ability of individuals and businesses to access public services in an efficient and cost-effective manner, the Federal Ministry of Information Technology has following features in its policy.

- In Pakistan, the use of computer applications in government organizations and services should be promoted to improve efficiency, transparency in operations and service delivery, as well as to establish and enable access to public information.
- Compulsory IT department in all provinces of Pakistan
- At least 2% of the annual budget must be spent on IT services in Pakistan
- Ensure that computer literacy must be mandatory for all new government employees
- Internet and intranet e-mail communications must be established for inter-office communication and the automation of a physical file system to a computerized file system, and must establish national databases.

The public services are to be provided through following three mediums when digital is implemented:

- E- services of government
- Vending Machines and ATMs; and
- Manually

Each year, some projects in the above categories are implemented. Practical achievements to date include:

- Paksat-1 successfully launched processes in January 2003 and the EOSS high-resolution Earth Observation satellite will be launched soon
- The first online portal for Pakistan, "Online Citizen", was launched with around 11,000 web pages available and more than 500 downloadable forms (Ghayur (2006).

A comprehensive study has been conducted to identify the several challenges facing e-government in several public sector organizations of Pakistan. The work process is fully transformed that brought changes in the administration setup of government due to intensive advancement in information and communication technologies. The implementation of e-government is a significant challenge because it requires a significant change in the existing work process. The different challenges are identified by conducting a case study such as technological infrastructure, organizational aspect and coordination with other organizations. The results of case study shows that the implementation of e-government is a big challenge, as the infrastructure of ICT and financial resources are limited and not available within organizations (Qaisar, & Ghufran, 2010)]. The significance of employing e-government is increasing nowadays in worldwide governments for delivering public services to the people. The e-government readiness index is a composite measurement of capacity and willingness of the countries for the use of government for ICT led development (Khan & et al, 2010). Pakistan is in the start of employing its e-governance and so far it seems to be heading towards Primary Rainbow E-Governance with Hierarchical Government Model (Ghayur, 2006). Many government services are provided online and each year it appears that new services are added (Heeks, 2001). Figure 3 and Figure 4 depicted Pakistan current state of e-governance and presents worldwide ranking of Pakistan in e-government according to United State survey 2018 (https://publicadministration.un.org/egovkb/en-us/Reports/UN-E-Government-Survey-2018).

## Current ICT in Pakistan

The index of readiness of the electronic government of Pakistan between 2003 and 2018 in the reports of the global survey on the state of readiness of the electronic government of the United Nations is presented in table 1. This is very clear that Pakistan's e-government readiness index is extreme below from world average and Pakistan has decreased its rank as compared to other countries of the world instead of improving its ranking. However in 2018 Pakistan ranking has been improved yet not comparable with its competitors. According to the United Nations e-Government Survey, measured for 192 countries (Arfeen

*Figure 3. Pakistan e-government development index rank*

*Figure 4. Pakistan e-participation index rank*

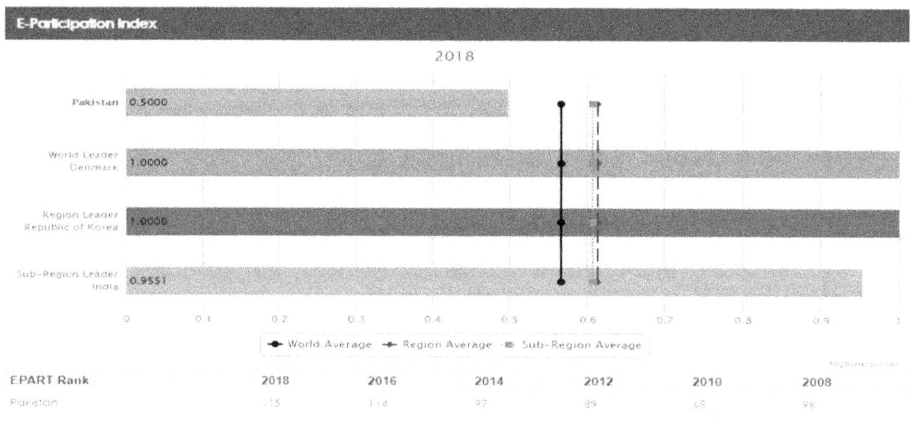

*Table 1. Pakistan's e-government readiness index*

| Year | Index | Rank | Rank Change | World Average |
|---|---|---|---|---|
| 2003 | 0.2475 | 137 | | 0.4020 |
| 2004 | 0.3042 | 122 | -15 | 0.413 |
| 2005 | 0 .2836 | 136 | +14 | 0.4267 |
| 2008 | 0.3160 | 131 | -5 | 0.4515 |
| 2010 | 0.2755 | 146 | +15 | 0.441 |
| 2012 | 0.2823 | 156 | +10 | 0.4882 |
| 2014 | 0.2580 | 158 | +2 | |
| 2016 | 0.2583 | 159 | +1 | |
| 2018 | 0.3566 | 148 | -11 | |

& Khan, 2009), Pakistan's ranking was 137 in 2003, which upgraded to 136 in 2005 and to 131 in 2008 (Un, 2008). On the other hand, in 2010 Pakistan dropped drastically its ranking by 15 ranks to 146. In same way, in 2012 its ranking further dropped by 10 ranks to 156. In 2014 and 2016 Pakistan remain unsuccessful in boasting up its ranking and dropped its ranks each year. Due to this fact Pakistan government paid special attention to provide better governance to their citizen and as a result in latest survey of United nations 2018 Pakistan improved its ranking by 11 points. The majority of the high scorers in the 2018 e-Government development index are Denmark, Australia, South Korea, New Zealand, Japan, France, Finland, Sweden, Singapore, United States and Canada, all well established and highly stable countries, who have the well enough financial resources to extend and rollout advanced e-Governance initiatives, create a promising environment for citizen rendezvous and empowerment. Pakistan can improve its ranking by focusing on and looking for ways to increase citizen participation. It is the deepest hope that, in the future, the Government of Pakistan will be able to improve the allocation of resources and responsibilities to promote the provision of quality services to the public sector organizations.

*Figure 5. E-government index of Pakistan*

*Figure 6. E-government rank of Pakistan*

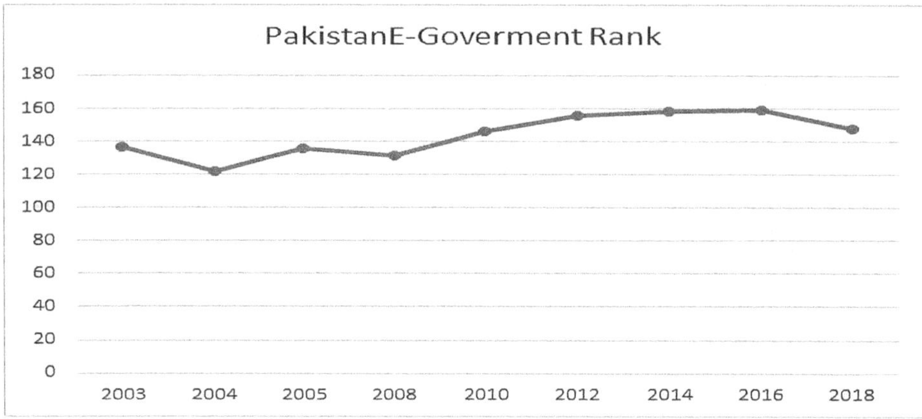


## ICT IN GOVERNMENT OF PAKISTAN

In Pakistan, the e-government was introduced in 2002. The e-government Directorate (EGD) was established in accordance with a decision of the Federal Cabinet in October 2002. The Government of Pakistan has converted the former Government Technology Commission to information in EGD as the Ministry of Information Technology. The main purpose of this portal is to bridge the gap between government and citizens by using personal computers, kiosks, telephones and other resources to immediately make the information available to them. Number of services are offered by EGD (www.e-governmentt.gov.pk) few of these are listed as follows [3]

- Federal Board of Revenue(FBR) provide e-tax payment
- Online Hajj applications processing
- Automation of Prime Minister Secretariat, Islamabad
- Disbursement of salary through ATMs
- Access to constitutional case laws at district Bar Associations online
- Online Higher Education Commission (HEC) Services
- Online procurement system
- Online management of planning, budget and finance
- Federal public service commission Online recruitment
- State office Automation
- E- Judiciary

In addition, Pakistan's National Database and Registration Authority (NADRA) (www.nadra.gov.pk) successfully provides identification, e-governance and secure documents that provide a number of distinct services, such as reducing identity theft, protecting the interests of citizens and facilitating the public. NADRA has gained international recognition for these services. Web acts as an anti-corruption tool in Pakistan in that it is used for different purposes, such as enumerating government officials guilty of a crime of corruption in order to deter people from committing similar crimes. Pakistan has also taken initiatives to digitize the courts and put in place an electronic justice system. In this regard, the Islamabad

High Court and Lahore High Court have taken the first step in introducing an online case management system. Mobile apps have been designed to help lawyers and complainants keep track of cases on their mobile phones. These app users will be able to search for cases using individual case numbers, party names, sponsor names, and lawyer names. The user can also consult the daily list of causes.

## ICT BASED PUBLIC SECTOR INITIATIVE

The investment of government in public sector information system leads towards the improvement in efficiency and the effectiveness of policy for attaining broader democratic moral values. New public management has applying these investments in several countries in the world by adopting different ideas and practices in the public sector. NPM proposes the public sector reforms to transform managerial and government practices. These reforms evaluate the socio political impact for adopting ICT in the public sector (Cordella & Bonina, 2012).

From the last decade, the government of Pakistan has initiated a number of e-Government projects. These projects are commenced in order to improve public's unified accessibility to information and citizens' gratification. In February 2005, the Directorate of E-Government launched an automation project for the Hajj wing of the Ministry of Religious Affairs of the Government of Pakistan. They wanted to make it a single online link to all major services, from candidate registration to hajj to data verification, group creation, survey and candidate selection, dissemination of information for selected candidates, from automatic printing of Hajj passports, to tracking pilgrims and their baggage, from planning to confirmation or changes in theft of hajjis during Hajj (Arfeen & Khan, 2009). Innovation in the public sector through the implementation of e-government has recently begun to gain momentum in developing countries, particularly in Pakistan. It deals with issues of governance, technology and culture that are very difficult to solve. (Arfeen & Kamal, 2014).

## DIGITAL GOVERNANCE TO THE CITIZENS

E-government provides different services to the citizens for electronic transaction with the government. These services vary according to the needs of the users, because of which different types of electronic administration have been developed. According to the World Bank, (2002); e-government delivery models can be briefly summarized as follows (Ilyas, 2016);

- Government to Citizens (G2C)
- Government to Businesses (G2B)
- Government to Employees (G2E)
- Government to Governments (G2G)
- Citizens to Governments (C2G)

G2C has been designed to create the smooth relationship between government and citizens. Through G2C citizens can easily access government information and services from all over the place, by use of various channels. As G2C is customer centric approach, here people can get all they need in just **"one stop"** i.e., place where multiple services are offered at one place. Here citizens can perform different operations that belong to multiple agencies without contacting individual agencies. A solo entrance point also boosts citizen involvement in the democratic process, as they can easily access administrative processes and more easily eloquent their needs with government representatives. While e-government initiatives improve the government's internal processes, e-citizens and e-services aim to improve the communication and quality of services between government and citizens. Finally, e-society initiatives focus on the government's interactions with the business community and civil society. As each area grows and develops, e-government initiatives may increasingly coincide (Ntulo & Otike, 2013).

Citizens can access government information and services anywhere in the world, at any time, via multiple channels. In addition to performing certain transactions, such as certifications, the payment of government taxes and the application for benefits, G2C facilitates the contact of citizens by overcoming temporal and geographic constraints, can in turn facilitate and increase the participation of citizens in government.

## BENEFITS OF ICT TO THE GOVERNMENT

Now a days, Information Communication Technologies (ICT) are used as the sensational treatments for curing a state from discomforts of corruption, mismanagement, poor governance, inflation, monopolies, business inactivity, high ratio of illiteracy and so on (Ghayur, 2006). The application of ICT is beneficial for both government and citizens. Government agencies get benefits in the terms of cost reduction and improved efficiency. On the other hand citizens obtain faster, more advantageous services from a supplementary vigilant and well-versed government. Additionally, e-government assurances to emulate the private sector by using ICT and providing very efficient, effective, translucent and reachable public services to citizens and businesses (Kayani & et al, 2011).

It is a fact that sector of ICT can contribute greatly to the national gross domestic product (GDP) of a country as proven by developments from other countries. Improved market competitiveness of a nation's products and services can be achieved by enabling ICT (Kundishora, 2010). On the top of that ICTs can influence positively on government and other sectors of the financial system as depicted in figure 6. In turn ICT can successfully support international economic amalgamation, improve living standards of citizens, and improve biodiversity utilization and management.

Information technology and communication is an essential to the everyday operations of federal and local government. ICT plays a key role in accelerating the flow of information and knowledge between government and citizens and in transforming how governments and citizens interact (Chinedu & Hussaini, 2018). ICT brings transparency and independence in government services. Today, in many countries, over 70% of taxpayers file taxes online, and many other transactions, such as driver's license renewal, parking ticket payment for government benefits management., the purchase of replacement ID cards for the purchase of online copies of personal documents for themselves and their families can be made online (Qaisar & Ghufran, 2010). ICT enable organization to make good and fast decisions. ICT plays a vital role in streamlining various government processes. Digital governance is enabled by ICT. In country like Pakistan, ICT has helped in making democracy vibrant and transparent. Figure 7 summaries the benefits of ICT to the government.

*Figure 7. Benefits of ICT to the government*

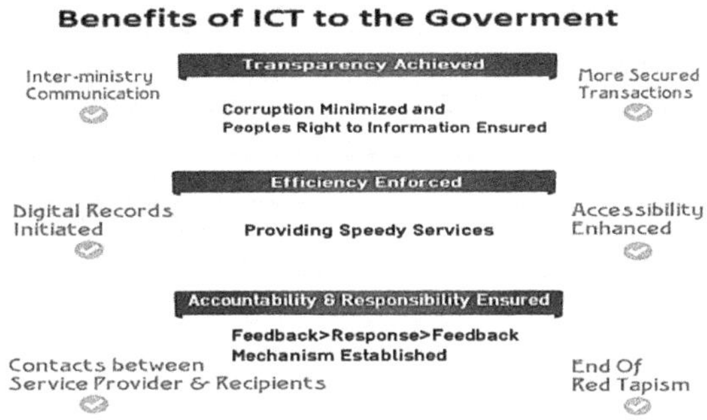


## ICT GOVERNANCE OPPORTUNITIES IN PAKISTAN

States and governments that, through the concept of e-government, seek to achieve a set of goals and objectives will benefit all individuals, institutions and societies. The crux of the discussion summed up with certain important tips to consider for successful e-government initiatives in Pakistan.

- **E-Assessment:** Initiatives must be launched to an e -readiness assessment to enlighten the stakeholders and user of e-services concerning the current state of affairs of technological infrastructure and increasing globalization and speeding competitive atmosphere and Strategies must be planned on the basis of assessment of human capital.
- **E-Voting:** Utill now voting in Pakistan has been done manually. There is a need to take initiative to launch e-voting system in Pakistan in order to make voting system more transparent.
- **E-Taxing:** Online payment of taxes may not only benefit government but also facilitate citizen to submit their taxes from a single online link. State Bank of Pakistan and Federal Board of Revenue have initiated the online collection of Government taxes and duties through single-link facility.
- **E-Population**: There is a need to initiate E-population system in Pakistan in order to improve data accuracy and facilitate citizen in registering their particulars online. E-population system are already implemented in developed countries that provide facilities to their citizen to register their full particular details online and perform searching and issuance of certificate copies from any Divisional Secretary Office, abridged need of translations, provide current information of citizens and share information of citizens between the institutions.
- Increasing wakefulness in public and private organizations through managing workshops, conferences and seminars in order to highlight real challenges opportunities and advantageous and implement the measure to revolutionize the ICT.
- Long-term is to prefer to meet the obstacles in the way of e-government adaptation and launch human capital development schemes via providing appropriate training and IT literacy in order to meet the coming technological innovativeness.

- The use of information and communication technology (ICT) has a considerable advantageous to Cost reduction and efficiency gains for private organizations and it benefits the major aspect of e-Government initiatives (Tapscott, 1996; Amit and Zott, 2001 and Malhotra, 2001).
- Quality of public Service Delivery to Businesses and Customers can turn down the dilemmas relating to violation of transparency, investment of hug timing in the administrative process, traditional manual work mechanism, high costs and miserable accessibility to authority.
- Cost Reduction and Efficiency Gains Researchers argue that ICT has considerable potential to contribute to efficiency gains and cost reductions for private organizations (Tapscott, 1996 and Malhotra, 2001).

## ICT GOVERNANCE ISSUES IN PAKISTAN

Although Pakistan is developing at a rapid pace, but the fact is that as a developing country, Pakistan faces many issues related to ICT governance few of which are listed below.

- Pakistan does not have required ICT infrastructure for e-governance.
- There are also "mindset gaps" involving overall resistance to change, data sharing and customer-orientation in public sector.

Pakistan requires a group of "e-champions" with the strategic vision of e-governance without which this whole initiative will become insignificant. The role of this group should be to put the e-governance on the top of national agenda, minimize the operational barriers and make it happen through channelizing all the resources into right direction.

## ICT GOVERNANCE CHALLENGES IN PAKISTAN

Governance structure in developing countries is usually stacked on excessive file work, lingering practices, time taking processes and a hierarchical approvals. There are numerous challenges that can delay progress of digital governance in underdeveloped countries (Ahmad & et al, 2013). The complexity and diversity of digital-governance initiatives infers the existence of several critical barriers to its implementation and management.

We can examine the challenges of ICT governance in Pakistan against the chronological and real-world background. The situation in Pakistan represents a distinct case where resident engagement every so often receives much less attention.

Major challenges of Pakistan may include

- Strategic challenges

Formulation of a digital governance policy is a crucial factor. This requires a strategic framework which covers all aspect of digitization.

- E-readiness (Adoption of change)

An innovation is less likely to be adopted. Adoptability to change is a major hindrance in the process of digitization in Pakistan. Lack of motivation, managerial lacunas, role ambiguity, and procedurals complexities adversely impact towards acceptance of change.

- Political consensus constraints

Thus, motivation of leadership is considered to be imperial factor that can result in success of ICT implementation. In government department's top management or decision making authorities are often politicians or their nominated personals. They work either for needs, or to support personal or political agendas. Often put unnecessary pressures to attain the benefits they are not entitled for.

- Inadequate resources

An effective implementation of digital governance requires strategic commitment and sound infrastructure. Often the staff of government organizations in Pakistan are either not capable or have low skill set. Inadequate technical resources, lack of capable staff and non-existence of polices for capacity building have been problems for years. Generally no policy subsists for the capacity building of existing staff, neither for the requirement of qualified employees.

- Corruption

Many underdeveloped countries like Pakistan where corruption is rampant, struggle a lot to achieve digital governance. Curbing corruption can result in more effectiveness of the digitization of governance process. Corruption control policies, procedures and measures are unlikely to exist in government organizations. Moreover, where the policy somehow subsists, is out ruled by the various factors like political influence, bribery, clerical loopholes and protracted process. As prone to high corruption Transparency International's (TI), ranks Pakistan at 33 on Corruption Perceptions Index (CPI) 2018, where range varies from zero (highly corrupt) to 100 (very clean).

- Social and cultural constraints

Social and cultural constraints can directly or indirectly impede the ICT governance in Pakistan. Social factors including gender, age, caste, poverty, literacy, class etc. whereas, cultural factors that can effect are norms, behavior, language and ethics. These factors are also backed by the hierarchies and decentralizing access weakens the spirit of revitalizing digital governance.

- Stakeholder Involvement

There is a need of positive environment that encourages stakeholders to participate in digital governance applications. Little formal policy making has occurred to encourage stakeholders to participate in the process. It seems that Leadership of Pakistan is much in disposed to encourage, guide, pursue or even require proactive consultation of stakeholders.

- Cybersecurity challenges

Cyber space is growing as this involves use of internet and telecommunication, invites hackers to misuse and disrupt use of cyberspace. One of the major issue all over world e-governments have been facing is cybersecurity challenges. Pakistan is a developing country and like other countries also facing cybersecurity issues as cybersecurity services are under development. There are fair chances of an attack on integrity, availability or confidentiality of the network or services of NADRA, the e-government and the financial markets of Pakistan by expert cybercriminals. Therefore, top priority of Pakistani government organizations must be the protection of confidential information. However, at present, Pakistan has no legal policy or law in place that can cautiously deal with the growing threat of cybercrime. The legal framework for criminal justice available in Pakistan is inadequate and poorly equipped to deal with the sophisticated cyber-age online threats (Awan & Memon,2016). In order to face this challenge, Pakistan needs to create a cybercrime unit to fight cybercrime.

Many other key challenges for E-Governance in Pakistan in addition to above mentioned challenges hinders the adoption of e-governance in Pakistan such as is the low literary rate of people, and the support of top management for resistance to change to digital processes, and the wobbly policies made by government. The management and deployment of huge ICTs in infrastructure create an issue, the possible intrusion of privacy, the insecurity of private data from intruders, the lesser collaboration between different organizations and the training of employees also constitute a challenge, therefore highly qualified professionals must foresee a monetary cost for the implementation and management of this system. Their absence also creates an obstacle to achieving this goal.

## VISION OF E-PAKISTAN

Vision must be clear and simple comprising of requirements and opportunities aligned with comprehensive strategy, implicating the consensus of stakeholders. The concept of E-Pakistan can be defined as the employment of internet for delivering e-services. An underdeveloped state like Pakistan has to put all of its effectors towards the transformation process. Pakistan as a state can emerged as a major player in the sphere by adopting E-Governance. The primary step towards e-Pakistan would be a transformation from manual to digital governance systems. It involves high digitalization standards and work processes. Significant progress can be achieved by determining the vision of E-Pakistan. With the development of world in field of ICT, Pakistan needs to catchup with challenges pertaining in the region & globally. Pakistan can undertake the state of upward surge by focusing on E-Governance, E- Health, E-Education, and as an agricultural country E-Agriculture. Keeping these fundamentals in view the vision of E-Pakistan can be described as "Establishing Access of all Government services at one-click, in a convenient manner, thereby improving reliability, efficiency, accountability, transparency, outreach, and networking capacity."

ICT has a significant role in achieving the goals of E-Pakistan. The vision can be implemented by formulation of strategic plan. During the development of the strategic plan, action against the critical factors should be prioritized as well. The strategic plan should cover all aspect of the actions, even the after effects of the plan should be considered. Effective approaches to implement and promote the principle of E-governance can provide the desired results. Moreover, to enhance access of citizen, ICT

*Figure 8. ICT strategic topology of Pakistan*

implementation program should be nation-wide. In Pakistan, major areas that need significant improvement are at remotest corner.

An empowered ICT Task Force with well-formed stated goals should be developed in Pakistan. The mandate of this force should be aimed to introduce ICT reforms in administrative and secretarial organizations of government. The taskforce should study, implement and introduce the projects by analyzing the working, environment and policies of organizations. A program for the awareness of stakeholders should also be designed with respect to their knowledge levels. Networked structures of public administration can develop a trust among the stakeholders. It also provides information, and knowledge required for a decision making. Existing assessment indicates that key to achieving this vision is to establish a relationship of trust among the stakeholders.

## FIVE/TEN YEARS ICT STRATEGIC PLAN FOR PAKISTAN

Strategy is a plan of action designed for achieving goals and objectives under conditions of uncertainty. In recent years, Pakistan's government has been struggling to take ICT initiatives to achieve the vision. An improve set of policies are promulgated and administrative practices has been started, however, there is still a lot to be done. In order to accomplish desired objective, the government of Pakistan has to aim for "building an ICT based knowledge driven society".

The vision and strategy of ICT should focus on people and not just only on technology. Mujahid, Y. H. (2002) has described the comprehensive approach adopted by Pakistan government in order to cover all short term and long term goals. The role assigned to ICT can be characterized in two ways ICT as production sector and ICT enabled services (Mujahid, 2002). The role of ICT strategies in Pakistan can be well understood as depicted in Figure 8 given below

The sectoral approach focuses on the development of a country's economy using information technology as a production or sector. In the short term, human resources as inputs and markets for absorbing outputs are critical to success. The other approach focuses on the long term goals and objectives of the country which obviously are an economic and prosperous Pakistan (Hameed, 2007).

Transparency international movement the leading global coalition against corruption has one of the national chapter named as TI Pakistan. This chapter is started working in Pakistan in 2002 from Karachi. It has its own board, advisory committee and member body that work together to plan the agenda against corruption according to the local context and it is registered as trust. It has the vision and values to reshape the anticorruption efforts. This organization has worked with several international and national organizations to reduce the corruption and proven itself one of the most capable anti-corruption organization. TI Pakistan will follow a strategic plan as a guiding document from the year 2017 to 2020. This strategy is align with the local context of Pakistan and will support to make relations to the largerTI movement and TI chapters. TI Pakistan will implement any initiative on the basis of next coming year's strategic policies. The TI strategic plan will helps the stakeholders of Pakistan to understand the organization better (Fund & Britain, 1997).

## SMART CITIES IN PAKISTAN

Smart cities has emerged as an urban development vision for gathering and integrating multiple information and communication technology and internet of things solution in order to manage the assets of city. The main aim of developing a smart city is to improve the life quality by adopting urban technologies for providing better services to residents to fulfill their needs. The benchmark definitions about smart city has been improved by (Caragliu & et al 2011) and his operational definition states: We believe that a smart city must be smart when investments in human and social capital and traditional (transport) and modern (ICT) communication infrastructures fuel sustainable economic growth and a high quality of life, with judicious management of natural resources through participatory governance. At present age in Pakistan the concept of smart city becomes a new idea and to understand what is still lacking in smart cities. However the eco-system, the thought leadership, the arrangement of institution and suggestions of public does not fully activated to achieve critical mass on the subject. Recently due to increasing urbanization and rising incomes trends shows that by 2025 the population of urban will increase from the recent 50 percent to nearly 60% by 2015 with 80% of growth coming from cities in Asia and Africa. The Pakistan's urban population has been increasing from 32% in 1998 to 40 percent today and if the trend continues it will reach up to 50% by 2025 (Vision, 2025). At present age 9 cities of Pakistan have populations over 1 million and 75 cities with population between 100, 000 to 1 million. The cities of Pakistan contribute 78% to the country GDP. One of a major problem faced by Pakistan is in urban influx between migrant and native populations as cities are unable to consume, provide shelter and employ rural migrants at a profit (Alavi & et al, 2018).

In this global word, smart cities has become an important milestone and considered as rudiment to achieve good governance. Pakistan has a lot of potential in I.T industry, and emerging as a major stakeholder in the world. A large pool of highly skilled professionals is present in Pakistan, which can empower the vision of smart cities concept. In the sphere of E-Pakistan the ICT and services remained urban-oriented. In some provinces, initiatives have already been taken towards the smart cities project. Provincial governments of Punjab and Khyber PakhtunKhwa (KPK) have started many projects to support the cause.

Establishment of E-KhidmatMarkaz in various cities of province Punjab. The public administration involves a number of departments, using different set of processes to provide services. To facilitate the public, an E-KhidmatMarkaz was designed to automate these process of public sector departments, by provision of service under one roof. The concept of One window service center is to integrate interlinked departmental process for provision of unhindered services. For this very purpose an autonomous body

"Punjab Information Technology board" has been constituted. The board aims to induce transparency and accountability while modernizing the process of Birth/Marriage/Death/Divorce/Domicile certificates, Vehicle Registration and Route Permits, Vehicle Transfer, Fine and Tax Collection, Driving License, and issuance/renewal of computerized national identity card.

Among many other services, the PITB spearhead for provision of system and solution along with infrastructure to other areas like education, health, law enforcement, legal sector and civil departments of province. The board has also initiated programs for increasing the digital literacy of the citizen's. As the Punjab's technology spearhead, PITBis effectively working towards following objectives:

- Decreasing distance to services.
- Extending service to remote areas
- Improvement of internal efficiency, accountability and transparency.
- Simplification of governmental procedures.
- Cost/Time Reduction and revenue generation.
- Achieving stakeholder's satisfaction.
- Introducing innovative and capacity building programs.
- Promotion culture of entrepreneurship

There is an urgent need to reduce the digital divide in the country in an integrated and comprehensive way. This is only possible when citizens are aware of e-government initiatives and how they can use them effectively in their day-to-day affairs.

## ICT BASED MODEL VILLAGE IN PAKISTAN

The concept of digital governance has also evolved with time. The concepts of ICT based Model village can be defined where resident are well aware of, and using ICT tools. As discussed before the major concern to achieve the vision of E-Pakistan is coverage of rural areas. The ICT Based Model Village stems from the goal to provide services to remote villages. Through provision of technology and service, the disadvantaged communities can uphold the economic development. With the emerging technologies now farmers are also using ICT tools for farming. A better crop can be conducting a soil test or cultivated crop by using the fertilizer recommendation software. Traditional society faces numerous challenges while transformation into ICT based modern society.

## DISCUSSION

With all of its benefits digital governance also expose the stakeholders involved in the process, resulting some perilous consequences as well. Policy and Regulation related issues are also considered as a major reason of failure of the digitization process. While formulation of strategic framework for the digital governance some critical barriers related to digitization can also cause complications. The factors like archiving, intellectual property rights, security, piracy, electronic signatures and transmission of information can affect the entire process.

The process of digital governance while providing e-service can also contributes towards the economy of Pakistan. As the economy of Pakistan is continuously declining, digital governance can play its role toward its stability. A sound decision making can only be possible it Leadership is aware of the current scenarios and can analyze the impact of decisions. With all of the shareholders on board, the digital governance can show its worth. A digital governance system should be able to provide reliable, efficient, cost effective and satisfying service to the stakeholders. A government that aims to improve digital governance emphasize not only on digitization and capability, but also towards performance monitoring and continuous improvements. Key to achieving this requirement to ensure that stakeholders use the e-services effectively. It would still be a big challenge for the Leadership of Pakistan to gather stakeholder, even if e-Pakistan initiative is implemented successfully. In Pakistan the Project related to the E-governance often got the low attention due to ill motivations, change resistance, and non-awareness. The overall progress is also affected by the lack of fundamental infrastructure. Recent studies of e-Government show that traditional public sector governance can be reformed by adopting a mixed approach. A national infrastructure development is required for computerization of government operations. It has been observed that practices related to E-Pakistan will enable a government in cost reductions and enhance efficiency of all government operations. The E-projects often involve trust, privacy, copy rights, digital signatures and identity management in the context of electronic governance.

## FUTURE OF ICT IN PAKISTAN

Information Communication Technology (ICT) has gained pragmatic development in last decades; the credit goes to the organizations such as the incumbent PTCL and Wateen Telecom that helped in plentiful growth ranging from availability of network to the broad infrastructures through deployment of fiber optic network across the country. Approximately more than 250 towns and cities are connected with each other through this extensive network. One of the major outcomes of this wide-ranging network is the provision of Mobile Financial Services, which reached up-to the transaction of worth Rs.3.76 billion via online banking. However, the future of ICT is largely dependent on the policies of Government for deregulation and consistency to bring foreign investors in the country

## CONCLUSION

The developing countries are based on the governance in which large amount of paperwork and file handling is the key. Such governance is time taking due to controls at many levels. The goal of any governments is to provide citizen welfare. The fast growing information and communication technologies (ICTs) has impacted the societies and changed the economic growth. Developing countries are focusing on electronic governance (e-governance) to become part of global world. The far impacts of e-governance are not yet clear but if not adopted can lead to less economic stability. Pakistan is also realizing that this new technology is essential for country's growth. To achieve this objective, he Planning Commission of Pakistan, the Department of Administrative Reform and Public Complaints, the Department of Information Communication Technology and other central ministries of the Center and various departments of state governments work together. Adopting technology will make citizens more active part of society.

At this time, we should motivate people for understanding of issues in good governance so that society as whole moves toward economic and social development. The media can be used for this purpose, a dedicated channel for information sharing, improve the norms and values by sharing experiences with each other, and a forum for intellectuals such as, academicians, business representatives and government policymakers can be useful . The government should be connected to regional networks, make use of resources so as to take maximum benefits. Moreover, we cannot ignore the role of private sector; a useful integrating with private sector in ICT can enhance the growth. Inventive partnership models between government, the private sector and non-governmental organizations (NGOs) can stimulate the economy and society of developing countries.

## FUTURE WORK

From the above discussions it is evident that E-government plays a vital role in improving citizen participation, transparency and overcome administrative bottlenecks. Developing countries like Pakistan are using digital technology to reform public service delivery. Pakistan Government should focus on improving IT infrastructure as being at the earlier stage of e-government services. Pakistan must implement policies and strategies that focus on utility, efficiency, IT infrastructure issues, sufficient assistance and guidance, esteem for user privacy, lack of awareness about the availability of e-government services and convenient access to e-government services. Advertising campaigns on the importance of using the Internet and e-government services should be strengthened. Additionally, Pakistan should focus on improving the literacy rate as one of the main impediment in adopting e-governance is low literacy rate. Therefore, in order to successfully diffuse e-government services, governments must understand citizen needs.

## REFERENCES

Ahmad, M. O., Markkula, J., & Oivo, M. (2013). *Factors affecting e-government adoption in Pakistan : a citizen' s perspective*. doi:10.1108/17506161311325378

Alavi, A. H., Jiao, P., Buttlar, W. G., & Lajnef, N. (2018). Internet of Things-enabled smart cities: State-of-the-art and future trends. *Measurement*, *129*, 589–606. doi:10.1016/j.measurement.2018.07.067

Arfeen, M., & Kamal, M. (2014). *Future of e-Government in Pakistan: A Case Study Approach*. Academic Press.

Arfeen, M. I., & Khan, N. (2009). Public sector innovation: Case study of e-government projects in Pakistan. *The Pakistan Development Review, 48*(4-II), 439.

Awan, J., & Memon, S. (2016). Threats of cyber security and challenges for Pakistan. In *International Conference on Cyber Warfare and Security* (p. 425). Academic Conferences International Limited.

Azam, A., Qiang, F., & Abdullah, M. I. (2013). Determinants of e-government services adoption in Pakistan: an integrated model. *Electronic Government, an International Journal, 10*(2), 105-124.

Beardsley, S. C., Enriquez, L., Bonini, S., Sandoval, S., & Brun, N. (2010). *Fostering the economic and social benefits of ICT*. The Global Information Technology Report 2009–2010: ICT for Sustainability

Caragliu, A., Del Bo, C., & Nijkamp, P. (2011). Smart cities in Europe. *Journal of Urban Technology, 18*(2).

Chandio, A., Haider, Z., Ahmed, S., Ali, M., & Ameen, I. (2019). E-government in Pakistan: Framework of opportunities and challenges. *GSJ, 6*(12).

Chinedu, P. U., & Hussaini, Y. (2018). ICT Deployment in Governance: A Potent Enabler of Accountability, Reliability and Transparency (ART) of Government Business. *Engineering Mathematics.*, *2*(2), 68–78.

Cordella, A., & Bonina, C. M. (2012). A public value perspective for ICT enabled public sector reforms: A theoretical reflection. *Government Information Quarterly*, *29*(4), 512–520. doi:10.1016/j.giq.2012.03.004

DPEPA. (2001). *Benchmarking E-government: A Global Perspective*. United Nations Division for Public Economics and Public Administration – American Society for Public Administration. Retrieved from http://unpan1.un.org/intradoc/groups/public/documents/un/ unpan021547.pdf/

Fund, H. L., & Britain, G. (1997). *Strategic Plan*. Environmental Defense Fund.

Ghayur, A. (2006). Towards good governance: Developing an e-government. *Pakistan Development Review*, *45*(4II), 1011–1025. doi:10.30541/v45i4IIpp.1011-1025

Gil-García, J. R., & Pardo, T. A. (2005). E-government success factors: Mapping practical tools to theoretical foundations. *Government Information Quarterly*, *22*(2), 187–216. doi:10.1016/j.giq.2005.02.001

Hameed, T. (2007). *ICT as an enabler of socio-economic development*. Academic Press.

Harfouche, A., & Robbin, A. (2015). E-Government implementation in developing countries. In From Information to Smart Society (pp. 315-327). Springer.

Heeks, R. (2001). *Understanding e-governance for development*. Manchester, UK: Institute for Development Policy and Management.

Ilyas, M. (2016). *E governance practices and models: Options for Pakistan. Institute for strategic studies, research and analysis*. ISSRA.

Kayani, M. B., Haq, M. E., Perwez, M. R., & Humayun, H. (2011). *Analyzing Barriers in e-Government Implementation in Pakistan*. Academic Press.

Khan, F., Khan, S., & Zhang, B. (2010). *E-Government Challenges in Developing Countries : A Case Study of Pakistan*. doi:10.1109/ICMeCG.2010.49

Kumar, R., Sachan, A., & Mukherjee, A. (2017, March). E-Government Adoption in India: a Mixed Method Approach. In *Proceedings of the 10th International Conference on Theory and Practice of Electronic Governance* (pp. 617-620). ACM. 10.1145/3047273.3047357

Kundishora, S. M. (2010). *The Role of Information and Communication Technology ICT) in Enhancing Local Economic Development and Poverty Reduction*. Harare: Zimbabwe Academic and Research Network.

Layne, K., & Lee, J. (2001). *Developing fully functional E-government : A four stage model*. Academic Press.

Misuraca, G., & Lusoli, W. (Eds.). (2010). *Envisioning digital europe 2030: Scenarios for ICT in future governance and policy modelling*. EUR 24614 EN, 12/2010. Retrieved from http://ipts.jrc.ec. europa. eu/publications/pub.cfm?id=3879

Misuraca, G., & Viscusi, G. (2015). Shaping public sector innovation theory : An interpretative framework for ICT-enabled governance innovation. *Electronic Commerce Research*, *15*(3), 303–322. doi:10.100710660-015-9184-5

Moor, M. J., & Norris, D. F. (2005). Does managerial orientation matter? The adoption of reinventing government and egovernmentat the municipal level. *Information Systems Journal*, 43–60.

Mujahid, Y. H. (2002). Digital opportunity initiative for Pakistan. *The Electronic Journal on Information Systems in Developing Countries*, *8*(1), 1–14. doi:10.1002/j.1681-4835.2002.tb00050.x

National Portal of India. (n.d.). Retrieved from https://india.gov.in/egovernance/national-e-governance-plan

Ntulo, G., & Otike, J. (2013). *E–Government: Its Role, Importance and Challenges*. School of Information Sciences. MoiUniversity.

PS. (2003). Retrieved from http://cns.miis.edu/research/space/pakistan/mil.htm

Qaisar, N., & Ghufran, H. (2010). *E-Government Challenges in Public Sector : A case study of Pakistan*. Academic Press.

Rehman, M. (2011). *Factors influencing the adoption of E-government in Pakistan*. Academic Press.

Rehman, M., Esichaikul, V., & Kamal, M. (2012). *Factors influencing e-government adoption in Pakistan*. doi:10.1108/17506161211251263

Sarker, N. I., Wu, M., Liu, R., & Ma, C. (2019). *Challenges and Opportunities for Information Resource Management for E-Governance in Bangladesh*. Springer International Publishing. doi:10.1007/978-3-319-93351-1_53

Shahghasemi, E., Tafazzoli, B., Akhavan, M., Mirani, G., & Khairkhah, T. (2013). *Electronic Government in Iran: A Case Study*. Online Journal of Social Sciences Research.

Song, H. J. (2004). *Building E-Governance through Reform* (Vol. 2). Ewha Womans University Press.

Tambouris, E., Gorilas, S., & Boukis, G. (2001). *Investigation of Electronic Government*. E-gov Project. Retrieved from http://academic.research.microsoft.com/Publication/104353/investigationof- electronic-government/

Tohidi, H. (2011). E-government and its different dimensions: Iran. *Procedia Computer Science*, *3*, 1101–1105. doi:10.1016/j.procs.2010.12.179

Un, E. (2008). *Government Survey 2008: From E-Government to Connected Governance/Dep. of economic and Social Affairs Div. for Publ. Administration and Development Management*. New York: United Nations.

Xu, Y. (2010). E-government and Governance. In *China*. Zhejiang Gongshang University School of Public Hangzhou P.R. China.

Zhou, X. (2004). E-government in China: A content analysis of national and provincial web sites. *Journal of Computer-Mediated Communication, 9*(4).

# Chapter 4
# Digital Governance in Rural Schools of Sri Lanka

**Kennedy Degaulle Gunawardana**
https://orcid.org/0000-0003-3381-9058
*University of Sri Jayewardenepura, Sri Lanka*

## ABSTRACT

*ICT has been used as a model for education in many parts of the world. At the end of the conflict in 2009, the Government of Sri Lanka embarked on an e-Sri Lanka initiative aimed at introducing ICT across several sectors. The education sector received substantial investment in ICT, but only a section of this sector was targeted. This study explores the barriers of introducing and implementing ICT. The analysis of the case studies together with the variables identified from the literature review provided the basis for developing the conceptual framework for the study. The target population for this study were teachers and principals in the rural schools of Trincomalee representing the rural population in Sri Lanka. A combined sample of 70 teachers and principals were considered for this survey across 20 schools in the district. The results indicate that internet access, resources, personnel, and security are essential for introducing and implementing ICT education in the rural areas.*

## INTRODUCTION

### Background to the Research

Sri Lanka's education sector is catered to by a mixture of nation-wide network of state supported, private and international schools. Primary enrolment of boys and girls is well above 90 per cent and secondary enrolment rate stands at over 80 per cent making the country one of the more successful nations in providing educational opportunities to all segments of its population (World Bank, 2011). However these statistics largely belie the severe vulnerability students' face in some of the more rural areas of the country.

For nearly three decades the country was plagued by a bitter civil war that claimed the lives of thousands and destroyed the livelihoods of many. In 2004, several coastal areas of the country were shattered by the Indian Ocean Tsunami that further worsened the economic conditions of the people.

DOI: 10.4018/978-1-7998-1851-9.ch004

In May 2009, the Sri Lankan Government declared the end to the conflict with nearly 300,000 civilians being housed in temporary camps waiting to be resettled back in their homes. The end of the war saw aid pouring into the island, creating a platform for post war development.

Since then the Government has embarked on mega infrastructure development projects; constructing roads, harbours and airports with a view to increasing economic development. In line with his election manifesto; the Mahinda Chinthanaya, (Department of National Planning Ministry of Finance and Planning, 2006) the President declared the year 2009 as the Year of English and IT as part of encouraging the creation of a knowledge society in Sri Lanka (Ministry of Education, 2009).

The Government introduced the e-Sri Lanka Initiative together with the Information and Communication Technology Agency (ICTA) of Sri Lanka to develop the economy of Sri Lanka, reduce poverty and improve the quality of life of the people (ICTA, 2009). Several projects were implemented under this initiative under various sectors.

The 'Nenasala' (Knowledge Centre) Project implemented by the Government together with the ICTA aims to provide Rural Knowledge Centres, e-Libraries, Distance and e-Learning Centres and Tsunami Camp Computer Kiosks. The initial phase of the project was implemented in 13 districts across the island. However this particular project had a selection criterion; where the selected division of the village needed to have a population of between 2000-5000 people, presence of a market with at least 15 wholesale vendors, presence of electricity and a type 2 school with at least 300 students (Nenasala, n.d.).

This meant that nearly 50 per cent of the schools (4,910) were not eligible to participate in the programme as they had less than 300 students (Ministry of Education, 2006).

The Ministry of Education together with the Asian Development Bank (ADB) implemented the Secondary Education Modernization Project (SEMP) aimed at connecting most of the secondary education schools and other related organizations on a wide-area network (WAN). The initial phase of the project was to connect 1000 schools, set up 100 computer resource centres and 8 provincial ICT centres. The services offered included IP-VPN technology, bandwidth connections, web hosting and filtering etc.

However in order for a school to join the project they needed to have the most basic infrastructure such as computers. This meant that most rural schools which did not possess this infrastructure were not part of the project.

Therefore a large number of rural schools lack not only the basic infrastructure for ICT education but basic educational requirements itself, such as teachers. Through the use of ICT these gaps can be fulfilled to a large extent depending on the type and nature of the problem at hand.

ICT has been used as a model for education in many parts of the world. Despite the developed countries exploiting its use exponentially, third world countries have been unable to keep up with the pace. (Valentine & Holloway, 2001) state that the potential of ICT to alleviate rural marginality is also recognized in education funding initiatives. The literature also suggests that ICT has been used in a variety of areas and to suit different purposes and needs of the culture (Vorakulpipat et al., 2010).

However the barriers vary from region to region and sometimes from country to country. Despite this the application and use of ICTs, have tremendous potential for improvements in every sector including education(Mansotraet al., 2009). On a broader perspective, researchers have found a few underlying reasons for the lack of ICT in the rural areas.

## Problem Statement

What are the barriers of introducing and implementing ICT education in the rural areas of Sri Lanka and how could these barriers be overcome?

## Objectives

1. To identify the barriers of introducing and implementing ICT education in the rural areas of Sri Lanka.
2. To find out a suitable medium of technology to implement ICT education in Sri Lanka.
3. To develop a framework for ICT education in the rural areas of Sri Lanka.
4. To estimate the cost / benefit analysis for sustainable development of ICT education in the rural areas of Sri Lanka.
5. To test the suitability of the framework in the Selected Rural Areas of Sri Lanka.

## Scope of the Project

1. The proposed research will be limited to the geographical boundaries of Sri Lanka.
2. The research covers and will be limited to the Sri Lankan secondary education sector supported by the Government of Sri Lanka.

# LITERATURE REVIEW

## Barriers of Introducing and Implementing ICT

(Raju, 2004) found that the **availability of computers**, **rate of teledensity**, **skill training** and **network connections** are fundamental barriers to the spread of access to ICTs and new technologies if applied in targeted ways can narrow the social divide and improve the standard of living in a wide range of health, education and economic outcomes.

**Shortage of ICT teachers** have proven to be a barrier in taking ICT education to the rural areas. (Dlodlo, 2009) points out that those who complete degree programmes in ICT disciplines tend to join the information technology industry rather than work in education. This attributes to differences in remuneration. (Mansotra, et al., 2009) also agree to the fact that lack of ICT trained teachers is a barrier. This is largely applicable even to Sri Lanka where those working in the IT industry are paid better remuneration than they would otherwise earn in education in the government school sector.

A majority of the government schools teach in the Sinhala and Tamil streams while a handful have begun teaching Advanced Level subjects in the English medium. While English is taught as a subject, the percentages of those who can converse and understand are vastly limited to those in the urban areas. (Dlodlo, 2009) identifies the **lack of ICT material in the local languages** as a barrier to taking ICT to the rural communities. Regionalization of software in areas where English is not the medium of education turns out to be much expensive (Mansotra, et al., 2009).

The **financial cost** of taking ICT to the rural villages is a significant barrier. Despite advancements in technology, ICT infrastructure still carries a level of investment which unless provided for, cannot be afforded by rural communities. The lack of exposure to ICT also discourages those in the rural areas of wanting to learn or spend on buying ICT equipment (Dlodlo, 2009).(Vorakulpipat, et al., 2010) found that budget and high IT skill were perceived as problems in implementing ICT for education in the rural areas.

**Access to proper hardware and software** also plays a crucial role in enabling ICT to reach the rural communities. With new piracy laws coming into effect, the cost of purchasing softwares turns out to be a huge investment. (Dlodlo, 2009) mentions that when proprietary software is used, it becomes too expensive for rural schools to purchase.(Vorakulpipat, et al., 2010) point out that a clear lack of computer hardware act as a major problem preventing rural communities access to the internet.

**Access to electricity** represents a significant barrier to ICT education in the rural areas. (Dlodlo, 2009), highlights that despite some areas having access to electricity, due to unemployment, the households find it difficult to pay the high cost of electricity. While she focuses on the ICT education aspect in homes in this particular point, this is also applicable to rural schools, which receive very limited funding from the central government.

## Suitable Options for ICT Implementation

Technology has revolutionized the landscape of education over the past few decades. From email to internet to video conferencing, access to information has literally reached our fingertips. This has led to 21$^{st}$ century being rightly called the information Age.

One of the sectors that have hugely benefitted from this technological revolution has been the education sector. It has helped to build bridges between countries and share information on a common platform, paving the way for collaborative learning.

In traditional education, teaching and learning is fixed, teaching tasks are carried out through such activities as preview, lecture, revision and exercises(Chen, et al., 2009). However with the advent of the internet, the boundaries have been shifted and newer benchmarks set.

Despite the advances in technology and its use in education, rural communities are still not receiving the benefits of it. Apart from the traditional barriers such as cost and trained ICT personnel, the type of technology also hinders the implementation of ICT education in the rural areas.

Due to the lack of proper infrastructure, dial-up connectionsare the most commonly used mode of connection to the Internet in rural schools. The cost of Internet access in this case is proportional to the amount of time the connection is up(Bhowmick, et al., n.d.). Therefore the **lack of a stable internet connection** in the area creates a technological barrier.

**Utilizing wireless communications** on rural and remote communities is forecasted to be enormous(Cebrail, et al., 2009). This is mainly due to the fact that telecommunication service providers are looking to cut down on costly installments of wireline communication infrastructure such as standard PSTN telephone lines. Also given that **mobile broadband devices** are growing in popularity, technologies such as WiMAX will be able to provide connectivity to the rural communities. (Cebrail, et al., 2009) propose **the introduction of Cognitive Radio and IEEE 802.22 Wireless Regional Access Network Technologies**which will enable educational opportunities for rural and remote areas. This technology helps to solve inefficient wireline solutions in rural and remote areas.

(liang, et al., 2008) suggest a more up and coming technology in the form of **IPTV** (internet protocal television) for e-learning in rural areas. They prove that the quality of distance learning will improve remarkably due to the availablility of interactive and individual education service system.

(Dlodlo, 2009) suggests the **setting up of a network of community-based ICT learning centers**. Instead of implementing ICT education in each and every school, a community based learning centre might act as a viable option in the rural communities where village folk are closely knitted. The centres can be powered by a viable electricity solution and if proven successful can be implemented in the other areas.

Under technology addressing the **issue of electricity** is deemed necessary. A significant number of villages in the rural areas of Sri Lanka do not have access to electricity while some have limited access where power is provided during a particular time frame of the day. Therefore the use of a **sustainable power supply component** is required if ICT education is to be used in the rural areas.

Rising fuel costs make the user of diesel powered generators rather costly. **Hybrid power systems** are more reliable in terms of supplying electricity whilst also ensuring a clean and inexpensive way without a dieselgenerator(Abdullah, et al., 2009).

## Sustainability

Sustainability is a key component stakeholders look for when implementing a particular project. ICT education project implementations **do not usually provide direct return on investments** but **increase the social and economic welfare** of the community. Since Governments are not primarily focused on profit-making, a majority of such projects are undertaken by them. Private sector companies undertake such projects as part of their CSR initiatives but not of the magnitude of what a Government would implement.

(Lallement, et al., 2006) points out that establishment and operation of tele-centers in marginal areas have to be treated as **social investment projects** with **subsidies justified on equity grounds**. It is therefore important for the Government or the implementing agency to give priority to the least cost and most efficient approaches and technologies.

Presently ICT educational projects are funded externally either through loans or grants from institutions such as the World Bank and ADB. Using funds to implement such ICT projects is worthwhile only if the rural communities are taught or a platform is created to sustain these projects once the grant is used up.

In Sri Lanka however this is not the case as educational projects are implemented and are left to the communities to sustain without providing them with guidance or a plan on how to go about it.

Sustainability will largely depend on the **level of infrastructure required** and the **future potential of the rural area**. If by providing ICT education, the Government is able to raise the standard of education and thus create a more knowledgeable workforce coming out of the rural communities, this will augur well for the future of the country.

Potential investors will look at implementing business ventures in areas which were previously inaccessible largely due to the war, but more importantly due to lack of the quality of labour available. This meant that new ventures were limited to the urban areas only.

These are aspects the Government can look at in its bid to take ICT education to the rural masses.

## REVIEW OF LOCAL LITERATURE

### The Secondary Education Sector in Sri Lanka

Since the end of the conflict in 2009, the Sri Lankan population has stabilized around the 20.3 million (Department of Census and Statistics, 2012) mark largely due to the peaceful atmosphere in the country and prospects of a booming economy. Many of those who fled the country during the course of the conflict have returned with the prospect of rebuilding and setting up base in their home country. However three years on, despite the advent of mega development projects, cost of living has risen considerably and the promise of a booming economy are far from being felt.

The Sri Lankan secondary education sector has encountered its fair share of problems. In 2011, the Ministry of Education released erroneous advanced level results due to a computational error in the calculating system (ColomboPage, 2011). Despite the mishaps, the Government has continued to fund the free education system in the country amounting to 2.0% of the GDP (CBSL, 2011).

According to a report by the Data Management Branch of the Ministry of Education, the number of Government schools amounted to 9,905(Ministry of Education, 2012). This was anincrease of 191 schools from the census in 2006 where there were 9,714 schools. This change could be attributed to the conflict when people fled the areas where the conflict intensified and schools were abandoned. Despite people returning to their villages, several schools in these areas have not returned to a functioning status.

On a more positive note there has been an increase in the number of 1AB and 1C schools (Figure 1) (Ministry of Education, 2012). This is a positive sign because it shows an increase in the number of schools offering Advanced Level subjects. This allows students to stay in their native villages and pursue their education. However an increase in these types of schools doesn't necessarily mean better standards of education.

Continuing with the rising trend, there has been an increase in teachers in all the schools. Between 2006 and 2012, the number of teachers has increased by nearly 10,000 (Figure 2) (Ministry of Education, 2012).

Furthermore 1, 2 and 3-9 teacher schools have dropped considerably over the period in concern (Figure 3) (Ministry of Education, 2012). This shows that more and more teachers are being sent to schools which are currently short of teachers. However this census does not show that teachers in some of the

*Figure 1. Schools by functional grade*

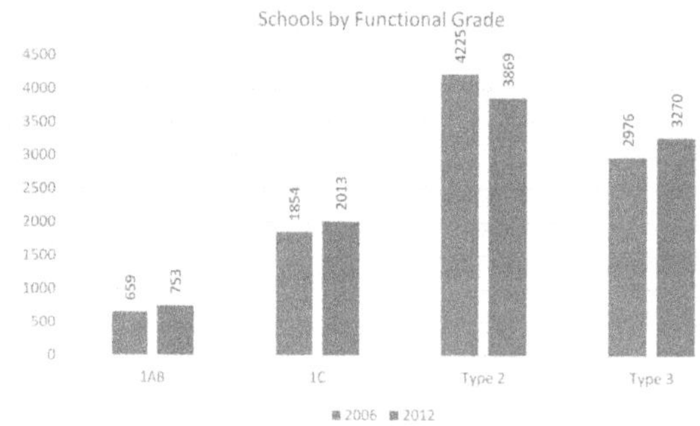

*Figure 2. Teachers by functional grade of schools*

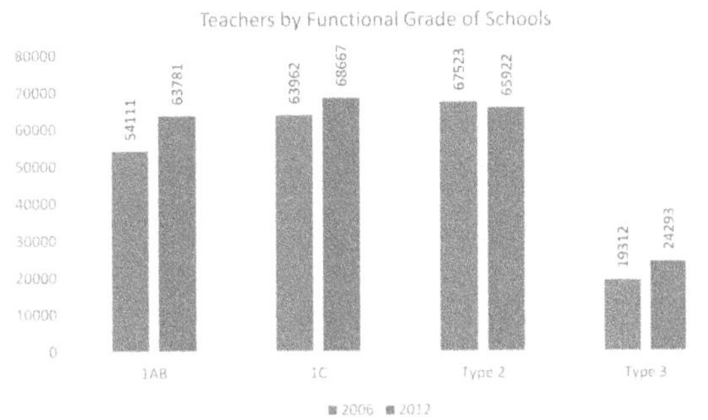


*Figure 3. Schools by teacher numbers*

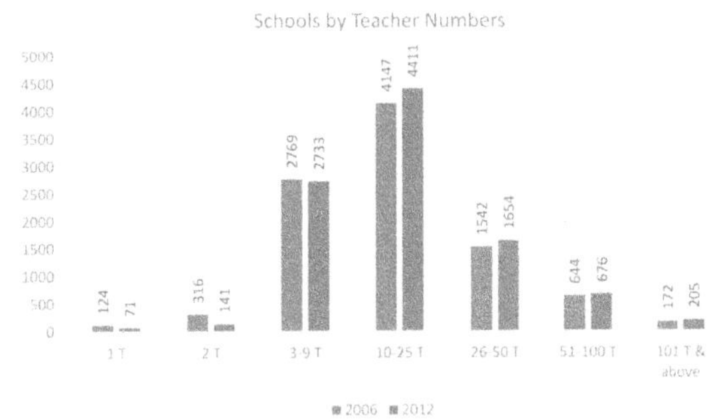


schools move to more prominent schools thus leaving 1 and 2 teacher schools without any teachers at all thus leading to closure of these schools.

According to the Figure 4, the number of graduate teachers has increased during the period 2006 – 2012 (Ministry of Education, 2012). While this increase can be taken as a positive sign, the drop in the number of trained teachers is damaging. Trained teachers are especially important for the lower grades where attention to students is paramount.

Between 2006 and 2012, entry into schools has increased approximately by a little over 160,000 (Figure 5) (Ministry of Education, 2012). This could be due to the fact that the height of the conflict was between 2006 and 2009 where more than 300,000 people including children were displaced in the North.

The student to teacher ratio has largely remained the same going down by one point overall during the period (Figure 6) (Ministry of Education, 2012). However this statistic belies the actual situation in the rural areas.

Despite the increase in Graduate teachers over the period, the student to graduate teacher ratio has reduced (Figure 7) (Ministry of Education, 2012). In 2006 the student to untrained teacher ratio stood at an alarming 514:1. But by 2012 this number had only reduced to 475. This is still a significantly high number considering the fact that there are trained teachers in the country. However teachers tend to request transfers to urban areas and avoid postings to rural and remote areas.

*Figure 4. Teachers by qualification*

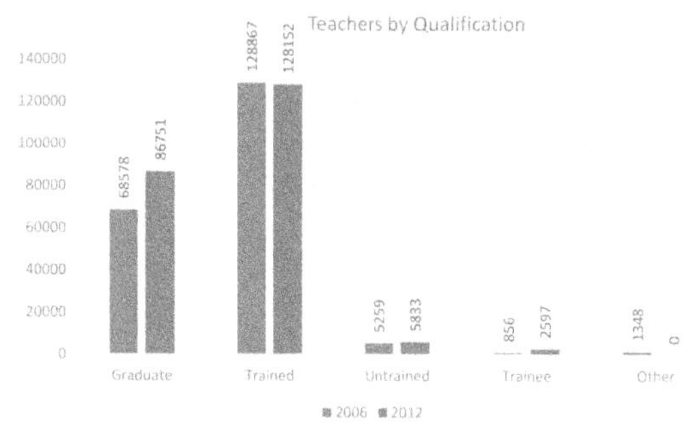


*Figure 5: Number of Students by Functional Grade of Schools*

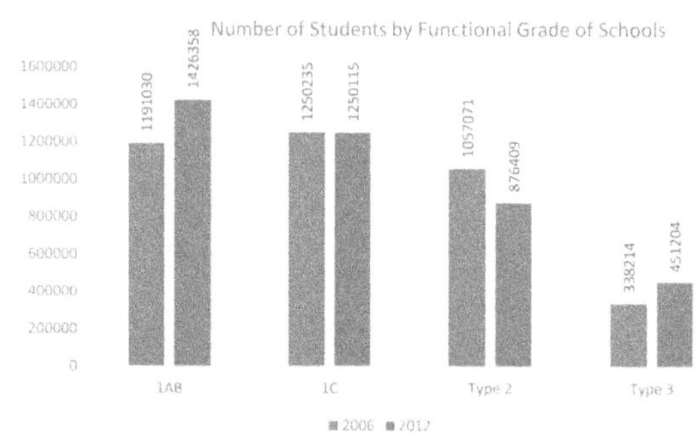


*Figure 6. Student to teacher ratio by functional grade of schools*

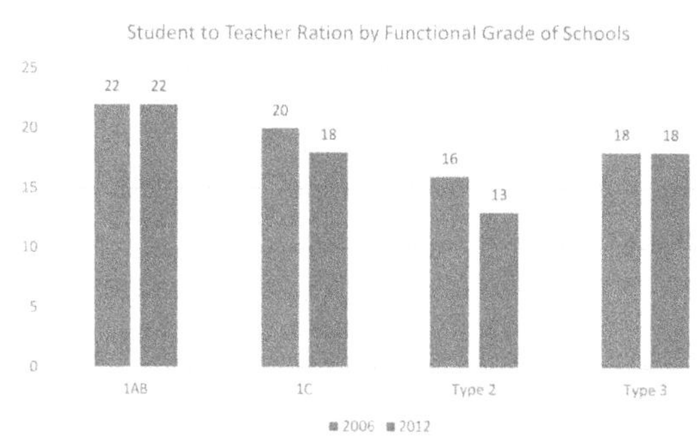


## ICT PROGRAMMES IN SRI LANKA

### e-Sri Lanka Programme

The Government of Sri Lanka launched "e-Sri Lanka", a national development initiative in 2003, with the aim of enhancing growth and equity through improved access and use of information communication technology. The project supports the empowerment of the rural poor, disadvantaged groups, women and

*Figure 7. Student to teacher ratio by teacher qualification*

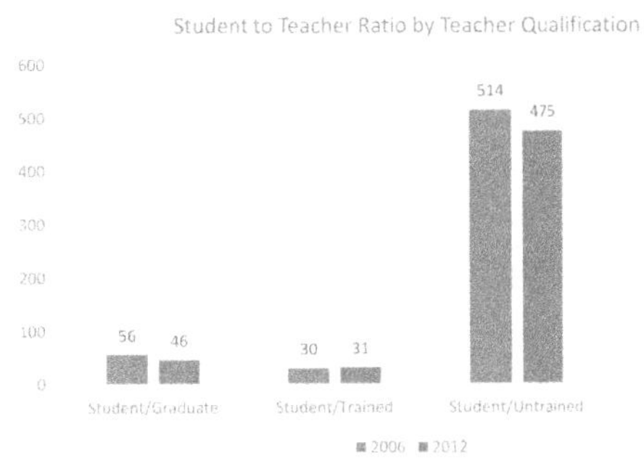


youth through increased and affordable access to information and communication tools (Skill International Pvt Ltd, 2010).

Under this initiative the Government set up the **Nenasala (Knowledge Centre) Project**. The Nenasala stands for —— Knowledge Centres or "Wisdom outlets"and aims to empower the population in rural areas of the country through affordable community access to ICT. The Nenasalas were established in rural areas starting in the South and scaled up to the North East, where connectivity is provided.

The programme targeted residents of small rural communities (e.g. farmers, rural youth) residing in small towns with between 2,000 and 5,000 people. The programme also aimed to provide parallel distance e-Learning and basic computer literacy training services to a broader population that also includes urban and semi-urban disadvantaged groups: e.g. students, small and micro enterprises, women, disabled and marginalized groups.

In order for a Grama Niladari division to be considered as a potential location for the establishment of a Nenasala it had to have:

- A population base between 2,000 and 5,000 people
- A fixed market with at least 15 wholesale vendors within a 5 km radius,
- Reliable energy, and
- A Type 2 school with at least 300 students

The project based its foundation on the lines of long term sustainability and set up three types of knowledge centres:

1. Nenasala – Rural Knowledge Centre – This type of knowledge centre involved the establishment of multi service community centres which provided access to internet, email, telephones, fax, photocopy and computer training classes.
2. E-Library Nenasala – This type of knowledge centre follows a reducedversion of the previous one together with a community driven model where some services are provided free while some are provided for a fee to enable sustainability of the centre.Computer based training (CBT) media is available to use off-line in English, Sinhala and Tamil as well as a large e-library of books and periodicals for the use of students of all ages.

3. Distance and E-Learning Centres –This type of knowledge centre will have distance and e-learning services inclusive of all infrastructure facilities such as video conferencing room, multi-media computer laboratory and a playback room.

The aim of the Nenasala project was to establish 1000 centres throughout the country which provided access to internet, email, telephone, fax, photocopy, computer training classes and other ICT services with a view to significantly increase the level of economic activity in rural communities, raise levels of computer literacy among children and adults and enable rural populations to access government information and services online. As at October 2010, only 629 centres have been established under this project. As a solution for the affordability gap, a voucher programme has been introduced to offer free access initially to the Nenasala services. The equipment and facilities supported by the Nenasala project were; 2-4 computers, Internet connectivity subsidy and ICT trained staff.

According to the evaluation report conducted on 237 of the 629 centres established so far, nearly 86% of the Nenasalas' were community, religious and individually owned while only 1.69% were owned by schools.

It was also noted that 59% of the users were students and 5% were school teachers. **Out of the 59% of students, 66.73% were secondary school students.** This statistic itself justifies the need for ICT facilities in schools located in the rural areas.

The survey also indicated that students in secondary education accessed the Nenasalas more frequently with 152 of them accessing it more than 4 times a week while 333 accessed it 2-3 times a week and 152 accessed the Nenasalas once a week. Those who accessed it once a week may have done so due to the need to travel a distance from their residence to the Nenasala and thus preferred to use it once a week.

In terms of duration of use at the Nenasala, students engaged in secondary education were found to use the Nenasalas for the longest periods of time. 180 students used it for 2-3hrs, 271 used it for 1-2hrs, 86 used it for 1/2-1hr and 70 used it for less than 1/2hr.

School teachers were also found to be using the Nenasalas for longer durations. 3 teachers used it for 3-5hrs, 12 used it for 2-3hrs, 20 used it for 1-2hrs, 9 used it for 1/2-1hr and 8 used it for less than 1/2hr.

Another significant highlight of the survey was that school work was reported the second highest use with 24.06% at the Nenasala behind ICT training which was 26.83%. This further necessitates the need for ICT facilities in the rural schools.

According to the conclusion of the report, the ownership and management of the project has an effect on the sustainability of the Nenasala. It was found that individual owner operated centres with profit sharing between owner and operator have better chance for financial sustainability. The data indicate that Nenasala centres are located in adequately spaced buildings that facilitate the housing of ICT equipment. This has contributed to the positive change in access to ICT facilities.

The report also points out that youth are the major group of beneficiaries and they access and use ICT facilities at a frequency of 3-4 times a week.

We can conclude that the Nenasala project has indeed made a positive impact on the community. However this project was focused more on improving access to the community and not primarily focused on students in the secondary sector.Furthermore the selection criteria for the Nenasala singled out many schools in the rural areas from having access to the Nenasala. Therefore there is a gap that needs to be addressed especially with such overwhelming numbers in favour of students using ICT facilities.

## Secondary Education Modernization Project

The Government of Sri Lanka requested the support of the Asian Development Bank (ADB) to modernize secondary education, as a means of promoting economic and social development and reducing disparities. To support this request the ADB designed the Secondary Education Modernization Project (SEMP). The project's objective was to modernize the secondary school system and improve educational quality, so as to prepare youth to compete in the modern global economy. Improvements in quality were expected to lead to higher pass rates, especially through modernization of the curriculum, improved access for disadvantaged students, and enhanced education management (Asian Development Bank, 2009).

Highlights of the ICT components are discussed below:

Multimedia units (MMUs) were included in the project design to modernize teaching. Classrooms in 2,169 schools (against an appraisal target of 2,500) were converted to MMUs. Schools were expected to provide a room, furniture and a teacher. Each MMU was equipped with a television, video cassette recorder or video compact disc player, radio, tape recorder, an overhead projector and a white board. The National Institute of Education prepared 45 provincial master trainers to train teachers in charge of MMUs and developed a handbook on multimedia education. Mission visits to schools revealed mixed success in the use of MMUs. Utilization ranged from equipment being kept in a storage room to MMUs that were booked for a large number of periods and had a library of multimedia materials available. Due to the shortage of classrooms in most schools, the room identified for MMU was typically small and could not accommodate large classes. A bigger problem, however, is that educational media technology has changed substantially since the project design was prepared. The present use of overhead projectors, video cassette recorders and audio tape players is rare. Computers and digital media projectors that provide superior quality presentations have replaced the older technology.

The project included provision of computer learning centres (CLCs) in 800 schools - a target that was later revised upward to 1,000 schools - to develop computer literacy and support computer-assisted learning (CAL) in core subjects. In total, 1,006 schools received (i) civil works to convert existing classrooms to CLCs; and (ii) furniture, air conditioners and 15–25 computers, depending on the total number of students enrolled. Funds to cover the recurrent operating costs for the first 2 years were provided by the project. Schools were encouraged to organize after-hour computer courses for a fee to generate revenue, primarily to offset utility costs and recurring maintenance after expiry of the supplier warranty. Schools were permitted to collect and retain fees for after-hour usage of the computers and to use the money to pay operating costs. At project completion, around 70% of schools were successful in recovering costs.

In its effort to promote sustainability the project introduced some innovative features.

The provision of computers and air conditioning units for the CLCs inevitably raised school operating costs, which include the cost of electricity, maintenance and repairs, and internet connectivity. The project covered these costs for the initial 2 years. Realizing that the absorption of these costs into the general education budget would have substantial financial implications, the project design included features to shift the responsibility of covering the operating costs from the Ministry of Education (MOE) to the schools. Schools were expected to devise strategies to recover operating costs by organizing after-hour computer courses for a fee. A significant milestone in this respect was the issuance of MOE Circular No. 2005/29 dated 30 September 2005, which permitted schools to collect and retain fees for after-hour usage of the computers and to use the money to pay operating costs, rather than having to transfer the funds to the central treasury(Ministry of Education, 2005). This gave the schools a sense of ownership and required principals and teachers to proactively apply management and marketing skills.

**SchoolNet** - The project provided for Internet connectivity for the CLCs once they were fully functioning and sustainable. Rather than letting individual schools take care of their internet connectivity, MOE used the opportunity to exploit economies of scale and established a wide-area network called SchoolNet. SchoolNet connects both the schools equipped with a CLC and other related organizations, such as 90 computer resource centres, 17 national colleges of education, MOE, the National Institute of Education, eight provincial ICTcentres and the central and regional project management offices. This resulted in substantial cost savings, standardization of access, and a better bargaining position with respect to the supplier.

In addition to network and internet connectivity for each of the institutions, SchoolNet also provides services for learning and teaching through its website (http://www.schoolnet.lk/). SchoolNet aims to improve interaction and information exchange between students and teachers from different schools to enhance the teaching and learning environment. Features include webmail for teachers and students, provisions for webpages of individual schools, and voice communication among schools through internet protocol phone.

SchoolNet also intends to become a one-stop resource page for schools in Sri Lanka. It currently provides announcements for various school competitions, such as for the annual educational software competition, which has become an important source for tapping software developed at the school level by students and teachers and replicating it for other schools. This approach, in combination with adopting software from various other sources, is a more efficient approach than the centralized in-house development of computer-assisted learning software by NIE. The aim is to make these learning materials available on SchoolNet, facilitating easy access by all schools.

## DESCRIPTIVE STATISTICS

### Quantitative Analysis

The data collected from the survey has been summarised below. The summarised data shows the sample data size grouped by Age, Gender, Medium of Instruction, Type of School and other parameters that were used in the questionnaire. The collected data shown below has been used in the processing of statistical results.

### Demographic Data

The survey targeted 20 schools in the district of Trincomalee. 50 teachers making represented 71 percent of the surveyed population. The remaining included the 20 principals from each of the schools.

A majority of those surveyed were between the ages of 21-30 and 31-40 indicating that the teaching population largely consists of young people.

There was a good balance of male to female representation in the survey. However 54.3 percent of the surveyed population were females.

Teachers in majority of the schools surveyed indicated that the medium of instruction was Tamil with 63 percent making up for this representation.

*Figure 8. Type of respondent*

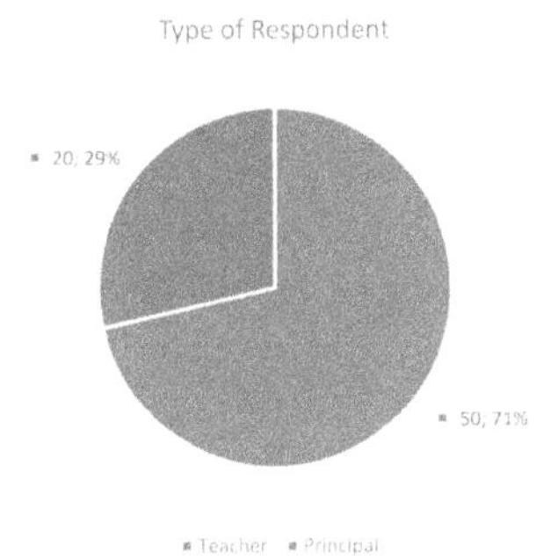


*Figure 9. Age of the respondent*

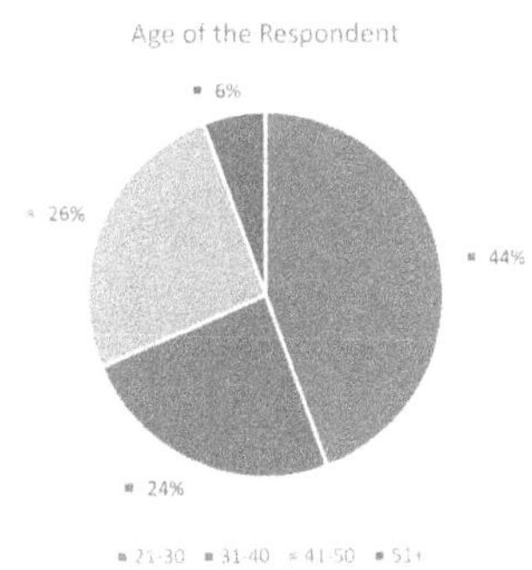


*Figure 10. Gender of the respondent*

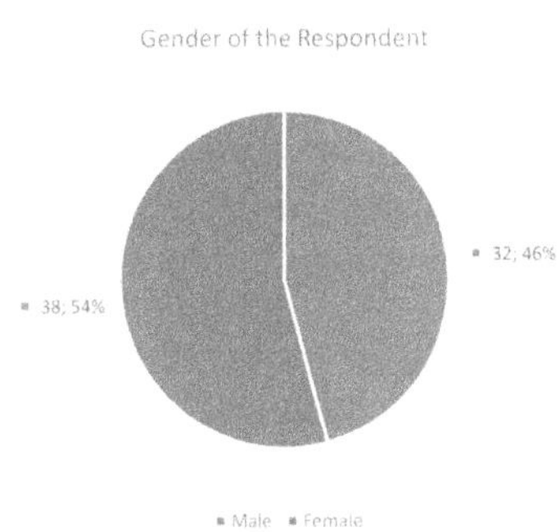


67 percent of the teachers surveyed were from 1C schools with 23 percent from 1AB schools. This is an appropriate representation since these schools offer the advanced level subjects apart from having higher student numbers.

Out of the surveyed population 14 percent were ICT teachers while a larger proportion was represented by Maths and Science teachers.

*Figure 11. Medium of instruction*

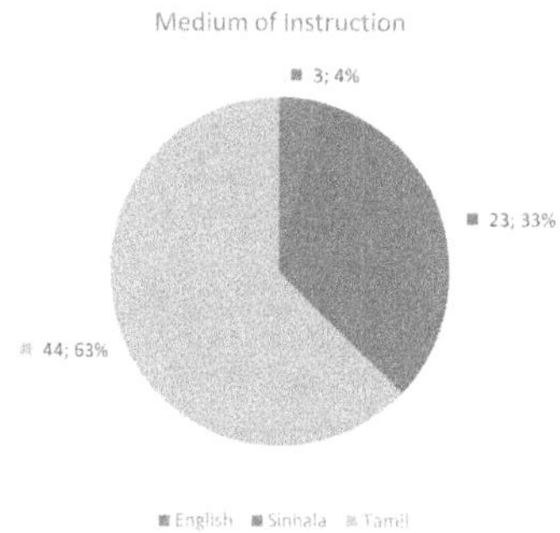


*Figure 12. Type of school*

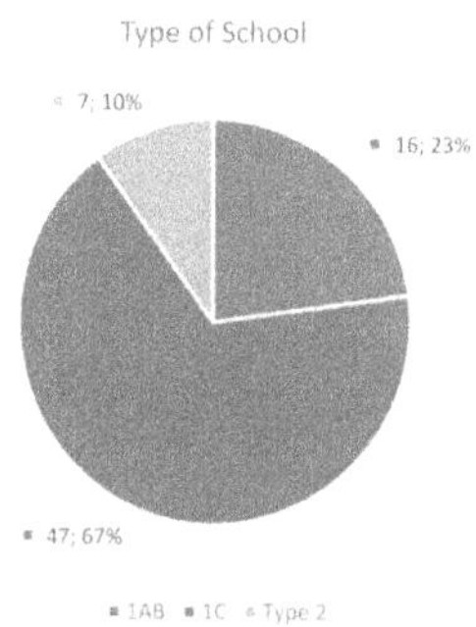


*Figure 13. Subject taught by the respondent*

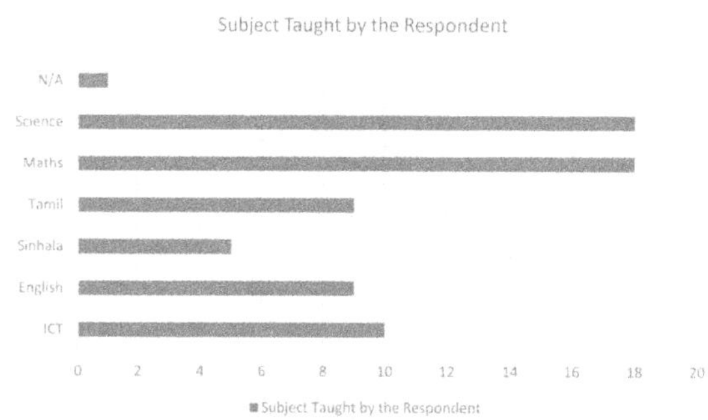


89 percent of the persons surveyed mentioned that they know how to use a computer. This is an important indicator especially if ICT is to be successful in these rural areas. Also it implies that persons in these areas have had access to computers. The respondents' knowledge of how to use a computer was further validated from the following questions:

*Figure 14. Do you know how to use a computer?*

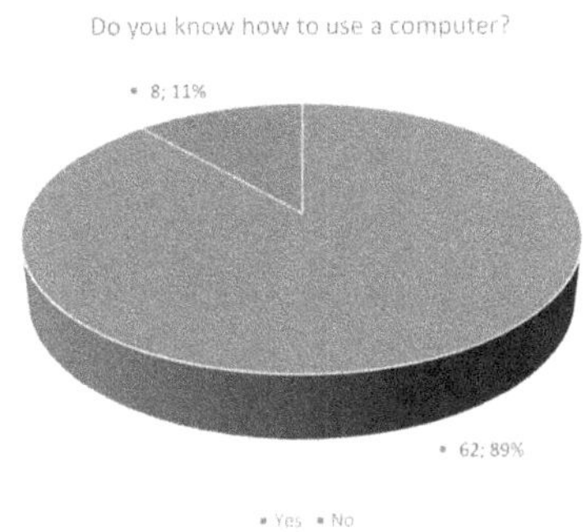


*Figure 15. Type a letter on a word document*

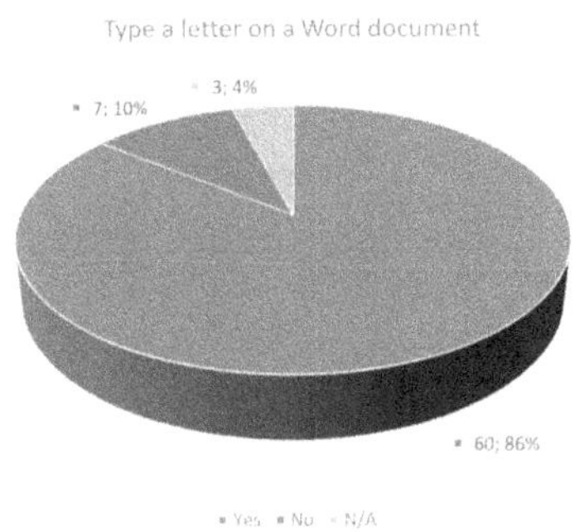


*Figure 16. Send an email*

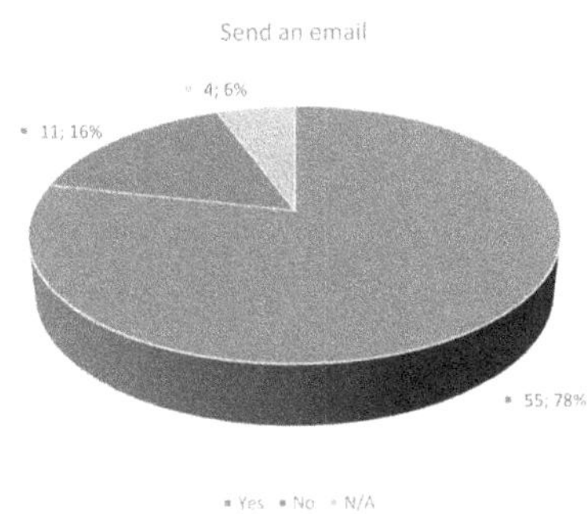


86 percent mentioned that they can type a letter on a word document.
78 percent mentioned that they can send out an email.
85.7 percent mentioned that they knew how to play a CD.
78 percent indicated that they knew how to browse the internet.

*Figure 17. Play a CD*

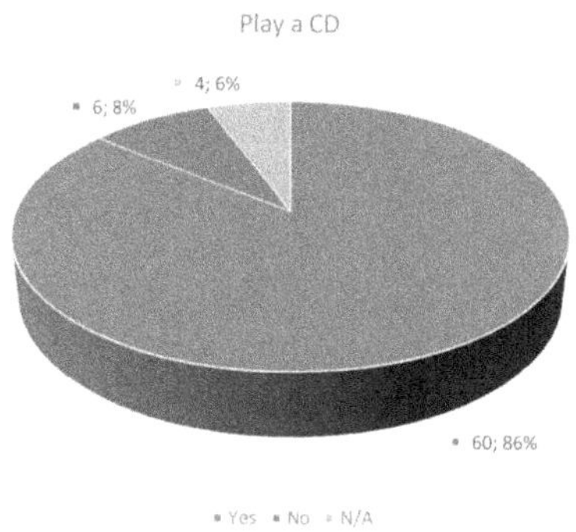


*Figure 18. Browse the internet*

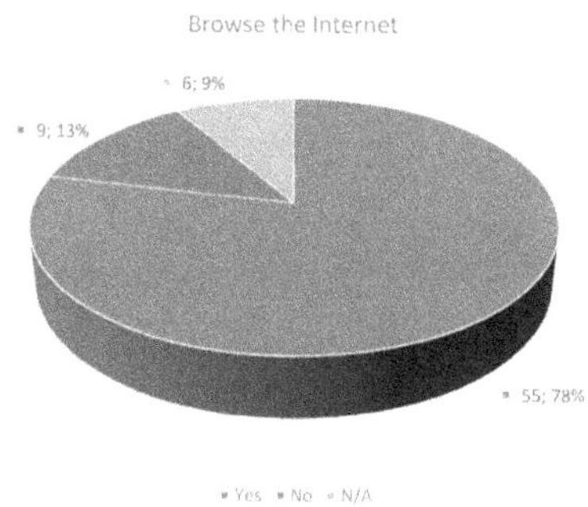


*Figure 19. Have you received computer training?*

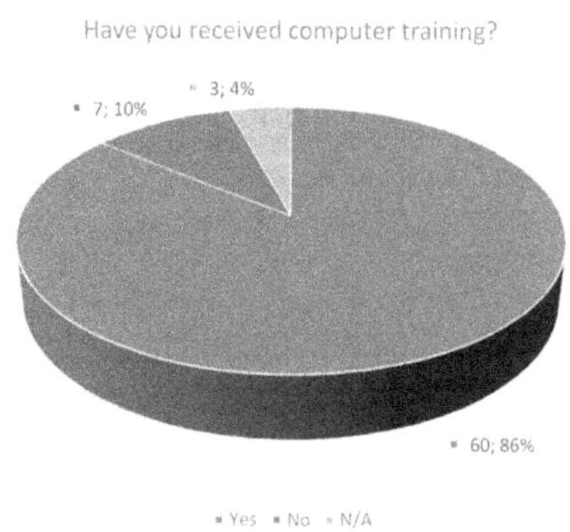


Respondents' were asked if they received some sort of computer training. 86 percent mentioned that they have had some sort of computer training before. The subsequent questions gather more data on how these trainings were obtained.

*Figure 20. Government sponsored*

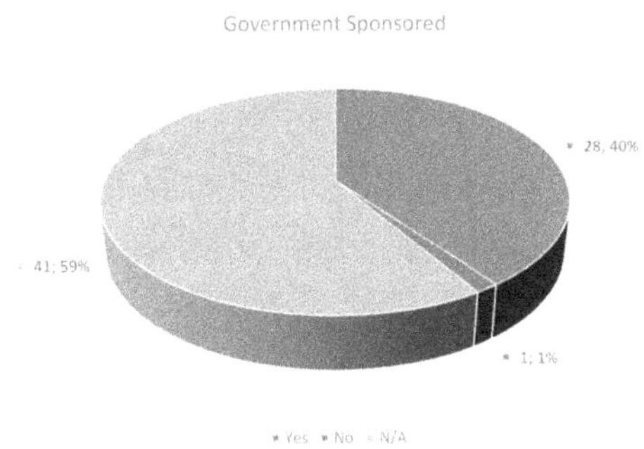


*Figure 21. Private class*

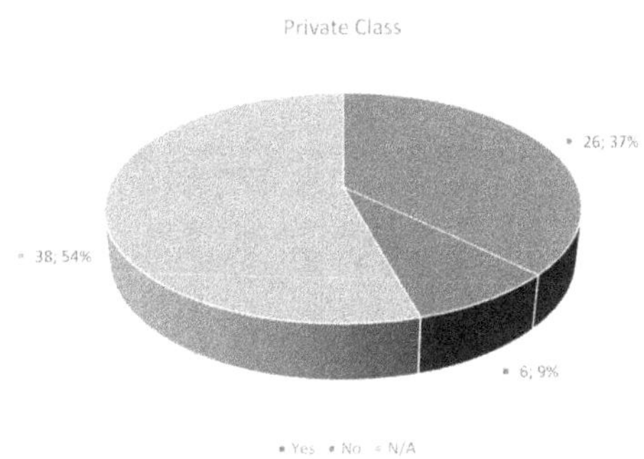


*Figure 22. University*

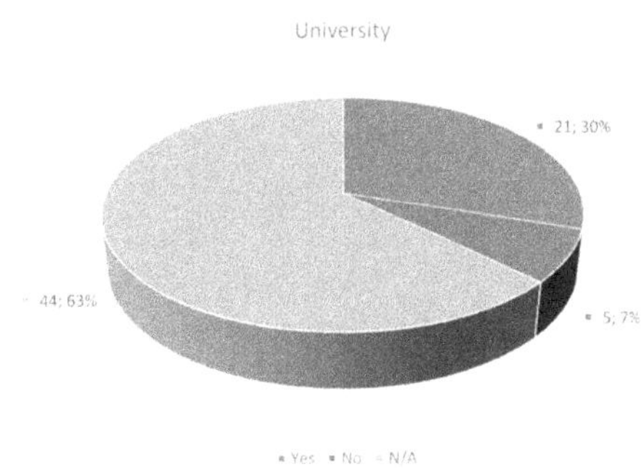


40 percent received training through a Government sponsored programme.

26 percent received training through a private class which is usually self-sponsored.

30 percent of the respondents mentioned that this training was through university. This could have been while they were following their undergraduate studies.

22.9 percent indicated receiving training in school.

*Figure 23. School*

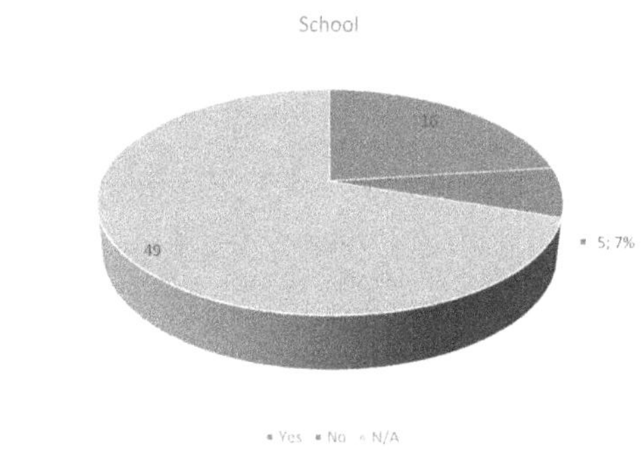


*Figure 24. Do you have an ICT related qualification?*

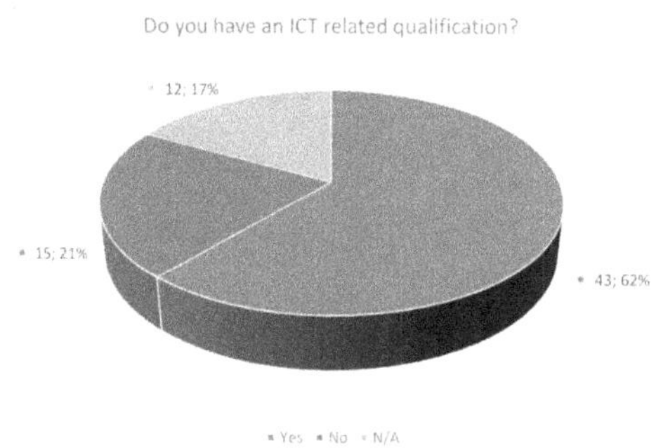


*Figure 25. Certificate course*

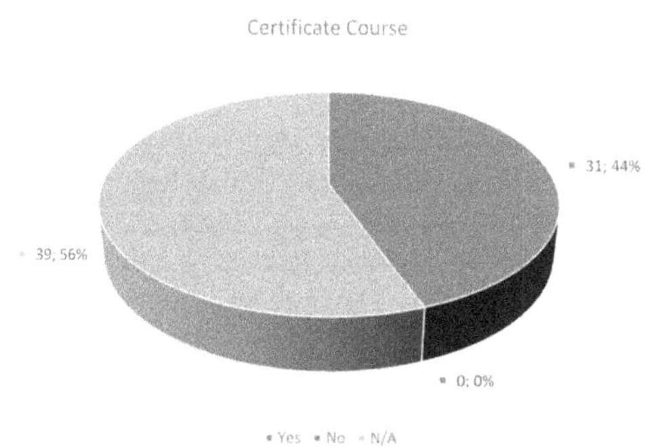


62 percent mentioned that they had an ICT related qualification. This is quite a significant figure indicating the interest in ICT. It also highlights the importance that is attached to it by teachers in rural areas. The subsequent questions were asked to deduce the type of ICT qualification.

44 percent mentioned that they have followed a certificate course.

22.9 percent possess a diploma in an ICT related qualification.

*Figure 26. Diploma*

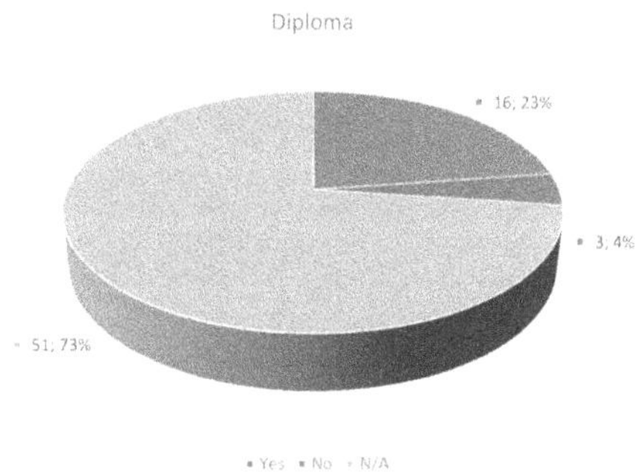


*Figure 27. Higher diploma*

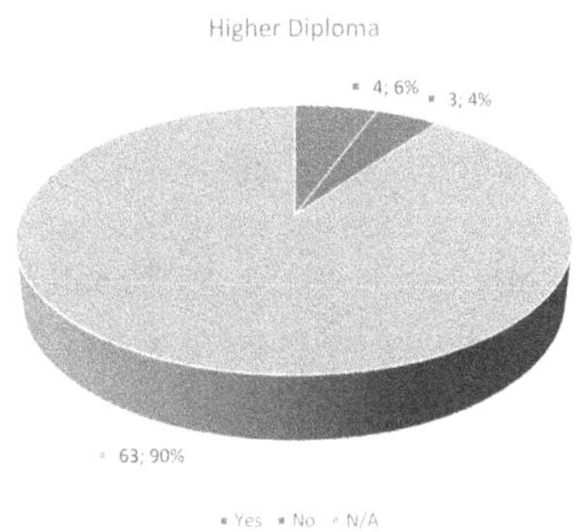


*Figure 28. Degree*

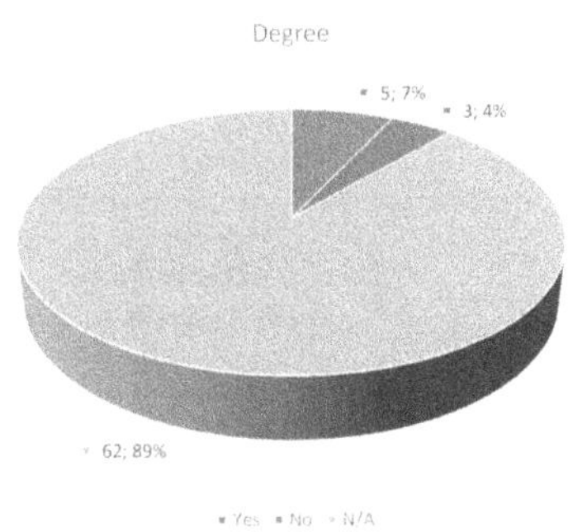


6 percent hold a qualification equivalent to a higher diploma.
7 percent hold degrees ICT.
There were no respondents' with postgraduate qualifications such as a masters in ICT.

*Figure 29. Masters*

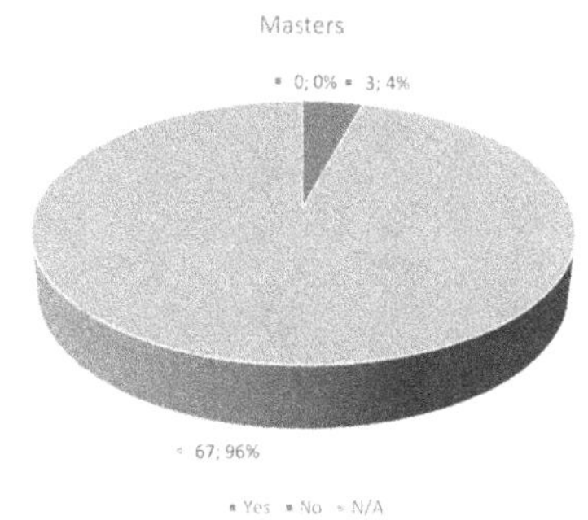


## PREFERENCE DATA

### Internet Access

Internet access was identified as a key component that could impact the introduction and implementation of ICT in the rural areas. The survey focused on gathering details on the availability of internet access in the chosen area, its importance and its usage. 55.7 percent strongly agreed that internet access is important for ICT education and another 32.9 percent agreed bringing it a cumulative total of 88.6 percent. Assessing the perceptions of internet use, 40 percent agreed that internet access distracts students from their work, while 32.9 percent remained neutral and 17.1 percent disagreed. 34.3 percent strongly agreed and 55.7 percent agreed that high speed connections allow more students to access the internet. Respondents also strongly agreed (28.6 percent) and agreed (48.6 percent) that the internet was used to obtain O/L and A/L results. 14.3 percent strongly agreed that internet connections were available in their area while 40 percent agreed to the same statement. However 22.9 percent each responded with neutral and disagreed to the statement.

| **Internet access is important for ICT education** | | | **Internet access distracts students from their work** | | **High speed connections allow more students to access the internet** | | **Internet is used to obtain O/L and A/L results** | | **Internet connections are readily available in your area** | |
|---|---|---|---|---|---|---|---|---|---|---|
| **Criteria** | **Frequency** | **Percent** | **Frequency** | **Percent** | **Frequency** | **Percent** | **Frequency** | **Percent** | **Frequency** | **Percent** |
| Strongly Disagree | 0 | 0.0 | 4 | 5.7 | 0 | 0.0 | 0 | 0.0 | 0 | 0.0 |
| Disagree | 4 | 5.7 | 12 | 17.1 | 0 | 0.0 | 0 | 0.0 | 16 | 22.9 |
| Neutral | 4 | 5.7 | 23 | 32.9 | 7 | 10.0 | 16 | 22.9 | 16 | 22.9 |
| Agree | 23 | 32.9 | 28 | 40.0 | 39 | 55.7 | 34 | 48.6 | 28 | 40.0 |
| Strongly Agree | 39 | 55.7 | 3 | 4.3 | 24 | 34.3 | 20 | 28.6 | 10 | 14.3 |
| Total | 70 | 100 | 70 | 100 | 70 | 100 | 70 | 100 | 70 | 100 |

| Purchasing new computers is expensive | | | Computers breakdown regularly | | Repairing computers is not expensive | | Softwares need to be purchased and updated regularly | | All computer peripherals must be provided by the Government sufficiently | |
|---|---|---|---|---|---|---|---|---|---|---|
| Criteria | Frequency | Percent | Frequency | Percent | Frequency | Percent | Frequency | Percent | Frequency | Percent |
| Strongly Disagree | 0 | 0.0 | 0 | 0.0 | 0 | 0.0 | 0 | 0.0 | 0 | 0.0 |
| Disagree | 0 | 0.0 | 22 | 31.4 | 22 | 31.4 | 11 | 15.7 | 12 | 17.1 |
| Neutral | 18 | 25.7 | 10 | 14.3 | 0 | 0.0 | 14 | 20.0 | 11 | 15.7 |
| Agree | 19 | 27.1 | 29 | 41.4 | 31 | 44.3 | 35 | 50.0 | 21 | 30.0 |
| Strongly Agree | 33 | 47.1 | 9 | 12.9 | 17 | 24.3 | 10 | 14.3 | 26 | 37.1 |
| Total | 70 | 100 | 70 | 100 | 70 | 100 | 70 | 100 | 70 | 100 |

## Resources

Resources play a significant part in the implementation of ICT. 47.1 percent strongly agreed that purchasing new computers is expensive. 41.4 percent agreed that computers breakdown regularly and 44.3 percent disagreed that repairing computers was not expensive, i.e. that it's expensive to repair. 50 percent agreed that softwares needed to be purchased and updated regularly. 67.1 percent (strongly agreed and agreed) responded that all computer peripherals must be provided by the Government sufficiently.

## Perceptions

Perceptions of using ICT were sought from the respondents to better understand their mind sets. 27.1 percent strongly agreed and agreed respectively on encouraging students to research on the internet. 34.3 percent agreed and 15.7 percent strongly agreed to use ICT resources in their teaching. Interestingly 28.6 percent strongly agreed to not use the computer or internet when they teach in class. 38.6 percent agreed that they were concerned that students will no longer read books if they use computers. A

| I encourage students to research on the internet | | | I am happy to make use of ICT resources in my teaching | | I prefer to not use the computer or the internet when I teach in class | | I am concerned that students will no longer read books if they use computers | | I feel that it students always learn through computers I won't have a job | |
|---|---|---|---|---|---|---|---|---|---|---|
| Criteria | Frequency | Percent | Frequency | Percent | Frequency | Percent | Frequency | Percent | Frequency | Percent |
| Strongly Disagree | 13 | 18.6 | 20 | 28.6 | 10 | 14.3 | 8 | 11.4 | 11 | 15.7 |
| Disagree | 7 | 10.0 | 7 | 10.0 | 12 | 17.1 | 4 | 5.7 | 21 | 30.0 |
| Neutral | 12 | 17.1 | 8 | 11.4 | 9 | 12.9 | 17 | 24.3 | 28 | 40.0 |
| Agree | 19 | 27.1 | 24 | 34.3 | 19 | 27.1 | 27 | 38.6 | 10 | 14.3 |
| Strongly Agree | 19 | 27.1 | 11 | 15.7 | 20 | 28.6 | 14 | 20.0 | 0 | 0.0 |
| Total | 70 | 100 | 70 | 100 | 70 | 100 | 70 | 100 | 70 | 100 |

| Any staff member can teach ICT | | | Qualified ICT instructors are hard to find | | ICT instructors choose to teach in rural schools | | One ICT instructor is adequate for the entire school | | Teachers who receive ICT training can teach ICT in the school | |
|---|---|---|---|---|---|---|---|---|---|---|
| Criteria | Frequency | Percent | Frequency | Percent | Frequency | Percent | Frequency | Percent | Frequency | Percent |
| Strongly Disagree | 18 | 25.7 | 7 | 10.0 | 7 | 10.0 | 3 | 4.3 | 0 | 0.0 |
| Disagree | 49 | 70.0 | 8 | 11.4 | 39 | 55.7 | 30 | 42.9 | 0 | 0.0 |
| Neutral | 0 | 0.0 | 0 | 0.0 | 10 | 14.3 | 3 | 4.3 | 4 | 5.7 |
| Agree | 0 | 0.0 | 48 | 68.6 | 11 | 15.7 | 30 | 42.9 | 39 | 55.7 |
| Strongly Agree | 3 | 4.3 | 7 | 10.0 | 3 | 4.3 | 4 | 5.7 | 27 | 38.6 |
| Total | 70 | 100 | 70 | 100 | 70 | 100 | 70 | 100 | 70 | 100 |

positive outcome emerged where only 14.3 percent indicating they felt if students always learn through computers would lead to losing their jobs.

## Personnel

70 percent disagreed that any staff member can teach ICT and 68.6 percent agreed that qualified ICT instructors are hard to find. Only 15.7 percent agreed that ICT instructors choose to teach in rural schools while 42.9 percent agreed that one ICT instructor is adequate for the entire school. 38.6 percent strongly agreed and 55.7 agreed that teachers who receive ICT training can teach ICT in the school.

## Security

The security of implementing ICT was also enquired. 17.1 percent strongly agreed that the computers were located in an open room while 31.4 percent disagreed. 44.3 percent agreed that the computers are located in a computer laboratory. 31.4 percent agreed that computers are safe from being stolen by outsiders while 28.6 percent disagreed. 57.1 percent noted that access to computers are closely monitored. 44.3 percent disagreed on the fact that computer peripherals are insured.

| The computers are located in an open room | | | The computers are located in a computer laboratory | | The computers are safe from being stolen by outsiders | | Access to computers are closely monitored | | The computer peripherals are insured | |
|---|---|---|---|---|---|---|---|---|---|---|
| Criteria | Frequency | Percent | Frequency | Percent | Frequency | Percent | Frequency | Percent | Frequency | Percent |
| Strongly Disagree | 8 | 11.4 | 0 | 0.0 | 5 | 7.1 | 0 | 0.0 | 10 | 14.3 |
| Disagree | 22 | 31.4 | 11 | 15.7 | 20 | 28.6 | 12 | 17.1 | 31 | 44.3 |
| Neutral | 17 | 24.3 | 11 | 15.7 | 12 | 17.1 | 10 | 14.3 | 20 | 28.6 |
| Agree | 11 | 15.7 | 31 | 44.3 | 22 | 31.4 | 40 | 57.1 | 5 | 7.1 |
| Strongly Agree | 12 | 17.1 | 17 | 24.3 | 11 | 15.7 | 8 | 11.4 | 4 | 5.7 |
| Total | 70 | 100 | 70 | 100 | 70 | 100 | 70 | 100 | 70 | 100 |

| | Using ICT for education significantly enhances my knowledge | | Using ICT for education improves interactive learning | | Using ICT in education makes people dependent on computers and technology | | Infrastructure required to use ICT in education is costly | | ICT in education can be sustained in the long run through effective use of resources | |
|---|---|---|---|---|---|---|---|---|---|---|
| **Criteria** | **Frequency** | **Percent** | **Frequency** | **Percent** | **Frequency** | **Percent** | **Frequency** | **Percent** | **Frequency** | **Percent** |
| Strongly Disagree | 0 | 0.0 | 0 | 0.0 | 0 | 0.0 | 0 | 0.0 | 0 | 0.0 |
| Disagree | 0 | 0.0 | 0 | 0.0 | 0 | 0.0 | 0 | 0.0 | 7 | 10.0 |
| Neutral | 7 | 10.0 | 7 | 10.0 | 42 | 60.0 | 14 | 20.0 | 14 | 20.0 |
| Agree | 42 | 60.0 | 35 | 50.0 | 28 | 40.0 | 42 | 60.0 | 42 | 60.0 |
| Strongly Agree | 21 | 30.0 | 28 | 40.0 | 0 | 0.0 | 14 | 20.0 | 7 | 10.0 |
| Total | 70 | 100 | 70 | 100 | 70 | 100 | 70 | 100 | 70 | 100 |

## Effective Use of ICT Education

60 percent agreed 30 percent strongly agreed that using ICT for education significantly enhances knowledge. 50 percent agreed and 40 percent strongly agreed that using ICT for education improves interactive learning. 40 percent noted that using ICT in education made people dependent on computers and technology. 60 percent agreed that infrastructure required to use ICT in education is costly. 60 percent agreed that ICT in education can be sustained in the long run through effective use of resources.

## CONCLUSION AND RECOMMENDATIONS

The present study was designed to identify the barriers of introducing and implementing ICT education in the rural areas of Sri Lanka and how could these barriers be overcome. In this investigation, the aim was to assess the key factors that were explored in the literature review and the subsequent case studies such as Internet Access, Resources, Perceptions, Personnel and Security.

The result of this investigation show that despite the fact that all the factors have an impact on introducing and implementing ICT education, Internet Access, Resources, Personnel and Security have greater impact on implementation.

The second major finding was that in the context of Sri Lanka, Perceptions were not considered an important factor among teachers and they were forthcoming in the implementation of ICT education.

The results of this research report support the idea that implementing ICT education is based less on perception but more on resources. In Sri Lanka it is evident that the rural population is willing to embrace ICT education and it is the duty of decision makers to channel resources to these places. The research also points out several success stories where rural schools have ideated their own models for sustaining the running costs involved in ICT education.

An artefact was designed and tested through a discussion with the ICT teacher or a senior pupil from the school. 25 schools schools were chosen at random from the 25 districts around the country to test the artefact. The favourable results imply that the artefact is acceptable.

Sustainable solutions have also been suggested for each of the variables identified and these will enable a cost effective approach to bringing effective ICT Education to the Rural Schools in Sri Lanka.

## FUTURE RESEARCH

This study mainly focused on the barriers of introducing and implementing ICT education in the rural areas of Sri Lanka. There is potential to further extend this research with the ultimate aim of looking how to maximize the benefit to students of using ICT in their education.

Future studies can also look at how ICT is being used in schools to fuel innovation and increasing the intellectual capacities of both students and teachers.

One interesting aspect to look at is the knowledge exchange between the city schools and the rural schools, if at all this could be carried out. This exchange of knowledge will help bridge the gaps and disparities between the urban and rural centres.

Finally large scale projects carried out by the Government need to look at more sustainable and holistic approaches to ensure effective use in the long run and this increasing the use of ICT in the rural areas of Sri Lanka.

### Case Study 1

Namagal Tamil Vidyalam is situated in the remote town of Thirukkadaloor in the district of Trincomalee approximately 270km from the capital Colombo. The school was established in 1950 and provides classes up to Ordinary Level (Grades 1 – 11); i.e. a Type II School. Subjects are taught primarily in Tamil whilst English is taught as a second language. The school has 154 pupils with 72 girls and 82 boys. There are 17 teachers which comprise of 14 ladies and 3 gents amounting to a teacher student ratio of 1:9. The school has four computers, all which are in good working condition. The computers were donated by the Government through the Ministry of Education however the school does not possess an internet connection and therefore has no access to email. Lack of funds to obtain and maintain a monthly connection was cited as the primary reason. The school also has one colour printer.

According to the Principal of the school, ICT was taught as a subject for one year, but a lack of interest in the second year resulted in the subjected being cancelled out. The school does not have a designated ICT instructor or teacher, but a combination of three teachers allow the students to use the computers during their periods of work covering subjects such as science, geography and art. The staff have received training for ICT from the Government and have attended several such programmes. The principal believes that continued ICT training for the teachers will improve their skill levels.

The parents of the students lack the financial capacity to fundraise or donate as they spend almost all their time on livelihood activities however there are a handful of students who have computers in their homes. The school enjoys a continuous supply of electricity and has a functioning solar power facility.

The principal of the school was willing to conduct ICT related classes for a nominal fee as a means to fund the cost of obtaining an internet connection but she had her reservations. The school was previously surrounded by Tamil households. During the riots of 1983 the residents fled and the subsequent conflict left many dead while some took up residency in other countries. The houses that were left behind were taken over by Sinhalese who settled in over a period of time. While the Sinhalese households lived in fear during the conflict, the absence of it has created a sense of fear within the school as the students are ethnic Tamils. There have been several reported cases of school property being vandalised and a lack of concrete action on the part of the authorities has only worsened the plight of those in the school. The principal also pointed out the presence of drug addicts in the vicinity of the school especially during after school hours and on weekends. This has discouraged teachers and students from engaging in any extracurricular activities or additional revision classes.

She was hopeful that providing the situation improves, they could provide additional classes during the weekend as this would provide them the opportunity to raise some funds to obtain and pay for a regular internet connection.

## Case Study 2

Nilaveli Tamil Maha Vidyalam is situated in the remote town of Kuchchaveli in the district of Trincomalee approximately 300km from the capital Colombo. The school was established in 1979 and provides classes up to Advanced Level (Grades 1 – 13); i.e. a Type 1C School. Subjects are taught primarily in Tamil whilst English and Sinhala are taught as second languages. The school has 800 pupils with 370 girls and 430 boys. There are 40 teachers which comprise of 27 ladies and 13 gents amounting to a teacher student ratio of 1:20. The school has 130 students in the Advanced Level stream following the Commerce, Science and Arts streams. The school has 23 computers, out of which 20 are in good working condition. The computers were donated by the Government through the Ministry of Education. The school is part of the e-village programme operated by ICTA and has internet access provided via a lower bandwidth leased line by Sri Lanka Telecom together with email access. The school also has its own website. The school also has three printers out of which two are in working condition.

According to the Principal of the school, ICT is taught as a subject and there are two dedicated staff with one of them being an IT Graduate. The staff have received training for ICT from the Government and have attended several such programmes. The principal believes that continued ICT training for the teachers will improve their skill levels. He points out that the use of computers is popular amongst the male teachers as opposed to the female teachers. Despite the leased line facility, the bandwidth is quite slow and the lack of an alternate service provider in the area has prevented them from switching internet connections. The school enjoys a continuous supply of electricity and has a functioning solar power facility.

While the school does not provide additional training for the students, there are a few who obtain ICT classes from private service providers andthere are a handful of students who have computers in their homes.

The school has been part of the e-village programme for the past three years. Through the programme they have provided ICT training to school-leavers who have not used computers before. The programme has encouraged and trained students to create basic websites and blogs. Websites have been created by the students for farmers and other livelihood enterprises as a marketing tool. The programme has also brought about creativity amongst the students with one of them creating a blog for the school library with the types of books that are available.

The principal was in favour of teaching subjects with the aid of computers but highlighted that some teachers had reservations. Reservations were two fold; one was that some of them were reluctant to use computers due to the lack of understanding while others feared for their job security. However he suggested that this could be addressed through an effective awareness mechanism and was aware of the benefits of using ICT as a medium for effective teaching and adding value to education.

As an option for raising funds to buy more computers and obtain a high speed internet line if available, the principal was willing to offer ICT related classes in the school for a nominal fee to outsiders. He pointed out that they had provided training before but it was done free of charge and it was a success, so there was no reason it would not succeed again even if a nominal fee was charged. However in order to provide training with internet access they would require a faster connection which presently is not possible.

## Case Study 3

An-Noor Tamil Vidyalam is situated in the remote town of Serunuwara in the district of Trincomalee approximately 280km from the capital Colombo. The school was established in 1971 and provides classes up to Ordinary Level (Grades 1 – 11); i.e. a Type II School. Subjects are taught primarily in Tamil whilst English is taught as a second language. The school has 256 pupils with 96 girls and 160 boys. There are 13 teachers which comprise of 4 ladies and 9 gents amounting to a teacher student ratio of 1:20. The school has four computers, out of which 3 are in good working condition. The computers were donated by United Nations Children's Fund (UNICEF). The computers have access to the internet through a mobile 3g broadband connection. The school also has one colour printer.

According to the principal one of the computer's is being used for office administrative purposes which leave only two working computers to be used among the students. ICT is not taught as a subject as there is no dedicated teacher or instructor to conduct classes. The staff have received training for ICT from the Government and have attended several such programmes. The principal believes that continued ICT training for the teachers will improve their skill levels. He points out that despite the trainings only a few teachers make use of the computers. The school enjoys a continuous supply of electricity and has a functioning solar power facility.

He pointed out that the parents of the students are daily wage labourers and had no financial capacity to fundraise for the school. Most of the parents were finding it difficult to pay a nominal monthly fee to the school for its upkeep. Students above Grade 5 have access to the two computers in school and use it mainly to draw artwork and design cards. They are not provided access to internet or email as the mobile 3g connection is used in the school office.He also noted that there are a handful of students who have computers in their homes.

A more encouraging fact was that the old boys of the school were very supportive of the development of the school and were willing to help out in any way to raise funds and contribute towards the development of the school. The principal is fully in favour of providing ICT related classes in the school for a nominal fee which would provide them the required funding to obtain internet access for the other computers in the school. However he pointed out that the lack of computers prevents them from carrying out any of these initiatives. He was fully supportive of embracing new technologies to improve the school's teaching methods which would benefit the students and the school as a whole.

## REFERENCES

Abdullah, M. (2009). Review and comparison study of hybrid diesel/solar/hydro/fuel cell energy schemes for a rural ICT Telecenter. *Energy*, *35*, 639–646.

Asian Development Bank. (2009). *Sri Lanka: Secondary Education Modernization Project.* Available at: http://www.adb.org/documents/sri-lanka-secondary-education-modernization-project

Bhowmick, P. (n.d.). *Samvidha: A ICT System for Personalized Offline Internet Access for Rural Schools.* Academic Press.

CBSL. (2011). *Annual Report 2011.* Available at: http://www.cbsl.gov.lk/pics_n_docs/10_pub/_docs/efr/annual_report/AR2011/English/content.htm

Cebrail, C., Tuncer, A., & Ozsashin, A. (2009). *A New Future Technology in Rural and Remote Education: IEEE 802.22 Cognitive Radio Based Wireless Regional Access Network (WRAN).* IEEE.

Chen, M., Li, G., & Feng, Z. (2009). Web-based Education Technology and Application. In *The Rural Basic Education*. IEEE.

ColomboPage. (2011). *Sri Lanka national exam results blunder due to a computer error - Education Minister.* Available at: http://www.colombopage.com/archive_11B/Dec29_1325183041JR.php

Department of Census and Statistics. (2012). *Census of Population and Housing.* Available at: http://www.statistics.gov.lk/PopHouSat/CPH2011/Pages/Activities/Reports/CPH_2012_5Per_Rpt.pdf

Department of National Planning Ministry of Finance and Planning. (2006). *Mahinda Chintana: Vision For A New Sri Lanka*. Colombo: State Printing Corporation.

Dlodlo, N. (2009). Access to ICT education for girls and women in rural South Africa: A case study. *Technology in Society*, *31*, 168–175. doi:10.1016/j.techsoc.2009.03.003

facts.co. (n.d.). *Sri Lanka Map - blank Political Sri Lanka map with cities.* Available at: http://srilanka-map.facts.co/srilankamapof/srilankamap.php

Garson, D. (2004). *What's a good value for Cronbach's Alpha.* Available at: http://www.pmean.com/04/CronbachAlpha.html

Gliem, J., & Gliem, R. (2003). *Calculating, Interpreting, And Reporting Cronbach's Alpha Reliability Coefficient For Likert-Type Scales.* Available at: https://scholarworks.iupui.edu/handle/1805/344

ICTA. (2009). *eSri Lanka.* Available at: http://www.icta.lk/en/e-sri-lanka.html

Lallement, D., Terrado, E., & Zhang, Y. (2006). Empowering information and communication technologies in isolated areas: Learning from the solar-net villages program in honduras. *Renewable & Sustainable Energy Reviews*, *10*(1), 46–53. doi:10.1016/j.rser.2004.08.002

Liang, Z., Quan, L., Jie, Y., & Juan, S. (2008). *Education IPTV for E-learning in Rural Area.* IEEE.

Mansotra, V., Sharma, V., & Sambyal, G. (2009). *IT Initiatives in Education 'A Case Study for J&K State*. IEEE.

Ministry of Education. (2005). *Ministry of Education - Circulars.* Available at: http://www.moe.gov.lk/english/images/stories/circulars/2005-29s.pdf

Ministry of Education. (2006). *School Census - Preliminary Report.* Statistics Branch.

Ministry of Education. (2009). *Year of English & IT.* Academic Press.

Ministry of Education. (2012). *Sri Lanka Education Information.* Colombo: Ministry of Education.

Nenasala. (n.d.). *Establishment of Nenasalas.* Available at: http://www.nanasala.lk/

Raju, K. (2004). A case for harnessing information technology for rural development. *Elseiver, 36*, 233-240.

Sekaran, U. (2006). *Research Methods for Business: A Skill Building Approach* (4th ed.). New Delhi: John Wiley & Sons Ltd.

Skill International Pvt Ltd. (2010). *Outcome Evaluation Report of Nenasala Project ICTA Sri Lanka.* Available at: http://www.icta.lk/index.php?option=com_phocadownload&view=category&download=17:nenasala-outcome-evaluation-final-report-october-2010&id=2:reports&Itemid=270&lang=en

Valentine, J., & Holloway, S. (2001). A window on the wider world? Rural children's use of information and communication technologies. *Journal of Rural Studies, 17*(4), 383–394. doi:10.1016/S0743-0167(01)00022-5

Vorakulpipat, C., Siwamogsatham, S., Kamolsook, A., & Jamchudjai, P. (2010). *An Empirical Study of IT Adoption in Rural Areas of Thailand.* IEEE.

World Bank. (2011). *Sri Lanka - Education in Sri Lanka.* Available at: http://www.worldbank.lk/WBSITE/EXTERNAL/COUNTRIES/SOUTHASIAEXT/SRILANKAEXTN/0,contentMDK:21635561~pagePK:141137~piPK:141127~theSitePK:233047,00.html

# Chapter 5
# Smart Cities and Digital Governance

**Mamoona Humayun**
*College of Computer and Information Science, Jouf University, Saudi Arabia*

**N. Z. Jhanjhi**
https://orcid.org/0000-0001-8116-4733
*Taylor's University, Malaysia*

**Malak Z. Alamri**
https://orcid.org/0000-0002-3484-0994
*Jouf University, Saudi Arabia*

**Azeem Khan**
*Taylor's University, Malaysia*

## ABSTRACT

*With the ubiquitous low-cost sensor devices and widespread use of IoT, the paradigm is shifted from urban areas towards a smart city. A smart city is an urban area that uses IoT technologies to collect data and manage resources efficiently. The vision is to improve the capabilities and to solve the citizens' problems (e.g., energy consumption, transportation, recycling, intelligent security, etc.) in an efficient way. A smart city is a multidimensional term including a smart economy, smart mobility, smart living, smart environment, smart people, and smart governance. Although the concept of a smart city is increasing and currently there exist many such cities in many developed countries, one of the key challenges faced by these cities is good governance. Smart cities need smart governance to run the city in a smarter way, and effective digital governance is a solution to this end. Digital governance refers to the use of digital technology in government practices.*

DOI: 10.4018/978-1-7998-1851-9.ch005

*Figure 1. Digital governance*

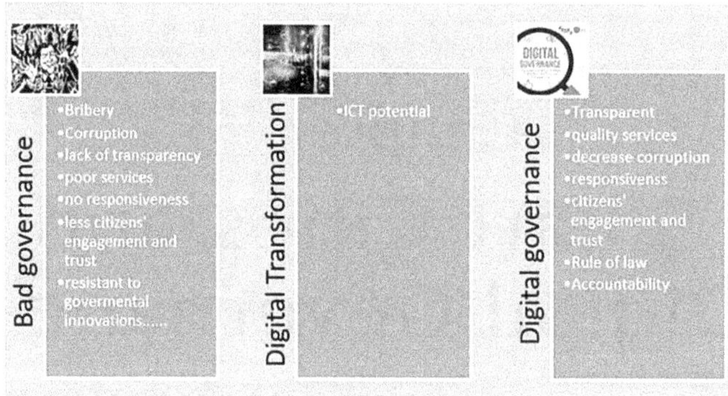


## INTRODUCTION

There exists no definitive explanation of smart city due to the wide variety of technologies incorporated in smart cities' infrastructure. However; the smart city is a designation given to the city/urban area which incorporates information and communication technologies (ICT) to increase the quality of services such as security, energy consumption, transportation, and other required utilities by reducing resource consumption and wastage of city resources (Abu-Matar, 2016; Musa, 2016). In order to effectively manage smart cities, smart administration is required that can be achieved through smart governance. Digital governance or smart governance are the two sides of the same coin. Digital governance includes the use of ICT in creating a progressive and effective government and public partnership so that all sections of a smart city become strengthen and integrated (Jucevičius, Patašienė, & Patašius, 2014; Nam & Pardo, 2011; K. Paskaleva, 2013).

Smart cities are very important segments of society whose sustainability depends on efficient digital governance. Digital governance of smart cities aims to make the system of these segments transparent so that a common citizen is well informed of all government information. In digital governance, government information does not remain limited to a particular community rather it is accessible by all sections of society (Coe, Paquet, & Roy, 2001; Meijer, Lips, & Chen, 2019).

Developed countries in the world are paying more attention to the development and governance of smart cities from more than a decade ago. Now, the developing world is also following the same trend. Cities are developed based on smart city model using latest ICT, and strong digital governance is used to strengthen this infrastructure (Chourabi et al., 2012; Mora, Bolici, & Deakin, 2017). This smart governance has given positive results across regions. Web-portals, mobile apps, online forums, and other online services have helped citizens to share their queries, experiences and suggestions to government authorities. Similarly, there exist apps that help citizens to report crimes and bribes happening in the society, complaints about a area, suggestions regarding government policies, tax portals, and security apps for public safety. All these ICT tools help in improving the governance of smart cities (Anand & Navío-Marco, 2018; Viale Pereira, Eibl, & Parycek, 2018; Webster & Leleux, 2018). Figure 1 shows the importance of digital governance for smart cities.

*Figure 2. E-government model*

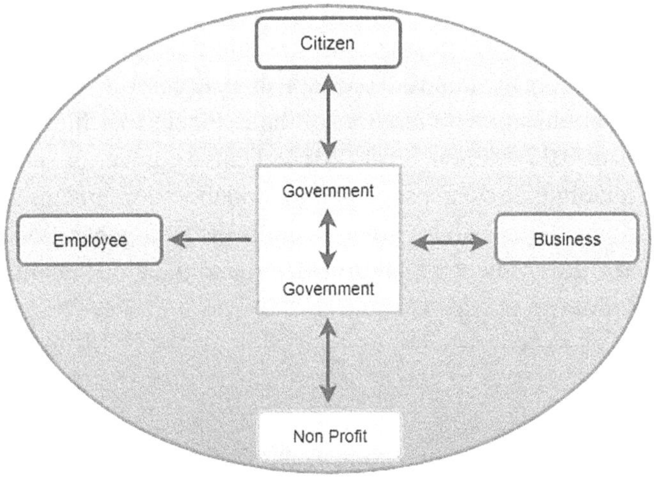


## LITERATURE REVIEW

Internet technology has gained much popularity in almost every field of life, as a result every sector of the economy needs to adopt information and communication technologies to accomplish their tasks in a best possible way (Ismagilova et al., 2019, Ivars-Baidal et al., 2019). In the current era of technology, individuals, as well as organizations, are dependent on technology for accomplishing their processes especially for communication and management. The rapid growth of ICT has given rise to the concept of smart city, and smart governance to effectively run these smart cities(David et al., 2018, Pereira et al., 2018). In order to get a better understanding of the topic under discussion, in the section below we define some important features of smart cities along with the role of digital governance

### E Government and Smart Government

E-government is a way adopted by current governments to use the latest ICTs especially web-bases applications to assist the citizens and organizations by providing them easy access to government information and required services. The aim of e-government is to facilitate the citizens, improving the standard and quality of services and providing them with the opportunities to participate in governmental processes (Fang, 2002, Jimenez et al., 2014, Scholl, 2005). E-governance brings citizens and government closer to each other by establishing citizens' trust on government. This trust help them in acceptance and adaptation of new rules and regulations (Carter and Bélanger, 2005). E-government brings government closer to the employee, citizens, nonprofit organizations and business. Thus, e-government have eight different dimensions namely government to government (G2G), government to business and business to government (G2B and B2G), Government to nonprofit and nonprofit to government (G2N and N2G), Government to citizens and citizens to government (G2C and C2G) and government to employees (Carter and Bélanger, 2005, Fang, 2002) as shown in figure 2.

Currently, many governments are transforming their e-government to the smart government due to rapid advancement in smart devices (Glybovets and Mohammad, 2017, Harsh and Ichalkaranje, 2015). The smart government refers to the use of smart devices for interconnecting government with citizens, employee, business and other organizations. To establish a successful smart government, the government need to know the communication preferences of their citizens to achieve effective communication (Glybovets and Mohammad, 2017, Mellouli et al., 2014). Unlike e-government, smart government focus on three key domains including; citizens participation, collaboration among government and citizens and information transparency. Hence, smart government is the next generation of e-government that is indirectly a citizen directed government which provides round the clock quality services to its citizens (Anthopoulos, 2017b, Mellouli et al., 2014).

## Service Improvement

Advancement in information and communication technologies especially the rapid growth of smart devices have transformed urban cities towards smart cities. This technological advancement has also transformed the style of governance towards digital governance. Good governance enables the government to be accountable in front of its citizens by providing them with adequate legal frameworks and effective processes to improve the quality of services (Pereira et al., 2018). Adaptation of digital technologies by public sector improve the lives of citizens and quality of services in the following way (Milakovich, 2012, Dunleavy et al., 2006, Asgarkhani, 2005, Barns, 2018, Lei and Tang, 2019, Sarker et al., 2018)

- Government make good relationship with citizens by improving citizens' participation
- Improve institutional capabilities so that they could provide efficient services
- Strict monitoring and evaluation of public sector institutions
- Implementation of digital technology in a strategic way
- Accountability
- Developing better trust relationship between government and citizens
- Provide convenient ways of information sharing
- Political awareness through improved digital infrastructure

## Sustainability and Resilience

The urban cities are transforming into smart cities rapidly, but this rapid growth comes with many challenges. Two important challenges that smart cities need to face are sustainability and resilience. Smart cities need sustainability in such a way that natural resources of city may be preserved, and an ecological balance could be maintained. The second important challenge faced by smart cities is resilience, it refers to the capacity of a smart city to recover quickly from a tough situation. Smart cities need to develop their capacities so that they could adapt to daily or occasional threats and need to choose a better strategy to face upcoming problems (Viitanen and Kingston, 2014, De Jong et al., 2015, Zheng et al., 2019, Almusaylim and Zaman, 2019). Sustainable development consists of three components: social development, economic development, and ecological/environmental development. The social component of sustainability refers to the citizens' growth and satisfaction. Economic component of sustainability refers to the absence of excessive fluctuations in the economy and environmental component of sustainability refers to the ecological balance. These three components are the pillars of sustainability and need to be maintained by smart cities, and it is only possible if good digital governance prevails in these smart cities (Ahvenniemi et al., 2017, Lombardi et al., 2012, Bătăgan, 2011). Figure 3 describes three pillars of sustainability

Resilience is a positive attitude toward disaster. Climate proofing, disaster planning and recovery, and citizens empowerment are some strategies that may be helpful in making a smart city resilient (Viitanen and Kingston, 2014, Suciu et al., 2013). The four important dimensions of resilience are economy, society, governance, and environment. Some factors that help to measure the economic resilience of a city include GDP growth rate, unemployment ratio, business startup, and failure rate, and working population. Similarly, a number of private and public sector organizations and their alliance, revenues earned, transparency and accountability are some factors that help to measure governance of a smart city. Further, the income level of citizens, gender ratio, poverty level, and population growth are factors that are helpful in measuring the society level resilience of a city. Factors that are helpful in measuring environmental resilience of a city include Population density, green areas, percentage of built-up areas, new development (Baron, 2012, Kaika, 2017, Kitchin, 2015).

## Citizen Engagement in and Evidence-Based Decision-Making Processes

Smart cities use modern technologies to improve quality of life and to provide better services to their citizens in urban space by keeping ideological balance. However; the smooth working of smart cities is only possible if citizens have a good liaison with the government and there exists trustworthy relationship between citizens and government. Citizens' engagement in decision-making processes is very important for successful implementation, working and executions of government projects and programs (Nalbandian et al., 2013, Yang and Callahan, 2007, Landsbergen Jr and Wolken Jr, 2001). Below are some points that highlight the role of citizens' engagement in evidence-based decision-making processes (Accordino, 2013, Matheus et al., 2018, Linders, 2012, Cunningham and Cunningham, 2006)

*Figure 3. Pillars of smart city's sustainability*

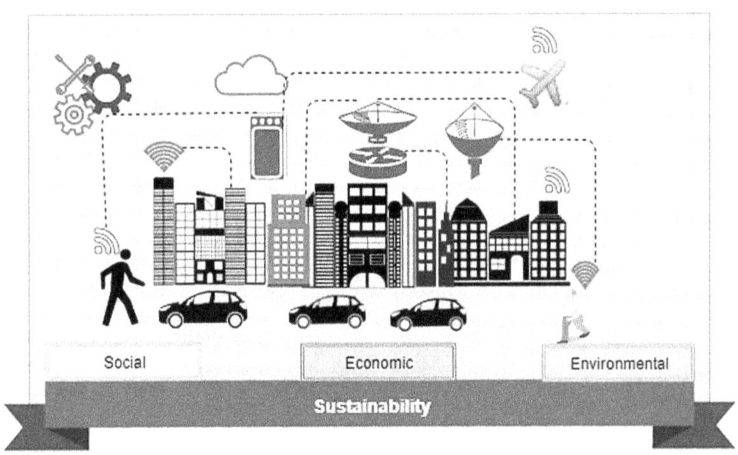

- The government can use citizens' experience and expertise to design better plans, services, and solutions
- Citizens help the government in data collection with the help of smart devices and social media apps
- Citizens are more aware of the problems existing in the cities and thus they can help the government in addressing these problems at early stages
- Active participation of citizens in the decision-making process promote the democratic environment and thus provide help in making city a sustainable and resilient one
- The trust among citizens and government help in reducing waste and pollution from the city
- Effective feedback from citizens help the government in refining its policies
- Online voting helps in making democratic decisions

## Collaborative, Open and Citizen-Centric Forms of Governance

Governance refers to the actions undertaken by a government through norms and laws for governing a state, organization or territory. Currently, the trend of governance is changing from traditional government setup towards collaborative, citizen-centered and open governance. This new trend of governance is adopted by many developed democratic states especially in their smart cities where citizens are in direct collaboration with government through smart devices and active social media apps (Chen, 2010, Linders et al., 2012, Ingrams et al., 2018). In the current era of technology, governance is based on transparency, collaboration and participation where data, processes and services are open and easily accessible by all citizens. Advancement and rapidly growing use of ICT has increased the capacity of civil society to make demands and offer the best solutions to the government. There are three important factors that make governance open, collaborative and citizen centric in smart cities; these factors include: inclusiveness, privacy and democratic responsibility. Inclusiveness refers to the inclusion of citizens in the decision making process. Privacy refers to the safety and security of individual data and information and democratic responsibility means that citizens need to participate actively in all city matters and should provide valuable feedback for the improvement of a city (Chatfield and AlAnazi, 2015, Nam, 2012).

## Smart City Dimensions

A smart sustainable city is multidimensional that uses the latest Information and communication technologies to improve its quality of life and services. Some key dimensions of a smart city include smart people, smart economy, smart governance, smart environment, smart mobility, and smart living (van Winden and van den Buuse, 2017, Anthopoulos, 2015, Albino et al., 2015). Smart city has smart people, who are open-minded, educated, and skilled and are willing to cooperate with the government in decision making processes. The life in smart cities is also smart where there are good facilities of health, education, housing, public safety and has an attraction for tourists. Another important dimension of a smart city is smart governance that is a transparent, responsiveness and citizen-centric governance where citizens are in good collaboration with the government and participate in decision making processes. The environment of smart city is also smart where there is less pollution and wastage, natural resources are protected to keep the city environment clean and government makes effective strategies for the protection and safety of the environment. The mobility in smart cities is also smart; there exist good transportation system, strong ICT infrastructure to provide national and international accessibility to the citizens. Finally is the economic dimension of smart city. Smart cities have smart economies where there is good private public partnership, the city export is good and have international reputation, government encourages entrepreneurship and thus GDP growth rate is good (Benevolo et al., 2016, Tian et al., 2018, Bibri, 2018). Figure 4 highlights key dimensions of smart city along with their features

## The Role of ICT in Smart Cities

A smart sustainable and resilient city uses ICT and other means to improve the standard of life and quality of services. ICT is a backbone of the smart city without that establishment and governance of a smart city is almost impossible. Following points highlight the role of ICT in smart cities (Ferro et al., 2013, Falconer and Mitchell, 2012, Neirotti et al., 2014, Bifulco et al., 2016)

- It acts as an enabler of smart city and interconnects government and citizens
- The planning and growth of ICT infrastructure is a prime focus of smart city governance
- ICT helps inefficient utilization of resources
- It helps to save cost and energy by interconnecting various node of society on a common infrastructure

*Figure 4. Dimensions of a smart city*

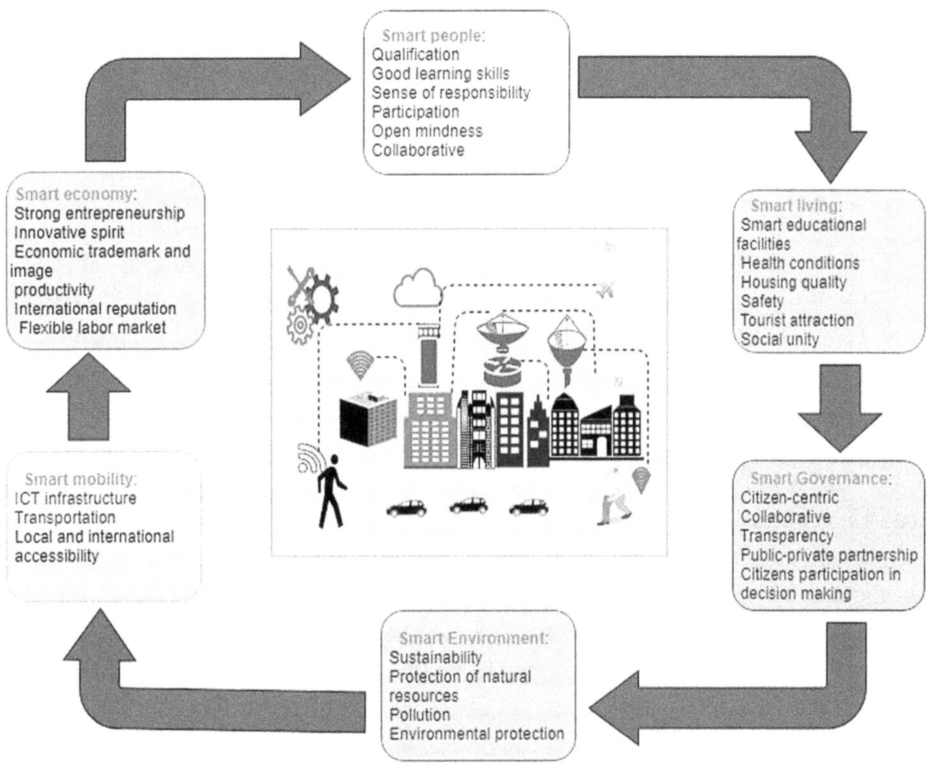

- Helps to streamline basic city infrastructure including transportation, water, telecommunication, and other utilities
- Provide ways to the citizens for collaboration and participation in decision-making processes
- Helps inefficient data collection and analysis
- ICT promotes knowledge sharing
- It helps in predicting natural disasters and helps in making useful strategies to save the city from these disasters
- Provides easy and timely access to information that helps in quick and effective decision making

## Security Challenges for Smart Cities

Smart city is a massively interconnected world that uses ICT infrastructure to provide better life and services to its citizens. However, this massively interconnected infrastructure is not free from potential harms if it became the target of cybersecurity attacks. Smart city infrastructure involves massive amount of data and information sharing that needs to deploy thousands of sensor nodes for sharing this big amount of data, the risk associated with this information sharing is: whether the data is protected from leakage, theft or cyber-attacks? The protection and control mechanism are sufficient enough or not? (Sookhak et al., 2018, Gottwalt and Karduck, 2018, Nayyar et al., 2019) In the section below, we discuss some of the key security requirements and challenges that are indispensable for a secure smart city

- Secure communication: an important component of a smart city is network communication that interconnects different units of a smart city for collecting, sharing and transferring information inside and outside of the city. Secure communication needs secure wired and wireless connection; smart cities need to adopt effective security mechanisms to provide secure communication. Some mechanisms of secure communication that may be adopted are: cryptographic methods, designing an effective key distribution system, creating secure communication channels and generating security credential for IOT devices (Sookhak et al., 2018, Alamri et al., 2019).
- Identification, authentication and access control is another security challenge faced by smart cities. There exist various authentication and access control protocols that may help in constructing a solid communication channel among a different unit of smart cities. Some of the well-known protocols for authentication and access control are IBE (identity-based encryption), RBAC (role based access control and ABE (attribute-based encryption) (Gottwalt and Karduck, 2018, Nayyar et al., 2019).
- Secure booting: Mostly IoT devices become the target of security attacks while booting; therefore secure booting is indispensable to protect smart cities from security attacks (Nayyar et al., 2019, Ahuja and Khosla, 2019).
- Security monitoring and controlling: Monitoring the network traffic for disclosure of potential attacks and taking proper measure to protect the city from these attacks is a challenging task. Smart cities must design a proper security monitoring, control, analysis and response mechanism based on international standards to protect the cities from potential harms (Sookhak et al., 2018, Gottwalt and Karduck, 2018, Nayyar et al., 2019).

## Communication Infrastructure for Smart Cities

Communication is a key subject of the smart city. The modern cities are now no longer dependent on a landline telephone or mobile communication. The modern cities especially the smart cities are based on new communication mechanism that has separate infrastructure which combines sensor networks, mobile communication and information networks. Further, the actors involved in smart cities communication infrastructure are not limited to individuals only; rather vehicles, buildings, streets, and other household devices are integrated into new communication infrastructure. The existing networks do not provide enough mechanisms to handle this huge data volume, therefore; there is a need to configure existing networks with latest ICT technologies to make it compatible with communication need of the city. The communication infrastructure of the smart city needs to provide timely availability of a data to its citizens (Zanella et al., 2014, Bakıcı et al., 2013, Balakrishna, 2012, Yaqoob et al., 2017).

## Transportation for Smart Cities

With the rapid growth of urbanization, the number of vehicles is also increasing rapidly. The road network construction is not increasing with the same pace due to which the problem of road congestion and road accidents is increasing, and it is becoming a core problem of urbanization. Smart cities adopt smart transportation solution that provides traffic data center, latest network equipment's and video investigation platform to overcome the problem of traffic congestion and to satisfy the citizens' need of transportation. This smart transportation system is based on ICT infrastructure that is used to build integrated traffic management system, proper signal control system, traffic flow monitoring and collection system, traffic guidance system and electronic police system (Menouar et al., 2017, Xiong et al., 2012, Liu et al., 2017).

Smart cities aim to provide standard life and quality of services to its citizens and transportation is one of the cores needs of citizens. Hence, an intelligent transportation system is important in smart city movement. Intelligent transportation system include regular monitoring of traffic patterns, monitoring of areas with high pedestrian movement, coordinating train timings, monitoring metro stations and much more to facilitate the citizens. Intelligent transportation must be the part of smart city initiatives especially when large events are hosted in a city that increases city traffic and security concerns (Zhu et al., 2016, Chen et al., 2017, Ding et al., 2017).

## Health at Smart Cities

In a traditional healthcare system, physician examines an individual based on the previous history and diagnosis and recommend medicines for the patient. The doctors mainly rely on their experience and training instead of using computing technologies. In the current era of technology, the trend is totally changed from traditional healthcare setup towards smart healthcare system. Smart cities heavily rely on sensors technology to predict environmental parameters such as temperature, humidity, traffic conditions, pollution, and power grid status. The value of these environmental parameters helps in understanding citizens' state at any given time. Further, the use of ICT for maintaining electronic health record and personals health records have helped a lot in providing a global view of the health status of smart cities' residents (Solanas et al., 2014, Boulos and Al-Shorbaji, 2014).

The rapid growth of smart devices especially mobile devices and body area network have provided the facility of self-monitoring to mobile owners by using mobile sensors and ICT. Now the sensors are not only limited to personal devices, rather they are being embedded in a physical environment. These sensors help in collecting data of environment that may be used to monitor changes in health status of citizens. Use of computing technologies to interconnect citizens with health providing institutions helped in overcoming geographical and temporal barriers. Hence it can be claimed that the use of IoT based health care system in smart cities have improved the efficiency of healthcare system a lot (Pramanik et al., 2017, Hussain et al., 2015).

Designing smart healthcare is not an easy task, it needs potential planning. Some of the important points that need to be included in smart healthcare system planning include a tool for population health analysis, timely delivery, and management of healthcare services, smart emergency, remote monitoring, and establishment of smart health clinics, digital hospital solutions, smart healthcare education, and training. All these services must be the important concern of any smart city government as citizens' health is very important for the smooth execution of all the activities (Letaifa, 2015, Mehmood et al., 2017).

### Education

Consider a scenario, John has successfully completed his intermediate education. He wants to get admission in Jouf University, Saudi Arabia. He opens the website of university from his mobile and applies in BS computer science program. He appears in online test and interview and finally qualifies for admission. John arrives at the university on his first day with the feelings of excitement and fear. He opens the university App from his mobile; this app tells him about his complete schedule of the day, the location of the classroom, the distance of classroom from his current position. John gest intimation on his mobile about all the upcoming events of the university, his total attendance, list of assignments that need to be submitted before mentioned a deadline, successful submission of assignments, online quizzes, marks of quizzes and assignments. John scans the lecture material using his smart phone....

All the facilities provided to john in the above scenario are due to smart education that is only possible if the latest ICT and IoT is used in education. In order to improve the quality of teaching and students' degree of understanding; smart cities need to take different initiatives. These initiatives include smart classrooms, enabling virtual education, use of digital media, and use of ICT, etc. (Winters, 2011, Lombardi et al., 2012).

## SMART CITIES RECENT STATICS REGION BASED

The main objective of a smart city is to improve the overall infrastructure of the city by providing quality life and efficient services to its Citizens. In the current era of technology, developed countries are putting a lot of efforts in making their cities smart. The developing countries are also not behind in this race. Now the question is that, what are the criteria to measure whether a city is smart or not? Mere implementation of quality ICT infrastructure is not enough. If the city is having very good ICT infrastructure and all facilities of life but it is not free from theft and pollution, nobody will like to live in this smart city. Hence, we can say that smart cities are judged across a number of parameters. Smart cities may be assessed and analyzed across nine dimensions, these dimensions include economy, human capital, technology infrastructure, social cohesion, environment, governance, international outreach, health, and transportation (Hall et al., 2000, Chourabi et al., 2012, Nam and Pardo, 2011).

If we analyze the recent statistics of smart cities: London has been declared as the smartest city in the world. The second and third in this race are New York and Paris respectively. Other than these, the list of Top 10 smart cities of the world includes: Tokyo, Singapore, Toronto, Seoul, Hong Kong, and Amsterdam. If we analyze the statistics of smart cities continent wise: America stands first in the category, then comes Europe having 12 cities which are among the top 25 smart cities of the world. Then comes North America having 6 smart cities followed by Asia having four cities and Oceania having three smart cities (Paskaleva, 2009, Giffinger and Gudrun, 2010).

## SMART CITIES CASE STUDY

A case study is a research methodology that model real-world situations to demonstrate the use of a product or service in a better way. In the section below, we provide two case studies of smart cities to give a better understanding of smart cities and digital governance

### London

According to the sixth edition of IESE Cities in Motion index 2019, London has been declared as the smartest city of the world. IESE Cities in Motion observed about 174 cities from 80 different countries, these cities were analyzed based on some key dimensions of smart cities. These dimensions include Human capital, economy, environment, social cohesion, governance, international outreach, technology infrastructure, urban planning, mobility, and transportation. In order to be ranked among smart cities of the world, the city not only needs to be focused only in the technology dimension, rather it should focus on above mentioned multiple dimensions of smart city vision. In 2019, London has ranked highest in almost all the above-mentioned dimensions of a smart city. In human capital, London hosts a maximum number of programmers than any other city of the world. London also hosts an open data platform that benefits more than 50,000 individuals, researchers and practitioners every month. London also innovated a lot in smart transportation by installing capsules and Heathrow pods that is an underground terminal used to connect citizens with Heathrow airport that is one among the busiest airports of the world. It has also started a mega project (Cross rail project) that will add a new train network among the existing networks of a city. Not only in human capital and transportation system, London is also growing pressure on education, healthcare, energy, pollution and other utilities of life and thus providing a quality lifestyle to its residents (Cowley et al., 2018, Batty, 2018, Joss et al., 2019).

### New York

According to IESE Cities in Motion index 2019, New York has been ranked second in the list of smart cities of the world due to its top place in various key dimensions of a smart city including economy, human capital, international outreach, urban planning, mobility and transportation, and technology infrastructure. However; it still lacks in social cohesion as compared to London. If we analyze the latest statistics of New York, it has the world's most important economic center that consists of more than 7000 high technology firms and provides excellent services, such as the free Wi-Fi service. The New York has upgraded lighting system to save energy. It has smart water metering system that provides automated meter reading to get the overall status of water consumption. Through this smart water metering system, citizens are notified of their water consumption four times a day so that they could control it. New York sanitation department is the largest in the world and is very famous for its waste management. Almost, 10,500 tons of waste is picked and disposed on a daily basis. Further, smart bins are available in the city that has built-in sensors to monitor the level of the trash, these trash bins also have a compactor that runs with the solar system and it allows a trash bin to accommodate five times more waste than the traditional bins. Different techniques are adopted by the NYC department of health and mental hygiene to control air pollution in the city. Active rebates are introduced in the city for converting the system from oil to natural gas. Further, smart monitoring helps in predicting the energy and other utility consumption. Solving the issues at an early stage is better and cost-effective than solving it after the crisis. New York has an efficient monitoring system to monitor and control energy and other utility consumption (Anthopoulos, 2017a, Li et al., 2016, Ben Khalifa et al., 2017, Indergaard, 2019).

## SMART CITIES IMPACT ON ECONOMIC GROWTH

One of the key factors that motivate cities to move towards smart vision is better economic growth. This is the reasons why countries around the world are investing so much in building smart cities. Below are some benefits of smart cities that are tied with economic growth (Shapiro, 2006, Visvizi et al., 2018, Artmann et al., 2019, Batabyal and Nijkamp, 2019)

- **Automation:** The vision of the smart city is to use the latest ICT technologies to perform different tasks of society. Automation helps in the economic growth of a city by saving labor cost and reducing errors.
- **Efficiency:** Most of the work done in smart cities is using sensors, unlike human these sensors do not require any rest. These sensors work 24/7, 365 days in a week to collect, analyze and monitor the data. These sensors also ensure that the resources of the city are utilized in an efficient way.
- **Risk Mitigation and Damage Reduction:** Smart cities use sensor technology for data collection and analysis, these sensors predict disasters while collecting the data of sensitive areas. Early detection of risk and disaster help the smart government in making useful strategies for mitigating these risks and thus reducing damage.
- **Quality Life:** The vision of a smart city is to provide a standard life to people, as a result, the people living in smart cities are healthy, educated and are satisfied with their life. This satisfaction of life makes them responsible citizens and they play their role to improve the city more.
- Full-time connectivity of Human and devices promote collaboration across regions. Connectivity helps individuals to collaborate on various projects and thus ultimately contribute well in the economic growth of a city.
- **Better Sustainability:** Smart cities have a better ecosystem that predicts uncertainties in advance and thus contribute well to the sustainability of a city that positively affects the economy of a smart city.

- **Smart Transportation:** Transportation plays a very important role in the economy; smart cities have a smart transportation system that helps in quick movement of goods and services. Smart transportation system reduces congestion and thus save the time of people. People can use this time for some productive activities and thus can contribute well to the economic growth of a city.
- **Smart Building:** Smart cities have smart buildings that are connected to each other and even generate electricity. This help in the sustainability of a city and positively affect the economic growth of a city.
- **Big Data:** Smart cities' sensors collect a huge amount of data that can be used in commerce.

The next section will discuss the conclusive remarks of the chapter, by highlighting the smart city benefits, ease of operations, sustainability, etc. it also provides a strong link among smart city and digital governance, which may uplift the life style and civic sense of the residents.

## DISCUSSION

A smart city is a label given to the city that uses information and communication technologies to improve the quality of life and services. A smart city has multi-dimensions and the aim is to promote the city in such a way that it provides a clean, sustainable and resilient environment to its citizens. The key dimensions of smart city include: Human capital, economy, environment, social cohesion, governance, international outreach, technology infrastructure, urban planning, mobility, and transportation. Smart cities use IoT sensors for data collection, analysis, and decision making. Mere implementing ICT and IoT sensors are not enough to make a city smart, it needs smart governance for efficient utilization of these resources. Smart governance/ digital governance is the use of technology to facilitate planning and decision-making process. Digital governance of a smart city is citizen-centric governance, it emphasizes citizens collaboration and participation in the decision making process. Apparently, it seems that the smart city is a tension free place to live and enjoy life but it is not like that. Without smart governance, there is a chance of theft, corruption, pollution, and lack of social cohesion that make the idea of smart city flop. Government makes a huge investment in developing a smart city, and good governance make it sustainable through proper monitoring and controlling. Smart city needs smart people, smart transportation system, smart healthcare system, latest ICT infrastructure and all this is possible only with the help of good digital governance in which there are collaborative and trustworthy relationship among government and citizens. If the citizens of smart cities are provided with smart education and health facilities and they feel comfort in their life, they will contribute well in the development of a city.

## FUTURE OF SMART CITIES

The idea of having a dream city (city having enough energy resources, quality education, better healthcare facilities, congestion-free transportation, pollution-free environment) is now not too far away from reality. The advancement of ICT and IoT sensors have made it easy to develop urban areas in such a way that it may become an ideal place to live in. A smart city is a solution towards this end, its vision is to use intelligent solutions for improving quality of life and efficiency of services with the help of good digital governance. However; smart cities need to face many challenges in the future. One of the big challenges is rapidly growing urbanization. The rapid growth of urbanization has put more strain on global environment and infrastructure in the presence of limited resources. The solution to these problems is the application of emerging technologies such as IoT and automation. Further, the private-public partnership is essential for the economic growth of smart cities. Further; the investment in various dimensions of the smart city needs to be made in a strategic way and with proper planning by prioritizing the different dimensions of a city. This strategic planning and prioritization of services will make the city smart in a real sense.

## CONCLUSION

ICT has become an integral part of the modern lifestyle that was otherwise unimaginable. All the segments of life including transportation, communication, security, healthcare, etc. are dependent on ICT. In the current era of technology, smart and sustainable cities are using ICT and other means to provide a quality life and efficient services to its citizens. However; in order to effectively manage and run these smart cities, good digital governance is inevitable. Digital governance helps in creating a collaborative environment, strengthening democracy, and citizens' participation and trust in government policies. The vision of digital or smart governance is to make the overall system transparent by providing citizens with government information. To conclude, the role of digital governance in smart cities have following important implications

- Digital governance focuses on decisions that are necessary for improving the quality of life in smart cities. These decisions may be in various dimensions (smart living, smart people, smart health, smart education, smart mobility, etc.)
- Use of technology to optimize services and quality of life
- A strong focus on citizens, establishing a trust relationship between government and citizens and involving them in the decision making process.
- A smart city has major benefit of 24/7 monitoring and assurance of high level security around the clock.

## REFERENCES

Accordino, F. (2013). The futurium—A foresight platform for evidence-based and participatory policy-making. *Philosophy & Technology*, *26*(3), 321–332. doi:10.100713347-013-0108-9

Ahuja, K., & Khosla, A. (2019). *Driving the Development, Management, and Sustainability of Cognitive Cities*. IGI Global. doi:10.4018/978-1-5225-8085-0

Ahvenniemi, H., Huovila, A., Pinto-Seppä, I., & Airaksinen, M. (2017). What are the differences between sustainable and smart cities? *Cities (London, England)*, *60*, 234–245. doi:10.1016/j.cities.2016.09.009

Alamri, M., Jhanjhi, N., & Humayun, M. (2019). Blockchain for Internet of Things (IoT) Research Issues Challenges & Future Directions: A Review. *Int. J. Comput. Sci. Netw. Secur*, *19*, 244–258.

Albino, V., Berardi, U., & Dangelico, R. M. (2015). Smart cities: Definitions, dimensions, performance, and initiatives. *Journal of Urban Technology*, *22*(1), 3–21. doi:10.1080/10630732.2014.942092

Almusaylim, Z. A., & Zaman, N. (2019). A review on smart home present state and challenges: Linked to context-awareness internet of things (IoT). *Wireless Networks*, *25*(6), 3193–3204. doi:10.100711276-018-1712-5

Anthopoulos, L. (2017a). Smart utopia VS smart reality: Learning by experience from 10 smart city cases. *Cities (London, England)*, *63*, 128–148. doi:10.1016/j.cities.2016.10.005

Anthopoulos, L. (2015). Defining smart city architecture for sustainability. *Proceedings of 14th electronic government and 7th electronic participation conference (IFIP2015),* 140-147.

Anthopoulos, L. G. (2017b). *Understanding Smart Cities: A tool for smart government or an industrial trick?* Springer. doi:10.1007/978-3-319-57015-0

Artmann, M., Kohler, M., Meinel, G., Gan, J., & Ioja, I.-C. (2019). How smart growth and green infrastructure can mutually support each other—A conceptual framework for compact and green cities. *Ecological Indicators*, *96*, 10–22. doi:10.1016/j.ecolind.2017.07.001

Asgarkhani, M. (2005). Digital government and its effectiveness in public management reform: A local government perspective. *Public Management Review*, *7*(3), 465–487. doi:10.1080/14719030500181227

Bakıcı, T., Almirall, E. & Wareham, J. (2013). A smart city initiative: the case of Barcelona. *Journal of the Knowledge Economy, 4*, 135-148.

Balakrishna, C. (2012). Enabling technologies for smart city services and applications. In *2012 sixth international conference on next generation mobile applications, services and technologies*. IEEE.

Barns, S. (2018). Smart cities and urban data platforms: Designing interfaces for smart governance. *City, Culture and Society, 12*, 5-12.

Baron, M. (2012). Do we need smart cities for resilience. *Journal of Economics & Management*, *10*, 32–46.

Batabyal, A. A., & Nijkamp, P. (2019). Creative capital, information and communication technologies, and economic growth in smart cities. *Economics of Innovation and New Technology*, *28*(2), 142–155. doi:10.1080/10438599.2018.1433587

Bătăgan, L. (2011). Smart cities and sustainability models. *Informações Econômicas*, *15*, 80–87.

Batty, M. (2018). *Inventing future cities*. MIT Press. doi:10.7551/mitpress/11923.001.0001

Ben Khalifa, M., Redondo, R. P. D., Vilas, A. F., & Rodríguez, S. S. (2017). Identifying urban crowds using geo-located Social media data: A Twitter experiment in New York City. *Journal of Intelligent Information Systems*, *48*(2), 287–308. doi:10.100710844-016-0411-x

Benevolo, C., Dameri, R. P., & D'Auria, B. (2016). *Smart mobility in smart city. Empowering Organizations*. Springer.

Bibri, S. E. (2018). A foundational framework for smart sustainable city development: Theoretical, disciplinary, and discursive dimensions and their synergies. *Sustainable Cities and Society*, *38*, 758–794. doi:10.1016/j.scs.2017.12.032

Bifulco, F., Tregua, M., Amitrano, C. C., & D'Auria, A. (2016). ICT and sustainability in smart cities management. *International Journal of Public Sector Management*, *29*(2), 132–147. doi:10.1108/IJPSM-07-2015-0132

Boulos, M. N. K., & Al-Shorbaji, N. M. (2014). *On the Internet of Things, smart cities and the WHO Healthy Cities*. BioMed Central.

Carter, L., & Bélanger, F. (2005). The utilization of e-government services: Citizen trust, innovation and acceptance factors. *Information Systems Journal*, *15*(1), 5–25. doi:10.1111/j.1365-2575.2005.00183.x

Chatfield, A. T., & Alanazi, J. (2015). Collaborative governance matters to e-government interoperability: An analysis of citizen-centric integrated interoperable e-government implementation in Saudi Arabia. *International Journal of Public Administration in the Digital Age*, *2*(3), 24–44. doi:10.4018/ijpada.2015070102

Chen, Y., Ardila-Gomez, A., & Frame, G. (2017). Achieving energy savings by intelligent transportation systems investments in the context of smart cities. *Transportation Research Part D, Transport and Environment*, *54*, 381–396. doi:10.1016/j.trd.2017.06.008

Chen, Y.-C. (2010). Citizen-centric e-government services: Understanding integrated citizen service information systems. *Social Science Computer Review*, *28*(4), 427–442. doi:10.1177/0894439309359050

Chourabi, H., Nam, T., Walker, S., Gil-Garcia, J. R., Mellouli, S., Nahon, K., . . . Scholl, H. J. Understanding smart cities: An integrative framework. *2012 45th Hawaii international conference on system sciences*, 2289-2297.

Cowley, R., Joss, S., & Dayot, Y. (2018). The smart city and its publics: Insights from across six UK cities. *Urban Research & Practice*, *11*(1), 53–77. doi:10.1080/17535069.2017.1293150

Cunningham, P., & Cunningham, M. (2006). eParticipation in policy-making: the research and the challenges. *Exploiting the knowledge economy: Issues, applications and case studies, 3*, 364.

David, N., Mcnutt, J. G., & Justice, J. B. (2018). *Smart cities, transparency, civic technology and reinventing government. In Smart Technologies for Smart Governments*. Springer.

De Jong, M., Joss, S., Schraven, D., Zhan, C., & Weijnen, M. (2015). Sustainable–smart–resilient–low carbon–eco–knowledge cities; making sense of a multitude of concepts promoting sustainable urbanization. *Journal of Cleaner Production*, *109*, 25–38. doi:10.1016/j.jclepro.2015.02.004

Ding, H., Zhang, C., Cai, Y., & Fang, Y. (2017). Smart cities on wheels: A newly emerging vehicular cognitive capability harvesting network for data transportation. *IEEE Wireless Communications*, *25*(2), 160–169. doi:10.1109/MWC.2017.1700151

Dunleavy, P., Margetts, H., Bastow, S., & Tinkler, J. (2006). New public management is dead—Long live digital-era governance. *Journal of Public Administration: Research and Theory*, *16*(3), 467–494. doi:10.1093/jopart/mui057

Falconer, G., & Mitchell, S. (2012). Smart city framework. *Cisco Internet Business Solutions Group, 12*, 2–10.

Fang, Z. (2002). E-government in digital era: concept, practice, and development. *International Journal of the Computer, the Internet and Management, 10*, 1-22.

Ferro, E., Caroleo, B., Leo, M., Osella, M., & Pautasso, E. The role of ICT in smart cities governance. *Proceedings of 13th international conference for E-democracy and open government. Donau-Universität Krems*, 133-145.

Giffinger, R. & Gudrun, H. (2010). Smart cities ranking: an effective instrument for the positioning of the cities? *ACE: Architecture, City and Environment, 4*, 7-26.

Glybovets, A. & Mohammad, A. (2017). E-Government Versus Smart Government: Jordan Versus The United States. *EUREKA: Social and Humanities*, 3-11.

Gottwalt, F., & Karduck, A. P. (2018). *Securing smart cities—a big data challenge. In Information Innovation Technology in Smart Cities*. Springer.

Hall, R. E., Bowerman, B., Braverman, J., Taylor, J., Todosow, H., & Von Wimmersperg, U. (2000). *The vision of a smart city*. Upton, NY: Brookhaven National Lab.

Harsh, A., & Ichalkaranje, N. (2015). *Transforming e-government to smart government: A South Australian perspective. In Intelligent Computing, Communication and Devices*. Springer.

Hussain, A., Wenbi, R., Da Silva, A. L., Nadher, M., & Mudhish, M. (2015). Health and emergency-care platform for the elderly and disabled people in the Smart City. *Journal of Systems and Software, 110*, 253–263. doi:10.1016/j.jss.2015.08.041

Indergaard, M. (2019). A developmental network city? Double embeddedness in New York. *Cambridge Journal of Regions, Economy and Society*, rsz013. doi:10.1093/cjres/rsz013

Ingrams, A., Melitski, J., Holzer, M. & Manoharan, A. P. (2018). Big data and open government. *E-Government and Information Technology Management: Concepts and Best Practices*, 183.

Ismagilova, E., Hughes, L., Dwivedi, Y. K., & Raman, K. R. (2019). Smart cities: Advances in research—An information systems perspective. *International Journal of Information Management*, *47*, 88–100. doi:10.1016/j.ijinfomgt.2019.01.004

Ivars-Baidal, J. A., Celdrán-Bernabeu, M. A., Mazón, J.-N., & Perles-Ivars, Á. F. (2019). Smart destinations and the evolution of ICTs: A new scenario for destination management? *Current Issues in Tourism*, *22*(13), 1581–1600. doi:10.1080/13683500.2017.1388771

Jimenez, C. E., Solanas, A., & Falcone, F. (2014). E-government interoperability: Linking open and smart government. *Computer*, *47*(10), 22–24. doi:10.1109/MC.2014.281

Joss, S., Sengers, F., Schraven, D., Caprotti, F., & Dayot, Y. (2019). The smart city as global discourse: Storylines and critical junctures across 27 cities. *Journal of Urban Technology*, *26*(1), 3–34. doi:10.1080/10630732.2018.1558387

Kaika, M. (2017). 'Don't call me resilient again!': The New Urban Agenda as immunology… or… what happens when communities refuse to be vaccinated with 'smart cities' and indicators. *Environment and Urbanization*, *29*(1), 89–102. doi:10.1177/0956247816684763

Kitchin, R. (2015). Making sense of smart cities: Addressing present shortcomings. *Cambridge Journal of Regions, Economy and Society*, *8*(1), 131–136. doi:10.1093/cjres/rsu027

Landsbergen, D. Jr, & Wolken, G. Jr. (2001). Realizing the promise: Government information systems and the fourth generation of information technology. *Public Administration Review*, *61*(2), 206–220. doi:10.1111/0033-3352.00023

Lei, T., & Tang, Y. (2019). Digital Governance Model for Big Data Era----Based on Typical Practices in Singapore. *Humanities and Social Sciences*, *7*(2), 76. doi:10.11648/j.hss.20190702.15

Letaifa, S. B. (2015). How to strategize smart cities: Revealing the SMART model. *Journal of Business Research*, *68*(7), 1414–1419. doi:10.1016/j.jbusres.2015.01.024

Li, M. Ch'ng, E., Chong, A. & See, S. (2016). The new eye of smart city: novel citizen sentiment analysis in twitter. *2016 International Conference on Audio, Language and Image Processing (ICALIP)*, 557-562. 10.1109/ICALIP.2016.7846617

Linders, D. (2012). From e-government to we-government: Defining a typology for citizen coproduction in the age of social media. *Government Information Quarterly*, *29*(4), 446–454. doi:10.1016/j.giq.2012.06.003

Linders, D., Wilson, S. C., & Bertot, J. C. (2012). *Open government as a vehicle for government transformation. In Public Sector Transformation through E-Government*. Routledge.

Liu, Y., Weng, X., Wan, J., Yue, X., Song, H., & Vasilakos, A. V. (2017). Exploring data validity in transportation systems for smart cities. *IEEE Communications Magazine*, *55*(5), 26–33. doi:10.1109/MCOM.2017.1600240

Lombardi, P., Giordano, S., Farouh, H., & Yousef, W. (2012). Modelling the smart city performance. *Innovation (Abingdon)*, *25*(2), 137–149. doi:10.1080/13511610.2012.660325

Matheus, R., Janssen, M., & Maheshwari, D. (2018). Data science empowering the public: Data-driven dashboards for transparent and accountable decision-making in smart cities. *Government Information Quarterly*, 101284. doi:10.1016/j.giq.2018.01.006

Mehmood, Y., Ahmad, F., Yaqoob, I., Adnane, A., Imran, M., & Guizani, S. (2017). Internet-of-things-based smart cities: Recent advances and challenges. *IEEE Communications Magazine*, *55*(9), 16–24. doi:10.1109/MCOM.2017.1600514

Mellouli, S., Luna-Reyes, L. F., & Zhang, J. (2014). Smart government, citizen participation and open data. *Information Polity*, *19*(1,2), 1–4. doi:10.3233/IP-140334

Menouar, H., Guvenc, I., Akkaya, K., Uluagac, A. S., Kadri, A., & Tuncer, A. (2017). UAV-enabled intelligent transportation systems for the smart city: Applications and challenges. *IEEE Communications Magazine*, *55*(3), 22–28. doi:10.1109/MCOM.2017.1600238CM

Milakovich, M. (2012). *Digital governance: New technologies for improving public service and participation*. Routledge. doi:10.4324/9780203815991

Nalbandian, J., O'Neill Jr, R., Michael Wilkes, J., & Kaufman, A. (2013). Contemporary challenges in local government: Evolving roles and responsibilities, structures, and processes. *Public Administration Review*, *73*(4), 567–574. doi:10.1111/puar.12059

Nam, T. (2012). Suggesting frameworks of citizen-sourcing via Government 2.0. *Government Information Quarterly*, *29*(1), 12–20. doi:10.1016/j.giq.2011.07.005

Nam, T., & Pardo, T. A. (2011). Conceptualizing smart city with dimensions of technology, people, and institutions. *Proceedings of the 12th annual international digital government research conference: digital government innovation in challenging times*, 282-291. 10.1145/2037556.2037602

Nayyar, A., Jain, R., Mahapatra, B., & Singh, A. (2019). *Cyber Security Challenges for Smart Cities. Driving the Development, Management, and Sustainability of Cognitive Cities*. IGI Global.

Neirotti, P., De Marco, A., Cagliano, A. C., Mangano, G., & Scorrano, F. (2014). Current trends in Smart City initiatives: Some stylised facts. *Cities (London, England)*, *38*, 25–36. doi:10.1016/j.cities.2013.12.010

Paskaleva, K. A. (2009). Enabling the smart city: The progress of city e-governance in Europe. *International Journal of Innovative Research and Development*, *1*(4), 405–422. doi:10.1504/IJIRD.2009.022730

Pereira, G. V., Parycek, P., Falco, E., & Kleinhans, R. (2018). Smart governance in the context of smart cities: A literature review. *Information Polity*, *23*(2), 143–162. doi:10.3233/IP-170067

Pramanik, M. I., Lau, R. Y., Demirkan, H., & Azad, M. A. K. (2017). Smart health: Big data enabled health paradigm within smart cities. *Expert Systems with Applications*, *87*, 370–383. doi:10.1016/j.eswa.2017.06.027

Sarker, M. N. I., Wu, M., & Hossin, M. A. (2018). Smart governance through bigdata: Digital transformation of public agencies. *2018 International Conference on Artificial Intelligence and Big Data (ICAIBD)*, 62-70. 10.1109/ICAIBD.2018.8396168

Scholl, H. J. (2005). Interoperability in e-Government: More than just smart middleware. *Proceedings of the 38th Annual Hawaii International Conference on System Sciences*, 123-123. 10.1109/HICSS.2005.336

Shapiro, J. M. (2006). Smart cities: Quality of life, productivity, and the growth effects of human capital. *The Review of Economics and Statistics*, *88*(2), 324–335. doi:10.1162/rest.88.2.324

Solanas, A., Patsakis, C., Conti, M., Vlachos, I. S., Ramos, V., Falcone, F., ... Perrea, D. N. (2014). Smart health: A context-aware health paradigm within smart cities. *IEEE Communications Magazine*, *52*(8), 74–81. doi:10.1109/MCOM.2014.6871673

Sookhak, M., Tang, H., He, Y., & Yu, F. R. (2018). Security and privacy of smart cities: A survey, research issues and challenges. *IEEE Communications Surveys and Tutorials*, *21*(2), 1718–1743. doi:10.1109/COMST.2018.2867288

Suciu, G., Vulpe, A., Halunga, S., Fratu, O., Todoran, G., & Suciu, V. (2013). Smart cities built on resilient cloud computing and secure internet of things. *2013 19th International Conference on Control Systems and Computer Science*, 513-518.

Tian, J., Li, H. & Chen, R. (2018). Individual Behavior Change under Smart City Environment—A Proposal of Smart Citizen Concept with Four Dimensions. *iConference 2018 Proceedings.*

Van Winden, W., & Van Den Buuse, D. (2017). Smart city pilot projects: Exploring the dimensions and conditions of scaling up. *Journal of Urban Technology*, *24*(4), 51–72. doi:10.1080/10630732.2017.1348884

Viitanen, J., & Kingston, R. (2014). Smart cities and green growth: Outsourcing democratic and environmental resilience to the global technology sector. *Environment & Planning A*, *46*(4), 803–819. doi:10.1068/a46242

Visvizi, A., Lytras, M. D., Damiani, E., & Mathkour, H. (2018). Policy making for smart cities: Innovation and social inclusive economic growth for sustainability. *Journal of Science and Technology Policy Management*, *9*(2), 126–133. doi:10.1108/JSTPM-07-2018-079

Winters, J. V. (2011). Why are smart cities growing? Who moves and who stays. *Journal of Regional Science*, *51*(2), 253–270. doi:10.1111/j.1467-9787.2010.00693.x

Xiong, Z., Sheng, H., Rong, W., & Cooper, D. E. (2012). Intelligent transportation systems for smart cities: A progress review. *Science China. Information Sciences*, *55*(12), 2908–2914. doi:10.100711432-012-4725-1

Yang, K., & Callahan, K. (2007). Citizen involvement efforts and bureaucratic responsiveness: Participatory values, stakeholder pressures, and administrative practicality. *Public Administration Review*, *67*(2), 249–264. doi:10.1111/j.1540-6210.2007.00711.x

Yaqoob, I., Hashem, I. A. T., Mehmood, Y., Gani, A., Mokhtar, S., & Guizani, S. (2017). Enabling communication technologies for smart cities. *IEEE Communications Magazine*, *55*(1), 112–120. doi:10.1109/MCOM.2017.1600232CM

Zanella, A., Bui, N., Castellani, A., Vangelista, L. & Zorzi, M. (2014). Internet of things for smart cities. *IEEE Internet of Things Journal, 1*, 22-32.

Zheng, L., Kwok, W.-M., Aquaro, V., & Qi, X. (2019). Digital Government, Smart Cities and Sustainable Development. *Proceedings of the 12th International Conference on Theory and Practice of Electronic Governance*, 291-301. 10.1145/3326365.3326403

Zhu, F., Li, Z., Chen, S., & Xiong, G. (2016). Parallel transportation management and control system and its applications in building smart cities. *IEEE Transactions on Intelligent Transportation Systems*, *17*(6), 1576–1585. doi:10.1109/TITS.2015.2506156

# Chapter 6
# From E-Government to Digital Government:
## The Public Value Quest in the Tunisian Public Administration

**Fatma Bouaziz**
*University of Sfax, Faculty of Economics and Management, Tunisia*

## ABSTRACT

*Information and communication technologies (ICTs) know a steady evolution that leads to the development of digital technologies. These affect human lives and organizations' activities. In particular, they have become ubiquitous in the public sector almost throughout the world. The extensive use of ICTs in public administration (PA) contributes to the emergence of digital government. It is perceived as a means of public value creation and represents a part of a paradigm shift in public management. This chapter intends to review the milestones of PA reforms in a developing country, namely Tunisia, to highlight how it is using ICTs to create public value. A content analysis of data on public sector reforms and e-government strategies is performed. The results show progressive ICTs deployment by Tunisian PA over time. They illustrate a search to public value creation through the digital government.*

## INTRODUCTION

Information and communication technologies (ICTs) have known continuous progress leading to the development of digital technologies and simultaneously digital government. These technologies were recognized for their potential to re-invent the public sector. Benefits result from their capacity to enhance the efficiency and effectiveness of public services, reduce costs, and improve communication, interaction, as well as collaboration between people and public administration (PA) (OECD, 2003; UN, 2002-2018; Gill Garcia et al., 2018). Moreover, it was assumed that they help to build trust, equity and transparency and to combat corruption (Gross Lopes et al., 2019; Valle-Cruz, 2019). Briefly, according

DOI: 10.4018/978-1-7998-1851-9.ch006

to these authors, ICTs and digital government have the capacity to transform the public sector and to lead to the creation of public value. The latter is the value created by PA through services, laws regulation and other actions. It refers to various benefits for society, which are determined by citizens' preferences (Kelly et al., 2002; OECD, 2014; Valle-Cruz, 2019).

Public value has been introduced in the sphere of public sector since the work done by Moore (1995). Recently, several authors have shown an interest in the public value and have agreed that ICTs and digital government are critical to its creation. This has led to a growing body of research, both theoretical and empirical, on the value of e-government (Twizeyimana and Andersson, 2019). The latter authors have classified the existing studies on the topic into three streams of research. The first stream is concerned with understanding the value of e-government. The second one seeks to hypothesize the value of e-government by developing methods for measuring e-government from a value perspective. The third stream is concerned with hypothesizing the value of e-government for developing methods for the design and implementation of e-government from a value perspective.

This chapter intends to review the evolution of the approaches of the public management and in parallel the progress of ICTs' use by the PA. It studies their contribution to the PA modernization and the public value creation. At the empirical level, the author will carry out a descriptive study based on a content analysis of secondary data related to the use of ICTs and digital government by Tunisian PA. The purpose is to highlight the progress of the ICTs' use by the Tunisian PA and to check how they are being used as a means of the PA modernization and public value creation.

Conducting this research may be justified at theoretical and managerial levels. The theoretical level is related to the fact that although the digital government was studied from a public value viewpoint, researches from a developing country's perspective remain scarce (Karunasena and Deng, 2012; Deng et al., 2018; Okong'o and Kyobe, 2018; Gross Lopes et al., 2019; Twizeyimana and Andersson, 2019). So, a contribution to reduce this theoretical gap on the topic may be made.

At a managerial level, mechanisms to measure the success of Tunisian e-government initiatives are not defined (Chaabouni and Kamoun, 2018). Yet, the public value perspective has been proved "appropriate" and "necessary" in evaluating the performance of e-government initiatives in developing countries (Karunasena and Deng, 2012; Deng et al., 2018). Therefore, this perspective is mobilized to identify the public values that are searched through Tunisian digital government initiatives. This may lead to dress a list of public values that projects managers can use as a diagnostic tool to assess their success and to justify the expenditures in a context of resources scarcity.

This chapter is organized as follows: the first section will review three approaches of public management, namely the traditional PA, the new public management and the public value management. Second and third sections describe the evolution to digital government and the concern for creating public value. The pursued method is highlighted in section four. Results about the evolution of ICTs' use and digital government in Tunisian PA and the quest for public value creation are exposed in section five.

## APPROACHES TO PUBLIC MANAGEMENT

When discussing the modernization of the PA, researchers (Kelly et al., 2002; Marcon, 2014) evoke three principal approaches. These are the traditional approach of PA, the new public management, and the public value management. These will be briefly exposed in the following.

## The Traditional Approach of Public Administration

Traditional approach of PA appears in the United States in the late 1900s and matured by the mid-twentieth century (Hughes, 2003; Bryson et al., 2014). This approach has been raised from a legal perspective. According to this approach, PA consists of a set of heterogeneous organisms that are created, managed and / or controlled by the public authorities and that fulfill the social functions of the state. These functions include the delivery of public services which are "activities which, for reasons of general interest, are considered by the states as being unable to be regulated and satisfied by the law of the market alone or not susceptible of private appropriation" (Gyon, 1997, p.141). The general interest is the principle of order that allows society to achieve integration and unity. It leads to satisfy the interest of the community. Indeed, meeting the needs of general interest is the main objective of the PA. It contributes to the specificity of public management compared to the management in private and economic organizations. In fact, PA is managed according to a set of principles based on a legal regime, the bureaucratic rationality, the monopolistic nature of the activity and the political constraint (Hughes, 2003; Sarker, 2006). Moreover, public management is determined by a set of abstract, impersonal and objective rules fixing the modes of management and action: the texts that define the division of tasks, the hierarchy of responsibilities, the conditions of employment and compensation, the allocation and use of resources, the choice of investments to be made, the methods of financing, etc.

The traditional approach of PA has been regarded as "the most successful" approach of public sector management (Hughes, 2003). However, since the late 1970s, this approach has become inadequate for the needs of modern societies, and has been-criticized (Sarker, 2006). In fact, the bureaucratic operating modes have been considered as synonymous to heaviness. They have inhibited responding to the new socio-economic constraints, the transformation of the environment and the pressure of international donors. Giving up with this approach to adopt new ones has become necessary (Hughes, 2003). In fact, in order to respond to national and international requirements and to enhance the competitiveness of the economy, the PA should firstly improve its performance by rationalizing its resources' exploitation and improving its services quality. As a result, a new approach to the PA based on the methods of private management has emerged. This is the managerial approach or the new public management.

## The New Public Management Approach

The new public management appeared in certain Anglo-Saxon countries during the 1980s and 1990s, and particularly in the United Kingdom (Pettigrew, 1997; Bryson et al., 2014). It describes a new way of studying and managing public sector organizations (Hughes, 2003). New public management, managerialism, neo-managerialism, public management and reinvention of the PA (Gow and Dufour, 2000) are expressions used to evoke the managerial approach to the PA. For Pettigrew (1997, p.113), the new public management "has been about fostering the intrusion of private sector models and individuals into the public sector". Its application to public management was based on the shortcomings of bureaucracy and its organizing principles.

According to the new public management, the functioning of the PA is based on a list of techniques such as privatization, contracting, decentralization, partnerships, results management and customer orientation (Osborne and Gaebler, 1992). Nevertheless, the new public management cannot be reduced to a set of administrative techniques. It is rather a transformation of the bureaucratic state to an entrepreneurial one that is supported by a cultural and behavioral changes in PA management (Gow and Dufour, 2000).

The new public management founds its rational on the private sector, which emphasizes clientelism (Gross Lopes et al., 2019). However, reducing citizens to customers may overshadow the democratic values of impartiality and equality and underestimate the collective preferences. Therefore, a contemporary alternative or reaction to new public management approach, namely the public value perspective, was promoted (Gross Lopes et al., 2019).

## The Public Value Management

The public value management, first articulated by Moore (1995), "places itself beyond traditional public administration-bureaucratic logic and the core aspect of the new public management approach, to take a 'post-competitive government' characterization, in which relationships are a dominant focus" (Marcon, 2014). It requires "thinking about government and public services less as machines or structures and more as "complex adaptive systems"; it distinguishes itself by its "capacity to analyze and understand the interconnections, interdependencies and interactions between complex issues" (Marcon, 2014).

Moore (1995, in Okong'o and Kyobe, 2018) defines public value as an equivalent of private sector's shareholder worth. He contends that the generation of public value is a core role of government. According to Kelly et al. (2002, p.4), "public value refers to the value created by a government through services, laws regulation and other actions". The value is determined by citizens' preferences refracted through the decisions of elected politicians (Kelly et al., 2002, pp. 3-4). This definition is also adopted in recent works on public value (Valle-Cruz, 2019).

Public value "refers to various benefits for society that may vary according to the perspective or the actors" (OECD, 2014, p.6). It flows from three sources, which are (Kearns, 2004):

- The delivery of high quality services shown through a series of factors such as their availability, the satisfaction of users, the perceived importance of the service and the fairness of its provision and finally its cost.
- The achievement of outcomes that are seen as desirable by the public such as improvements in health, reduced poverty or environmental improvements.
- Trust in public institutions which makes citizens more likely both to accept government action and to have a sense of association with it.

The table 1 characterizes some differences between the three approaches of public management.

Nowadays, public value is an important issue for governments around the world and it is a difficult commitment for public servants toward citizens (Valle-Cruz, 2019). The extensive use of digital technologies and the development of digital government can facilitate creating public value by PA. In fact, it becomes hard to imagine any government function or governance process that does not involve extensive use of information and technology (Gil-Garcia et al., 2018). Moreover, the e-government policies are more assessed in terms of their ability to increase the PA capacity of producing public value (Kearns, 2004; Deng et al., 2018; OECD, 2018; Valle-Cruz, 2019). The latter will be discussed in the following section.

*Table 1. Approaches to public management*

| | Traditional public management | New public management | Public value |
|---|---|---|---|
| **Public interest** | Defined by politicians/ experts | Aggregation of individual preferences, demonstrated by customer choice | Individual and public preferences (resulting from public deliberation) |
| **Performance objective** | Managing inputs | Managing inputs and outputs | Multiple objectives<br>-service outputs<br>-satisfaction<br>-outcomes<br>Maintaining<br>-Trust/ legitimacy |
| **Dominant model of accountability** | Upwards through departements to politicians and through them to praliament | Upwards through performance contracts; sometimes outwards to customers through market mechanisms | Multiple<br>-citizens as overseers of government<br>-customers as users<br>-taxpayers as funders |
| **Prefered system for delivery** | Hierarchical department or self-regulating prefession | Private sector or tightly defined arms-length public agency | Menu of alternatives selected pragmatically (public sector agencies, private companies, Community interest companies, community groups as well as increasing role for user choice) |
| **Approach to public service ethos** | Public sector as monopoly on service ethos, and all public bodies have it | Speetical of public sector ethos (leads to inefficiency and empire building) – favous customer service | No one sector has a monopoly on ethos, and no one ethos always appropriate. As a valuable resource it needs to be carefully managed |
| **Role for public participation** | Limited to voting in elections and pressure on elected representatives | Limited – apart from use of customer satisfaction surveys | Crucial- multi-faceted (customers, citizens, key stakeholders) |
| **Goal of managers** | Respond to political direction | Meet agreed performance targets | Respond to citizen/ user preferences, renew mandate and trust through guaranteeing quality services |

Source: Kelly et al. (2002). Creating public value: An analytical framework for public service reform. http://www.allamreform.hu/letoltheto/kozfeladatok/kulfoldi/public_value2.pdf

## FROM E-GOVERNMENT TO DIGITAL GOVERNMENT AND THE CONCERN FOR CREATING PUBLIC VALUE

The steady evolution of ICTs leads to the development of digital or emerging technologies and simultaneously digital government. Gil-Garcia et al. (2018) and Valle-Cruz (2019) give some examples of digital technologies. These are mobile technologies, social media, GPS, the Internet of things, wireless technology, sensor networks, artificial intelligence, Big Data, nanotechnology, biotechnology, cognitive systems, renewable energy technologies and biometric software. Globally, these technologies refer to "ICTs, including the Internet, mobile technologies and devices, as well as data analytics used to improve the generation, collection, exchange, aggregation, combination, analysis, access, searchability and presentation of digital content, including for the development of services and apps" (OECD, 2014, p.6).

Statistics published by the UN (2002-2018) show an extensive use of digital technologies at the different levels and branches of governments around the world, and this is labeled as digital government.

Zwass (2006) defines digital government as "government facilitated by the internet-web and allied information and telecommunication technologies". The author distinguishes between two principal domains of digital government, which are e-government and e-democracy.

E-government has been introduced in the public sphere since the years 1990s. It has been defined as "the use by the governments of ICTs, and particularly the Internet, as a tool to achieve better government" (OECD, 2014, p.6). It refers to services provision and information dissemination through ICTs by PA to their stakeholders (other public organizations, citizens, businesses, and civil society). Therefore, it is understood as the use of ICTs (Internet, Intranet, EDI, decision support systems) by PA in their information processes, communication, transaction and decision-making. By involving government agencies, citizens, businesses and civil society in electronic relationships, the objective is to improve the efficiency and effectiveness of public organizations and the delivery of public services (Bouaziz and Chaabouni, 2013). Accordingly, e-government addresses the needs of all levels of government as providers of services (Zwass, 2006). Efficiency and access to information and transactional services are the primary concerns. It also includes the use of electronic means to ensure security, safety and the citizens' well-being. For that purpose, e-government is based on electronic interactions between PA, citizens, businesses, employees, and actors in civil society. All of these interactions are focusing on the use of ICTs to provide information and services and to realize internal administration activities, management and decision-making.

The second domain of digital government, i.e. e-democracy, implies "the use of ICTs to support the process of broad participation of citizens and of those who represent others, in debates, political discourses, surfacing and elaboration of societal issues, emergence of representation, polling, and voting" (Zwass, 2006). This domain includes activities of e-democracy, e-voting and political participation online, which are related to e-governance. In fact, digital government is "a concept that has broadened in scope from an early focus on the use of ICT for government administration to the more recent notion that information and technology influence administration, management, and governance" (Gil-Garcia et al., 2018, p.634). E-governance is "the public sector's use of information and communication technologies with the aim of improving information and service delivery, encouraging citizen participation in the decision-making process and making government more accountable, transparent, and effective" (UNESCO, 2011). This definition is adopted by Gil-Garcia et al. (2018, p.634) to define digital government.

In this work, we align with the definition of the digital government proposed by the OECD (2014). It is considered as "the use of digital technologies, as an integrated part of governments' modernization strategies, to create public value. It relies on a digital government ecosystem that is comprised of government actors, non-governmental organizations, businesses, citizens' associations and individuals, which supports the production of and access to data, services and content through interactions with the government" (OCDE, 2014, p.6). In this definition two major ideas are specified:

- The first one is the idea about the contribution of digital government to the creation of public value. Thanks to digital technologies, the PA can better respond to stakeholders' requests and preferences, thus creating public value. This gives more specification to the concept of "better benefits" highlighted in the definition of e-government.
- The second idea is the alignment between digital government strategy and modernization strategy that focuses on the public value creation. To lead to the creation of public value, the digital gov-

ernment should be inscribed in a strategy of PA transformation based on the abilities of digital technologies. In fact, the public value creation is firstly highlighted by Moore (1995) as a paradigm shift of public organizations. Moreover, digital government is understood as "a phenomenon" that "involves new styles of leadership, new decision-making processes, different ways of organizing and delivering services, and new concepts of citizenship" (Gil-Garcia et al., 2018, p.634).

Thus, the evolution towards the digital government fits more under a paradigm shift of PA modernization, and change of management than an automatic application of ICTs to existent processes, structures, and operational modes. In fact, these technologies are disruptive in organizations and nations, changing traditional forms and altering the status quo (Valle-Cruz, 2019). They are capable of changing the ways of communication, the interaction between users and organizations, facilitating communication and interactions, as well as reducing time and costs. Therefore, the governments' use of ICTs should be combined with organizational changes to improve the structures and operations of PA and to be able to create public value.

## CREATING PUBLIC VALUE THROUGH DIGITAL GOVERNMENT

In 2001, Schedler and Sharf have claimed, "modernizing the state without e-government is not thinkable anymore, both in theory and in practice". This has always been the case since it becomes hard to imagine any government function or governance process that does not involve extensive use of information and technology (Gil-Garcia et al., 2018). ICTs and digital government have become a key element of public sector modernization around the world. It is a continuous process that began in the 1970s and 1980s with centralized computing. At that time, the first applications of ICTs enabled the automation of work, the improvement of data processing capacities thanks to storage and calculation capacities and, consequently, the reduction of costs of administrative procedures.

Today, ICTs are adding communication capabilities that promote connectivity, interactivity and collaboration between individuals and organizations. The evolution of the potential of ICTs leads to the development of e-government. ICTs, mainly the Internet and the World Wide Web, add a new "e" dimension to the government agencies (UN, 2002, p.4). ICTs change the ways of communicating, working, consuming and learning of individuals and organizations. By this, they have the potential to be a driver of economic and social development. Moreover, the implementation of applications, such as telecommunication networks and Internet standards, contributes to replacing PA management with new forms of public management and makes it possible to go beyond bureaucratic organization (Schedler and Sharf, 2001; Hughes, 2003). In addition, e-government has the potential to support the transformation of PA into an efficient and effective system (OECD, 2003, UN, 2004). It improves the efficiency of public services, contributes to the achievement of economic development goals, and establishes trust between governments and citizens (OECD, 2003, UN, 2004). Accordingly, e-government has the potential to transform the PA into a "better PA" (OECD, 2003, p. 24), which creates public value.

The concept of public value is increasingly becoming an innovative driver in modern e-government endeavors. Authors are unanimous about the fact that e-government leads to the creation of public value (Kearns, 2004; Karunasena and Deng, 2012; Deng et al., 2018; Okong'o and Kyobe, 2018; Gross Lopes et al., 2019; Valle-Cruz, 2019; Twizeyimana and Anderson, 2019). Public value of e-government is the impact of e-government in respect to government operations, actions, policies, and services for citizens

(Twizeyimana and Andersson, 2019). It is the ability of e-government to provide improved efficiency in government, improved services to citizens, and social values such as inclusion, democracy, transparency and participation (Twizeyimana and Andersson, 2019). Accordingly, ICTs and e-government encompass strategic goals that cover efficiency, effectiveness and economy issues, and include political and social objectives such as trust in government, social inclusion, community regeneration, community well-being and sustainability. Kearns (2004) contends that public value can be used to aid decision making, to assess performance and, in the e-government context, to provide a bridge between technology and wider policy communities. ICTs are drivers of change in organizations. They generate public value because they lead to better service delivery, better interaction between public servants and citizens, better transparency and efficiency as well as less corruption (Gross Lopes et al., 2019; Valle-Cruz, 2019). This may have impacts on trust, engagement and public value. A summary of public value dimensions and sub-dimensions resulting from e-government is made in table 2.

*Table 2. Public value dimensions and sub-dimensions*

| Authors | Public value dimensions and sub-dimensions |
|---|---|
| Kearns (2004) | The provision of services that are widely used; Increased levels of user satisfaction with services; Increased information and choice available to service users; Greater focus on the services that the public believes to be most important; Increased focus of new and innovative services towards those most in need; Reduced costs of service provision; Improved delivery of outcomes; A contribution to improve levels of trust between citizens and public institutions |
| Karunasena and Deng (2012) | Quality of information; delivery of e-services; user-orientation of e-gov service delivery; efficiency of government organizations; responsiveness; government organizations' contribution to environmental sustainability |
| Deng et al. (2018) | Delivery of quality public services: quality of information, functionalities of the e-services, user orientation;<br>Effectiveness of public organizations: organizational efficiency, openness;<br>Achievement of socially desirable outcomes: equity, self-development, trust, environmental sustainability |
| Okong'o and Kyobe (2018) | Accuracy and up-to- date information,<br>online communication and online access to government document,<br>user friendly web and none technology savvy web,<br>data sharing and lean staffing,<br>online filing of complaints and online display of government data,<br>online enquiries and online case tracking / follow up, access by the marginalized,<br>security of government information and credibility of information exchanged,<br>providing information on upcoming public policies and online participation in policy making, and<br>limiting duplication of resources and enabling recycling of consumables. |
| Twizeyimana and Andersson (2019) | Improved public services; Improved administrative efficiency; Open government (OG) capabilities; Improved ethical behavior and professionalism; Improved social value and well-being; Improved trust and confidence in government |
| Gross Lopes et al. (2019) | Collaborative process: Dialogue with society; Collaboration Citizen ; empowerment<br>Governmental readiness: Governmental transparency; Collaboration readiness<br>E-service design: User orientation, Ease of use; Usability<br>Citizen trust: Trust in the technology; Trust in the government<br>Creation of public value: Convenience, Service experience, Attitude |
| Valle Cruz (2019) | Anti-corruption strategies, Open data, Access to information, Data privacy |

## METHOD

The purpose of this chapter is to highlight the progress of the ICTs' use by the Tunisian PA and to check how they are being used as a means of the PA modernization and public value creation. In this regard, "the depth of insights that can be gained from qualitative studies is important for the field" (Bélanger and Carter, 2012, p.375).

Therefore, in order to achieve the purpose of the chapter, it would be useful to conduct qualitative research based on a retrospective analysis of technology-driven reforms of public sector and content analysis of e-government strategies. The Tunisian PA reform is studied in a chronological way through secondary data sources. The key ones are:

- The sections dealing with the PA modernization contained in the quinquennial economic and social development plans (ESDP) (1982-1986 until 2016-2020).
- The document of the e-strategy program 2009-2014.
- The document related to "Smart-Gov 2020".

A content analysis of the selected documents is conducted. According to Aktouf (1987), the content analysis is a detailed study technique of the document contents. Its aim is to extract from them the meaning, associations and intentions, not directly perceivable by simply reading them. Miles and Huberman (2004) have defined three steps of content analysis. The first step is categorization. It deals with classifying or codifying the elements of the analyzed documents under diverse categories for understanding the meaning of the message (Miles and Huberman, 2004). In this step, the data are divided into homogeneous units to recreate, retrospectively, a structured document on the subject to study. As advocated by Miles and Huberman (2004), a list of codes to assign to the segments of data is created in advance. To this end, the dimensions and sub-dimensions of e-government public value proposed by Deng et al. (2018) (table 1) were adopted. They were retained because they result from a validated factors framework leading to public value of e-government, from the public services delivery perspective and from the public organizations efficiency. This framework was initially elaborated by Karunasena and Deng (2012). It was the first of its kind to examine public value of e-government by considering all the dimensions of public value creation in the context of developing countries (Deng et al., 2018). It was also validated by Okong'o and Kyobe (2018). To identify the key public values through the selected documents, checklist matrices were used. These provide a way to organize and display the data according to a key theme or variable. In addition, during data analysis, the researcher allows categories to emerge from the data. The second step of content analysis is inference. In this step, the author examines the coded documents and the checklist matrices to infer the public values. The third step of data analysis is interpretation. This third step is to interpret the results based on the presence or absence of certain codes, taking into consideration the studied context.

## RESULTS

### Evolution of ICTs Use and Digital Government in Tunisian Public Administration

The Figure 1 shows the evolution of three indexes, namely e-government development index (EGDI - series 1), online services index (OSI - series 2) and e-participation index (EPI - series 3), performed by the Tunisian PA. It is elaborated based on data displayed in the UN reports on e-government for the period 2002-2018.

The E-Participation Index (EPI) is based on: (i) e-information – availability of online information; (ii) e-consultation – online public consultations, and (iii) e-decision-making – directly involving citizens in decision processes. The UN surveys assesse the availability of e-participation tools on national government portals for each of the above criteria.

The EGDI assesses e-government development at the national level. It is a composite index based on the weighted average of three normalized indices (UN reports). These indices are the following:

- One-third is derived from a Telecommunication Infrastructure Index (TII) based on data provided by the International Telecommunications Union (ITU),
- One-third from a Human Capital Index (HCI) based on data provided by the United Nations Educational, Scientific and Cultural Organization (UNESCO), and
- One-third from the Online Service Index (OSI) based on data collected from an independent survey questionnaire, conducted by UNDESA, which assesses the national online presence of all 193 United Nations Member States. The survey questionnaire assesses a number of features related to online service delivery, including whole-of-government approaches, open government data, e-participation, multi-channel service delivery, mobile services, usage uptake, digital divide as well as innovative partnerships through the use of ICTs.

*Figure 1. The evolution of the digital government indexes*

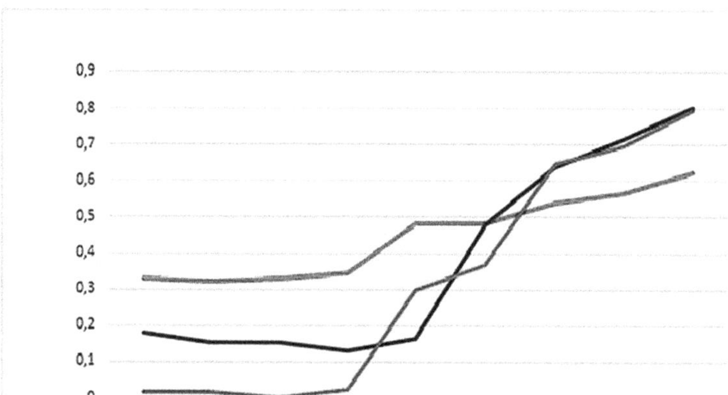


| | 2003 | 2004 | 2005 | 2008 | 2010 | 2012 | 2014 | 2016 | 2018 |
|---|---|---|---|---|---|---|---|---|---|
| Série 1 | 0,329 | 0,3227 | 0,331 | 0,3458 | 0,4826 | 0,4833 | 0,539 | 0,5682 | 0,6254 |
| Série 2 | 0,179 | 0,154 | 0,1538 | 0,1304 | 0,1641 | 0,4771 | 0,6378 | 0,7174 | 0,8056 |
| Série 3 | 0,017 | 0,0164 | 0 | 0,0227 | 0,3 | 0,3684 | 0,6471 | 0,6949 | 0,7978 |

The figure 1 reveals that there was a steady progress in the three indexes over time. Improvements in EGDI and in providing online services have been steady over the last eight years (from 2010 to 2018). Improvements in EGDI were driven by significant improvements in OSI. It is also the case for EPI, which has a significant development. Enhancements in the values of EGDI, OSI and EPI lead Tunisia to be among the countries that have high indexes. They show a great interest of Tunisian PA to the use of ICTs and digital government with its two domains, namely e-government and e-democracy. In fact, important and steady improvements in the three indicators have been performed since the year 2010. It is worth noting that since 2010, Tunisia has entered a phase of democratic transition. This may explain the greater use of ICTs to benefit from their advantages in this regard. This interest is also the result of efforts, actions and programs engaged through several years to modernize the PA functioning. Indeed, according to Chaabouni and Kamoun (2018), ICTs have transformed sectors of the Tunisian economy and administration over the last 20 years. This digital transformation affects Tunisian central, regional and local PA, where applications are implemented to get closer to the citizens, to facilitate and accelerate the delivery of services, to strengthen the trust of citizens and to improve organizational performance (Chaabouni and Kamoun, 2018). This choice is in line with the country's strategy for the development of e-government. In fact, several actions and projects have been engaged by the PA over the last few years, such as e-finance, TradeNet, e-CNSS, e-justice, e-local government, etc. (African Development Bank, 2017). Moreover, some performance indicators about digital technologies and e-government that are given in the report of the African Development Bank (2017) show a continuous emphasis from the government to enhance these indicators. The target is to have a leading position among the Arab and African countries by 2021 in the domain and to implement the "Digital Tunisia 2020" National Strategic Plan. These indicators are summarized in Table 3.

## The Quest for Public Value Creation by the Digital Government in Tunisian Public Administration

In Tunisia, the PA has a crucial role as a means of economic and social development. To fulfill this role, a process of modernization has been engaged since the 1970s. Three checklists (Tables 4, 5, 6 appendix) were elaborated to display data contained in the documents related to Tunisian PA modernization and e-government. The dimensions and sub-dimensions of public value proposed by Deng et al. (2018) were used as a first list of categories to code the data. The content of the three checklists shows that the modernization of the Tunisian PA is based on the following major axes:

- The adoption of organizational solutions through the re-design of PA. These organizational solutions are the decentralization and redistribution of powers between central, regional and local levels, the creation of new functions (administrative mediator, supervising citizen, administrative control committees) and the allocation of the human, material and financial resources required to fulfill the new responsibilities.
- The modernization of the human resources management system by the revision of the statutes of the civil service, the revision of the remuneration, the promotion, the evaluation and the training systems.
- The adoption of managerial solutions such as management by objectives, quality management approach.
- The improvement of the quality of public services.

*Table 3. Performance Indicators in the domain of digital government and digital technologies*

| Indicator | Baseline situation | Target |
|---|---|---|
| **Digital vision of the government (NRI-GEF index, Scale: 1-7)** | 3.58 (2016) | 5.85 (2021) |
| **Digital ranking of Tunisia** | World: 81 (2015)<br>Africa: 4 (2015)<br>Arab countries: 9 (2015) | World: 40 (2021)<br>Africa: 1 (2021)<br>Arab countries: 3 (2021) |
| **Availability of online administrative services (United Nations index, Scale: 0-1)** | 0.63 (2014) | 0.77 (2021) |
| **Percentage of households with access to online government services** | 18.2% (2015) | 100% (2021) |
| **Number of digital technology jobs created per year** | 7,500 (2014) | 25,000 (2021) |
| **Use of e-participation platforms by citizens (UN DESA index, Scale: 0-1)** | 0.64 (2014) | 0.85 (2021) |
| **Successful promotion of e-government (WEF-NRI index, scale: 1-7)** | 4.14 (2014) | 5.6 (2021) |
| **Number of interoperable IS established to guarantee the development of e-government** | 0 (2016) | 10 (2021) |
| **Number of computerized/ online government services** | 200 | 500 (including 300 accessible online) |
| **Number of administrative sites connected to the government network (RNIA)** | 550 | 1.600 |

Source: African Development Bank. (2017). Support project for the implementation of the "digital Tunisia 2020" National Strategic Plan. https://www.afdb.org/fileadmin/uploads/afdb/Documents/Project-and-Operations/Tunisia_-_Support_Project_for_the_Implementation_of_the_%E2%80%9CDigital_Tunisia_2020%E2%80%9D_National_Strategic_Plan.pdf, p.8.

- The simplification of administrative procedures for citizens, businesses and Tunisian as well as foreign investors;
- The better reception and information through creating offices of relations with the citizens, administrative information via correspondence and simplified telephone numbers.
- The modernization of the PA by the use of computers, ICTs and the development of e-government programs.

The axes of administrative reform are mentioned in all the analyzed ESDP. Among them, ICTs are a key pillar of the PA modernization. The sixth ESDP (1982-1986) emphasizes the rationalization of the administrative operations through the office automation and the use of computers. This leading role of ICTs in the modernization of the PA is consolidated over time and through all the ESDP. ICTs are considered as a tool for the implementation of a modern, "communicating and informational" PA. They are used to improve the internal processes of the PA, to inform citizens and to deliver higher quality public services. Moreover, an e-government program has been launched since 2002. It includes all projects undertaken by PA to put information and forms online, provide public e-services and develop specific web sites and portals (Circular of prime minister 02-07-2003). Training actions that improve the abilities of human resources to use ICTs has been planned.

Thus, greater interest to ICTs applications and e-government projects has been made since 2002. ICTs were more considered as a means for an open PA, citizen-need-centered than an automation tool of internal administrative processes and work (10th ESDP, 2002-2006).

Moreover, the e-strategy program (2009-2014) focuses on the improvement of the relationships with citizens, the development of firms' competitiveness, and the achievement of high levels of users' satisfaction. This will be achieved through the simplification and dematerialization of existent procedures, the implementation of e-services, the integration of actual information systems, the reinforcement of the human resources capacities and the certified training. These objectives are more reinforced over the program "Smart Gov 2020" (December 2016). This program seeks to transform the PA to a more performing and open administration, providing accessible, simple and high quality services to citizens and companies, and actively contributing to development (Doc. Smart-Gov 2020). Based on the desired vision of the administration, the "Smart Gov 2020" program was articulated around four major axes, where digital technologies and e-government projects are a priority. These axes are:

- Integration of the PA through interoperability of actual information systems, sharing of infrastructure and electronic data exchange.
- The opening of the PA via a framework of transparency, reuse of data and citizen participation.
- User oriented PA by putting online simple, end-to-end and user-oriented services.
- The appropriation of the digital technologies: by securing the digital transformation of the administration, generalizing the use of digital technologies and establishing digital trust within the administration and with its users.

The analysis exposed above reveals the presence of the three dimensions of public value, namely delivery of quality public services, effectiveness of public organizations, and achievement of outcomes. Organizational and managerial reforms and the adoption of ICTs in internal administrative processes and external relationships with citizens and businesses were simultaneously planned and used to achieve these dimensions. While the creation of public value is not clearly mentioned, its dimensions and sub dimensions proposed by Deng et al. (2018) were identified, except the sub-dimensions equity and environmental sustainability. These two sub-dimensions were mentioned in the analyzed data. Yet, there is no indication that they will be achieved through ICTs' use and digital government. Findings are summarized in table 7.

It follows that digital technologies and digital government are deployed for Tunisian PA modernization and that this modernization aligns with the approach of public value management. Results reveal that the deployment of digital technologies (web 2.0, government cloud, internet, mobile applications, m-services, Intranet, government portal, e-petition, (doc. e-strategy, 2009-2014; doc. Smart Gov 2020) is a pillar of this modernization. Moreover, it seems that digital technologies are considered as a tool to achieve transparency and to combat corruption. These two values were not mentioned among the dimensions and sub-dimensions of Deng et al. (2018). They were added to the list of categories after their emergence through content analysis. These public values were cited by Gross Lopes et al. (2019) and Valle Cruz (2019) as they could be achieved through e-government (table 1).

Besides, findings demonstrate an evolution of the use of ICTs by Tunisian PA over time. They reveal the existence of three major periods of ICTs adoption as described in figure 2: 1982-2001, 2002-2009 and 2010-2020.

*Table 7. Results' summary*

| Public value dimensions | Public value sub – dimensions | Evaluation |
|---|---|---|
| **Effectiveness of public organizations** | Organizational efficiency<br>Openness<br>Responsiveness | ✓<br>✓<br>✓ |
| **Delivery of quality public services** | Quality of information<br>Functionalities of e-services<br>User orientation | ✓<br>✓<br>✓ |
| **Achievement of outcomes** | Equity<br>Self- development<br>Trust<br>Participatory democracy<br>Environmental sustainability<br>**Transparency and anti-corruption** | It is not mentioned if it will be achieved through ICTs<br>✓<br>✓<br>✓<br>It is not mentioned if it will be achieved through ICTs<br>✓ |

*Figure 2. The evolution of ICTs use in Tunisian public administration*

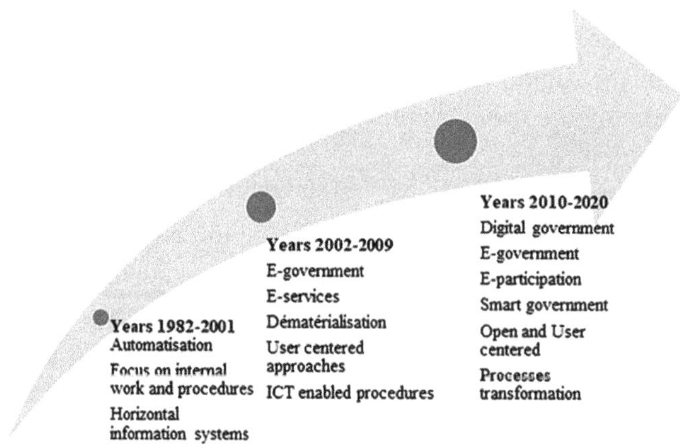


For the first period, ICTs are used as a tool of automation, focusing on internal work and procedures. Horizontal information systems are also promoted. For the second period, ICTs are more used to adopt e-government applications based on a citizen-centered approach. Dematerialization of procedures covers those related to internal processes and external relationships. In the third period, digital government is more promoted. E-government, e-participation and smart government are more used to allow open and user- centered administration and processes transformations.

## CONCLUSION

The purpose of this chapter was to highlight the progress of the ICTs' use by the Tunisian PA and to check how they are being used as a means of the PA modernization and public value creation. The contribution lies in the study of ICTs for the creation of public value in a developing country, namely Tunisia. To achieve this purpose, content analysis of secondary data related to PA modernization and e-government was performed based on the framework of public value of Deng et al. (2018). It leads to the identification of all the dimensions and sub-dimensions proposed by these authors, except of equity and environmental sustainability. Although these two sub-dimensions of the dimension "outcomes achievement" were mentioned, there is no indication that they will be achieved through ICTs and e-government programs. Moreover, it was shown that Tunisian e-government programs are perceived as a tool to enhance transparency and combat corruption. These were not considered in the framework of Deng et al. (2018), which allows its enrichment.

This chapter is a first step of a research on the PA modernization and public value creation through digital government in Tunisian PA. Despite its limitations since it focuses on the analysis of secondary data, it offers a first investigation of the digital government impacts from a public value creation perspective. At a managerial level, the results can contribute to a better understanding of these impacts by Tunisian public managers and allow better decision making about the digital government projects perceived in terms of creating public value. The list of public values studied in this chapter may serve as a diagnostic tool for digital government initiatives. It can provide a useful tool for evaluating their performance.

Future research should be conducted with Tunisian public managers for the triangulation of the data and better verification of the contribution of digital government to the PA modernization and the creation of public value.

## REFERENCES

African Development Bank. (2017). *Support project for the implementation of the "digital Tunisia 2020" National Strategic Plan*. Retrieved from https://www.afdb.org/fileadmin/uploads/afdb/Documents/Project-and-Operations/Tunisia_-_Support_Project_for_the_Implementation_of_the_%E2%80%9CDigital_Tunisia_2020%E2%80%9D_National_Strategic_Plan.pdf

Aktouf, O. (1987). *Méthodologie des sciences sociales et approche qualitative des organisations*. Québec: PUQ.

Bélanger, F., & Carter, L. (2012). Digitizing government interactions with constituents: An historical review of e-government research in information systems. *Journal of the Association for Information Systems*, *13*(5), 363–394. doi:10.17705/1jais.00295

Bouaziz, F., & Chaabouni, J. (2013). Criteria for assessing the success of e-government projects. In J. R. Cil-Carcia (Eds.), E-Government Success Factors and Measures: Theories, Concepts, and Methodologies, (pp. 273-288). IGI Global. doi:10.4018/978-1-4666-4058-0.ch015

Bryson, J. M., Crosby, B. C., & Bloomberg, L. (2014). Public Value Governance: Moving Beyond Traditional Public Administration and the New Public Management. *Public Administration Review*, *74*(4), 445–456. doi:10.1111/puar.12238

Chaabouni, J., & Kamoun, F. (2018). *Usage des applications en ligne dans les municipalités & perspectives de villes intelligentes*. Retrieved from http://www.aegid-foundation.org/events/archives/3/telechargements/rapport-integre-usages-des-applications-developpees-pour-les-municipalites.pdf

Deng, H., Karunasena, K., & Xu, W. (2018). Evaluating the Performance of e-Government in Developing Countries: A Public Value Perspective. *Internet Research*, *28*(1), 169–190. doi:10.1108/IntR-10-2016-0296

Gayon, C. (1997). De la managérialisation du secteur public à un véritable métissage public-privé. *Revue Française de Gestion*, 115-120.

Gil-Garcia, J. R., Dawes, S. S., & Pardo, T. A. (2018). Digital government and public management research: Finding the crossroads. *Public Management Review*, *20*(5), 633–646. doi:10.1080/14719037.2017.1327181

Gow, J., & Dufour, C. (2000). Le nouveau management public est-il un paradigme ? cela a-t-il de l'importance? *Revue Internationale des Sciences Administratives, 4.*

Gross Lopes, K. M., Macadar, M.A., & Luciano, E.M. (2019). Key drivers for public value creation enhancing the adoption of electronic public services by citizens. *International Journal of Public Sector Management*. doi:10.1108/IJPSM-03-2018-0081

Hughes, O. (2003). *Public management and administration: an introduction* (3rd ed.). Creative Print and Design.

Karunasena, K., & Deng, H. (2012). Critical factors for evaluating the public value of e-government in Sri Lanka. *Government Information Quarterly*, *29*(1), 76–84. doi:10.1016/j.giq.2011.04.005

Kearns, I. (2004). *Public Value and E-Government*. Institute for Public Policy Research.

Kelly, G., Mulgan, G., & Muers, S. (2002). *Creating public value: An analytical framework for public service reform*. Retrieved from http://www.allamreform.hu/letoltheto/kozfeladatok/kulfoldi/public_value2.pdf

Marcon, G. (2014). Public Value Theory in the Context of Public Sector Modernization. In Public Value Management, Measurement and Reporting. 323-351. doi:10.1108/S2051-663020140000003014

Miles, M. B., & Huberman, A. M. (2004). *Analyse des données qualitatives* (2nd ed.). De Boeck.

Moore, M. H. (1995). *Creating public value: Strategic management in government*. London: Harvard University Press.

OECD. (2003). *The e-government imperative. OECD e-government studies*. Retrieved from www1.worlbank.org/publicsector/egov/E-governmentImperative.pdf

OECD. (2014). *Recommendation of the Council on Digital Government Strategies*. Retrieved from http://www.oecd.org/gov/digital-government/Recommendation-digital-government-strategies.pdf

Okong'o, K., & Kyobe, M. (2018). Empirical Examination of E-Government in Developing Countries and its Value in Kenya's Public Service. *The Electronic Journal Information Systems Evaluation*, *21*(1), 35–45.

Osborn, D., & Gaebler, T. (1992). *Reinventing government. How the entrepreneurial spirit is transforming the public sector*. Addison-Wesley.

Pettigrew, A. (1997, Sept.). Le new public management conduit à un nouveau modèle hybride public-privé. *Revue Française de Gestion*, 113-120.

Sarker, A. E. (2006). New public management in developing countries: An analysis of success and failure with particular reference to Singapore and Bangladesh. *International Journal of Public Sector Management, 19*(2), 180–203. doi:10.1108/09513550610650437

Schedler, K., & Sharf, M.C. (2001). Exploring the interrelation between electronic government and the new public management: a managerial framework for electronic government. *13ième conférence Zurich.*

TAP. (2017). *Tunisie digitale 2020: seuls 5% des projets programmés ont été réalisés.* Retrieved from https://www.huffpostmaghreb.com/2017/02/16/tunisie-digitale_n_14796854.html

Twizeyimana, J. D., & Andersson, A. (2019). The public value of E-Government - A literature review. *Government Information Quarterly, 36*(2), 167–178. doi:10.1016/j.giq.2019.01.001

United Nations. (2002). *Benchmarking e-government: A Global perspective, assessing the progress of the UN member states.* Retrieved from www.golconferenceca/presentations/eGovernment_UN.pdf

United Nations. (2003). *World Public Sector Report 2003: E-Government at the Crossroads.* New York: UN.

United Nations. (2004). *Global e-government readiness report 2004: Towards access for Opportunity.* New York: UN. Retrieved from http://unpan1.un.org/intradoc/groups/public/documents/un/unpan019207.pdf

United Nations E-Government Survey. (2008). *From e-Government to Connected. Governance.* Author.

United Nations E-Government Survey. (2012). *E-Government for the People.* Author.

United Nations E-government Survey. (2014). E-*government for the future we want.* Author.

United Nations E-government Survey. (2016). *E-government in support of sustainable development.* Author.

United Nations E-government Survey. (2018). *Gearing e-government to support Transformation towards Sustainable and resilient societies.* Author.

Valle-Cruz, D. (2019). Public value of e-government services through emerging technologies. *International Journal of Public Sector Management, 32*(5), 530–545. doi:10.1108/IJPSM-03-2018-0072

Zwass, V. (2006). The web-internet compound as the infrastructure of digital government. *Business Process Management Journal, 12*(1), 7–12. doi:10.1108/14637150610643715

## ADDITIONAL READING

Andersen, L. B., Jørgensen, T. B., Kjeldsen, A. M., Pedersen, L. H., & Vrangbæk, K. (2012). Public Value Dimensions: Developing and Testing a Multi-Dimensional Classification. *International Journal of Public Administration, 35*(11), 715–728. doi:10.1080/01900692.2012.670843

Guthrie, J., & Russo, S. (2014). Public Value Management: Challenge of Defining, Measuring and Reporting for Public Services. In Public Value Management, Measurement and Reporting Studies in Public and Non-Profit Governance, (3), 3-17. doi:10.1108/S2051-663020140000003001

Hazlett, S. A., & Hill, F. (2003). E-government: The realities of using IT to transform the public sector, *Managing Service Quality. International Journal (Toronto, Ont.)*, *13*(6), 445–452. doi:10.1108/09604520310506504

Knox, C., & Janenova, S. (2019). The e-government paradox in post-Soviet countries. *International Journal of Public Sector Management*, *32*(6), 600–615. doi:10.1108/IJPSM-08-2018-0173

Moon, M. J. (2002). The evolution of E-government among municipalities: Rhetoric or reality? *Public Administration Review*, *62*(4), 424–443. doi:10.1111/0033-3352.00196

Omar, K., Scheepers, H., & Stockdale, R. (2011). e-Government service quality assessed through the public value lens. In M. Janssen et al. (Eds.), EGOV 2011, LNCS 6846, 431–440, IFIP International Federation for Information Processing.

Pang, M.-S., Lee, G., & DeLone, W. H. (2014). IT Resources, Organizational Capabilities, and Value Creation in Public-Sector Organizations: A Public-Value Management Perspective. *Journal of Information Technology*, *29*(3), 187–205. doi:10.1057/jit.2014.2

Savoldelli, A., Misuraca, G., & Codagnone, C. (2013). Measuring the Public value of e-Government: The eGEP2.0 model, *Electronic. Journal of E-Government*, *11*(1), 373–388.

Scott, M., DeLone, W., & Golden, W. (2016). Measuring e-Government success: A public value approach. *European Journal of Information Systems*, *25*(3), 187–208. doi:10.1057/ejis.2015.11

Siddiquee, N. A. (2016). E-government and transformation of service delivery in developing countries: The Bangladesh experience and lessons, *Transforming Government: People. Process and Policy*, *10*(3), 368–390. doi:10.1108/TG-09-2015-0039

## KEY TERMS AND DEFINITIONS

**Digital Government:** A concept that has broadened in scope from an early focus on the use of ICT for government administration to the more recent notion that information and technology influence administration, management, and governance.

**Digital Technologies:** These are mobile technologies, social media, GPS, the Internet of things, wireless technology, sensor networks, artificial intelligence, Big Data, nanotechnology, biotechnology, cognitive systems, renewable energy technologies and biometric software.

**E-Democracy:** The use of ICTs to support the process of broad participation of citizens, and of those of them who represent others, in the debate, political discourse, surfacing and elaboration of societal issues, emergence of representation, polling, and voting.

**E-Governance:** The public sector's use of ICTs with the aim of improving information and service delivery, encouraging citizen participation in the decision-making process and making government more accountable, transparent, and effective.

**E-Government:** Refers to services provision and information dissemination through ICTs by PA to their stakeholders (other public organizations, citizens, businesses, and civil society). It is the use of ICTs by PA in their processes of information, communication, transaction and decision-making.

**New Public Management:** A new way of studying and managing public sector organizations focusing on the intrusion of private sector models and individuals into the public sector.

**Public Value:** The value created by PA through services, laws regulation and other actions. It refers to various benefits for society, which are determined by citizens' preferences.

*Table 4. Checklist matrix of the public value dimension, namely effectiveness of public organizations*

| *Public Value Sub -dimensions* | *Sentences extracted from the analyzed documents* | *Comments of the researcher* |
|---|---|---|
| **Organizational efficiency**<br>**Openness**<br>**Responsiveness** | *Modern, efficient, transparent PA (ESDP 82-86)<br>*Extended automation to rationalize administrative processes (ESDP 82-86)<br>*Decentralization (ESDP 87-91)<br>*Modernization of services by the use of office automation and computers (ESDP 87-91)<br>*Re-design of organizational structures, Simplification of procedures, Quality circles (ESDP 92-96)<br>*Modernization of the work tools through the use of ICTs, the deployment of horizontal systems such as ADAB, INSAF, SICAD and other computer applications (ESDP 92-96)<br>*More decentralization and organizational re-design (ESDP 97-2001)<br>*Re-design of administrative structure (ESDP 2002-2006)<br>*Generalization of ICT use by setting up a secure internal network in all administrative structures, developing a national information system and strengthening the electronic exchange of administrative documents (ESDP 2002-2006)<br>*Implementation of management quality systems and certification to ISO Standards (ESDP 2002-2006)<br>*Creation of one-stop shops (ESDP 2002-2006)<br>*Quality management system in the public sector (ESDP 2007-2011)<br>*Internal process rethinking, procedures dematerialization by taking advantage of digital technologies as well as resources and infrastructures rationalization (Smart-Gov 2020).<br>*Open public administration (ESDP 82-86)<br>*Improvement of relationships between PA and citizens through the simplification of procedures, the enhancement of the information activities, the modernization of the services and the rationalization of forms (ESDP 87-91)<br>*Improvement of the relationships between the PA and the citizens via the consolidation of information and orientation offices<br>*Development of SICAD (ESDP 97-2001)<br>*Improvement of the reception, information and orientation activities for better relationships with citizens (ESDP 2002-2006)<br>*Creation of offices of relationships with citizens to respond to their requests and provide more quickly the best services (ESDP 2002-2006)<br>*The establishment of an open government through the adoption of open government principles by PA, etc. (ESDP 2016-2020).<br>*Open Data Support and Development by the development of a legal framework for public consultation, development of a national system for public procurement, support of the communication function in the administration (ESDP 2016-2020).<br>*A performing and open administration, providing accessible, simple and high quality services to citizens and companies, and actively contributing to development (Smart-Gov 2020).<br>*Improvement of the relationships between the PA and the citizens via the consolidation of information and orientation offices<br>*Development of SICAD (ESDP 97-2001)<br>*Improvement of the reception, information and orientation activities for better relationships with citizens (ESDP 2002-2006)<br>*Creation of offices of relationships with citizens to respond to their requests and provide more quickly the best services (ESDP 2002-2006)<br>*More attention to the citizen's needs and concerns (ESDP 2007-2011) | Both organizational solutions and ICTs solutions are proposed as a means to realize the dimension entitled effectiveness of public organizations. The three sub-dimensions, namely organizational efficiency, openness and responsiveness are concerned with these solutions. |

*Table 5. Checklist matrix of the public value dimension, namely Delivery of quality public services*

| Public Value Sub -dimensions | Sentences extracted from the analyzed documents | Comments of the researcher |
|---|---|---|
| Quality of information<br>Functionalities of e-services<br>User orientation | *Quality of Public services (ESDP 82-86)<br>*Improvement of services quality (ESDP 97-2001)<br>*Improvement of public services quality (ESDP 2002-2006)<br>*Internal and external information availability (ESDP 2002-2006)<br>*Use of ICT to better inform and guide citizens (ESDP 2002-2006)<br>*A performing and open administration, providing accessible, simple and high quality services to citizens and companies, and actively contributing to development (Smart-Gov 2020).<br>*Creation of an information system and a web site to diffuse information about procedures and legislative texts (ESDP 2002-2006)<br>*Use of ICT to better inform and guide citizens (ESDP 2002-2006)<br>*Definition of a program of public e-services (ESDP 2002-2006)<br>*More attention to the citizen's needs and concerns (ESDP 2007-2011)<br>*Support of the administration's contribution to improve the business climate (ESDP 2007-2011)<br>*A performing and open administration, providing accessible, simple and high quality services to citizens and companies, and actively contributing to economic and social development (Smart-Gov 2020),<br>*Development of end-to-end, accessible, multi-channel online services centered on the needs of the citizen, the company and the administration. (Smart-Gov 2020)<br>*Modernization of the State's information systems to enable a better user orientation (Smart-Gov 2020) | Through the documents, emphasis on the quality of public services is made.<br>The three sub-dimensions, namely quality of information, functionalities of e-services and user orientation, are highlighted. |

*Table 6. Checklist matrix of the public value dimension, namely Achievement of outcomes*

| Public Value Sub -dimensions | Sentences extracted from the analyzed documents | Comments of the researcher |
|---|---|---|
| Equity<br>Self- development<br>Trust<br>Participatory democracy<br>Environmental sustainability<br>Transparency and anti-corruption | *Services provided to citizens on the basis of equity (ESDP 82-86)<br>*Modernization of human resources management (ESDP 87-91)<br>*Training Programs of Human resources to be able to use computers and office tools (ESDP 92-96)<br>*HRM modernization, training of HR (ESDP 97-2001)<br>*Acquisition of specialist skills in the field of computer science and spread of digital culture and training of agents in the computer field with the aim of setting up a modern communicating and computerized administration (ESDP 2002-2006)<br>*Enhancement of the administrative development indicators to international standards (ESDP 2007-2011)<br>*Human resources management and training to use ICT (ESDP 2007-2011)<br>* Improvement of HR potentials and capacities (ESDP 2016-2020)<br>* Affectation of human resources to the e-government projects and training and certification (doc. E-strategy 2009-2014)<br>*Generalization of ICT use by setting up a secure internal network in all administrative structures, developing a national information system and strengthening the electronic exchange of administrative documents (ESDP 2002-2006)<br>*Strengthening of digital trust in administrative online services (Smart-Gov 2020)<br>*Citizen participation (ESDP 87-91)<br>*Use of tools and applications for dialog with citizens (Web 2.0, participative tools) (doc. E-strategy 2009-2014)<br>*The establishment of an open government through the adoption of the principles of open government by PA, the adoption of a transparency based system, the citizen participation in the definition of public policies, the prevention of corruption (ESDP 2016-2020).<br>*Implementation of a citizen participation framework that promotes participatory democracy (Smart-Gov 2020),<br>*More informed citizens, more active in public life, and more easily communicating with the administration, thanks to the use of digital technologies (Smart-Gov 2020)<br>*Save of energy, phone, paper and staff (ESDP 87-91)<br>*Cost reduction and expenditures rationalization (ESDP 2007-2011)<br>*Combat of interventionism, favoritism, bribery and all other forms of corruption (ESDP 82-86)<br>*Modern, efficient, transparent public administration (ESDP 82-86)<br>*Definition of responsibilities for more transparency (ESDP 2002-2006)<br>*The promotion of e-government is of great importance given to the role of ICT in achieving the goals, with transparency of the administrative services and better coordination (ESDP 2007-2011)<br>*The establishment of an open government through the adoption of open government principles by PA, the adoption of a transparency based system, the citizen participation in the definition of public policies and the prevention of corruption (ESDP 2016-2020). | The sub dimension equity has been advanced since the sixth ESDP (1982-1986). There is no indication that this dimension will be attained with ICTs.<br>There is a focus on the need to modernize the human resources management, to acquire the specialized skills in ICTs and to train the users to be able to use digital technologies.<br>The idea of citizen participation is not recent. It is more based on digital technologies.<br>Indicators of environmental sustainability are present. However, there is no indication that they will be achieved by ICTs and e-government.<br>The Tunisian PA seeks to combat corruption and enhance transparency. The role of e-government in improving transparency is also highlighted. |

# Chapter 7
# The Role of IoT in Digital Governance

**Azeem Khan**
*Taylor's University, Malaysia*

**N. Z. Jhanjhi**
https://orcid.org/0000-0001-8116-4733
*Taylor's University, Malaysia*

**Mamoona Humayun**
*Jouf University, Saudi Arabia*

**Muneer Ahmad**
*University of Malaya, Malaysia*

## ABSTRACT

*The acronym IoT stands for internet of things. The IoT ecosystem can be envisioned as a set of physical electronic devices embedded with intelligence, connected through a network, enabling them to collect and exchange data, and allowing these devices to be sensed and handled remotely between the physical and cyber worlds. The devices connected through the internet has been influencing all walks of our life ranging from individual, societal, educational, industrial, entrepreneurial, and related to governance as well. As we are connected and surrounded with a plethora of connected smart devices, it seems there is a great risk of security and privacy in several aspects, such as device authentication, data theft, device manipulation, data falsification, etc., to name a few. Hence, the current chapter has been undertaken to explore and comprehend the security and privacy related implications, opportunities, future directions, and challenges involved in implementing digital governance with IoT.*

DOI: 10.4018/978-1-7998-1851-9.ch007

## INTRODUCTION

The origin of communication over Internet was perceived as human to human interaction but with the advent of IoT, the concept of communication over Internet has changed tremendously as now the communication encompasses from beings to things that is Human to Machines, Machine to Machine, and vice versa. With the advancement of IoT it seems that this technology is moving towards an automated society where everything and everyone will be connected (R. Khan, Khan, Zaheer, & Khan, 2012). IoT is the major technological revolution in recent human history which has changed the directions of computing comprising several physical objects with unique identity and pervasive connection. In this technological continuum everything around us like smart vehicles, smart gadgets, smart home appliances, smart wearables etc., are all interconnected to share their data with each other, as well as pass their data to the cloud. IoT ecosystem is expanding every moment based on the existing network environments with embedded intelligence. IoT is one of the fastest growing area in the history of computing with a projected increase of 50 billion devices by the end of 2020, which is 80 times greater compared to the population growth as illustrated in Figure 1.

Digital Governance DG or Electronic-Governance EG is projected as the digital presence of an organization encompassing its means of communications with the external world, these means of communication can be websites, blogs, social media channels, portals and all other internet related products and services. Thus, DG is defined as the use of Information and Communications Technology ICT to exchange, integrate and enable communication to dissipate information involving numerous standalone systems and services between government and businesses, government and citizens, and several other back office processes and interactions with in the entire governments operational framework. With the advent and advancement of ICT many manual processes with in the government sector have been automated to improve the quality of service in public sector to name a few are registration of documents online, online payment provisions for various bills, taxes, online registrations of assets belonging to individuals and business organizations, etc. thereby enabling governments to function smoothly with pace and efficiency for the public welfare. ICT now has extended its horizon with an extension of IoT which has provisioned automation for several systems that are facilitating civil life to a large extent to name a few of IoT enabled DG systems are: parking systems allowing public know the parking updates,

*Figure 1. Projected growth of IoT*
*Source: (Zahrah A. Almusaylim & Zaman, 2018)*

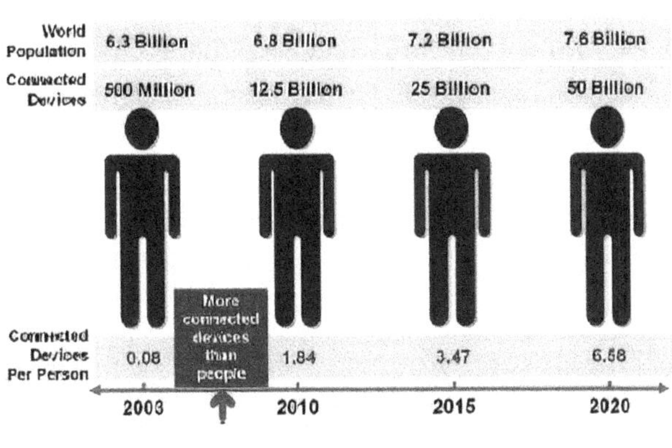

weather systems to update on weather, traffic control systems to update citizens with the current status of the traffic which are allowing people to plan their travel, security checks at airports, electricity billing systems, automated electronic passports, etc., that are being helpful to manage and administer all these automated systems. IoT in DG is playing a very vital role in improving the quality of life. Hence, to improve and facilitate the systems governments all over the world are investing and willing to invest for developing secure, effective and efficient systems that could facilitate public welfare and smooth governance of the state. Automated systems created by IoT has not only influenced individuals and industries but also, they have greatly impacted the whole governance processes as well globally.

IoT expediates electronic governance by connecting physical objects with a unique IP address forming a network of physical objects which are embedded with intelligence and benefiting organizations for connected with governance to function smoothly thereby enabling digital governance mediated with technology comprising relationship between citizens and governments, development of policy, communication and expression of public will. IoT in DG also provides enough quality data which can be used by organizations to improve their decision making process which in turn can help them to grow with the pace required by the markets. (Brous & Janssen, 2015)

Thus from the above discussion, it is obvious that IoT in DG encompasses whole arena of our life ranging from individual, societal, industrial, educational, entrepreneurial, entertainment, automotive, cloud services etc., which helps in automating the whole governing process digitally thereby enabling the whole system to function smoothly with the help of IoT devices connected through the seamless network of smart gadgets and tightens together several areas to be addressed specifically security(K. Hussain, N. Jhanjhi, H. Mati-ur-Rahman, J. Hussain, & M. H. Islam, 2019), connectivity, efficient energy consumption, etc., IoT in DG requires energy efficiency, security, interoperability with software applications. Thus, it can be comprehended that IoT provides a rich ground for collaborative, interdisciplinary research. From security and privacy perspective, IoT in DG is perceptible to several security attacks to name a few are ransomware, phishing and identity theft(Kok, Abdullah, JhanJhi, & Supramaniam, 2019). Security, privacy and safety are the major concerns in this whole IoT in DG landscape encompassing Smart Transportation, Smart City, Smart House, Smart Phone, Smart School, Smart Healthcare, Smart Industries, Smart Agriculture, etc. As depicted in Fig 2.

Figure 2 gives us a general overview of IoT DG topography which relates to the real-world objects, and those objects are embedded with intelligence in the whole system enabling whole system with an ability to make decisions smartly. The whole scenario though seems to be smartly connected, but there are many threats and vulnerabilities which compel us to reflect upon how secure are the systems in this whole spectrum of interconnected smart devices(Zahrah A Almusaylim & Jhanjhi, 2019). To understand these underlying security intricacies at the very outset we need to address the fundamental question about security and privacy in IoT DG. IoT DG Security addresses the issues such as: privacy, secure communication over existing infrastructure, safety in terms of data for every connected user and to ensure uninterrupted seamless delivery of services stated by the established IoT DG ecosystem.

The subsequent sections are further divided into main sections and their sub sections discussing the research work done so far through exhaustive literature review highlighting taxonomy, opportunities, challenges, future directions, critical issues with a few case studies that are already implement with IoT in digital governance.

*Figure 2. IoT Landscape*
*Source: (Al-Fuqaha, Guizani, Mohammadi, Aledhari, & Ayyash)*

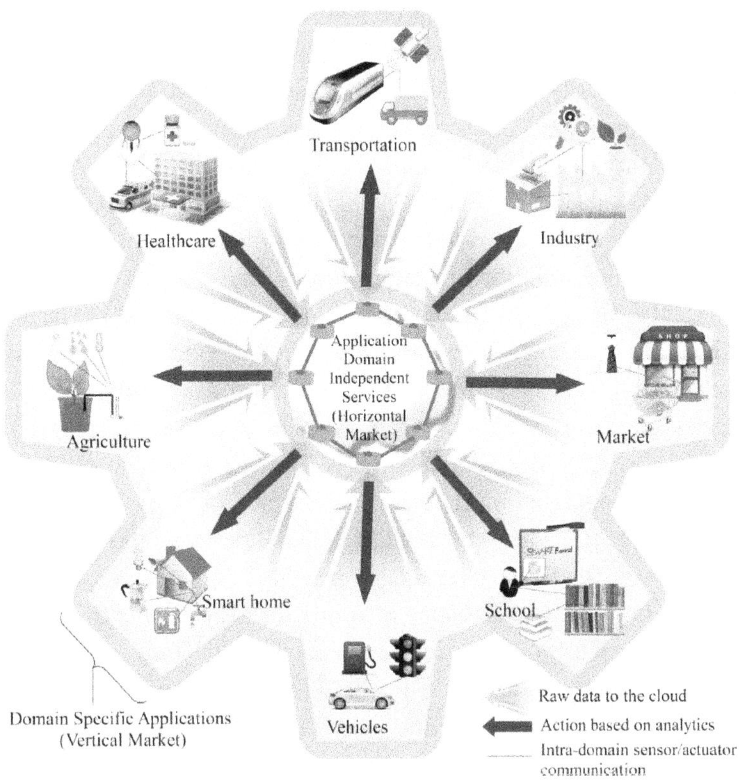


## LITERATURE REVIEW

Earlier only computers were connected for the transmission of data but now we see with the advent of IoT not only desktop computers but also many smart physical things such as smart vehicles, smart cameras, smart home appliances, and industrial equipment etc., are now connected to a seamless network of networks. Within few years it is expected the scale may increase to 20 billion devices currently it is said that we already have 6 billion devices. With all these connected devices there is a serious concern about how secure they are in this whole connected Internet of Things ecosystem. During recent years it was observed that numerous security cameras were attacked with DDoS to hamper the security and safety over the internet, which caused serious concerns over social media(Naik & Maral, 2017). As perceived from general perspective Internet of Things is not just related to software instead it comprises a whole system of intelligently connected physical devices. This whole ecosystem of hardware, software, cloud infrastructures are prone to several security and privacy issues. This whole ecosystem though seems to be serene and aesthetic superficially in fact they are still in their nascent stage and are inclined towards many security loopholes and are in need to address them with the optimized security solutions (Almulhim & Zaman, 2018; Maria Almulhim, 2019). The following synopsis gives us a brief overview of the research work carried by various researchers in this area with their significant contributions.

Alaba et al (Alaba et al., 2017) discussed current threats and security issues prevailing in an entire IoT Landscape. They elaborated their discussion on security taxonomy of IoT from layered perspective with an emphasis on Architecture, Communication and Application's context. Rizvi et al (Rizvi, Kurtz, Pfeffer, & Rizvi, 2018) presented latest trends in IoT by recognizing critical domains, the security necessities, challenges and security solutions currently faced by IoT. Secure Communication between IoT devices and servers is considered very crucial over a network, to accomplish this security concern single password authentication mechanism is implemented which is vulnerable to many security threats to address this issue Shah & Venketesan (Shah & Venkatesan, 2018) presented a mechanism for mutual authentication between IoT system and cloud. Apart from many security threats the primary threats creeping into IoT ecosystem are tampering of data, cloning of devices connected to IoT, unauthorized control of device, DoS, DDoS, exposure to sensitive data etc., Naik & Vikas (Naik & Maral, 2017) addressed two among those aforesaid security issues namely device cloning and exposure to sensitive data. As the devices connected to IoT ecosystem are heterogeneous and are under continuously vulnerable to security attacks that may sabotage the network in terms of confidentiality, integrity and authenticity of citizens to address this Sridhar & Smys(Sridhar & Smys, 2017) proposed Intelligent security framework with double and mutual authentication mechanisms as depicted in Fig 4., these mechanisms can lessen the traffic over the network resulting improvement in terms of performance and decreased consumption of bandwidth. Das et al (Das, Zeadally, & He, 2018) provided an extensive overview on IoT protocols encompassing authentication, key management and securing the privacy of the citizens. Implementation of IoT applications is simplified using IoT frameworks which are defined as a set of protocols, standards and a set of guiding rules. Ammar et al (Ammar, Russello, & Crispo, 2018) surveyed a total of eight IoT frameworks and elucidated their architecture, hardware compatibility, security characteristics and essentials of developing third-party smart applications.

Stergiou et al (Stergiou, Psannis, Gupta, & Ishibashi, 2018) discussed three different systems and integration of these systems. Which comprises of cloud computing, IoT and Big Data. They proposed installation of a security wall in between internet and cloud server for a better security management. Tiwari & Gupta (Tewari & Gupta, 2018) discussed security concerns between IoT and Traditional net-

*Figure 3. Taxonomy of IoT security*
*Source: (Alaba, Othman, Hashem, & Alotaibi, 2017)*

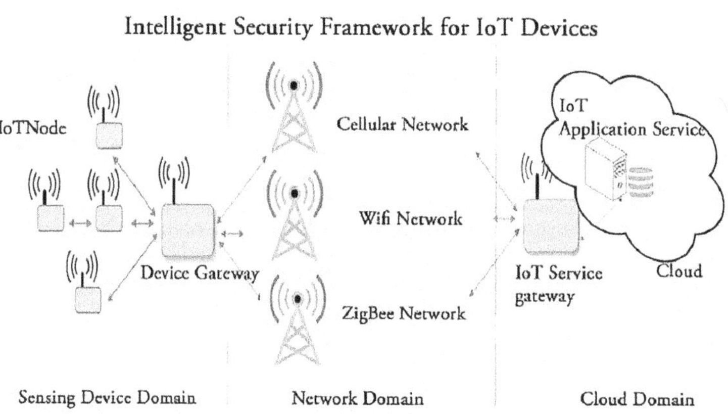

*Figure 4. Intelligent security framework*
*Source: (Sridhar & Smys, 2017)*

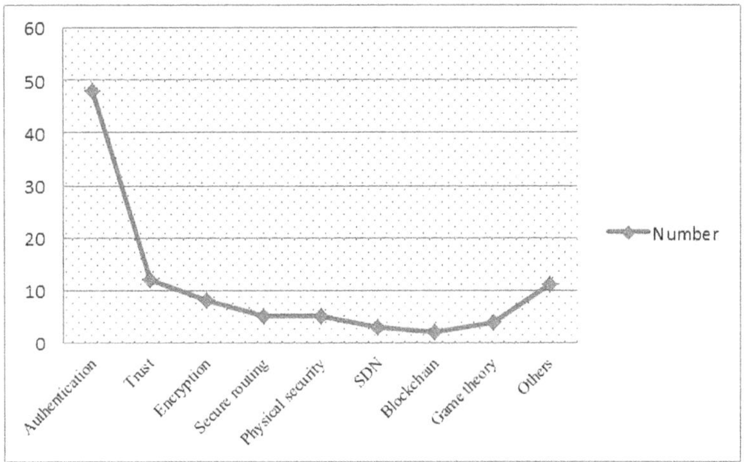


work. Mardiana & Haslina (Hassan, 2018) provided an exhaustive analysis of IoT Security in the recent years ranging from 2016 until 2018, there exhaustive findings are depicted in Figure 5.

Mavropoulos et al (Mavropoulos, Mouratidis, Fish, & Panaousis, 2018) discussed analysis of security mechanism for smart public transport system using APPARATUS which is a framework for security and it also facilitates the analysis of IoT systems from security perspective. Sha et al (Sha, Wei, Yang, Wang, & Shi, 2018) discussed security challenges and summarized three architectural designs which could pave the way as a reference for further research for security protocols and designing of algorithms to address those challenges. Mocrii et al (Mocrii, Chen, & Musilek, 2018) presented a review an IoT based smart houses and discussed extensively their architecture, the software being used, communications technology and last but not least they also addressed issue of privacy and security in IoT based smart homes. Martino et al (Di Martino et al., 2018) discussed a survey on IoT reference architectures, their security and their interoperability. Banerjee et al(Banerjee, Lee, & Choo, 2018) presented a survey on IoT security solutions ranging from year 2016. They also proposed block chain technology to share datasets of IoT among relevant stakeholders. Chifor et al (Chifor, Bica, Patriciu, & Pop, 2018) presented authentication mechanism with FastID Online protocol for citizens to authenticate their devices. They also proposed an authorization stack which is a light weight protocol for smart home IoT applications. Bu et al (Bu, Isakov, & Kinsy, 2018) presented Threshold Security Sharing TSS mechanism acronymized as TSS. TSS splits the information to be exchanged into chunks in whole IoT landscape and then is shared by all the devices in the system. By doing this it ensures that the information being exchanged is reliable, robust and secure. Bhoyar et al (Bhoyar, Sahare, Dhok, & Deshmukh, 2018) presented a comprehensive review on security challenges over communication technologies with in IoT. They discussed at length the comparison within IoT. They elucidated elaborately wireless technology at three different layers namely physical, media access control and network layer. Chakib Bekara (Bekara, 2014) investigated the security issues such as Identity Spoofing, eavesdropping, data tampering, authorization, etc., and challenges such as scalability, Mobility, deployment, latency, etc., for an IoT based Smart Grid SG. Kouicem et al (Kouicem, Bouabdallah, & Lakhlef, 2018) presented a comprehensive top down survey with an emphasis on Software Defined Networks(SDN) and Blockchain Technology. Hellaoui et al

*Figure 5. IoT security publications 2016 – 2018*
*Source: (Hassan, 2018)*

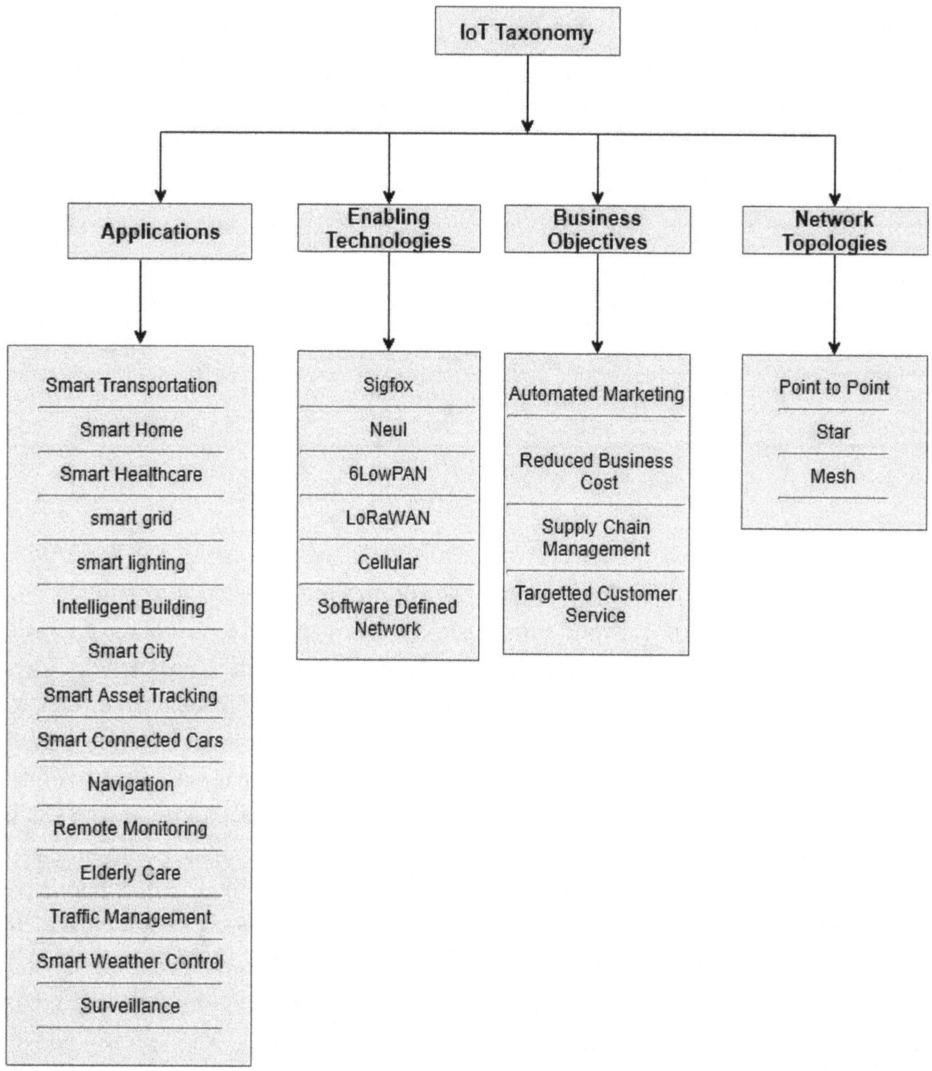


(Hellaoui, Koudil, & Bouabdallah, 2017) presented survey on energy-efficiency used in IoT security services. They also proposed taxonomy for energy saving mechanisms in security protocols. Khan et al (N. Khan, Sakib, Jerin, Quader, & Chakrabarty, 2017) presented analysis of several algorithms with their performances that can be implemented on IoT devices, they also discussed some critical cases and comparative analysis for better understanding. Their study also gave an overview of several algorithms implemented in IoT. The results presented can help researchers to better understand and design of IoT. Mathur et al (Mathur et al., 2017) presented security solutions that can safeguard IoT from DoS and unauthorized access. Sadique et al (Sadique, Rahmani, & Johannesson, 2018) discussed security challenges in IoT from applications perspective. Alem & Masud (Čolaković & Hadžialić, 2018) presented a comprehensive review on challenges, open research issues and enabling technologies in an IoT ecosystem.

Hanan (Elazhary, 2018) presented technologies related to the IoT such as the Internet of Nano Things IoNT, and the Internet of Underwater Things IoUT. Yang et al (Yang, Li, & Sun, 2018) investigated the characteristics of IoT devices in three layers of network layers namely, application, transport and network layer respectively, taking advantage of these characteristics and with the network protocols, based on neural network algorithms they created fingerprints for IoT devices. These device fingerprints were utilized to find out 15.3 million devices connected with network which paved the way for them to examine thoroughly distribution characteristics in cyberspace. Jose & VijayaLakshmi (Jose & Vijyalakshmi, 2018) emphasized their discussion on most significant IoT security issues and algorithms pertaining to them. Sfar et al (Sfar, Natalizio, Challal, & Chtourou, 2018) presented an overview of the IoT security roadmap and they presented a case to highlight the components and interactions of cognitive and systematic approach. Ray (Ray, 2016) presented an in-depth analysis about the existing IoT cloud service providers with their advantages and disadvantages with a greater detail. Subsequent section explores detailed IoT based applications for Digital Governance.

## DETAILED TAXONOMY OF IOT BASED APPLICATIONS IN DG

This section discusses the detailed taxonomy of IoT in DG comprising technologies that enable IoT to function, describing its applications in various domains, various categories of network topologies involved which are enabling communications among these devices and lastly objectives of IoT which are to be attained by the businesses with the implementation of IoT in DG. Following below are the explanations for the detailed taxonomy of IoT in DG as illustrated in Fig 6.

*Figure 6. Taxonomy of internet of things*

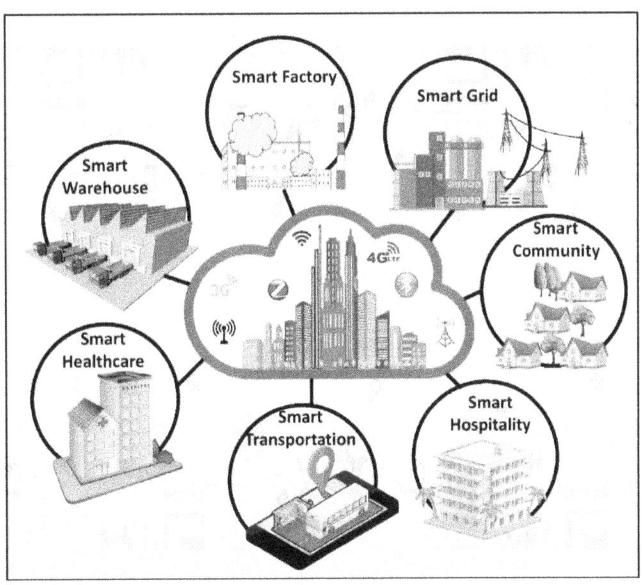

## APPLICATIONS

There are numerous applications of IoT in DG, succinctly they can summarized as smart home, smart transportation, smart healthcare, smart grid, smart lighting, intelligent building, smart city, smart asset tracking, smart connected cars, navigation, remote monitoring, elderly care, traffic management, smart weather control and surveillance. Smart transportation helps in automation of whole logistics from the carrying goods from source to delivering at the destination enabling quality of service. Smart home enables citizens to monitor and control their homes remotely. Smart grid enabled with smart meters avails citizens to manage their consumption cost of energy with optimal usage of power. smart lighting helps governance of cities, saving power by consuming only the amount of power required, in this way all other applications such as smart asset tracking, connected cars, navigation, intelligent buildings, weather control, surveillance, elderly care and smart traffic management are enabling and facilitating citizens by improving the quality of life.

## ENABLING TECHNOLOGIES

For the aforesaid applications, the enabling technologies are: Sigfox, neul, 6LowPAN, LoRaWAN, Cellular and Software Defined Networks(SDN). As the whole IoT landscape is comprised of heterogenous connected smart devices,to enable communication between these interconnected devices different communication and networking technologies such as low-power personal area network (6LowPAN), Neul, Sigfox, low-range wireless area network (LoRaWAN), cellular and Software Defined Networks(SDN) are used. Low-power objects such as electricity meters and smartwatches are connected through Sigfox. Neul is weightless wide range networking technology. Emerging networking technology to route network traffic intelligently and to eliminate bottlenecks without larger strain on network is through software defined networks(SDN). LoRaWAN is designed to target wide area networks and supports low-cost mobile bidirectional communication in IoT with strengthened security. cellular (GSM/3G/4G) communication capabilities are used to support the long-distance operations of IoT applications. It is considered as ideal for the sensor-based low-bandwidth-data projects. 6LowPAN is an IP-based network protocol that defines new encapsulation and header compression mechanisms. It can be used in multiple communications platforms such as Wi-Fi, 802.15.4, and sub-1GHz ISM.

## BUSINESS OBJECTIVES

Aforesaid applications with the above discussed technologies facilitate the following business objectives: Automated marketing, reduced business cost, supply chain management targeted customer service. IoT can benefit businesses by making whole process of marketing automated thereby reducing business costs. It also facilitates ease in supply chain management and also by helping in knowing the targeting customers for businesses.

## NETWORK TOPOLOGIES

IoT network topologies are categorized into three types namely point to point, mesh and star. Direct connection is established between two nodes in point-to-point, all the devices in a network are connected through a centralized hub in star topology, lastly all nodes are connected to each other in a mesh topology. The capability of these topologies is estimated through following attributes in orderly mentioned: latency, throughput, scalability, range and fault resiliency.

## IOT OPPORTUNITIES IN DIGITAL GOVERNANCE

The application of Information & Communication Technology(ICT) in governance is transforming the relationship between citizens, businesses, various government departments are enabling the implementation of e-governance possible and are facilitating in making whole operational process of governance operational process efficient and speedy. e-governance intends to improve the quality of life for its citizens, availing local economy, transport, traffic management, environment and interaction with the government. IoT opportunities in DG are enormous to name a few are in the following domains: smart health care, smart economy, smart homes, smart cities, traffic management, smart lighting, smart wastage management, smart water management in urban areas, smart parking in cities, disaster management in case of emergencies, etc., These can be a means of delivering values to citizens by securing privacy.

As there are many opportunities for citizens residing in towns and villages in cities comparatively many are migrating to cities thereby increasing the population in cities enormously. According to a study 3% of the earth's land area is occupied by cities and they are consuming 75% of natural resources, producing 60 – 80% of global greenhouse gas emissions. It seems that in near future there will be considerable increase in urbanization and their impact on environment will also increase considerably as of today their impact is 54% which may increase to 66% by 2050. (O'Dwyer, Pan, Acha, & Shah, 2019)

To accommodate this human migration flooded towards cities, and to make city administration more efficient and manageable it is incumbent and inevitable for the city administrators to take help of ICT infrastructure to make cities smart, neat, clean and green. To achieve this objective IoT plays a vital role in enhancing the ICT infrastructure making the cities smart by facilitating the governments in many ways to name a few: waste management, avoiding traffic congestion, addressing pollution level which is rampant in almost all cities of the world, smart parking, smart homes, etc. which would be hassle to manage using traditional systems. Smart cities enable citizens with updates on traffic at their fingertips enabling them to schedule their travels avoiding traffic congestion, updating weather conditions, pollution levels, automating public transport system by the improving better transport facilities through smart transit thereby enabling the whole system of administration in cities for smooth governance.

Apart from other opportunities, data that is generated by plethora of these connected devices is also a big opportunity that can be used in a better way by implementing data analytics and data mining so that data can be a useful tool in making intelligent decision that can benefit citizens at large. To accommodate this opportunity there is a need to come up with new techniques to address this issue as the traditional ways of analyzing the data being generated is obsolete now.

*Figure 7. Smart city IoT application domains*
*Source: (Curzon, Almehmadi, & El-Khatib, 2019)*

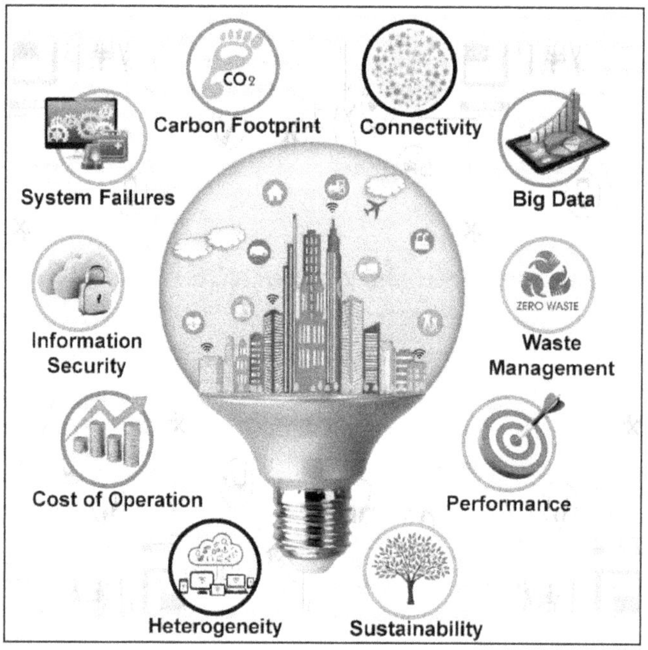


The sole objective of smart cities is to enable communication between physical devices connected through ICT, social media and business infrastructure to improve the overall intelligence in city operations to avail better decision making. One of the opportunity of IoT is in smart governance of Smart Cities which utilizes Wireless Sensor Networks for applications such as data collection for city metrics on pollution or noise levels around the city, managing waste collection and controlling street lights [50].

## CHALLENGES LINKED WITH IOT FOR DIGITAL GOVERNANCE

Though there are many opportunities (Maher Omar Alshammari, 2017) with implementation of IoT in DG, there are numerous inherent challenges associated with their implementation, to name a few they can be summarized as: cost, heterogeneity, security and privacy, storage, technology, standards, regulatory standards, scalability and sustainability. (Silva, Khan, & Han, 2018)

## COST

As there is growing interest by many stakeholders in implementing IoT in many facets of life. Many Governments globally are very keen towards putting IoT in DG due to its numerous benefits for the welfare of the citizens in many aspects. In implementing IoT in DG there are many challenges involved first and foremost challenge among them is the matter of cost. The cost that is incurred in implementing IoT systems across the places that may yield good profits for the governments in terms of productivity and can contribute to the increase in overall GDP of the country. Cost which is an essential ingredient in implementing this whole automated systems can be further classified into three types viz., design, implementation and operational cost. Design cost is the cost that is incurred to design the whole infrastructure of IoT in digital governance, implementation cost is the cost that is incurred to implement the whole IoT infrastructure that comprise whole plethora of hardware and its installations apart from it the other cost which is operational cost is the cost that is incurred once the whole system goes live and put to practice. Hence any organization which is inclined towards implementation of IoT in governance should consider this challenge and come up with strategies to that can help in optimizing the cost at all levels.

## HETEROGENEITY

The second significant challenge which cannot be undermined is the heterogenous nature of the devices that are involved in this whole network. As there are different vendors supplying devices for this arena of physical devices, compatibility among them is one of the issue and a great challenge which if not addressed can lead to chaos of the whole system. As the devices connected with the plethora of IoT infrastructure are comprised of multiple vendors, hence integrating them at the application level and proper functioning of these interconnected devices is another challenge which needs to be addressed.

## SECURITY AND PRIVACY

Though there are many security mechanisms (Malak Alamri, 2019) implemented at both infrastructure and information level(K. Hussain, N. Z. Jhanjhi, H. Mati-ur-Rahman, J. Hussain, & M. H. Islam, 2019), but they are not addressing the problems that are in rise hence there is a need to address this challenge as every now and then there are security breaches that are reported at large. In the past there has been reported such as hacking of drug infusion pumps, cameras, smart fridges, baby monitors, thermostats, and even the radio in a car that are signifying a security nightmare being caused by the future of IoT and this has been a concern that has drawn the attention of prominent tech firms and government agencies across the world. As in the near future IoT will become an inseparable part of our lives. Security concerns will not be restricted on safeguarding sensitive information and assets. But our very own lives and health can become a target of IoT hack attacks(S. J. Hussain et al., 2019).

## CONNECTIVITY

Uninterrupted connection of internet to the devices and networks will be one of the biggest challenges for the future of IoT in DG, currently centralized, server/client paradigm is in use to authenticate, authorize and connect different nodes in a network, which is not sufficient for future IoT infrastructures and ecosystems involved in the whole communication process, because every now and then there are thousands of devices are that are being connected to this seamless network. When these connected device networks grow to join billions and hundreds of billions of devices, centralized systems will turn into a bottleneck thereby requiring huge investments and expenditure in maintaining IT hardware including cloud servers that can handle such large amounts of data enabling information exchange, and entire systems connected can go down if the server becomes unavailable.

Though there are many technologies that are availing communication between the devices but still there is a problem of connectivity among the devices especially when these devices are connected to the vehicles as this could involve the communication during motion which, if is not maintained stable then there could be situation which can create hassle for the entire system as this produces communication lag which can result in time delay that can affect the business to a large extent incurring losses hence there is a need to address this challenge as well.

## STORAGE

As the data that is being generated by these zillions of devices connected together are producing tons and tons of bytes of data which needs to be stored, apart from storing this data there is need to how to make this data useful so that it can benefit the citizens at large; for instance data being generated by the devices connected to automobiles, when this devices are in motion it means that there are no traffic hurdles; but if these vehicles are stand still or moving with less speeds connotes the traffic situation on roads; this data being sent to cloud via cloud servers enable citizens connected to this seamless network informed about the traffic situations at various places in cities connected to the network thereby enables the citizens of that region to plan their travel efficiently, apart from this disaster management can also be made efficient with the data being generated by this devices by creating applications which can avail

*Figure 8. Challenges associated with IoT (Silva et al., 2018)*

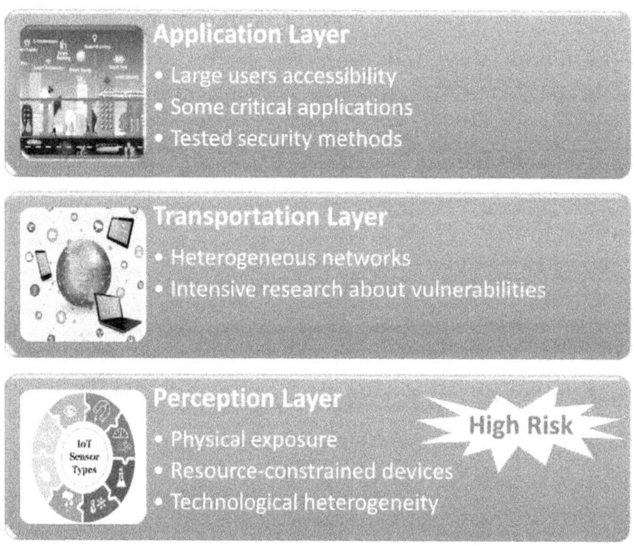

disaster management in times of emergencies which can be related to health of financial and monetary matters, thus apart from aforesaid many other domains can also be explored. Hence apart from new storage mechanisms, there is a need to develop systems that can use the data being generated as well as proper storage mechanisms to store this data.

## TECHNOLOGY

As there are many parties involved into the development of the systems that work collaboratively to make the whole IoT infrastructure work efficiently. There are times where we can experience system failures this can due to problems with in the devices connected to IoT network or it can be at the application level. Hence failure of systems either at device level of application level can be considered due to the failure of technology hence proper functioning of technology is also a challenge as the failure of the systems connected with these technologies if they don't function well, then they bring whole businesses connected to these whole IoT landscape falls under risk. Hence we can conclude that this challenge also needs to be addressed for the proper implementation and functioning of IoT in digital governance. (Banafa, March 14, 2017; Kim, Ramos, & Mohammed, 2017)

## STANDARDS

Standards include network protocols, communication protocols, and data-aggregation standards which needs to be addressed as the growing networks will make the existing protocols and data-integration standards obsolete hence, there is a need to address all these activities of handling, processing and storing the data collected from the sensors connected to this seamless network of devices. Without proper standards this whole plethora of devices cannot function properly, hence maintaining proper standards within different stakeholders involved in this whole system to work is itself a being challenge which in turn requires enormous efforts and interest among different groups and parties involved in this whole spectrum.

## REGULATORY STANDARDS

There is also a need for Regulatory standards at government level, to control data markets which are missing at a bigger picture of this IoT landscape in DG, especially As there are numerous stakeholders involved in the whole network of IoT devices, which in turn are creating enormous of data, as the data is now treated as the new oil that is discovered which is used by third parties which may be recognized or unrecognized, these parties may take the data without the consent of the owner and may misuse it or sell to other interested parties which can become a threat to the individual or society at large, hence if proper security checks are not implemented at various levels of IoT device regulation then there are chances that this whole system may go futile. Hence there should be proper regulation by the government agencies with clear procedures for products and services to preserve data, to properly use the data with the citizens consent, only by this way there can be data consistency and data preservation of an individual, society thereby whole governing process. if this is implemented properly at government level then the whole system will work properly resulting the whole system to be a boon rather than bane. (Banafa, March 14, 2017)

## SCALABILITY

As the network of the physical devices are increasing considerably, there is a dire need to accommodate these devices to the existing networks. It is a significant factor that facilitates the devices connected to the IoT landscape with the flexibility to meet the changing demands as well as environmental conditions which contributes to efficiency, competitiveness and quality. The significance of scalability underlies in productive utilization of available resources. Scalability comprises of networks, businesses, software and hardware.

## SUSTAINABILITY

If we consider whole IoT implemented system, it may involve cost at various levels at design level, implementation level, operational level and maintenance level. For instance if we consider the infrastructure of smart cities there we see that this system works fine but if we look at the income that this whole plethora of devices are generating if they are unable to cope up with the maintenance of the whole system in turn if there are overhead expenses on the authorities then it will be a great bottleneck for the whole system to sustain hence there is need to address this challenge and efforts should be made to develop a system that can sustain on its own apart from creating overhead expenses at different levels, if it occurs frequently then it may create a situation where the whole system may cease to function resulting in chaos. Hence, we can conclude this is one of the significant challenge that needs to be addressed in implementation of IoT in DG.

## FUTURE DIRECTIONS OF IOT FOR DIGITAL GOVERNANCE

The implementation of IoT in DG in future depends on decentralizing IoT networks, Fragment of its tasks can be moved to the edge and fog computing models wherein smart devices connected to IoT hubs can take charge of mission-critical operations and cloud servers can take responsibility of gathering data and data-analytics. The rapid growth of IoT based devices in DG brings numerous benefits to businesses and consumers. Those benefits can take different forms for citizens, businesses and for governments. Consumers can travel more efficiently by avoiding traffic jams based on traffic reported by other vehicles connected to these plethora of devices into a seamless network. They also help consumers to save money by reducing energy usage and insurance premiums. Consumers can be healthier, safer and more independent by using wearable devices that provide feedback on health especially for elderly people living at homes they facilitate with remote monitoring of their health and safety. by studying how customers behave; Businesses can provide better products and services; businesses can also discover needs for

*Figure 9. IoT system risks*
*Source: (Frustaci, Pace, Aloi, & Fortino, 2018)*

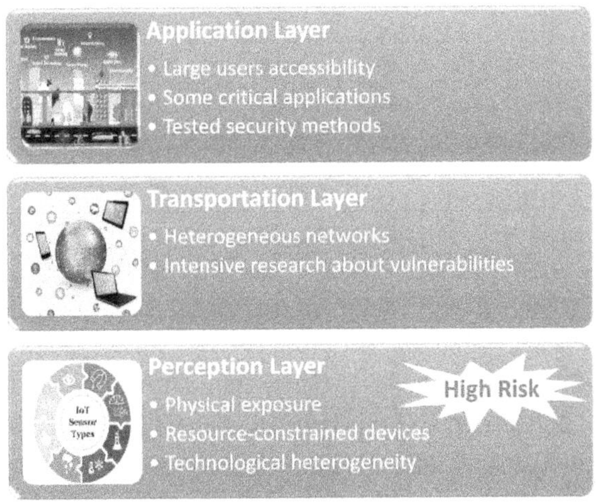


new products or services thereby increasing their revenues which in turn can help the nations to create more employment opportunities thereby reducing the problem of unemployment for citizens residing in a country. long-term health care costs can be reduced for citizens and governments to invest less to maintain and manage the whole health sector with better remote support for the elderly in their own homes. Road safety can be improved based on data gathered from thousands of drivers. The efficiency of street lighting can be improved by dimming lights on empty roads (Pundir, Sharma, & Singh, 2016).

## CRITICAL ISSUES OF IOT IN DG

As depicted in figure 11. It is quite obvious that among the layers of IoT, the lower level which is called the perception layer is at the highest security risk level because of the deployment of physical devices and their exposure to open environments apart from it, it also has a very large hardware limitations due to technological heterogeneity that limit the implementation of effective security measures (Hussain, Hussain, Jhanjhi, & Humayun, 2019). Transportation layer compared to perception layer is at the lower risk level due to the known drawbacks of standard wireless data transfer technologies, as well as known threats in access networks. Lastly the application layer has an unpredictable level of risk based on the implemented application. In addition, compared to the perception layer, application layer has more mature technology, less threats, and tested security methods.

## CASE STUDIES

This section briefly discusses a summary of the case study carried out by researchers in irrigation domain.

Table 1 exhibits a case study carried out in Rayalaseema of Andhra Pradesh, India, is an analysis conducted on the production of Rice, Wheat, Cotton and Ground nut crops in an area of 10 cents each. Half of this experiment was carried out using IoT controlled cultivation and the other half without IoT but

*Table 1. Comparison of various crops using manual and IoT system*

| Crop | Watering | | Usage of pest | | Soil strength | | Informing authorities | | Production | |
|---|---|---|---|---|---|---|---|---|---|---|
| | Manual | IoT | Manual | IoT | Manual | IoT | Manual | IoT | Manual | IoT |
| Rice | 4.03 | 6.45 | 3.33 | 6.58 | 4.78 | 7.32 | 5.43 | 7.32 | 6.43 | 7.85 |
| Wheat | 4.82 | 7.02 | 3.78 | 6.89 | 5.14 | 8.33 | 3.69 | 5.87 | 5.77 | 7.96 |
| Cotton | 2.78 | 4.92 | 4.52 | 7.05 | 5.89 | 6.99 | 4.27 | 5.98 | 6.98 | 8.32 |
| Ground nut | 2.52 | 4.28 | 3.96 | 5.43 | 4.22 | 7.38 | 4.21 | 6.87 | 5.96 | 7.02 |

Source: (Kumar, 2017)

from the results it was found that the former was better than latter. The data collected was concentrated on five constraints namely: Moisture in soil with proper watering, Pest controlling, Strength of the soil which constitutes minerals and salts of the soil, reporting concerned authorities timely about disorders found in the crops and lastly the harvest of the crop. Sensors were used to record the proportion of soil moisture and humidity by generic water flow. vibration and pressure sensors were used to record pest control. magnetic, pressure sensors were used to record the minerals and salts of the soil. programming was done using Embedded Development Kits EDK and the timely reporting was done through Global System for Mobile Communication GSM to the respective authorities. Apart from the aforesaid sensors, weight detecting sensors were also used in collecting the data.

This clearly indicated the proved advantages with and without the implementation of IoT in agricultural sector. Apart from the case study, the results proved that this approach can be further extended to other business sectors enabling increase in production thereby increasing the economy of the nation with an environment friendly approach. (Kumar, 2017)

## DISCUSSION

From the above sections, it is obvious that IoT in DG has many advantages helping the citizens with a quality of life with an environment friendly approach. IoT in DG has many benefits ranging from automating manual clerical work with online documentation which enables citizens to save a lot time and cost, automating machines with driver less cars, automating health checkups with smart wearables, automating the whole governing process with includes administration at both regional and federal levels.

As there is a rapid growth of urbanization there is a huge shift of citizens from rural are to urban areas thereby increasing the population in cities creating a challenge for the governance of these cities, hence this problems seems to be minimizing gradually with the implementation of technology in cities resulting smart cities which encompasses the whole city infrastructure and resources automated helping citizens to travel smoothly with smart transit, smart transportation, proper control over homes is made available with smart homes. Elderly people can be monitored remotely if they are under medical supervision their health records are updated on clouds via smart wearables thereby enabling efficient health management with the help of smart health devices. Buildings are also automated with the concept of intelligent buildings, business are getting great benefit enable them to grow and enjoy more benefits in terms of profits with the automation of whole supply chain management from source of acquiring raw materials to the end till the goods are delivered to customers at home with the help of smart tracking systems which are helping in controlling the inventories and managing them efficiently for better profits. Apart from it IoT is also used in various public sectors to increase the productivity of crops, efficient usage of resources available for the organizations, in power sector with the advent of smart meters power consumption have been minimized considerably due to on time feedback given by the systems connected with IoT involved in DG. IoT in DG is also being helpful in city surveillance.

It is clear that Internet of Things in DG aims to integrate seamlessly both physical and digital worlds in one single ecosystem that makes up a new intelligent era of Internet helping the governments in improving the quality of life of the citizens with a good participation of efficient governance in terms of interaction between Government to Citizens G2C, Government to Businesses G2B and Governments to Governments G2G within the federal and state governments of a nation. From security aspect of IoT in DG, from figure 5 and 6 it is evident that though there is considerable research work being carried out on IoT security issues in DG but still it is not enough and it is not able to cater the need of exponentially growing IoT landscape in DG leaving a huge gap between the growth and creeping security issues. From Figure 5 it is obvious that the amount of research work focused from security perspective is mostly on Authentication only and other areas of security such as Trust, Encryption, Secure mining, Physical Security, SDN, Blockchain, Game theory and others are still in their nascent stage and are in need to address them abundantly. From Figure 6 it is obvious that under Authentication security mechanism only Lightweight Authentication mechanism has gained much attention by the researchers leaving behind other security mechanisms providing a huge vacuum to address. To understand this and to gain an in-depth acquaintance with these intricate issues an effort has been made to encompass IoT, their infrastructure, security concerns in this whole plethora of connected physical devices and networks. This research also aims to elaborate and elucidate researchers regarding different possible IoT security issues, leaving room for many unanswered questions which still needs attention by the researchers.

Though there are numerous advantages and benefits with IoT, to state emphatically IoT is a boon for the governments as it helps in automation of the systems that were running on traditional information systems. There is no doubt that these new systems running on IoT are facilitating the whole process of governance there is still dearth of proper security issues that needs to be addressed to name a few security concerns and data theft as the data is being sold by third parties without the consent of the owner of the data. This has created a great concern for the individuals as well as organizations working on automated systems automated by IoT in DG to name a few such security issues are ransomware, phishing and identity theft. All these issues were addressed, and specific solutions have been discussed above. Thus, we can conclude though there are many advantages associated with IoT in DG there are also underlying threats with which the whole stakeholders associated with this network are to be aware and take wise and conscious decisions which can contribute to the wellbeing of citizens and for the purpose of efficient governance as well.

## CONCLUSION

Applications of IoT in DG encompasses several areas, which can be broadly classified as commercial, infrastructural, industrial, pertaining to consumers, and also related to managing, administering and coordinating various units involved for the smooth running of a state with digital governance. The application of IoT for carrying out services within the entire Government framework includes National

Defense, managing smart cities, building an ecosystem for water supply, city planning, and control. In present era of technological advancement IoT has been regarded as key potential area gaining much attention by researchers from multiple disciplines. Since it is growing rapidly and is witnessing exponential growth, influencing all walks of our life. There has been rapid increase in IoT landscape, and we have witnessed tremendous change in the whole computing paradigm as it is evident from the history of computing. Though the paradigm of communication has been changing availing space for many more physical devices to connect over the internet, there is still space to find answers related to security as the whole IoT ecosystem is continuously growing; simultaneously it is prone towards several security threats and attacks which are evident from literature. These attacks may be harmful and may sabotage the whole network, making it futile. Thus, this research undertaking aims to present an outline with the present-day research trends from safety and security perspective, which can provide base for future researchers to proceed further.

## FUTURE WORK

From the above discussions it is evident that as the future needs predicts more human dependencies on machines, hence there is need for the governments and regulatory bodies to come up with policies and cyber laws with its strict implementations for future IoT framework in DG which is expected to function effectively, if it is not handled properly then whole technology which is a boon for the whole human society may turn into a bane, thereby causing havoc in the whole IoT landscape. Hence it is the responsibility of all stakeholders to participate actively and responsibly to protect whole Technological Landscape.

## REFERENCES

Al-Fuqaha, A., Guizani, M., Mohammadi, M., Aledhari, M., & Ayyash, M. (2015, 4). Internet of things: A survey on enabling technologies, protocols, and applications. *IEEE Communications Surveys and Tutorials*, *17*(4), 2347–2376. doi:10.1109/COMST.2015.2444095

Alaba, F. A., Othman, M., Hashem, I. A. T., & Alotaibi, F. (2017). Internet of Things security: A survey. *Journal of Network and Computer Applications*, *88*, 10–28. doi:10.1016/j.jnca.2017.04.002

Almulhim, M., & Zaman, N. (2018). *Proposing secure and lightweight authentication scheme for IoT based E-health applications.* Paper presented at the 2018 20th International Conference on Advanced Communication Technology (ICACT).

Almusaylim, Z. A., & Jhanjhi, N. (2019). Comprehensive Review: Privacy Protection of User in Location-Aware Services of Mobile Cloud Computing. *Wireless Personal Communications*, 1–24.

Almusaylim, Z. A., & Zaman, N. (2018). A review on smart home present state and challenges: Linked to context-awareness internet of things (IoT). *Wireless Networks*. doi:10.100711276-018-1712-5

Ammar, M., Russello, G., & Crispo, B. (2018). Internet of Things: A survey on the security of IoT frameworks. *Journal of Information Security and Applications*, *38*, 8–27. doi:10.1016/j.jisa.2017.11.002

Banafa, A. (2017). *Three Major Challenges Facing IoT*. Retrieved from https://iot.ieee.org/newsletter/march-2017/three-major-challenges-facing-iot.html

Banerjee, M., Lee, J., & Choo, K.-K. R. (2018). A blockchain future for internet of things security: A position paper. *Digital Communications and Networks*, *4*(3), 149–160. doi:10.1016/j.dcan.2017.10.006

Bekara, C. (2014). Security Issues and Challenges for the IoT-based Smart Grid. *Procedia Computer Science*, *34*, 532–537. doi:10.1016/j.procs.2014.07.064

Bhoyar, P., Sahare, P., Dhok, S., & Deshmukh, R. (2018). Communication Technologies and Security Challenges for Internet of Things: A Comprehensive Review. *AEÜ. International Journal of Electronics and Communications*.

Brous, P., & Janssen, M. (2015). *Advancing e-Government Using the Internet of Things: A Systematic Review of Benefits*. doi:10.1007/978-3-319-22479-4_12

Bu, L., Isakov, M., & Kinsy, M. A. (2018). A secure and robust scheme for sharing confidential information in IoT systems. *Ad Hoc Networks*. doi:10.1016/j.adhoc.2018.09.007

Chifor, B.-C., Bica, I., Patriciu, V.-V., & Pop, F. (2018). A security authorization scheme for smart home Internet of Things devices. *Future Generation Computer Systems*, *86*, 740–749. doi:10.1016/j.future.2017.05.048

Čolaković, A., & Hadžialić, M. (2018). Internet of Things (IoT): A review of enabling technologies, challenges, and open research issues. *Computer Networks*.

Curzon, J., Almehmadi, A., & El-Khatib, K. (2019). A survey of privacy enhancing technologies for smart cities. *Pervasive and Mobile Computing*, *55*, 76–95. doi:10.1016/j.pmcj.2019.03.001

Das, A. K., Zeadally, S., & He, D. (2018). Taxonomy and analysis of security protocols for Internet of Things. *Future Generation Computer Systems*, *89*, 110–125. doi:10.1016/j.future.2018.06.027

Di Martino, B., Rak, M., Ficco, M., Esposito, A., Maisto, S., & Nacchia, S. (2018). Internet of things reference architectures, security and interoperability: A survey. *Internet of Things*, *1*, 99–112. doi:10.1016/j.iot.2018.08.008

Elazhary, H. (2018). Internet of Things (IoT), mobile cloud, cloudlet, mobile IoT, IoT cloud, fog, mobile edge, and edge emerging computing paradigms: Disambiguation and research directions. *Journal of Network and Computer Applications*.

Frustaci, M., Pace, P., Aloi, G., & Fortino, G. (2018). Evaluating critical security issues of the IoT world: Present and Future challenges. *IEEE Internet of Things Journal*, *5*(4), 2483–2495. doi:10.1109/JIOT.2017.2767291

Hassan, W. H. (2018). Current research on Internet of Things (IoT) security: A survey. *Computer Networks*.

Hellaoui, H., Koudil, M., & Bouabdallah, A. (2017). Energy-efficient mechanisms in security of the internet of things: A survey. *Computer Networks*, *127*, 173–189. doi:10.1016/j.comnet.2017.08.006

Hussain, K., Hussain, S. J., Jhanjhi, N., & Humayun, M. (2019). *SYN Flood Attack Detection based on Bayes Estimator (SFADBE) For MANET.* Paper presented at the 2019 International Conference on Computer and Information Sciences (ICCIS). 10.1109/ICCISci.2019.8716416

Hussain, K., Jhanjhi, N., Mati-ur-Rahman, H., Hussain, J., & Islam, M. H. (2019). Using a systematic framework to critically analyze proposed smart card based two factor authentication schemes. *Journal of King Saud University-Computer and Information Sciences.*

Hussain, K., Jhanjhi, N. Z., Mati-ur-Rahman, H., Hussain, J., & Islam, M. H. (2019). Using a systematic framework to critically analyze proposed smart card based two factor authentication schemes. *Journal of King Saud University - Computer and Information Sciences.* doi:10.1016/j.jksuci.2019.01.015

Hussain, S. J., Ahmed, U., Liaquat, H., Mir, S., Jhanjhi, N., & Humayun, M. (2019). *IMIAD: Intelligent Malware Identification for Android Platform.* Paper presented at the 2019 International Conference on Computer and Information Sciences (ICCIS). 10.1109/ICCISci.2019.8716471

Jose, D. V., & Vijyalakshmi, A. (2018). An Overview of Security in Internet of Things. *Procedia Computer Science, 143*, 744–748. doi:10.1016/j.procs.2018.10.439

Khan, N., Sakib, N., Jerin, I., Quader, S., & Chakrabarty, A. (2017). *Performance analysis of security algorithms for IoT devices.* Paper presented at the Humanitarian Technology Conference (R10-HTC), 2017 IEEE Region 10. 10.1109/R10-HTC.2017.8288923

Khan, R., Khan, S. U., Zaheer, R., & Khan, S. (2012). *Future internet: the internet of things architecture, possible applications and key challenges.* Paper presented at the 2012 10th international conference on frontiers of information technology. 10.1109/FIT.2012.53

Kim, T., Ramos, C., & Mohammed, S. (2017). Smart City and IoT. *Future Generation Computer Systems, 76*, 159–162. doi:10.1016/j.future.2017.03.034

Kok, S., Abdullah, A., Jhanjhi, N. Z., & Supramaniam, M. (2019). Prevention of Crypto-Ransomware Using a Pre-Encryption Detection Algorithm. *Computers, 8*(4), 79. doi:10.3390/computers8040079

Kouicem, D. E., Bouabdallah, A., & Lakhlef, H. (2018). Internet of things security: A top-down survey. *Computer Networks, 141*, 199–221. doi:10.1016/j.comnet.2018.03.012

Kumar, S. P. (2017). Internet of Things for sophisticated e-governance: A special focus on agricultural sector. *International Journal of Trend in Research and Development.*

Maher Omar Alshammari, A. A. A. N. Z. (2017). Internet of Things (IoT): Charity Automation. *(IJACSA). International Journal of Advanced Computer Science and Applications, 8*(2), 166–170.

Malak Alamri, N. J. M. H. (2019). Blockchain for Internet of Things (IoT) Research Issues Challenges & Future Directions: A Review. *IJCSNS International Journal of Computer Science and Network Security, 19*(5), 244–258.

Maria Almulhim, N. I. N. Z. (2019). A Lightweight and Secure Authentication Scheme for IoT Based E-Health Applications. *IJCSNS International Journal of Computer Science and Network Security, 19*(1), 107–120.

Mathur, A., Newe, T., Elgenaidi, W., Rao, M., Dooly, G., & Toal, D. (2017). A secure end-to-end IoT solution. *Sensors and Actuators. A, Physical*, *263*, 291–299. doi:10.1016/j.sna.2017.06.019

Mavropoulos, O., Mouratidis, H., Fish, A., & Panaousis, E. (2018). Apparatus: A framework for security analysis in internet of things systems. *Ad Hoc Networks*. doi:10.1016/j.adhoc.2018.08.013

Mocrii, D., Chen, Y., & Musilek, P. (2018). IoT-based smart homes: A review of system architecture, software, communications, privacy and security. *Internet of Things*, *1-2*, 81–98. doi:10.1016/j.iot.2018.08.009

Naik, S., & Maral, V. (2017). *Cyber security—IoT*. Paper presented at the Recent Trends in Electronics, Information & Communication Technology (RTEICT), 2017 2nd IEEE International Conference on.

O'Dwyer, E., Pan, I., Acha, S., & Shah, N. (2019). Smart energy systems for sustainable smart cities: Current developments, trends and future directions. *Applied Energy*, *237*, 581–597. doi:10.1016/j.apenergy.2019.01.024

Pundir, Y., Sharma, N., & Singh, Y. (2016). Internet of things (IoT): Challenges and future directions. *International Journal of Advanced Research in Computer and Communication Engineering*, *5*(3), 960–964.

Ray, P. P. (2016). A survey of IoT cloud platforms. *Future Computing and Informatics Journal*, *1*(1-2), 35–46. doi:10.1016/j.fcij.2017.02.001

Rizvi, S., Kurtz, A., Pfeffer, J., & Rizvi, M. (2018). *Securing the Internet of Things (IoT): A Security Taxonomy for IoT*. Paper presented at the 2018 17th IEEE International Conference On Trust, Security And Privacy In Computing And Communications/12th IEEE International Conference On Big Data Science And Engineering (TrustCom/BigDataSE).

Sadique, K. M., Rahmani, R., & Johannesson, P. (2018). Towards Security on Internet of Things: Applications and Challenges in Technology. *Procedia Computer Science*, *141*, 199–206. doi:10.1016/j.procs.2018.10.168

Sfar, A. R., Natalizio, E., Challal, Y., & Chtourou, Z. (2018). A roadmap for security challenges in the Internet of Things. *Digital Communications and Networks*, *4*(2), 118–137. doi:10.1016/j.dcan.2017.04.003

Sha, K., Wei, W., Yang, T. A., Wang, Z., & Shi, W. (2018). On security challenges and open issues in Internet of Things. *Future Generation Computer Systems*, *83*, 326–337. doi:10.1016/j.future.2018.01.059

Shah, T., & Venkatesan, S. (2018). *Authentication of IoT Device and IoT Server Using Secure Vaults*. Paper presented at the 2018 17th IEEE International Conference On Trust, Security And Privacy In Computing And Communications/12th IEEE International Conference On Big Data Science And Engineering (TrustCom/BigDataSE). 10.1109/TrustCom/BigDataSE.2018.00117

Silva, B. N., Khan, M., & Han, K. (2018). Towards sustainable smart cities: A review of trends, architectures, components, and open challenges in smart cities. *Sustainable Cities and Society*, *38*, 697–713. doi:10.1016/j.scs.2018.01.053

Sridhar, S., & Smys, S. (2017). *Intelligent security framework for iot devices cryptography based end-to-end security architecture*. Paper presented at the Inventive Systems and Control (ICISC), 2017 International Conference on. 10.1109/ICISC.2017.8068718

Stergiou, C., Psannis, K. E., Gupta, B. B., & Ishibashi, Y. (2018). Security, Privacy & Efficiency of Sustainable Cloud Computing for Big Data & IoT. *Sustainable Computing: Informatics and Systems*.

Tewari, A., & Gupta, B. (2018). Security, privacy and trust of different layers in Internet-of-Things (IoTs) framework. *Future Generation Computer Systems*. doi:10.1016/j.future.2018.04.027

Yang, K., Li, Q., & Sun, L. (2018). Towards Automatic Fingerprinting of IoT Devices in the Cyberspace. *Computer Networks*.

# Chapter 8
# Success Stories of ICT Implementation in Saudi Arabia

**Saira Muzafar**
*King Faisal University, Saudi Arabia*

**N. Z. Jhanjhi**
https://orcid.org/0000-0001-8116-4733
*Taylor's University, Malaysia*

## ABSTRACT

*Information and communication technology (ICT) has a great role in the development, growth, and economy of the country. ICT refers to any system, device, or product that can store, manipulate, and receive and transmit digital data effectively. ICT brings clear differences in the economy of countries wherever it is successfully implemented. It becomes a reason of timely, accurate, and efficient operations and becomes a source of user satisfaction by providing them fast, better, easy service around the clock. As timely access of information and communication plays a vital role in the success of any organization either public or private, many developed countries are already competing in digitalization. It is not only convenience, but it is the prime ingredient of good governance and an indispensable part of modern life. This chapter discusses the successful implementation of ICT infrastructures in the context of good governance and builds a vibrant society, the main agenda of Vision 2030 of KSA.*

## INTRODUCTION

E-Government has been deploy by many countries all over the world due to its significant role in delivering transparent and timely government services to the citizens. The government departments in which technology is penetrated and practiced are more efficient and significantly affect country's international competitive power (Al-Mushayt, Haq & Perwej, 2009).

The government of Saudi Arabia is constantly working to modernize the ICT infrastructure in the country and achieve 71% Digital maturity level of e-government services in 2018, which is a great improvement as compared to 58% in 2016. Kingdom's Digital Transformation Program is one of the

DOI: 10.4018/978-1-7998-1851-9.ch008

principal goals to achieve Vision 2030 objectives. Saudi Arabia's National Vision 2030 aims to build digital government to boost the kingdom's economy based on fourth industrial revolution and establish a vibrant digital society by turning the country into a globally competitive ICT hub. Vision 2030 has 96 strategic objectives based on 13 Vision Realization Programs, all promising to improve the citizen's lifestyle.

The Saudi Ministry of Communications and Information Technology (MCIT) has taken several initiatives to infuse technology in government infrastructure and in 2018 Saudi Arabia ranked seventh worldwide by World Economic Forum (WEF) for efficient government spending index which is a key measure of global competitiveness. At present Saudi National Portal offer more 2,500, government services and the m-government alone addresses almost 130 applications for 70 government agencies. In the ongoing journey of digital transformation, KSA has achieved several milestones to serve citizens and attract foreign investors. Following section discusses some major e-projects deployed by KSA government to achieve goals of digital transformation (Saudi Vision 2030).

## MAJOR E-GOVERNMENT PROJECTS IN KSA

### Yesser Program

One of the important initiative under MCIT is to launch "Yesser Program" which covers three main domains G2G, G2C and G2B. Yesser Program aims to establish and manage most e-government services and products to provide fast and easy services to the people (e-Government Program "Yesser").

Main products of "Yesser" are as follows:

### Saudi Portal

Saudi e-Government Portal provides broad online services to citizens, residents, businesses, visitors and external organizations efficiently. In addition, to provide e-services the portal also provides directory of government agencies and links to Saudi laws and regulations (Saudi Portal).

### National Contact Center (Amer)

Amer Program launched in support of e-government strategic goals. It is the National Contact Center of Yesser program, designed to provide technical support, assistance and prompt advice to users. The center supports all government agencies and provides information about e-services to general public and to respond inquiries in both Arabic and English language by using variety of communication channels such as email, phone 9both landline and mobile), social networks, Fax, website, SMS, post etc.

The main objectives of the program is to address queries related to e-government issues and procedures, facilitate e-communication between individuals and organizations, reduce the load on service providers and to improve the response time (National Contact Center (Amer)).

## e-Government Mobile Services (Ma'ak)

Ma'ak is the formal application of e-government program launched by Yesser. The main objective is to create common e-government infrastructure. The application provides e-services manual, information and financial supports about mobile e-government projects

Other notable services include Yesser Data Center, Government Cloud Computing, Government Service Observatory, National Enterprise Architecture (NEA), Single Sign-on (SSO), Government SMS Gateway, e-Correspondence System, Government Service Bus (GSB), Government Secure Network (GSN) etc (e-Government Program "Yesser").

## MERAS PLATFORM

Meras launched by Tayseer Committee in 2017 as part of Saudi Vision 2030 to support businesses, it links 18 government and private sector entities to one-stop shops for investors. It is a unified e-service portal providing more than 40 e-services and significantly reduce transaction time from 81 days to 24 hours throughout the kingdom, where business owner have access to major business services such as commercial registration, chamber of commerce membership, zakat and tax returns, social insurance etc. Figure 1 shows the mechanism of Meras application (Meras Platform).

*Figure 1. Roadmap of meras application (meras platform)*

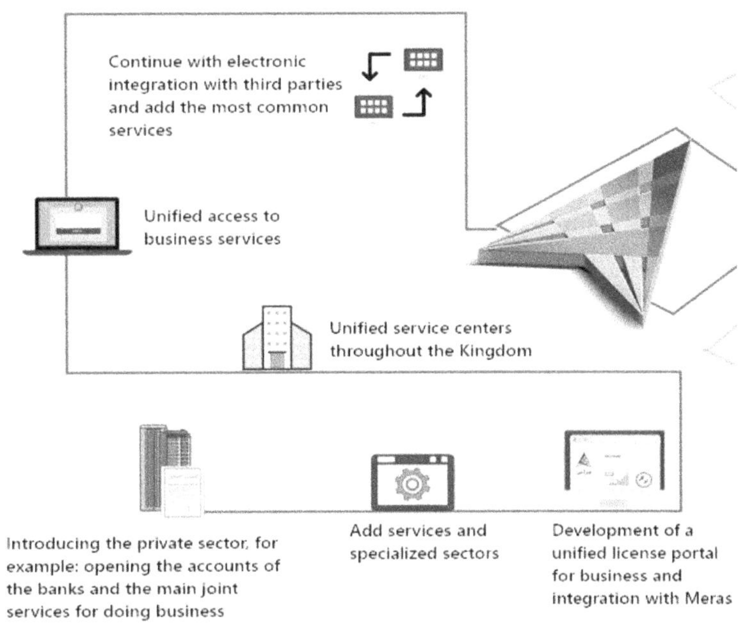

## DIGITIZING HEALTHCARE

Saudi Arabia has been using information technology in healthcare industry since 1970 by launching country's first Hospital Information System (HIS) in King Faisal Specialized hospital since then ministry of health (MOH) continuously taking steps to improve the health care sector. However, on broader spectrum the government started the operational phase of the national e-health plan in 2008 (Moghaddasi et al. 2018). Since then the government has been taking several steps to infuse technology and promote telemedicine in health care system. Following are some of the smartphone apps are launched by MOH in this context

## MOH APPS FOR SMARTPHONES

### Mawid Service

Mawid is an e-appointment app launched by Ministry of Health in 2017 with the aim to standardize management and reservation of patient appointments in government hospitals 8 million appointments reported in 2018 by this e-service. All user holding absher accounts can schedule or cancel appointments at primary health care centers. Beneficiaries can download a smartphone app or get the services through portal (Ministry of Health, MOH Apps for Smartphones).

### Seha App

Seha App provides online medical consultations from MOH's accredited doctors remotely. The App is specifically significant for rural population reported 260 thousand medical consultations in 2018 (Ministry of Health, MOH Apps for Smartphones).

### Ashanak App

This app designed to serve MOH staff members, through this app user can apply for promotions, register themselves for trips and can follow up the requests. Further, the app advertises MOH's mandate jobs and publishes a monthly online magazine call named "Wareed" to keep inform the MOH staff about activities and events (Ministry of Health, MOH Apps for Smartphones).

### Mawared App

Mawared is a single channel Enterprise Resource Planning (ERP) app facilitate employee self-services administrative tasks such as apply for leave application and office recommencing etc (Ministry of Health, MOH Apps for Smartphones).

## DIGITAL TRANSFORMATION IN JUSTICE SERVICES

Introduction of digital technology in court system promote paperless less environment and fast exchange of data and case files (Upadhyay, 2015). In liue of fast justice and streamline court related activities in the first quarter of 2019, an innovation platform launched by Saudi Ministry of Justice (MoJ) to boost and simplify the judicial process, documentation and to promote paperless environment which saves annually SAR 60 million. MoJ claimed 40% increase in output by going digital (Saudi Vision 2030). The key services transformed to digital services are to inform the defendant and the issuance of visit, custody and evictions order etc (Arab News, 2019).

Also, MoJ provides several e-services divided into two main categories of Courts and Notary through its official portal to access fast justice. The main court services include e-file enforcement application, e-file statement of claim, e-file probate application, e-marriage contract. Users can request several services through this platform such as request for judiciary custody, request to see the casefile, request for appeal or reconsideration and many other (Ministry of Justice, e-services).

The key online Notary services include submit and verify power of attorney (PoA), this service is the great ease for all Saudi citizens generally and specifically those living abroad. Other important e-notary services related to property registration, verfication and conveyance. (Ministry of Justice, e-services)

## FINTECH SAUDI

FinTech is an innovative business model in financial sector as financial stability is indispensable for the growth of any country, requires creating a transparent technology enabled financial system as a part of economic growth (Philippon. 2016).

One of the major strategic goal of Vision 2030 is to make the country a role model in economic and financial field this objective requires constant development in finance and banking sector to enhance socio-economic development and to support the growth of other economic sectors. Although online banking, credit and ATM cards have been already successfully practiced all over the kingdom however increased demand of secure systems and customers' financial services requires to introduce new technical trends such as Blockchain an Artificial Intelligence leads to the provision of fintech services, which promises to rise the financial services industry in Saudi Arabia.

Hence, Saudi Arabian Monetary Authority (SAMA) launched Financial Technology i.e. Fintech Saudi in April 2018 with the aim to create job opportunities by supporting small and medium enterprises, promote entrepreneurship, increase participation of private sector and to enhance financial technology (fintech) with the rise of digital transactions among banks, institutions, fintech international companies and partners. Further, government is very keen to raise awareness, increase the knowledge and expertise of citizens about the program's objectives by launching different campaigns (Saudi Arabian Monetary Authority, 2018).

## ABSHER PLATFORM

Absher is an civil affairs e-portal and app launched by Saudi Arabian Ministry of Interior (MoI) to facilitate Saudi citizens and expatriates to obtain MoI services and information at one click. Absher linked more than 130 government services to expedite the request process time with ease and comfort. MoI claimed absher as a one channel service for more than 12 million users.

In order to activate their accounts users can register themselves using website followed by submitting personal details and biometrics through self-service kiosks or by phone banking.

Absher allows users to apply for jobs, Hajj permits, visa-services, reduce time from 8 days to one day for renewal of several documents such as passports, residents' card, and driving license. Also, users can report for missing documents and request replacement of missing documents through portal as well as smartphone app. It is hassle-free and user friendly service which saves time and efforts significantly. (Ministry of Interior Saudi Arabia, Absher)

## SADAD-ONLINE PAYMENT SERVICE

Sadad is an online utility payment system launched by Saudi Arabian Monetary Agency (SAMA) to streamline bill payment process through different channels of nominated banks in kingdom. Sadad links local banks and commercial organizations through all available channels such as ATM, SMS service and through online or mobile banking. This service launched in October 2004 as the national electronic bill presentment and payment (EBPP). As of October 2019 the paid amounts exceeded SR 880 million (Saudi National Portal, The SADAD Payment System).

## FIRST 3D HOUSE BUILT

Use of additive manufacturing (AM) or 3-dimensional technology (3D) is the great revolution in the construction sector, which significantly reduces laborious on site work, construction time and environmental hazards. The technology is highly customized and already become popular in many countries such as Singapore and Hong Kong. Because of significant, work done off-site in a controlled environment, quicker and cheaper construction is possible with minimum waste (Paul et al., 2018). To prove as a leader of technology in the region, Saudi Aribia's Ministry of Housing announced in the last quarter of 2018, the successful building of first house using 3D printing technology in capital city of Riyad by one of the leading Dutch companies- Cybe. The projects' main objective linked with National Transition program 2020 and Vision 2030 as part of using cost effective, innovative, eco-friendly and latest technology for future construction (*Al Arabiya English, 2018).*

## ELECTRONIC READING OF WATER METERS

Saudi Arabia has mainly two water sources sea water and deep wells, in order to deal with this problem government invested in sea water desalination plants however water distribution without any loss is also big challenge for the government, to overcome this problem the government installed electronic water meters. Electronic water meters are compact in size with high precision and interference capability (*United States Patent No. US10240961B2).*

The National Water Company (NWC) takes initiative to install electronic meters for bill process automation, track water consumption data electronically and transmit readings using radio technology. The project aims to reduce water loss, optimize data systems, monitor and make sure networks must operate as per schedules.

These electronic meters based on ultrasonic and magnetic technology, tends to send readings to data centers installed in residential areas, which then forward data to the main data center linked to the e-billing system. The smart meters are designed to achieve high-level accuracy by recording data every 15 seconds and able to forward it even in rough climatic conditions. Moreover, smart meters send alerts in case of leakage, unusual high water consumption and meter damage.

This initiative is also support the National Transformation Program 2020 and Saudi Vision 2030 not only minimize operating cost but to maximize customer satisfaction (NWC).

## ETIMAD FINANCIAL PORTAL

In early 2018, Ministry of Finance (MoF) introduced an electronic platform to provide unified e-procurement services for both government and private sector. It is an financial interlinking platform deals with more than 450 government facilities (Ministry of Communication and Information Technology). The main objective of Etimad is to expedite MoF processes with ease, accuracy and minimized administrative cost.

Etimad allows government bodies to select appropriate vendors, display budget for project, and to manage tenders, budgets, contracts are few of key facilities. Private Organizations can submit business proposals; apply for tenders and financial claims through Etimad. The portal made to achieve Vision 2030 goals to increase private sector involvement by focusing transparency and to avoid procedural delay between MoF and stakeholders (Ministry of Finance).

### e-Procurement- Monafasat Portal

Traditional procurement systems costly and are proved to be inefficient due to lack of transparency and non-availability of timely information. Hence, in order to streamline business processes effectively and reduce cost and time of operations e-procurement is the best solution (Gunasekaran et al., 2009).

To meet the need of the hour KSA government launched Monafasat portal. **Monafasat is the global level advanced National Tendering and Procurement** system, which is an essential part of e-government. It is unified platform caters all government agencies for transparent and standard online procurement activities. The main objective of Monafasat platform is to adopt e-government for the sake of transparent procurement at national level, portal major services are to automate and standardize procurement activities, reduce labor cost, save time, and maximize economic return, control financial and administrative corruption, enforce audit system, fast and efficient data exchange between government and private sector etc. The portal also allow contractors to search tenders announced by government and submit their proposals, this practice boost quality and positive competition among users (Saudi National Portal, e-Procurement)

## HADAF- SAUDI HUMAN RESOURCES GATEWAY

Hadaf is an electronic service launched by Saudi Ministry of Labor and Social Development department to supports Saudi entrepreneurs under Human Resources Development Fund (HRDF). The main objective is to provide finances to train Saudi citizens for the sake of qualified and competitive national workforce in the private sector. The portal links the establishments with training organizations, receives the job requests, upskill the already employed personnel's and connect the students and job seekers through-matching mechanism to the targeted companies and jobs. Moreover, this interactive portal supports small businesses, university start-ups and entrepreneurs during all phases of project establishment.

The portal targets variety of stakeholders including Human Resources Development Fund (HRDF), Ministry of Labor and Social Development, International Labor Organization for Arab States, Foreign Investors, Companies and organizations, scholarship holders, HR practitioners, experts and students (Saudi Human Resources Gateway)

## HAJJ HACKATHON- HOLY EVENT MODERNIZATION

Hackathon is a digital innovation, these are events in which programmers and computer specialists gathered for software development in a short time period (Briscoe and Mulligan, 2014). To stimulate innovation under the objective of Vision 2030 the Saudi Federation for Cybersecurity, Programming and Drones with collaboration of Google launched Hajj Hackathon in the western Saudi city of Jeddah in July 2018.

The purpose of this event is to modernize the holy pilgrimage by improving several services including accommodation, transportation, health, food sanitation, crowd management, financial solutions, traffic control, waste management etc. It was biggest hackathon in the Middle East in which thousands of programmer, software developers, computer and IT enthusiasts from more than 100 countries participated, and gains the Guinness World Record recognition proves the country's technical leadership in the region (Bently, 2019)

## FEKRATECH

FekraTech is an interactive and creative platform launched by National Digitization Unit (NDU) under the supervision of the Ministry of Economy and Planning, and Ministry of Communications and Information Technology in support of digital transformation and innovation. The purpose of FekraTech is to encourage both citizens and residents to contribute in national digital transformation program. This e-service enables users to propose solutions, ideas and submit projects to overcome the posted challenges in different sectors over the platform (Saudi Vision 2030. FekraTech).

## SAUDI POST

Saudi Post is the government operated postal system functioning under the Ministry of Communications and Information Technology with the aim to keep up the strategic development of e-government, they launched the National Address Project System which considered as the first project of its kind in the entire Arab world. The unified and efficient national postal addressing system covers all regions, provinces, cities and villages of the Kingdom by using Geographic Information System (GIS) (Saudi Post, National Address).

Saudi Post offers delivery of documents and parcels in several domains as follows:

1. **Wasel:** This e-service allow user to transfer the incoming mail from the mailbox at the offices of the postal agencies to the beneficiary's website Saudi National Portal, Wasel).
2. **Wasel International:** It is a unique service provided by Saudi Post Corporation, brings the global electronic markets to Saudi Post Corporation customers. Wasel Worldwide provides customers with a free redeeming address in the United States, so that customers can use it as a rechargeable address in order to facilitate the purchase process from American electronic stores (Saudi Post, Wasel International).
3. **Jamaee:** Jamaee particularly designed to send admission related documents of candidates accepted in public universities without any charge, which eliminates overcrowding in university premises. This service provides a great comfort for students by removing the burden of travelling just to submit documents in order to complete admission process in universities located in areas far from their homes. Students have to submit the documents to the nearest branch of Saudi Post and can track the delivery of documents by online tracking service (Saudi Post, Jamaee).
4. **Mureeh:** Mureeh means convenient in Arabic, this service dedicated for speedy delivery of documents to government organizations. The service is extremely reliable and highly organized, aims to facilitate citizens by saving their time, money and efforts for delivering necessary documents to concerned public institutions and provides on-line tracking system for document follow up ((EMS, Mureeh).

## NEOM-SAUDI ARABIA'S FIRST SMART CITY

Saudi Crown Prince Mohammed bin Salman announced the launch of NEOM project as Saudi Arabia's first smart city in 2017, which is a cross border city located in the northwestern province of Tabuk. 'Neom" is derived from two words Greek work ***neo*** means new and letter ***m*** abbreviated from Arabic word ***M***ostaqbal means future (*Al Arbiya, 2017*). The planned city is another initiative of Vision 2030 digital transformation program, which aims to become a global center of innovation, trade and tourism.

The former CEO and chairperson of Alcoa Inc. and former president and CEO of Siemens AG hired for the direction of Neom development. The initial budget of this mega project is $500 billion invested by foreign investors and Saudi Public Investment Fund. The first phase of project expected to complete in 2025.

The concept of NEOM has six main areas:

1. **Prioritizing Humans**: The city will aim to create an "idyllic society" with comfortable and modern living conditions.
2. **Healthy Living and Transport:** Advanced technologies will used to create "an unprecedented transportation infrastructure".
3. **Automated Services:** 'E-government' practiced at the fullest.
4. **Digitization:** It will provide high-speed internet access and e-learning facilities to all citizens. Planning also includes many functions performed by Robots such as logistics, security and home delivery.
5. **Sustainability:** The city planned to power by wind and solar power plants and its buildings will "have a net zero carbon footprint."
6. **Innovative Construction:** "It will encourage new techniques and materials to ensure it can meet its future requirements" (Saudi Arabia: ready to lead)

## DISCUSSION

E-governance means smart governance to exchange information and deliver services electronically, which reduce cost of operation, expedite processes and facilitate the citizens to access public services in a convenient, effective and efficient manner. Hence, Saudi Arabia's government is quite promising to boost smart governance at all levels of public operations and promote a citizen-centric governance. The country's digital maturity level of e-government is constantly rising and the government introducing several e-projects to achieve digital transformation under Vision 2030 program diligently and presently it is in between design and implement phase of digital economy which puts country at cusp of its economic revolution. However, just implementation of technical projects is not sufficient it requires constant maintenance, review and user training to infuse and diffuse technology in all departments at the fullest. Although government is keen to prepare a technical workforce of citizens by launching different training programs, spending huge money on advanced technical solutions and outsourcing technical experts, still there is a more room and time to be digitized completely.

## CONCLUSION

It is evident that the IT sector of Saudi Arabia is rapidly evolving with the promising and clear goals of Vision 2030 programs to become a leader of innovation and technology in the Arab World. Use of modern technologies and infusion of e-services at government institutions plays a significant role to aligned business strategies, create new business, improves quality of life by enhancing citizen centric services and this results the tangible benefits of e-government. We conclude that by the end of 2030, the country would fully achieve the goal of digital transformation in terms of e-governance, which further increases the efficiency & effectiveness, and also contributes to country's economy. It is clear based on the current implementation of ICT technology in Saudi Arabia that the country economy is growing along with its participating in up lifting the daily life standards of the people.

## FUTURE WORK

The key pillar of the Saudi Arabia's Vision 2030 is digital transformation and its benefits are visible in all sectors. Electronic transformation of business processes results fast and efficient information exchange among stakeholders which is the prime requirement of e-government. Saudi Arabia is going through its digital transformation journey and the government is quite determined to achieve Vision 2030 goals effectively. Further research should seek to measure the success factor of newly developed systems, user satisfaction ratio, insights for security, maintenance and the level of infusion and diffusion at all levels and departments.

## REFERENCES

Absher Official Webpage. (n.d.). *Ministry of the Interior of Saudi Arabia*. Retrieved 28 October 2019 from https://bit.ly/2r2Gt9g

Al-Mushayt, O., Haq, K., & Perwej, Y. (2009). Electronic government in Saudi Arabia: A positive revolution in the peninsula. *International Transactions in Applied Sciences*, *1*(1), 87–98.

Bently, D. (2019, July). Hajj Hackers Are Bringing Tech to the Mecca Pilgrimage. *Fortune Magazine*. Retrieved from https://fortune.com/2019/07/27/modern-pilgrim-hajj-mecca/

Briscoe, G., & Mulligan, C. (2014). *Digital Innovation: The Hackathon Phenomenon*. London: Creativeworks London/QMUL.

Cheng, X., & Deak, J. G. (2019). *United States Patent No. US10240961B2*. Retrieved from https://patents.google.com/patent/US10240961B2/en

e-Government mobile services (Ma'ak). (n.d.). Retrieved September 30, 2019 from https://www.yesser.gov.sa/EN/BuildingBlocks/Pages/M-gov.aspx

e-Government Program "Yesser". (n.d.). Retrieved September 30, 2019 from https://www.yesser.gov.sa/EN/Pages/default.aspx

Express Mail Service. (EMS). (n.d.). *Mureeh*. Retrieved 25 October 2019 from https://www.ems.com.sa/en/services/mureeh

Gunasekaran, A., McGaughey, R. E., Ngai, E. W. T., & Rai, B. K. (2009). E-Procurement adoption in the Southcoast SMEs. *International Journal of Production Economics*, *122*(1), 161–175. doi:10.1016/j.ijpe.2009.05.013

Melin, U., & Axelsson, K. (2009). Managing e-service development - comparing two e-government case studies, Transforming Government. People. *Process and Policy*, *3*(3), 248–270.

Ministry of Communication and Information Technology. (n.d.). *Saudi Arabia Affirms Leading Position In Digital Transformation In Effort To Drive Global Economy In Argentina's G20 Summit*. Retrieved October 19, 2019 from https://www.mcit.gov.sa/en/media-center/news/99864

Ministry of Finance. (n.d.). *Ministry of Finance Launches Etimad - the Digital Services Platform*. Retrieved October 19, 2019 from https://www.mof.gov.sa/en/MediaCenter/news/Pages/news_20012017.aspx

Ministry of Health. (n.d.). *MOH Apps for Smartphones*. Retrieved October 20, 2019 from https://www.moh.gov.sa/en/Support/Pages/MobileApp.aspx

Ministry of Justice. (n.d.). *E-services*. Retrieved October 4, 2019 from https://www.moj.gov.sa/English/eServices/Pages/default.aspx

Moghaddasi, H., Mohammadpour, A., Bouraghi, H., Azizi, A., & Mazaherilaghab, H. (2018). Hospital Information Systems: The status and approaches in selected countries of the Middle East. *Electronic Physician*, *10*(5), 6829–6835. doi:10.19082/6829 PMID:29997768

National Contact Center (Amer). (n.d.). Retrieved September 30, 2019 from https://www.yesser.gov.sa/EN/BuildingBlocks/Pages/national_e-Government_contact_center.aspx

National Water Company (NWC). (n.d.). Retrieved 21 October 2019 from https://www.nwc.com.sa/English/OurCompany/MediaCenter/NewsandEvents/News/Pages/smart-meters-in-the-regions-of-the-Kingdom-by-the.aspx

Paul, S. C., Tay, Y. W. D., Panda, B., & Tan, M. J. (2018). Fresh and hardened properties of 3D printable cementitious materials for building and construction. *Archives of Civil and Mechanical Engineering*, *18*(1), 311–319. doi:10.1016/j.acme.2017.02.008

Philippon, T. (2016). *The FinTech opportunity*. NBER Working Paper No.22476.

Platform, M. (n.d.). Retrieved September 17, 2019 from https://meras.gov.sa/en/about/

Portal, S. (n.d.). Retrieved September 30, 2019 from https://www.yesser.gov.sa/EN/BuildingBlocks/Pages/saudi_portal.aspx

Post, S. (n.d.a). *Jamaee*. Retrieved 25 October 2019 from https://sp.com.sa/en/government-services/jamaee/

Post, S. (n.d.b). *National Address*. Retrieved 25 October 2019 from https://sp.com.sa/en/national-address/

Post, S. (n.d.c). *Wasel International*. Retrieved 25 October 2019 from http://www.waselalami.com.sa/ar/overview

Saudi Arabia Builds First House Using 3d Printing Technology in the Middle East. (2018, November 5). *Al Arabiya English*. Retrieved 21 October 2019 from http://english.alarabiya.net/en/business/technology/2018/11/05/Saudi-Arabia-builds-first-house-using-3D-printing-technology-in-the-Middle-East.html

Saudi Arabia: ready to lead the world in smart city development? (2018, August 12). Retrieved October 2, 2019 from https://www.alj.com/en/perspective/saudi-arabia-ready-lead-world-smart-city-development/

Saudi Arabian Monetary Authority. (2018). *Saudi Arabian monetary authority launches fintechsaudi with the objective to make the kingdom a pioneer in the financial technology sector*. (Retrieved October 24, 2019 from http://www.sama.gov.sa/en-US/News/Pages/news30042018.aspx

Saudi Human Resources Gateway. (n.d.). Retrieved 23 October 2019 from https://hrdf.org.sa/Program/406/Saudi_Human_Resources_Gateway?bc=264

Saudi justice ministry launches 15 new e-services to streamline procedures. (2019, September 20). *Arab News*. Retrieved October 2, 2019 from https://www.arabnews.com/node/1557051/saudi-arabia

Saudi National Portal. (n.d.). *e-Procurement*. Retrieved 17 October 2019 from https://www.saudi.gov.sa/wps/portal/snp/pages/e-Procurement

Saudi National Portal. (n.d.). *Wasel*. Retrieved 25 October 2019 from https://www.saudi.gov.sa/wps/portal/snp/government/servicedetails/9746

Saudi National Portal, The SADAD Payment System. (n.d.). Retrieved 21 October 2019 from https://www.saudi.gov.sa/wps/portal/snp/pages/ePayment

Saudi Vision 2030. (n.d.a). *FekraTech*. Retrieved October 20, 2019 from https://fekratech.gov.sa/faq?lang=en

Saudi Vision 2030. (n.d.b). Retrieved October 20, 2019 from https://vision2030.gov.sa/en/vision-progress

Upadhyay, M. H. (2015). E-Courts in India and E-Judiciary in India. *International Multidisciplinary Research Journal, 7637*(7), 2–5.

What does Saudi Arabia's mega project 'NEOM' actually stand for? (2017, October 24). *Al Arbiya*. Retrieved October 2, 2019 from http://english.alarabiya.net/en/business/economy/2017/10/24/What-does-NEOM-mean-.html

# Chapter 9
# Information and Communication Technology (ICT) for Intelligent Transportation Systems (ITS)

**Robithoh Annur**
*Universiti Tunku Abdul Rahman, Perak, Malaysia*

**Vasaki Ponnusamy**
*Universiti Tunku Abdul Rahman, Perak, Malaysia*

## ABSTRACT

*Transportation is an essential part of human activity that allows people to move from one place to other places. Modern transportation systems refer to intelligent transportation systems (ITS). It has been proven that information and communication technology (ICT) has a significant role to improve the quality and security transportation services. ITS can be described as an integration of computer, electronics, and communication technologies and management strategies that can increase the efficiency and safety of transportation. The application is a platform that manages the vehicles, drivers, passengers, road operators, and managers in an environment that is suitable for everyone to communicate and interact. This chapter presents the history of transportation systems and the role of ICT to bring ITS in different transportation modes. The main motivation of the chapter is to survey the various technology-enabled ITS systems that have been in place, analyzing their strengths and providing recommendations for such ITS systems in developing countries to improve the governance.*

DOI: 10.4018/978-1-7998-1851-9.ch009

## INTRODUCTION

Transport or transportation is derived from the Latin trans ("across") and portare ("to carry") which means the movement of people and goods from one place to another. In human history, peoples' feet were the first transportation mode. Animal such as horse, cow, camel, and others were utilized to become transportation mode. People then started the development of traditional transportation modes that operated without machine with a very simple vehicle like a cart. After the invention of machine, a lot of various types of modern vehicles were developed.

The transportation development is triggered by the fact that people need transportation to support their activities, for example going to work or going to school. Moreover, other sectors such as trading and other economic business need transportation to transport huge amounts of goods from one to other places. These needs have already been addressed by the availability of many transportation modes, that is not only land transportation but also water transportation mode and air transportation mode.

The increasing number of people who travel, the number of vehicles and the various transportation modes cause many problems in transportation such as traffic congestion, traffic accident and pollution due to gas emission. These problems further lead the need of intelligent transportation systems (ITS). In the development of ITS, information and communication technology (ICT) plays important role in presenting a safe, secure, effective and efficient transportation system. ICT allows people to save their travel time travel with useful traffic information and good traffic management.

This chapter presents overview on ITS and a survey on the various technology enables ITS systems. Best transportation system with technology enabled applications in developed countries like Japan, Sweden and South Korea provides will also be presented.

## ITS AND ITS DEVELOPENT

According to United Nation, the world population recently is about 7.6 billion and his population will increase by 83 million people every year (UN, 2017). In transportation sector, this means that the number of people who mobile also increases. This is indicated by the fact that the number of cars also increases incrementally. This condition will create problem such as pollution and traffic congestion. Pollution can decrease the convenience for the road users. Traffic congestion will cause further complicated problems such as creating stress and losing time for the road user. In business, traffic congestion means also losing revenue. Building new roads might be the solution but in fact this is not a good solution as this will invite people to buy cars. Moreover, the space of the earth is constant. Other conventional solution may not really solve the problems. So, a smart solution namely Intelligent Transport Systems (ITS) are currently being looked at by most countries.

The concept of ITS has actually been initiated since 1980s. It combines high technology and improvements in information systems, communication, sensors, and advanced mathematical methods with the conventional world of surface transportation infrastructure (Sussman, 2005). Later, according to the European Union (EU) Directive 2010/40/EU, ITS is a system in which information and communication technologies are applied in the field of road transport to manage infrastructure, vehicles, users (passengers and decision makers), traffic flows and mobility within cities, and to interface with other modes of transport.

*Figure 1. Intelligent transportation systems (ITS)*
*Source: (ETSI, 2013)*

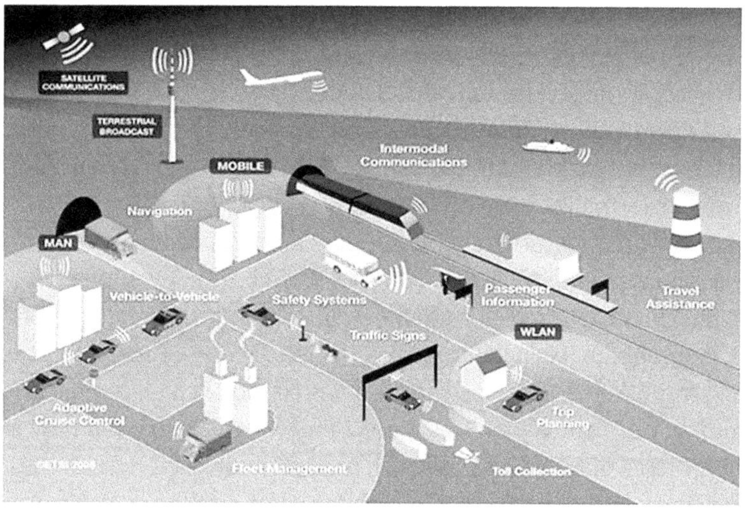


*Figure 2. The element of intelligent transportation system*
*Source: (Samadi et al, 2012)*

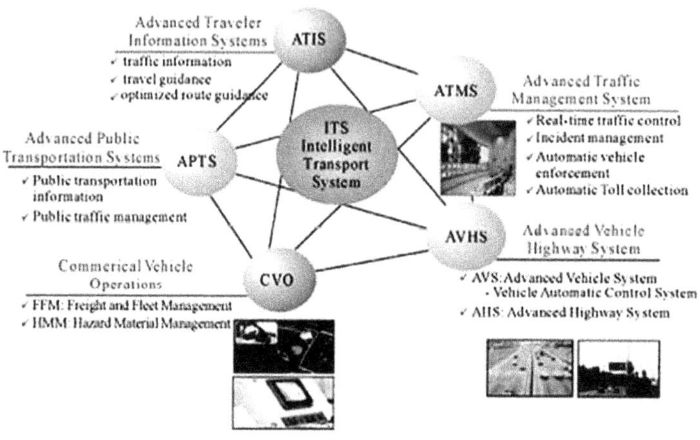


ITS is commonly devided into six elements: Advanced Traffic Management Systems (ATMS), Advanced Traveler Information Systems (ATIS), Advanced Vehicle Control Systems (AVCS), Commercial Vehicle Operations (CVO), Advanced Public Transportation Systems (APTS), and Advanced Rural Transportation Systems (ARTS) (Sussman, 2005).

## 1. Advanced Traffic Management Systems (ATMS)

ATMS integrates sensing system, communications, and data-processing technologies aiming to control traffic and reduce traffic congestion by improving the efficiency of utilization of existing infrastructures

(Shahgholian, 2018). Its implementation is based on four sections namely traffic information, traffic assignment, traffic optimization, and traffic prediction. Traffic information can be in the form of congestion maps that are collected in real time from camera, sensors, global positioning system tracking devices that provide the average speed, number of vehicle, occupancy of each lane of road. The data from these devices forms the basis for the development of the congestion map which relate flow speed and occupancy over a specific road link. Traffic prediction is performed based on the data provided by the traffic information. Traffic flow and traffic congestion are predicted in each link or route. The results of this process is useful for traffic assignment and traffic optimization, where route choices can be provided to vehicles or specified transportation system to improve the efficiency of the roads or links. Further, traffic control solutions can be performed to avoid traffic congestion.

## 2. Advanced Traveler Information Systems (ATIS)

Advanced Traveler Information System (ATIS) provides information about the transportation system to travelers. The information includes multi-modal trip planning, accident location, road conditions, weather forecast, route guidance services and advisory functions (Kem et al., 2017; Levinson, 2003; Ezell, 2010). These systems themselves also suggest the routes to take to avoid congestions, while taking the shortest and the fastest route as far as possible. Routing is a key problem in ATIS. Information can be provided both to drivers and to transit users and even to people before a trip to help them decide what mode they should use. ATIS have also been proven to be a factor affecting travel choice behavior of commuters (Sun et al., 2014).

## 3. Advanced Vehicle Control Systems (AVCS)

AVCS is a form of an enhancement of control capabilities and road driving safety by exploiting sensor networks on board and intervehicle communications, collaboratively use the information collected by individual vehicles. This system provides an intelligent vehicle with automatic functions, such as adaptive cruise control, night-vision sensors, radar, and automatic braking, to identify obstacles and adjacent vehicles and to prevent collisions with them. The AVCS will interface with the ATMS to provide automatic vehicle operations.

## 4. Commercial Vehicle Operations (CVO)

VCO is a subsystem of ITS which covers transportation busssines like the private operators of trucks, buses, vans and taxis to provide an operation with improved productivity and high efficiency.

## 5. Advanced Public Transportation Systems (APTS)

APTS aim to improve the safety of users and increase the efficiency of public transportation by implementing *transportation* management and information technologies.

These systems deal with the provision of real-time information for public transportation, automatic vehicle location (AVL) and electronic fare payments such as the use of smart-cards. APTS have made public transportation more attractive to the public as they provide enhanced visibility into timelines of buses, i.e. arrival and departure times (Ezell, 2010).

## 6. Advanced Rural Transportation Systems (ARTS)

The special economic constraints of relatively low-density roads and the question of how ITS technologies can be applied in this environment are challenges that are being undertaken by many rural states (Sussman, 2005).

## TECHNOLOGIES IN ITS

Data communication is the key implementation of ITS. As depicted in Figure 1, Radio frequency system services and specialized technologies are used to performed data communication in ITS. In ITS, there is a concept namely vehicular communication that allows intelligent data communication of vehicles. There 5 types of vehicle communication in ITS (CAAT,2018):

1. Vehicle to Infrastructure (V2I): The technology that allows data communication between vehicle and transportation infrastructure. Data are generated by the infrastructures and captured by the vehicle to provide information to the users.
2. Vehicle to Vehicle (V2V): The technology that is able to exchange information between vehicles. The information includes speed and position of surrounding vehicle. Vehicles are equipped with hardware and software to generate, capture and exchange the information. The main aim of V2V communication is to prevent accident between vehicles as the information can alert the divers allowing the drivers to take action to avoid accident.
3. Vehicle to Cloud (V2C): The technology that allows vehicles to get direct access to cloud system for information related to traffic information.
4. Vehicle to Pedestrian (V2P): The technology that enables communication between vehicles and pedestrians about information such as their position to anticipate and identify if there is a possibility of crash. This will help to improve pedestrian safety and comfort.
5. Vehicle to Everything (V2X): The technology that interconnects vehicles and its environment for safer and better driving experience.

To perform the aforementioned 5 types of communication, several technologies are widely used in ITS that will be discussed in this section.

## WIRELESS TECHNOLOGIES

### 1. Wireless Local Area Network (WLAN)

Wireless Local Area Network (WLAN) is a short range of wireless communication system (~ 250 m). It is commonly created with Wi-Fi technology that supports IEEE standard (IEEE 802.11a/b/g/n/ac). These Wi-Fi technologies operate at 2.4 GHz and 5 GHz with data rate varying from 2Mbps to 6.93 Gbps and use unlicensed spectrum to provide network access. Besides its high data rate, Wi-Fi is also reliable and very robust against disasters. It is then could be used in vehicles in support of Intelligent Transportation Systems to perform V2V direct communication (Ricci, 2010).

## 2. Wireless Access for Vehicular Environments (WAVE)

Wireless Access for Vehicular Environments (WAVE) is as of now considered as the most encouraging innovation for vehicular network. This technology supports IEEE 802.11p that is basically an approved amendment to the IEEE 802.11 standard. This will use the physical layer of IEEE 802.11a and quality of service enhancements of IEEE 802.11e. This operates on the frequency band of 5.850–5.925 GHz with data rate of 6–27 Mbs/s and coverage area of up to 1 km (IEEE standard 2010, Eichler, 2007; Xiang, 2009). WAVE is the core part of Dedicated Short-Range Communications (DSRC) which is a suite of standards at the heart of the communication of vehicular safety messages. In ITS, WAVE can be used to support high speed data communication between vehicles (V2V) or between the vehicles and the roadside infrastructure (V2I). It can provide Internet access, real time traffic information, vehicle safety services that finally can improves the safety of the transportation and reduces the traffic congestion.

## 3. Cellular Networks

Cellular network technologies are another potential solution to be used in ITS, besides DSRC, that can support V2I or V2X communications (3GPP, 2016). The technologies include GSM/GPRS at 800MHz and 1.8GHz based on GSM standard, UMTS at 800MHz and 2GHz based on 3GPP standard, LTE, LTE Advance. Cellular networks can provide communications with large coverage range, moderate delay and high-capacity. Moreover, compared to other wireless technologies, the infrastructure of cellular networks is now widely deployed in every country. To support ITS, 3GPP has developed specifications to enable the use of LTE mobile networks to provide connectivity between vehicles, roadside infrastructure and pedestrians inside and around the connected vehicles. This specification is named as 3GPP-supported Long-Term Evolution (LTE), with Proximity Services (ProSe) that can support for device-to-device (D2D) communications. Release-14 of this specification introduce a new D2D interface for enhanced vehicular networks which purposely design for high density network traffic and high speed vehicles (relative speeds up to 500 kph).

## 4. Worldwide Interoperability for Microwave Access (WiMAX)

Worldwide Interoperability for Microwave Access (WiMAX), based on the IEEE 802.16 standard, is a telecommunications technology aimed at providing wireless data over long distance in a variety of ways, from point-to-point links to full mobile cellular type access. Mobile WiMAX, defined by IEEE 802.16e, provides support for mobility. WIMAX offers portable connectivity, high-speed data, and a wireless alternative to cable/DSL for last mile broadband access. Moreover, this potential technology can also be used with cellular networks as a layover to increase capacity. Like cellular networks, WiMAX can be used for V2I or I2I long-range communication to provide, for example, high-speed Internet access to mobile users.

## 5. Bluetooth technology

Bluetooth is a wireless technology standard for exchanging data over short distances using short-wavelength microwave transmissions in the ISM band from 2400–2483.5MHz (including guard bands). This was standardized as IEEE 802.15.1. The range of each radio is approximately 10m but can be extended to 100m with an optional amplifier. Bluetooth has a lower data rate than Wi-Fi (less than 3 Mb/s). This technology is intended to provide cable free connection between different portable and/or fixed electronic devices. The main characteristics of Bluetooth technology include low power, low cost, low complexity, and robustness. Additionally, it can be used for mobile and wireless access to LANs, and Internet over mobile and other existing networks, where the backbone is wired but the interface is free to move. Bluetooth can be used for invehicle nomadic devices that provide cable free interface for onboard applications.

## 6. ZigBee technology

ZigBee technology supports IEEE 802.15.4 and is designed as a low-cost and flexible wireless medium. This operates in 2.4–2.4835 GHz and 868–915 MHz. It offers High throughput and low latency, low power consumption, reliability, and interoperability with small to medium data rates (20–40 kb/s @ 868–915 MHz and 250 kb/s @ 2.4 GHz.).

## 7. Infrared (IR)

Infrared (IR) is most suitable for applications requiring lane-specific communication due to its highly directional nature, increased data transfer rate (1 Mb/s) up to 100 m, and maximum data reliability. Furthermore, there is no need for licenses and agreements with providers, and no restrictions on bandwidth allocation. Main characteristics of this kind of wireless optical communication is physically secure data transfer, Line-of-Sight (LOS) and very low bit error rate (BER) that makes it very efficient. The main reason for using infrared had been wireless data transfer over the "last one meter" using point and shoot principles. Thus, it has been implemented in portable devices such as mobile phones, laptops, cameras, printers, medical devices and many more. Infrared technology has been used successfully in several ITS projects in Malaysia (electronic toll collection), Japan (Vehicle Information and Communication System [VICS]), and Germany (Truck Tolling Scheme), among others. It can also be used for line of sight communication between RSUs and vehicles while far-infrared can be used for video surveillance.

## 8. Radio-frequency identification (RFID)

Radio-frequency identification (RFID) is the wireless non-contact use of radio-frequency electromagnetic fields to transfer data, for the purposes of automatically identifying and tracking objects. RFID usually consist of two parts, first is tag and the other one is an interrogator/ reader. The tags contain electronically stored information. Some tags are categorized as passive tags that are powered by and read at short ranges (a few meters). Others use a local power source such as a battery, and may operate at hundreds of meters. RFID system is now widely used in ITS for electronic ticketing system, electronic toll card system, and luggage tracking and identification in air transportation system.

## 9. Near Field Communication Technology (NFC)

Near Field Communication (NFC) is a short-range wireless technology that enables the communication between devices over a distance of less than 10 cm. The NFC standard is defined by ISO/IEC 18092. Near Field Communication Technology. NFC builds upon RFID systems by allowing two-way communication between endpoints that can both convey and receive data. Communication is also possible between an NFC device and an unpowered NFC chip, called a "tag". This technology is mostly applied in Mobile payment and contactless smart cards payment (Coskun, Ok and Ozdenizci, 2011). In ITS, it used in smart card for toll and smart ticket token in public transportation.

### 10. Global Positioning System (GPS)

Global Positioning System (GPS) is a worldwide satellite navigation system formed by satellites revolving around the earth transferring radio signals to earth destined receivers (Thilagavathi and Rao,nd). Predic et al., (2007) stated that the AVL systems which were more reliable and economic then, were those using satellite-based GPS to determine the geographical location of the vehicle and cellular networks such as GPRS and EDGE for transmission of the data to control centers. GPS technology is open to use without limitations and depends on the type of receivers available (Hofmann-Wellenhof, Lichtenegger and Collins, 2013). In public transportation, GPS has been used for AVL, Vehicle tracking, in-route guidance and in-vehicle navigation (Setiawan et al., 2012). GPS technologies are also used in trip planning. Trip planning platforms use GPS data in order to provide people wishing to travel with accurate data. The data is then processed and used to calculate the link.

## SENSING TECHNOLOGIES

Accurate traffic information is crucial in ITS implementation that can be provided with the utilization of sensor technologies. Nowadays, there are various types of sensors to allow gathering the necessary traffic information used ITS. In (G. S. Tewolde, 2012; Guerrero-Ibáñez, 2018), the sensor is categorized based on its placement into in-vehicle sensor and in-road sensor.

### In-Vehicle Sensor

In-vehicle sensors are attached in the vehicle to improve the driving experience and for special purposes depending on the type of the sensor. This includes pressure sensor, fuel sensor, ultrasonic sensor inertial sensor, exhaust gas sensor, infrared sensor, rotation-rate sensors, wheel speed sensors, etc (Niewels, et.al, 2013).

In modern automotive industry, these sensors can be used to form a subsystem such as Radio detection and ranging (RADAR), Adaptive cruise control (ACC), Collision avoidance system (CAS), Lane change assistant (LCA), LiDAR (Light Detection and Ranging), Sound navigation and ranging (SONAR). Radar is a detection technique that uses electromagnetic wave to determine the position (range or angle) and relative velocity of objects. ACC is intelligent cruise control system using radar sensor that allows vehicles to automatically adjust the speed to keep safe distance from another vehicle ahead and accelerate again to a preset velocity. CAS is vehicle safety system by defining a warning strategy to the driver when the vehicle is in danger position and immediate action should be taken. This aim to prevent or reduce the severity of a collision. LCA is a driver assistance system that monitors the areas behind the vehicle and the blind spot of the driver by using radar sensors. It will generate a warning signal as the driver moves to change line if another vehicle is rapidly approaching on the neighbor lane. LiDAR is a sensor technology that can simultaneously provide the position of people or other objects surrounding the vehicles. It becomes the main component of a self-driving car or an autonomous car (or any robot) to observe the world with a few special characteristics such as continuous 360-degree visibility and highly accurate depth information (Guerrero-Ibáñez, 2018). SONAR is able to measure distances and velocities of objects by using sound propagation. It is normally an ultrasonic sensor that is mounted inside of the bumper in the front and rear of a vehicle to determine distance and position information of objects surround the vehicle and issues a warning if necessary.

### In-road sensor

Information regarding traffic flow, road condition and weather are very useful for road users. To obtain those information, some sensors are normally attached in the surface of pavement or roadway and some sensors are mounted over the center of the roadway or to the side of the roadway as depicted in Figure and Figure. In-road sensors include pneumatic road tube sensor, inductive loop detector (ILD), magnetic sensors, piezoelectric sensors, microwave radar, infrared sensors, ultrasonic sensors, passive acoustic array sensors, etc.

## IMPLEMENTATION OF ICT IN ITS IN DIFFERENT TRANSPORTATION MODE

### 1. Electronic Road Pricing (ERP) Technology

Electronic Road Pricing (ERP) is a technology to for traffic and congestion management. It aims to charge vehicles for the use of the road at times and at places of congestion. Unlike toll road systems, a road pricing system charges vehicles for usage of the road in accordance with the congestion they are causing (Menon & Guttikunda, 2010). This ERP system has been successfully used to reduce traffic in metropolitan cities such as London, Singapore, and Stockholm as a government transport strategy to provide a good transport planning, good provision of road network (Rasetyono, 2017; Rizal, 2017). This system has proven to reduce the rate of vehicles entering the limited area effectively as well as increasing the rate of vehicles that use the highway around the limited area thus increasing the usefulness of the road.

ERP consists of an in-vehicle unit (IU), antennas, vehicle detectors, surveillance cameras and lights mounted atop gantries, IC cards and central computers (Menon & Guttikunda, 2010). The IU is equipped with a stored-value card or credit card for ERP charge payment. Every time a vechicle passes through an ERP gantry, ERP charges are deducted from stored-value card or credit card via short-range radio communication.

### 2. Automatic license plate recognition (ALPR) Technology

ALPR which is also known as Automatic number plate recognition (ANPR) is a vehicle identification system by the plate number using an advanced machine vision technology. This is applied in ITS as an electronic congestion pricing systems for traffic management and congestion control. The ALPR system mainly consists of High-definition infrared digital camera(s) to capture the plate number, processor with character recognition software to perform plate recognition and database system.

*Figure 3. Piezoelectric traffic counter by the side of the road*
*Source: (Windmill, 2019)*

*Figure 4. Radar traffic sensor system n a31 highway (north-east of Italy, Vicenza-Rovigo)*

## 3. Autonomous vehicle (AV)

Autonomous vehicle or self-driving vehicle is an adavance technology in transportation that allows the vehicle to move and guide by itself with capability to sense the environment with little or no human conduction. The development of AV shows the advancement of embedded ICT in the automotive industry. This aims to provides but autonomous cars can truly minimize accidents by increasing the precision and reducing human intervention (one of the causes of traffic accidents), reduce congestion thus improve the traffic flow, and also reduce vehicle carbon emissions (Agharwal, 2018).

Acording to US National Highway Traffic Safety Administration (NHTSA), there are 5 different level of automation in AV (Anderson, 2016):

1) Level 0: The human driver is in complete control of all functions of the car.
2) Level 1: One function is automated.
3) Level 2: More than one function is automated at the same time (e.g., steering and acceleration), but the driver must remain constantly attentive.
4) Level 3: The driving functions are sufficiently automated that the driver can safely engage in other activities.
5) Level 4: The car can drive itself without a human driver.

## 4. Integrated Transport System for Public Transport

Travel with personalized car is more convenient for people, but it will raise problems such as congestion and pollution. Public transportation is a solution to meet the travel demands. However, many people do not want to use public transportation due to some issue like convenience, travel time and safety. According a survey conducted in 2010 in Europe, people will take public transportation with some conditions, for example: there is an easy mechanism to transfer between bus, train, car, tram, convenient payment method, good information in transport schedule. These reasons are basically common everywhere in the world. This become motivations of the public transportation operators worldwide to provide a convenient integrated transportation to attract people using public transportation. Nowadays, public transportation has provided IT'S THE following enabled facilities for convenient and better journey experience for the passengers.

1. Route and destination display

Route and destination display system using LED and LCD technologies has already applied in public vehicles such as trains and buses and stations or stops. In the vehicles, the signs are put on back, front, inside or sides of the vehicles showing the last destination and also the next station or next stop. This display is useful for the passenger to take the correct destination, and further will help them know when they are approaching the destination. In some vehicles, the display system is supported by the announcement by machine set by the driver. This will much more useful to make passenger realized their destination. For bus systems, this display system is also supported by the stop request button for passenger to the sign to driver that they want to stop in the next stop. Route display destination is normally shown in a simplified route map that mostly in displayed inside the vehicles and the stations. An example of bus route and destination display is given in Figure 5. This display will help passengers to take the correct bus and can estimate their waiting time.

2. Mobile and web-based applications for public transportation

Since these view years, web application and mobile application have been widely used for public transportation for various services. Trip schedule, available destination and ticketing are the most services. These allows passengers to easily plan their trip if they suddenly want to change the plan, it is will not be big issue as they can obtain the information very easily. In terms of ticketing, ticket can be generated online supported with barcode or QR code that can be easily saved in their mobile phone. This will save their time for queuing in the ticket counter or ticket machine. For transportation that needs check in such as scheduled air plane, check in can be conveniently done in the website or in the mobile applications together with boarding pass generation.

3. Trip planning

Trip planning is normally web-based application or mobile application provided by transportation provider or transportation authorities to help travelers planning their journey real time. This is very useful features especially for those who use multimodal transportation services where travelers just need to enter their location and destination, the application will provide all possible trips including the transportation mode, the travel time and the fare that the travelers need to pay. It is possible that the application will provide some trip alternatives for the traveler to choose from to make sure such a faster, safer, cheaper and comfortable ways of public transportation.

4. Smart card and electronic fare payments system

In modern transportation mode, cash is no longer accepted for fare payment. Token system has been widely used to replace cash as this system can increase the traveler convenience. This does not require the traveler to pay with exact change. As the technology developed, smart card has come as a new payment method electronically. This payment method is mostly based on RFID or NFC technology. The smart card is basically RFID or NFC tag to store the data. With this smart card, payment tracking is possible. This will also allow the traveler to monitor the available deposit and the validity of the amount in the card.

5. Automatic Vehicle Location (AVL)

AVL is basically a vehicle tracking systems that provide the geographic location of a vehicle with GPS as the main technology. This is widely used as a tool to manage emergency vehicles, service vehicles, and public transportation such as buses and trains. AVL provides real-time information about the location of vehicles that makes the user to monitor and update the current location. With the advance development in GPS technology, internet and wireless technologies and the Internet, AVL can be integrated with two-way radio and web-based application or mobile application to show the status of public transport vehicles. This system has been adopted by many transportation providers in the world. AVL systems have different functionalities dealing with the different types of data acquired from the moving vehicles. These systems have different components assigned to calculating the geographical information of buses and reporting the data to the control center using wireless internet technology.

Figure 6 shows an example of mobile application of AVL that is used in campus bus system in Chulalongkorn University, Thailand. In this system passenger can view the lay out of the route provided. Different route has different color, and the bus color is the same as the route color. The application will show the movement of the buses in real time. This will help the passenger to know when their bus will reach their location. So it can prevent the passenger to waste their time for waiting for the bus or prevent the passenger to miss the bus.

*Figure 5. Bus route and destination display in Jakarta, Indonesia*
*Source: (Vertanews, 2019; Tribunnews, 2019)*

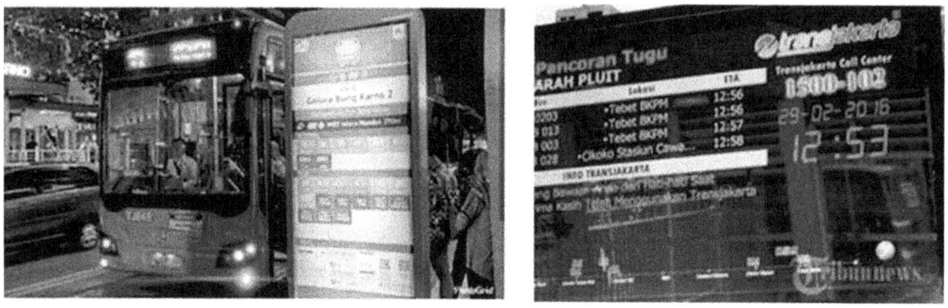


*Figure 6. Chula Pop Bus, a real-time bus vehicle tracking used in Chulalongkorn University*

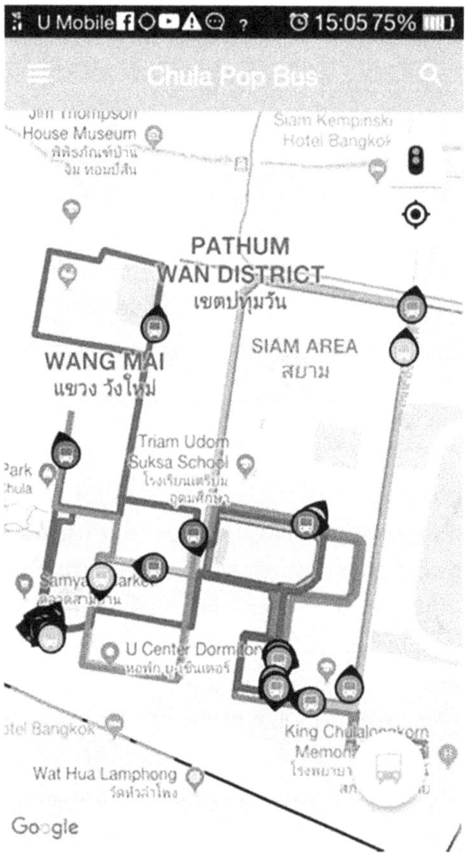

## ITS DEPLOYMENT

### Challenges and opportunities in ITS Deployment

ITS involves some elements such as people as user, government agency, and other stake holders that each of them has significant role to achieve a success ITS deployment. This section will discuss the challenges and opportunities in ITS deployment that affected by the aforementioned elements. The future transportation system that we are anticipating should be smart (Frost & Sullivan, 2008), whereby it should be able to address the needs anticipated from the transportation systems in the 21st century. It should be able to predict what is to be expected by the people and address the needs and also able to predict the future trends. Frost and Sullivan identified four main challenges that are foreseen in the current transportation system (Figure 7). The new smarter transportation systems should be able to address these challenges in order to make it efficient, safe and eco-friendly. Pojani, and Stead, 2015 have identified nine important elements as something crucial for sustainable urban transportation system in developing countries mainly in cities. They are listed as below:

1. **Road Infrastructure:** An effort to reduce congestion by introducing flyovers, tunnels, parallel routes, public transportation and later new road constructions are seen as some of the latest developments to improve road infrastructure.
2. **Rail Based Public Transport:** Railway systems construction consideration varies between cities, metro system and between metro lines. For some developing countries rail systems may not be a viable solution due to economic and practical reasons.

*Figure 7. Challenges for a smarter system Source: Frost and Sullivan (2010)*

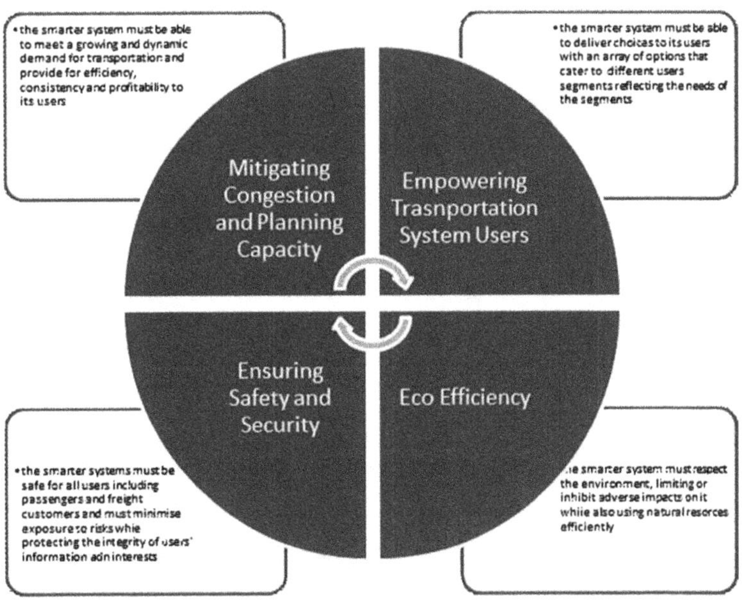

3. **Road Based Systems:** Road-based systems are the best public transportation systems for various services such education, employment and public. For some of the developing countries these amenities within reachable distances by walking or cycling and especially for some whereby automobile access is limited and not affordable. Color coded bus lanes are some of the advancements that we can see in the latest bus systems. Moreover, BRT systems are seen as some improvements in the bus systems by taking the ideas from the rail transit.
4. **Support for Non-Motorized Travel Modes:** Some of the developed and developing countries like Europe, Asia and Africa are promoting walking and cycling as alternative mode of transport. On the hand some politicians are alse seeing these modes of transportation as backward approach and may jeopardise their political status.
5. **Technological Solutions:** Employing ICT in promoting ITS to reduce pollution, accidents, congestion and energy consumption is seen as another alternative (Chowdhury and Sadek,2003) (Kulmala 2010). Electric cars and biofuels or other alternative fuels are also seen as some of the latest development in this. In some of East Asia, Eastern Europe and Latin America ITS has gone into the mode of deploying CCTVs, GPS tracking systems, electronic mode of ticketing, payment, bus management and toll systems have been introduced.
6. **Awareness-Raising Campaigns:** Urban transportation system usefulness can be cultivated by raising awareness through education, persuasion and campaigns although this may not be the most effective method (Shah and Ahn, 2006)
7. **Pricing Mechanism**: Pricing mechanism with the increased use of public transportation is crucial especially when it comes to private cars and other alternative transportation besides public transportation.
8. **Vehicle Access Restriction:** Some cities have been promoting on car banning or car rationing mechanism for reducing congestion (Mahendra 2008)
9. **Control of Land Uses:** This is control the overdeveloping the land for transportation purposes and also to cater for smaller cities whereby proper planning can be done (Jenks and Burgess, 2000).

Vaa et al. (2007) performed a thorough review on the effect of ITS pertaining to road safety issues. The evaluations were performed by categorizing into different aspects as "behavior effects on accidents, the accident study from real traffics, accident types and effects on accidents were analyzed. Various types of IT based system such as driver assistance system, vehicle information system, vehicle data collection systems and other telematics systems were analyzed. There are various accidents cases such as antilocking brake systems and electronic stability control (ESC) were analyzed and ESC seems to perform well in reducing number of accidents. driver Emotion and behavior is also important when it comes to ITS and number of accidents.

## ITS DEPLOYMENT IN DEVELOPED AND DEVELOPING COUNTRIES

ITS plays an important role in a nation's economic, social and environmental welfare. As shown in Figure 8, an urban transportation system should address these three main factors while focusing on the adoption of IT based systems. Under the economic umbrella, the efficiency in mobility, local economic develop-

*Figure 8. Sustainable urban transport system goals*
*Source: Litman (2014)*

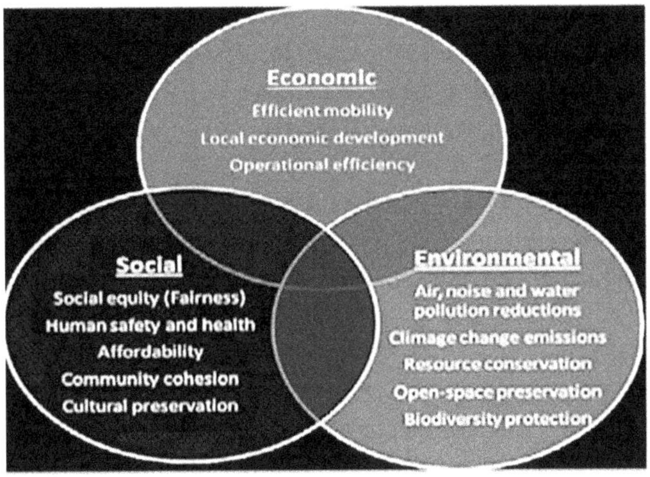


ment and operational efficiency should be addressed. In the social aspects, social equity, human safety and health, affordability, community cohesion and cultural preservation is important. Whereas in the environmental domain, reduction of pollution, resource conservation, biodiversity protection, open-space preservation and climate change emission should be carefully looked into. Any countries that plans to mimic the successful ITS adoption should also look into these factors without blindly implementing the system. The following section highlights some of the success stories of ITS implementation and their contributions in ensuring the following three factors are taken into consideration. The case studies are taken from some of the popular cities from the developed and developing countries which are efficiently operating ITS. (Bandauko et al. 2018)

## ITS System in Hong Kong

Hong Kong is one of the most popular cities in Asia with highly dense population with about seven million people living within an area that covers 1 100 square kilometers and yet has one of the best economic progress. Hong Kong is seen of having one of the most efficient public transport networks in the world. (Audenhove et al, 2014). According to OECD, Asia sees a very progressive economic growth that exceeds the growth of other world regions. And this projected that the growth may triple in year to come reaching 2032. Hong Kong adopts the Bus Rapid Transit (BRT) which emphasizes on the rapid bus movement and security of the bus ways. This smart transportation system provides threefold benefits in terms of social, environmental and economic. The system also helps to manage the public transportation systems. The transportation system in Hong Kong is a multimodal system which includes all types of rail systems that consists of minibus, ferries, taxis and trams. The system is a well consolidated system that make uses of the Octopus smart card systems that can be used for all modes of transportation systems and can also be used for parking payment facilities.

Hong Kong has been the top in the 2011 and 2013 urban mobility index mainly in the areas of "financial attractiveness of the public transport, share of public transport in modal split, smart card penetration, number of vehicles per capita, traffic safety, climate impact of transport and public sector initiatives". The rail road system in Hong Kong is one of the most prominent system in the city since the geography of the city is very suitable and also due to the huge numbers of users using the network. The rail system was initially constructed to integrate the rail public transport supply and the construction site. This huge contribution in the public transportation system in Hong Kong with the main investor Mass Transport Railway Corporation Limited (MTR) has dramatically reduced the car ownership ratio with 73 per every 1000 citizens. The public transportation development with the urban development is well supported by the Octopus Card IT initiatives. (Bandauko et al. 2018)

## ITS System in Europe

The European Commission plays an important role in the European transportation systems by encouraging all regions to have an "integrated Intelligent Transport System" that covers mobility and all modes of transportation. The system should address the local market and customer needs (European Commission, 2013). The main purpose of ITS adoption in the European cities is to provide "efficiency in availability of travel information, traffic and demand management, smart ticketing and urban logistics" (Transport Research and Innovation Portal, 2014). Some of the major contributions of the ITS in Europe are the reductions in greenhouse gas emissions, improved mobility and managed to reduce the congestion. The ITS in Europe combines the usage of CCTV and the ANPR technologies for proper transport management, improving safety and efficiency as well as reducing energy consumption and effects to the environment. The users benefit tremendously in terms of planning their travel, current network status and also the well-integrated system (European Commission, 2011a). With this, Stockholm becomes a good role model on the adoption of ITS in the European countries.

Stockholm achieves the second rank in the mobility index after Hong Kong (Audenhove et al, 2014). The geographical location of Stockholm poses some challenges in terms of access to the city and traffic control. The 2013 mobility index shows that stockhold outperforms in the following aspects: "a sustainable modal split with environmentally friendly modes, increase of shared cars per million citizens from 138 in 2011 to 400 in 2013, decrease in transport related fatalities from 21 per million in 2011 to 9 per million in 2013, decrease of carbon dioxide emissions from 1430kg per capita in 2011to 1348kg per capita in 2013", (Audenhove et al, 2014). The cycle path in Stockholm is the third best in terms of density with 4041 km lanes per 1000square km. The initiatives made by Stockholm to climb the mobility index ranking has made the transportation system smart with high speed lines and feeder lines. This has also improved the quality and capacity of all transport modes, while providing efficient reall time updates on the bus arrival times, traffic updates and also the cycle path network. Stockholm also introduced the permanent congestion charge in 2007 which led to reduction in traffic rates and congestion in the city. The report by IBM corporation (2010) stated that the traffic has been reduced by 18% and carbon dioxide emissions by 14% in the city centre and this has also improved the transportation ridership by 80 000 passengers per day (Bandauko et al. 2018)

*Figure 9. The inner-city trunk line routes of Stockholm*
*Source: (Odet & Gerasimo, 2016)*

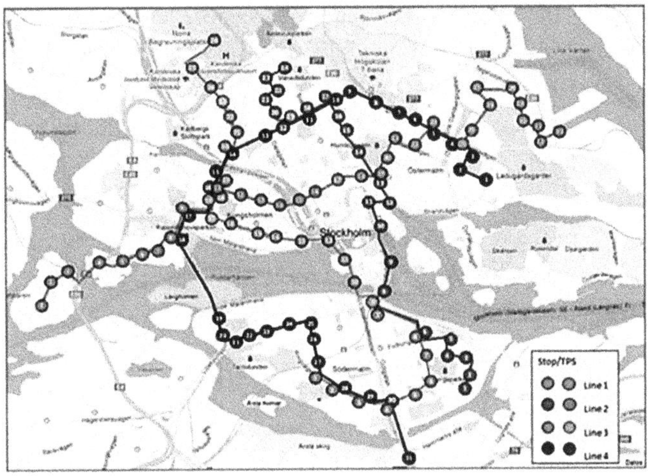


## ITS SYSTEM IN LATIN AMERICA

### Curitiba, Brazil

Some of the shortcomings in the Latin American traditional transportation system has led to the emergent of ITS by the adopting the BRT system (UN-Habitat, 2012). Curitiba and Lima offers one of the best examples on the implementation of the BRT to achieve the goal of smart cities and improving efficiency in managing transport systems. Curitiba provides one of the BRT in the world and it has great BRT model of bus ways (Figure 10).

*Figure 10. The exclusive bus ways in Curitiba*
*Source: (Wright, L. 2001)*

The system commutes about 2 million people per day with having 11 100 passengers one way on the busiest way and days. The bus usage is about 70% with the total estimated $1.5 million in BRT development. The bus system consists of the trunk and feeder bus system whereby passengers commute between the bus way vehicles, feeder routes and the inter-district links without additional charges. The private bus systems are under the supervision of the municipal and they have a "common color coding system". The Curitiba bus system has some unique characteristics such as "physically separated median bus lanes flanked by two local service streets, 26 mid-route and end-of line terminals for transfer among bus takers, an integrated bus fare structure, tube stations with off-vehicle fare collection and platform boarding of buses, direct express service in the parallel one-way arterials and distinctly colored bi-articulated buses along the bus way each with five doors and designed for level boarding at tube stations " (Figure 11&12). Curitiba BRT is one of the best urban transportation system and has become a model for other nations

*Figure 11. The modern station design and high end buses in Curitiba*
*Source: (Wright, L. 2001)*

*Figure 12. A car-free pedestrian zone in Curitiba*
*Source: (Wright, L. 2001)*

to benchmark with. Although the rate of car ownership is still high in Curitiba, many motorists still prefer using the public transport system due to the efficiency. Almost 28% of passengers have travelled by automobile before switching their decision to use public transport and this has helped to reduce the fuel consumption by 27 million litres per annum. The fuel consumption rate in Curitiba has reduced by 30% and the air pollution rate in Curitiba is one of the lowest in Brazil (Bandauko et al. 2018).

## Lima, Peru

The city of Lima has a population rate of 8.4 million (Audenhove et al, 2014) and it is the 5th biggest city in Peru. There are about 30 000 buses and 200 000 to 300 000 taxis that running across 600 routes within the city. Most of the buses and taxis are operated by privately owned individuals and there is comparatively higher competition between the bus and taxi operators. The transport system in Lima was initially very much fragmented with poor performance and has become a threat in terms of environmental and safety aspects. This shortcoming has greatly motivated towards the innovation in smart urban transportation system in the city by adopting the BRT system by combining the EL Metropolitano and the metro line known as Tren Electrico. The urban transport system helped to reduce the vehicle number in the city to 139 per 1000 people and this has tremendously reduced the traffic. The bus lanes are separated from the rest of the vehicles similar to what is implemented in Curitiba, Brazil (Figure 11). The metro lines in Lima operates for about 22km above ground with 16 stations. Besides its efficient BRT system and the Metro Line, Lima implements a regulatory framework in which financial incentives were given to those who can off road the oldest buses, vehicles that causes high pollution and least secure buses. This effort was to improve the efficiency and modernization of the road system as observed by Singapore and other modern nations (Bandauko et al. 2018).

The summary presented in the earlier section shows some of the success stories of ITS deployment in some well-known cities in the world. The cities like Stockholm, Curitiba, Lima and Hong Kong good examples on how a smart transportation system should be implemented. Table 1 shows some of the unique features that is applied in some of the developed countries and Table 2 presents similar outcomes but on different features. One good example of another vase study that could be looked into is the Brisbane BRT system as shown in Figure 13. Figure 14 shows a good example on how a modern bus looks like in ITS deployment. Another new dimension of intelligent transportation is by utilizing biometric systems. Ilhéus city of Bahia State deployed facial recognition system for ticket management so that unique identity of the user is used for ticket purchase.

*Table 1. ITS system facilities with specific features*

| FEATURE | US/ CANADA | AUSTRALIA /EUROPE | SOUTH AMERICA | TOTAL | % (OF 29 SYSTEMS) |
|---|---|---|---|---|---|
| Running Way | 13 | 5 | 6 | 24 | 83 |
| Stations | 12 | 4 | 3 | 19 | 65 |
| Distinctive Vehicles | 7 | 1 | 3 | 11 | 38 |
| Off-Vehicle Fare Collection | 1 | 0 | 3 | 4 | 14 |
| ITS | 7 | 1 | 3 | 11 | 38 |
| Frequent All-Day Service | 11 | 5 | 6 | 22 | 76 |
| **TOTAL SYSTEMS** | **17** | **6** | **6** | **29** | **100** |

Source: (Levinson et al, 2003)

*Table 2. Application of Intelligent Transportation System (ITS) Technologies (Selected Systems)*

| CITY | SYSTEM | AUTOMATIC VEHICLE LOCATION (AVL) | TELEPHONE INFO / STATIONS | PASSENGER INFORMATION AUTOMATED STATION ANNOUNCEMENTS ON VEHICLE | REAL TIME INFO AT STATIONS | TRAFFIC SIGNAL PRIORITIES |
|---|---|---|---|---|---|---|
| **U.S./CANADA** | | | | | | |
| Boston | Silver Line | X | X | X | X | |
| Charlotte | Independence Corridor | X | | | | |
| Cleveland | Euclid Ave | | | | | X |
| Hartford | Hartford-New Britain Busway | X | X | X | | |
| Los Angeles | Wilshire-Whittier & Ventura BRT | X | | | | X |
| Miami | Miami-S. Dade Busway | | X | | | Removed |
| Ottawa | Transitway | X | X | X | X | |
| Pittsburgh | South-East-West Busways | | X | Some buses | Selected locations | |
| Vancouver | Broadway and Richmond "B Lines" | X | | | X | X |
| **AUSTRALIA** | | | | | | |
| Brisbane | South East Busway | X | | X | X | |
| Sydney | Liverpool Parramatta BRT | X | | | X | |
| **EUROPE** | | | | | | |
| Rouen | Optically Guided Bus | X | | X | | X |
| **SOUTH AMERICA** | | | | | | |
| Bogotá | TransMilenio | X | | | | |
| Curitiba | Median Busway System | | | X | | |

Notes: (1) GPS and Control Center
Source: Individual Case Studies

Source: (Levinson et al, 2003)

*Figure 13. Griffith University BRT station, Brisbane Australia*
*Source: (Levinson et al, 2003)*

*Figure 14. Irisbus Civis Vehicle, Rouen, France*
*Source: (Levinson et al, 2003)*

## Public Transportation and ITS

ITS in the past have been focusing on the planning, control and management of transportation systems. Hence public transportation authorities are not trying to adopt ITS for planning, maintenance and operations (Ezell, 2010). Advanced traveler information system (ATIS) is now being employed in public transportation system to provide real-time information on schedules, accidents, routes and delays. Public transportation systems are now utilizing information and communication to provide passengers with accurate and reliable information (Oded & Gerasimos, 2016). There are various such real time bus tracking systems are available utilizing advanced data collection (ADC)methods llike automatic vehicle location (AVL) and automatic passengers counts (APC). As such in the following section, public transportation specifically the bus tracking system utilizing some of the features mentioned in some of the popular cities in the world is discussed. As discussed earlier, buses contribute to the main sector in the public transportation system.

*Figure 15. Ilhéus city (Bahia State) deployed facial recognition in 2012*
*Source: (Di Pasquele et al. 2016)*

## Bus Tracking System in Chicago

The Chicago Transit Authority (CTA) is a web application for public transportation system. CTA also have a mobile app that support all platforms but this mobile app is not official create by CTA itself but created by third parties. CTA was created by state legislation and operate on Oct. 1, 1947. This system is implemented by the municipal corporation that provides public transportation service (bus and heavy rail) in the City of Chicago which covers 35 surrounding suburbs, in Midwestern United States. This CTA Bus Tracker is a real-time bus tracking system that uses Global Positioning System (GPS) devices to positioning CTA buses. The users use text message to get estimated bus arrival time and on top of that a mobile application available. This system has its setbacks as well, whereby the users need to purchase ticket on specific place manually and no automatic alert is present on the phone application. Users need to send text message manually in order to get estimated time arrival. The system does not support time-to-time update information, require user to refresh manually

In this system (Figure 16), all buses have their own GPS installed and GPS will help to locate the bus. The query is sent to the server so that the server can generate the required information and will be sent to the database for storage. When users require certain bus information such as estimated time arrival of time, bus route, bus destination, the request will be sent to the server and the server will respond by sending the information to the website as well as to the customer service center and application as shown in Figure 17 to let their customers know.

## Bus Tracking System in New York

According to Larry Penner, 2015, Metropolitan Transportation Authority (MTA) initiated the Metropolitan Commuter Transportation Authority (MCTA) and it was a reoperation of the bankrupt rail road system. MCTA was chartered by the State Legislature in 1965 and became MTA in 1968 when took over by New York City Transit Authority. This is North America's largest transportation network that consists of 233 bus routes, in New York City. The system uses GPS to locate the MTA buses and a mobile ap-

*Figure 16. A mobile application for CTA bus system*

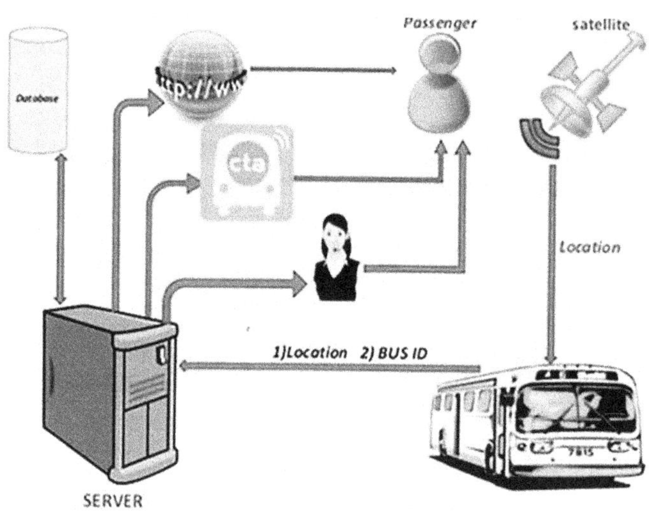

*Figure 17. A sample application on the mobile phone*

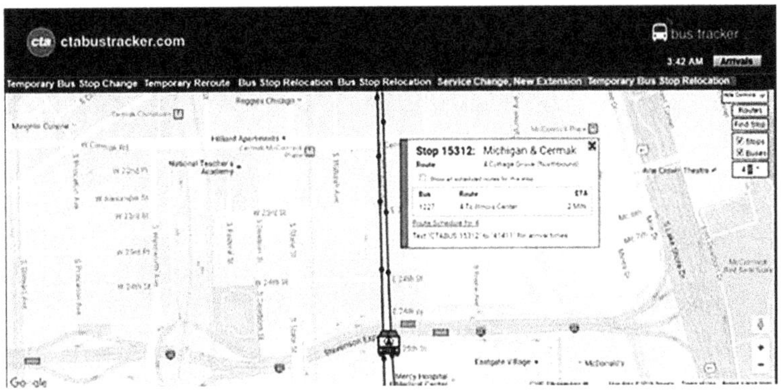


plication (Figure 18) is available for users to purchase online tickets. Guide dogs are allowed on MTA public transportation accordance with the Americans with Disabilities Act of 1990 (ADA). The systems has its setback whereby with limited routes that only operates 47 local routes in the Bronx, Brooklyn, and Queens, and 35 express bus routes between Manhattan and the Bronx, Brooklyn, or Queens. The system does not support real time information update and require users to refresh manually

## Bus Tracking System in Canada

Oakville Transit is the public transportation provider in Oakville, Ontario, Canada since 1972 and new Oakville Transit facility was completed in March 2011. It is a department of the town and a member of the Canadian Urban Transit Association. The system uses GPS data to predict the arrival time of bus with real time information update. The system provides different categories, such as school specials, senior specials, and regular schedules. It facilitates online selling fares according to the categories and offers different discount prices based on the categories. There is mobile application available with notification that can alert user about the bus arrivals. Users can also use text messages to get bus arrival time with real time information update. The system only provided limited routes and can only show one bus in particular bus stop (Figure 19).

*Figure 18 MTA System in New York*

*Figure 19. Tracking system in Canada*

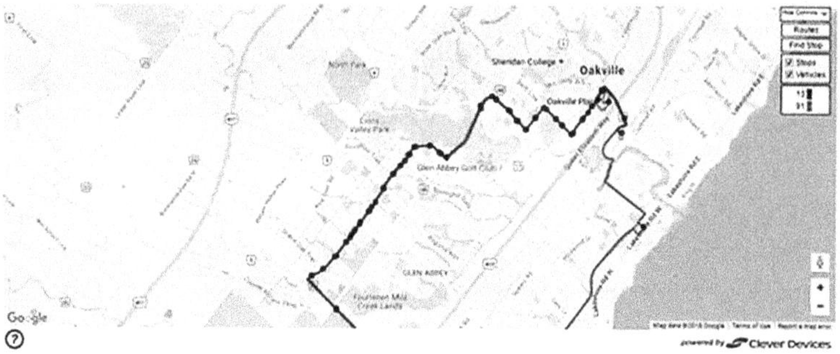


## Bus Tracking System in Singapore

MyTransport.SG was created by the Land Transport Authority (LTA) of Singapore. This is a portal base web application and this system provide the essential tools for users, such as interactive maps to get around Singapore. The system provides information and eServices for all land transport users. It provides real time information on bus arrival timings for all bus stops. This is a one stop arena to discover Singapore land transport related mobile applications. It provides traffic news so that users will be able to know the situation of the road. The system can display multiple buses that will arrive in particular stop (Figure 20). Nevertheless, the system does not display real time bus location on the map and users only know the estimated arrival time of the buses as there is no real-time map to refer to.

## Intelligent Parking Management System and ITS

Intelligent parking management system can be another contributing factor to the ITS to make it much more Efficicnt. Vianna ct al. 2004 statcs thc ITS can bc intcgrated into parking management system to utilize the features and information available in the ITS. Using the telematics, the parking system can be designed for various services such as parking guidance, better management of available parking spaces, proper road and operation control. A lot of effort has been taken into consideration in designing am intelligent parking management system. Some of the systems proposed are discussed in the following section.

This work by Patil et al., 2018 is an Intelligent Parking System utilizing Raspberry Pie, Android Application, QR Code and Infrared(IR) Sensor. The design of this system is to detect the availability of parking space through the IR sensor. The database of the application will be updated periodically and the result will be displayed through the Android application. Parking space can be reserved and reservation is confirmed through the user scanning the parking's QR Code. This proposed system provides parking monitoring system with low depletion, also it is easy to be conducted and is not cost heavy. It can be implemented by multi-level parking lot and open space parking lot. The only sensor used to implement automatic update of parking space is the IR Sensor. Although the proposed solution does not require high-end smart parking sensor, the proposed model is able to theoretically implement smart parking.

The work by Lekurwale *et al.,* 2016 proposes a smart parking system to lessen the time required to search for a parking in colleges. The operation of this system is on the platform of a mobile phone and it provides a graphical display of the empty parking spaces to the user so that it is possible for the users to reserve a parking space. (Lekurwale *et al.,* 2016) The system proposed by this paper also includes

*Figure 20. Tracking SYSTEM in Singapore*

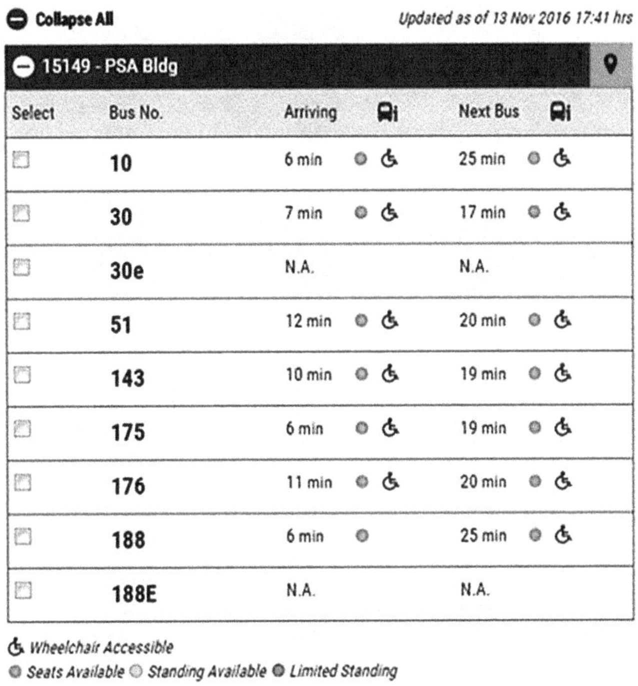


*Table 3. A Summary of Bus Tracking System*

| Function\System | CTA | MTA | Oakville Transit | MyTransport.sg |
|---|---|---|---|---|
| Real-time Tracking | YES | YES | YES | YES |
| Estimated next bus arrive | NO | NO | NO | YES |
| Mobile Application | YES | NO | YES | YES |
| ETA | YES | YES | YES | YES |
| Mobile Notification | YES | YES | YES | NO |
| Time-to-time Update Info | NO | NO | YES | YES |

functions such as reserving a parking space and a navigation system to navigate user to the parking space. QR Code is printed and displayed on each parking space to be scanned upon arrival and departure to update the database regarding the availability of the parking space. The only hardware addition required to implement this system is the printed QR Code. No sensor, gate, light indicator or other hardware is implemented in this system. This system includes a navigation function to navigate user to the parking space. As for the reserved function, reservation can only be feasibly implemented when all users share the same parking system. Since there is a possibility that not all students use the system, the more

*Figure 21. An example of intelligent parking system*

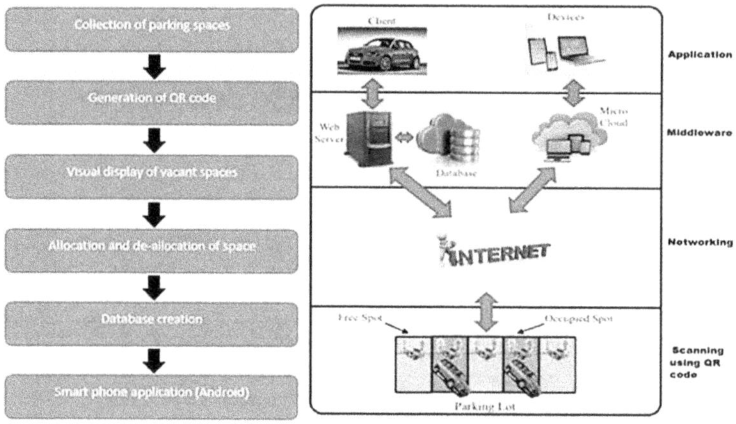


feasible method is to assign a person to place an 'Occupied' sign on the reserved parking space. With more resources, a gate system can be implemented to all parking space. Reserved parking space would require user to scan the QR code on the gate. After the system detects that the user is the valid user, the gate would be unlocked.

## System Architecture

Another work by Yang, Portilla & Riesgo 2012 proposes an implementation of smart parking service based on wireless sensor networks. In the design of this system, wireless sensor networks are used to detect the availability of a parking space. Each parking space is monitored using wireless sensor and the result is updated to the database periodically. Users will receive result which is shown on their mobile application based on the data stored in the database. (Yang, Portilla & Riesgo 2012). One of the strengths of this system is that user will receive real-time update on the availability of all parking space within the parking lot. Since a wireless sensor is deployed to monitor each parking space, the result displayed to user is very accurate. Besides, this system is fairly self-sustainably without needing human intervention. This allows the system to save cost in terms of human resources.

This is a smart parking system Soh et al., 2010 implemented through the integration between SMS services and car parking system. User may send a request through the system's mobile application to request for information on parking space availability and the system will reply the user via SMS. This SMS will contain information such as estimated time to reach, estimated time to reach, cost, entrance number and parking lot ID. (Soh et al., 2010) If the user fails to acquire the parking space replied in the SMS, they may request for the parking space information again. Charges will be charged for every request. User are allocated to the nearest parking space which is allocated after the BFS algorithm calculates the result. This allows user to save time and is able to decrease the chances of the parking space being taken away by other passing vehicle.

## ITS System Using IoT

Cardoso et al. 2013 developed a mobile application by deploying ITS using IoT devices. The system was constructed using a low-cost readily available Internet platform known as BeagleBone for controlling locally. The system is able to interpret signals from sensors, actuators by controlling the automatic gate over the TCP/IP network (Figure).

## SUMMARY

ICT technologies have been implemented in ITS to provide safe, convenient and efficient transportation system. The development of internet technologies, wireless technologies and sensing technologies make possibility to implement the technologies in the vehicles, or other transportation infrastructure. The indication of the successful use of ICT can be seen that transportation system is more organized, travel time is shorter, waiting time is shorter, real-time information is widely available. For further result, people will move to use public transportation and will reduce the number of vehicles on the road. To achieve this condition, development and consistent promotion of ITS is required. This further needs serious attention from government agencies and transportation stakeholder to make a good ITS planning in their territory. Moreover, these related agencies must provide the necessary ITS infrastructure. Countries that have successful ITS implementation is Japan, Singapore and some countries in Europe. In these countries travelling becomes very convenient and efficient. Besides the support from the related stake holders, education for people of awareness of ITS is also main key of the successful ITS implementation.

*Figure 22. IoT enabled ITS architecture*

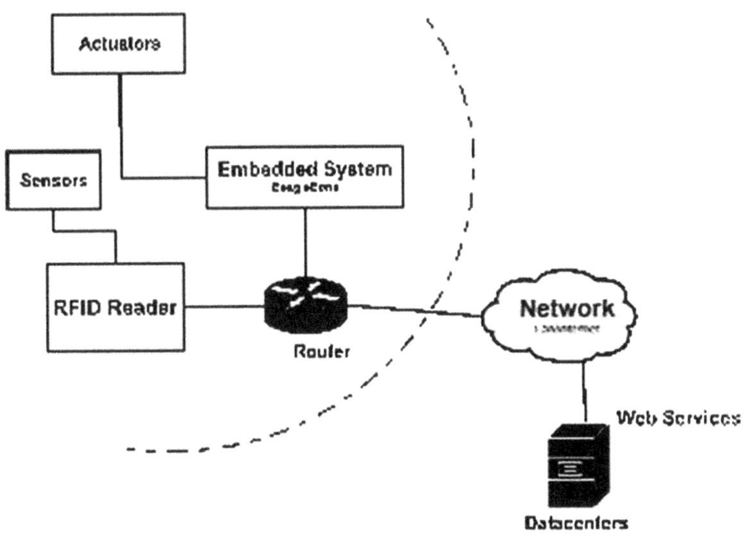

## REFERENCES

Agarwal, P., & Alam, M. A. (2018). Use of ICT for sustainable transportation. *28th International Conference on Future Environment and Energy (ICFEE 2018) IOP Publishing IOP Conf. Series: Earth and Environmental Science, 150*(2018). doi:10.1088/1755-1315/150/1/012032

Anderson, J. M., Kalra, N., Stanley, K. D., Sorensen, P., Samaras, C., & Oluwatola, T. A. (2016). *Autonomous Vehicle Technology: A Guide for Policymakers.* Santa Monica, CA: RAND Corporation. Retrieved from https://www.rand.org/pubs/research_reports/RR443-2.html

Bandauko, E., Bobo, T., & Mandisvika, G. (2018). Towards Smart Urban Transportation System in Harare, Zimbabwe. In Intelligent Transportation and Planning: Breakthroughs in Research and Practice (pp. 962-978). IGI Global.

Camillo, E. (2013). *A brief history of the evolution of transportation over time*. Academic Press.

Cardoso, R. M., Mastelari, N., & Bassora, M. F. (2013). Internet of things architecture in the context of intelligent transportation systems—a case study towards a web-based application deployment. In *22nd International Congress of Mechanical Engineering (COBEM 2013)* (pp. 7751-7760). Academic Press.

Cats, O., & Loutos, G. (2016). Real-Time Bus Arrival Information System: An Empirical Evaluation. *Journal of Intelligent Transport Systems*, *20*(2), 138–151. doi:10.1080/15472450.2015.1011638

Chowdhury, M., & Sadek, A. (2003). *Fundamentals of Intelligent Transportation Systems Planning*. Norwood, MA: Artech House.

Di Pasquale, G., dos Santos, A. S., Leal, A. G., & Tozzi, M. (2016). Innovative public transport in Europe, Asia and Latin America: A survey of recent implementations. *Transportation Research Procedia, 14*, 3284–3293. doi:10.1016/j.trpro.2016.05.276

ETSI. (2013). Retrieved from https://www.etsi.org/newsroom/news/728-2013-12-press-release-its-interop-event

Ezell, S. (2010) Explaining International IT Application Leadership: Intelligent Transport Systems. *The Information Technology and Innovation Forum*, 1-45.

Ezell, S. (2010). *Explaining international IT application leadership: Intelligent transportation systems*. Washington, DC: ITIF—The Information Technology & Innovation Foundation.

FTA. (2000). US Department of Transport: Federal Transit Agency. Staff, CAAT. Automated and Connected Vehicles. autocaat.org.

Guerrero-Ibáñez, J., Zeadally, S., & Contreras-Castillo, J. (2018). Sensor Technologies for Intelligent Transportation Systems. *Sensors*, *18*(4).

Jenks, M., & Burgess, R. (2000). *Compact Cities: Sustainable Urban Forms for Developing Countries*. London, UK: Spon Press.

Kem, O., Balbo, F., & Zimmermann, A. (2016). A Distributed Approach to Constructing Travel Solutions by Exploiting Web Resources. *2016 IEEE/WIC/ACM International Conference on Web Intelligence (WI)*, 713-716. 10.1109/WI.2016.0128

Kulmala, R. (2010). Ex-Ante Assessment of the Safety Effects of Intelligent Transport Systems. *Accident; Analysis and Prevention*, *42*(4), 1359–1369. doi:10.1016/j.aap.2010.03.001 PMID:20441853

Levinson, D., & Huo, H. (2003). Effectiveness of VMS Using Empirical Loop Detector Data. Transportation Research Board Conference.

Levinson, H., Zimmerman, S., Clinger, J., Rutherford, S., Smith, R. L., Cracknell, J., & Soberman, R. (2003). Bus rapid transit. *Transit Cooperative Research Program Report, 90.*

Mahendra, A. (2008). Vehicle Restrictions in Four Latin American Cities: Is Congestion Pricing Possible? *Transport Reviews*, *28*(1), 105–133. doi:10.1080/01441640701458265

Menon, G., & Guttikunda, S. (2010). *Electronic Road Pricing: Experience & Lessons from Singapore.* SIM-air Working Paper Series: 33-2010.

Niewels, F., Knoop, S., Jordan, R. A., & Michalke, T. (2013). *In-Vehicle Sensors. Encyclopedia of Automotive Engineering*. John Wiley & Sons.

Pojani, D., & Stead, D. (2015). Sustainable urban transport in the developing world: Beyond megacities. *Sustainability*, *7*(6), 7784–7805. doi:10.3390u7067784

Rasetyono, A. P. (2017). *IPTEK: Solusi Komprehensif Atasi Kemacetan Lalu-Lintas.* Retrieved from Kementerian Riset, Teknologi dan Pendidikan Tinggi (Kemenristekdikti) Republik Indonesia: http://dikti.go.id/iptek-solusi-komprehensif-atasi-kemacetan-lalu-lintas

Ricci, F. (2010). Mobile Recommender Systems. *Information Technology & Tourism*, *3*(12), 205–231. doi:10.3727/109830511X12978702284390

Rizal, M., Maulina, E., Purnomo, M., & Febrian, A. (2017). Electronic Road Pricing (ERP): A Systematic Mapping Study. *International Research Conference on Economic and Business 2017.*

Samadi, S., Rad, A., Kazemi, F., & Jafarian, H. (2012). Performance Evaluation of Intelligent Adaptive Traffic Control Systems: A Case Study. *Journal of Transportation Technologies*, *2*(3), 248–259. doi:10.4236/jtts.2012.23027

Shah, A., & Ahn, B. (2006). Intelligent Transportation Systems: Is it a Compatible Tool for Developing Countries? *Journal of Advanced Transportation*, *40*(3), 289–294. doi:10.1002/atr.5670400304

Shahgholian, M., & Gharavian, D. (2018). *Advanced Traffic Management Systems: An Overview and A Development Strategy*. Available: https://arxiv.org/abs/1810.02530

Sun, H., Wu, J., Wang, W., & Gao, Z. (2014). Reliability-based traffic network design with advanced traveller information systems. *Information Sciences*, *287*, 121–130. doi:10.1016/j.ins.2014.07.036

Sussman, J. M. (2005). *Perspectives on Intelligent Transportation Systems*. Springer.

Tewolde, G. S. (2012). Sensor and network technology for intelligent transportation systems. *2012 IEEE International Conference on Electro/Information Technology*, 1-7. 10.1109/EIT.2012.6220735

Tribunnews. (2019). Retrieved from https://www.tribunnews.com/images/regional/view/1640382/layar-informasi-busway

UN. (2017). Retrieved from https://www.un.org/development/desa/en/news/population/world-population-prospects-2017.html

Ursamajor. (2015). Retrieved from https://ursamajor.its-platform.eu/highlights/new-radar-traffic-sensor-system-along-a31-italy-ne

Vaa, T., Penttinen, M., & Spyropoulou, I. (2007). Intelligent transport systems and effects on road traffic accidents: State of the art. *IET Intelligent Transport Systems*, *1*(2), 81–88. doi:10.1049/iet-its:20060081

Vertanews. (2019). Retrieved from http://www.vertanews.id/homepage/read/9014/Integritas+Transportasi+Antarmoda+Jakarta+Menuju+Kota+Global

Vianna, M. M. B., da Silva Portugal, L., & Balassiano, R. (2004). Intelligent transportation systems and parking management: Implementation potential in a Brazilian city. *Cities (London, England)*, *21*(2), 137–148. doi:10.1016/j.cities.2004.01.001

Windmill. (2019). Retrieved from https://www.windmill.co.uk/vehicle-sensing.html

Wright, L. (2001, May). Latin American busways: Moving people rather than cars. *Natural Resources Forum*, *25*(2), 121–134. doi:10.1111/j.1477-8947.2001.tb00754.x

Xiang, W., Gozalvez, J., Niu, N., Altintas, O., & Ekici, E. (2009). Wireless Access in Vehicular Environments. *EURASIP Journal on Wireless Communications and Networking*, *2009*(1), 576217. doi:10.1155/2009/576217

# Chapter 10
# The Evaluation of Social Media Practice in Malaysia by Small and Medium Enterprises

**Noraniza Binti Md Jani**
*Universiti Teknikal Malaysia Melaka, Malaysia*

**Mohd Hafiz Bin Zakaria**
*Universiti Teknikal Malaysia Melaka, Malaysia*

## ABSTRACT

*The adoption of social media by large businesses has influenced small medium enterprises (SMEs) to be part of this phenomenon. Evaluating the results of the online efforts is a prerequisite to the success of social media adoption. As more branding activities move online, marketers are faced with the realisation that brands and businesses are not always relevant in all the social media facets. The study is aimed at solving the question of how SMEs in Malaysia are evaluating their online marketing efforts. The quantitative survey was deployed across six states in Peninsular Malaysia to get feedback from 125 active firms from different size, sector, year of establishment, and period of use of social media. The pattern of all these participants conducting the evaluation phase on the social media platform is descriptively examined from quantitative findings. It is used to formulate new information into literature and body of knowledge. Research implications in the practical and social form are also discussed.*

## ORGANIZATION BACKGROUND

The online community-based sensitivity that relies heavily on reviews and recommendations for product use, as well as sharing experiences with the brand, increasingly focusing the firm's surveillance on social media gateways. Profitable firms keep track of the ongoing practice of enforcing the benefits of strategic management (Peters et al., 2013), firms' performance (Wang & Kim, 2017), online reputation firms (Garcia-Madariaga & Rodriguez-Rivera, 2017), and brand presence (Keegan & Rowley, 2017). Further, in a strategic marketing plan, the brand will seek commercialization through a combination of

DOI: 10.4018/978-1-7998-1851-9.ch010

the evaluation process to the overall activities carried out through this growing channel. The majority of businesses, including small-scale ventures, are paying close attention to this practice (Doherty, 2015; Marzouk, 2016; Pan & You, 2017). As a consequence, a large number of marketers and executives has either began or increased the evaluation programmes in the space of their brand sites.

Recent studies have completed important business goals with social media evaluation. According to Keegan and Rowley (2017), social media evaluation is a process for measuring, clarifying and reporting on the success of social media initiatives beginning with identifying social media marketing objectives, managing the proper metrics selection, involving quantitative and qualitative data collection, and producing insights outlined into format reports to determine the success of coordination with a firm's business objectives, as well as corporate goals. When firms employ social media, basically they set up the marketing objective in accordance with the adoption. At that point, they should list the metrics to gauge all business-related activities via social media. To facilitate practitioners in managing a successful evaluation program, Lovett (2011) has classified a large number of metrics into four main categories: foundational metrics, business value metrics, outcome metrics and counting metrics. Each has a special function according to the usage boundaries. Some marketers recognize metrics as indicators. On one level, businesses are compulsory to employ Key Performance Indicators (KPIs), high-level metrics that help firms understand how they are doing against their target objectives and are outcome-oriented. The stable KPIs signalize the properly aligned marketing strategies with the business mission and corporate goals (Sterne, 2010; Lovett, 2011; Jeffrey, 2013).

Small Medium Enterprises (SMEs) in Malaysia were chosen to be subjects because of the appreciation on this group that continued to contribute significantly as a national economic player (SME Corporation Malaysia, 2018). They are supporter of large firms as suppliers and buyers of products in many service capacities, thereby enhancing the economic relationship between and within the industry. Despite the challenging environment, evidence shows that SMEs have continued to expand at a higher rate than large firms (SME Corporation Malaysia, 2017).

## SETTING THE STAGE

This study has captured four issues that accompany the practice of social media evaluation among SMEs. First, the existence of evaluation practices, the second is the proper stages on performing the evaluation, the third is whether the practices succeeded in achieving the objective of social media adoption, and lastly, whether it positively helped the management's decision in attaining business missions and corporate goals. This resulted in reviewing a number of publications in a well-positioned journal and scrutinized official reports issued by government agencies as well as by industry experts.

The initiative to evaluate the cyber activities usually raises questions about the extent to which they are effective and in line with the established business mission. Recent evidence found that social media was practiced without the establishment of a well-organized and good strategy (Effing & Spil, 2016). The majority of SMEs are still hesitant to launch marketing efforts in the online medium due to cost problems and lack of confidence to innovate (Digital News Asia, 2016; SME Corporation Malaysia, 2017). On the contrary, firms that have embraced digital marketing in promotional activities, are either trapped in deciding to extend or forward it with limited resources constraints. When further detailing the problem of social media evaluation, previous studies reported that businesses were impressed with the influx and inconsistent of existing metrics offered by many publishers and vendors (Madison, 2012; Barger & Labrecque, 2013), yet they can't find actionable insights on what firms solving for and resultantly fail to achieve expected objectives (Kaushik, 2011; Romaniuk, 2012).

More research is needed for future scholars and businesses alike to benefit from this adoption exercise. This is because until recently strategic social media research is still limited (Keegan and Rowley, 2017) and there is relatively little empirical data from previous studies that identify human attitudes with social media at the organization level (Reuter, Ludwig, Kaufhold, & Spielhofer, 2016). From the perspective of developing countries, there are only a handful of tasks in observing the effects of social media in marketing avenue (Oztamur & Karakadilar, 2014; Hassan, Nadzim, & Shiratuddin, 2015; SME Corporation Malaysia, 2017). Given the clear picture of the aforementioned issues, this study is presented to provide better understanding on social media evaluation practices among SMEs in Malaysia.

A considerable amount of literature has been published on the proposition of frameworks that used to evaluate social media marketing activities. Some study about a distinct multidimensional framework which is parallel with fast technology evolution (Pan, & You, 2017) while there are authors revise the measurement guidelines for the right dashboard metrics (Kaushik, 2011; Peters, Chen, Kaplan, Ognibeni, & Pauwels, 2013; Bartholomew, 2013). The other reviews have delineated the importance on impact, innovation, dialog, advocacy, and Return On Investment (Lovett, 2011; Ruhi, 2014; McCann & Barlow, 2015) as uncontrolled data has been shaped in the form of user-generated content (Drula, 2012; Schultz & Peltier, 2013) and poses a challenge to firms to build effective instruments for measuring online customer engagement with brands (Karpinska-Krakowiak, 2014) which has remained unresolved in marketing practices (Marzouk, 2016). Despite the social media measurement is often seen as closely prevalent on larger firms (Owyang, 2010; Peters et al., 2013), scholars have the responsibilities to discover the adaptable assessment in small industries that enable firms' strategic implementation of this nascent technology. Various studies have been implemented to identify human attitudes with social media at the organisation level.

SMEs who are performing the branded-content posts in the social media page have limited clue on its effectiveness particularly if the pre-planned metrics are used. Branded-site audience will alternately like, share, mention, comment or do other activities according to respective posts. On the basis that SME marketers are inexperienced, knowledgeable in minimum measurement evaluation techniques and have financial limitations (Stockdale, Ahmed, & Scheepers, 2012; Effing, & Spil, 2016), they always appreciate these easy and visible metrics such as provided by Facebook Insights or Google Analytics as this will be a good tangible measures for improving ads contents (Wiesel, Pauwels, & Arts, 2011). However, they might be exclude the effective measurement techniques that reflect the business value (Stockdale, Ahmed, & Scheepers, 2012) and firm performance (Oztamur & Karakadilar, 2014).

Although scholars assert that the awareness of firms to engage in social media evaluation mainly social media has become increasingly demanding (Hoffman & Fodor, 2010; Oztamur & Karakadilar, 2014; Aichner & Jacob, 2015), most found the management are trapped in the unclear direction of the results in which cannot be used to generate important decisions. In a more bigger picture, firms usually advanced a detailed framework on maintaining customers by improving or restructuring the strategy in leveraging digital platform (Felix, Rauschnabel, & Hinsch, 2017; Keegan & Rowley, 2017). This is a good advice for SMEs to survive in a fierce and fluctuating environment. The results of the social media evaluation must coincide with the objectives of marketing and corporate goals previously targeted.

## CASE DESCRIPTION

The quantitative approach was deemed appropriate for this study to provide a clearer picture of how SMEs in Malaysia are practicing their SNS assessments specifically in dealing with customers involved in brand websites. Therefore, the survey was used by using the cross-sectional time horizons in a set of questions. The population group has been chosen from Malaysian SMEs who were involved in online brand and product promotion through SNS and has initiated their own corporate page in Johor, Melaka, Negeri Sembilan, Selangor, Putrajaya, and Kuala Lumpur. Firms in this area were chosen because of the rapid establishment rate and represented 50.8% of total SMEs in Malaysia (SME Corporation Malaysia, 2018). According to preliminary planning, the profile is provided by SME Corporation Malaysia with a diversified sector and industry that earns MyScore recognition from 2 to 5 stars.

In developing the sound questionnaire design, this study has followed the focus areas proposed by Sekaran and Bougie (2016). Wording principles such as the content and purpose of the questions, the correct language, the type and sequencing of the questions are particularly important area in minimising bias. The second area concentrates on the measurement principles including the properly-used scales mechanisms, and the assessment of validity and reliability to test the goodness of data obtained. The questionnaire also should be appropriately and neatly organized in terms of questions, instructions and alignment, hence, the final focused area has addressed the issue of general appearance. For these three areas emphasized above, this study has received a review of four academic experts in the field of study, as well as two language experts during pre and post pilot studies. The level of appropriateness of the questionnaire was also tested by industry practitioners. Eleven firms who are not included in the actual sample have responded to a full set of questionnaires and provided good feedback to help improve the questions.

The questionnaire was piloted in quantitative phases to 30 SMEs in terms of obtaining valuable feedback. Content and formatting improvements are made to items after analysis of pilot data to achieve the target quality and better level of clarity. When the form is completed, its dissemination is carried out in stages through: 1) sending via email using profile information supplied by SME Corporation Malaysia and 2) facing digital marketers on premises in all six surveyed states. SMEs who received the questionnaire either on-line or by hand were tracked on social media and messenger applications in the effort to obtain data. After five challenging months, a total of 125 forms were successfully completed and the participants were rewarded as promised.

A descriptive analysis finally takes place to produce the statistics with the focus on SME profiles, SNS utilisation and their current evaluation practices in daily routines. In detail, quantitative data has been analysed using Google Spreadsheets to generate frequencies for needed variables and are visualised to demonstrate the charts and tables. Using this type of analysis, this paper expects to achieve the aspiration to uncover the practice of evaluation carried out by SMEs recently in the delimitation of six states.

### SME Profile

125 SMEs were involved in this study. 52.00% of these respondents are micro enterprises, 33.60% are small businesses and residues (14.40%) are medium scale firms. These statistics have verified that micro is the largest group, small firms is the second, and medium firms are minorities. SMEs from the services

*Table 1. Demographic of SME respondents*

| Item | Scale | Frequency | Percentage |
|---|---|---|---|
| Size | Micro enterprise | 65 | 52.00 |
| | Small enterprise | 42 | 33.60 |
| | Medium enterprise | 18 | 14.40 |
| Total | | 125 | 100.00 |
| Sector | Services | 91 | 72.80 |
| | Manufacturing | 29 | 23.20 |
| | Construction | 5 | 4.00 |
| Total | | 125 | 100.00 |
| Year of establishment | < 2 years | 16 | 15.40 |
| | 2 – 6 years | 35 | 33.70 |
| | 6 – 10 years | 27 | 26.00 |
| | > 10 years | 26 | 25.00 |
| Total | | 125 | 100.00 |

sector dominated this group (72.80%) while the other two sectors namely manufacturing and construction filled the remaining percentage (27.20%). The breakdown of participants by size and sector resembles the facts in the report by SME Corporation Malaysia. The majority of these participants have been in business for 2 to 6 years, making up 33.70%. Others exceed 6 years (51.00%), while the minority is still new (15.40%). Table 1 demonstrates the overall demographic information.

## Results on Social Media Adoption

From the aspect of social media marketing use on products and services, a large number of SMEs are still at the level of adjustment. This can be seen from the resulting statistics where 36.00% are new adopters to this platform (2 years or less), while 45.60% are between 2 and 6 years. SMEs that have long been on social media avenue are 18.40% only. As SMEs are small, the number of marketing-led staff on social media platforms is also small. Table 2 shows that on average, these operations are done by business owners themselves (24.55%) and marketers group of 2 to 4 people (23.35%), believed to be from medium-sized firms. The involved staff unit also learns self-app training exercises without any formal training (57.60%) for savings mainly on costs and other resources.

Respondents proved the monopoly of two major channels of social media as a digital marketing medium comprised of Facebook, and Instagram. Facebook is still a superior channel among social media category with the most frequency usage by marketing executives in their respective SMEs (55.19%). Instagram, a fast-moving medium for young people, follows the second frequency of use (24.03%). The rate of use on other social media channels is not prominent and cannot rival the two well-known mediums. Table 2 showcases the two highest frequency scores dominated by Facebook and followed by Instagram.

*Table 2. Social media marketing use*

| Item | Scale | Frequency | Percentage |
|---|---|---|---|
| Duration of use | < 2 years | 45 | 36.00 |
| | 2 – 6 years | 57 | 45.60 |
| | 6 – 10 years | 10 | 8.00 |
| | > 10 years | 13 | 10.40 |
| Total | | 125 | 100.0 |
| Staff strength in social media marketing | > 7 staffs | 7 | 4.19 |
| | 5 - 7 staffs | 7 | 4.19 |
| | 2 - 4 staffs | 39 | 23.35 |
| | 1 staff | 31 | 18.56 |
| | No staff, the owner will do it | 41 | 24.55 |
| Total | | 125 | 100.0 |
| Method of social media training | In-house training | 18 | 14.40 |
| | External training | 5 | 4.00 |
| | Both in-house and external training | 30 | 24.00 |
| | Self-learning | 72 | 57.60 |
| | In-house training | 18 | 14.40 |
| Total | | 125 | 100.00 |
| Frequently used channel | Facebook | 85 | 55.19 |
| | Instagram | 37 | 24.03 |
| | Twitter | 1 | 0.65 |
| | Google Plus | 14 | 9.09 |
| | LinkedIn | 3 | 1.95 |
| | Pinterest | 2 | 1.30 |
| | LinkedIn | 1 | 0.65 |
| | Pinterest | 5 | 3.25 |
| | Others | 6 | 3.90 |
| Total | | 154 | 100.00 |

## Do SMEs Evaluating Social Media Marketing Activities?

While digital marketing assessments have proved to be significant, it is surprising that practice among industry players is discouraging. Further inquiries in the survey to 125 SMEs find that only 79 or 63.20% evaluate their marketing efforts on the platform. 15 firms or 12.00% directly respond to no evaluation activities while 31 firms or 24.80% need time to implement them in the future. All the 46 SMEs were asked about their reasons for refusal (at least for now) to evaluate marketing efforts at social media. Through the closed options provided, many SME representatives put reason for *lack of time* (23.17%)

and *lack of knowledge* (18.29%) as the main cause. *Lack of cost* (13.40%) and *lack of guidelines* (10.98%) became the next reason they did not conform to the importance of digital marketing evaluation. Other reasons are not as much impressive as the percentage is small. Table 3 has details for reference.

## Results on Social Media Evaluation

Social media evaluation statistics are shown through Figure 1, which illustrates a form such as normal distribution. Most SMEs are moderate in conducting the evaluation (29.10%). Quite surprisingly, frequent rates (22.83%) and rarely (24.10%) practicing it are more or less the same. In the meantime, there are marketers who are too often evaluating (11.40%), on the contrary, some are not concerned with this practice (12.70%).

Since the social media evaluation is a complicated process and impact directly on the firm's decision, the owner is the most appropriate individual to undertake this responsibility. This is evident in the percentage of SME owners (45.70%) that are much higher than other positions in Figure 2. The managers and administrators of social media are in second place (15.00%) and third (13.40%) because as a human

*Table 3. Reason for not practicing social media evaluation*

| Reason for not practicing social media evaluation | Percentage |
|---|---|
| Lack of knowledge | 18.29 |
| Lack of guidelines | 10.98 |
| Lack of cost | 13.40 |
| Lack of time | 23.17 |
| Poor in technology skills | 9.76 |
| Not concern | 6.10 |
| The practice is not a necessity | 4.88 |
| New business, still no plans at this moment | 6.10 |
| Others | 7.32 |
| Total | 100.00 |

*Figure 1. Frequency of social media evaluation*

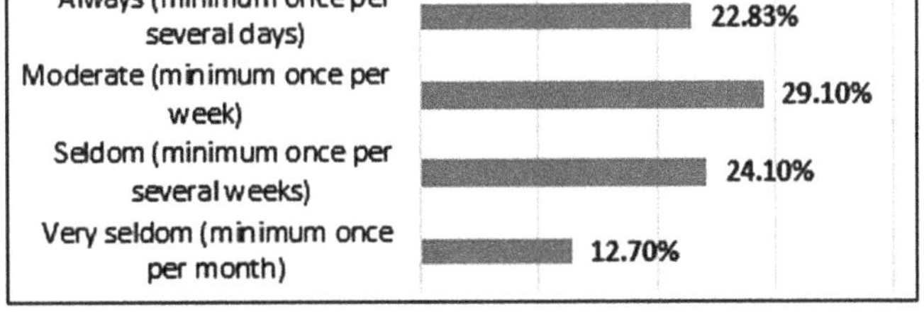

*Figure 2. Responsible person for social media evaluation*

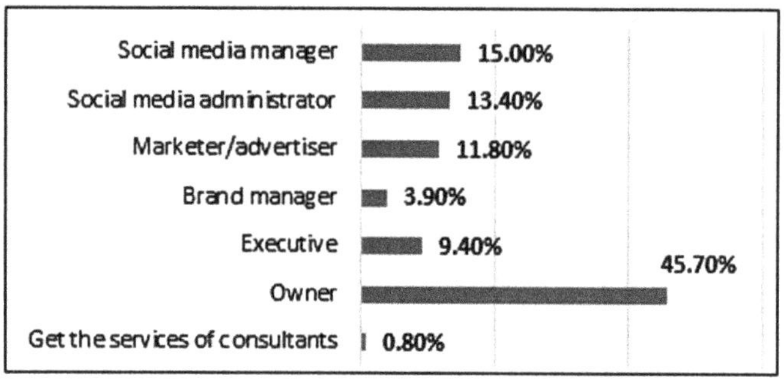


resources directly responsible with this application, they are required to administrate marketing activities at any time. Small and medium-sized firms often have many staff acting as marketers of goods and services. These are believed to form 11.80% and 9.40% of the figure on display.

Of the 79 firms, 60.80% confirmed their marketing success in social media acceptance. In fact, some of them (7.60%) have benchmarked the engagement with this app as very successful. Unfortunately, another 36.70% is uncertain of their success, and 2.50% say they rarely see what should be derived from investments in digital platforms. Figure 3 contains details on the percentage.

The outcome of the measurement efforts has identified several achieved business goals. Most SME respondents have selected service improvement with online customers (38.50%) as the primary goal achieved in social media marketing. Once a post is made, two-way communication starts to take effect and firms benefit from it to provide the best feedback on queries and comments on the brand page. Revenue growth is the second business goal that SME typically achieves in the context of the same question (34.20%). Other goals are less prominent and can be regarded as unfamiliar with the purpose of directly operating a business on a digital platform. More specifically, SME calls for increased sales (22.50%) as the most successful marketing goal when dealing with social media. Increased customer engagement followed in second place with 17.90%, while promoting customer satisfaction in third place (12.80%) followed by instil brand awareness (12.40%). It is known that the use of social media is aimed at reaching as many prospects, hence the brand awareness item gets the fourth option by SME (27.00%). The results from the above illustrations are shown in Figure 4 and Figure 5.

*Figure 3. Marketing success at social media*

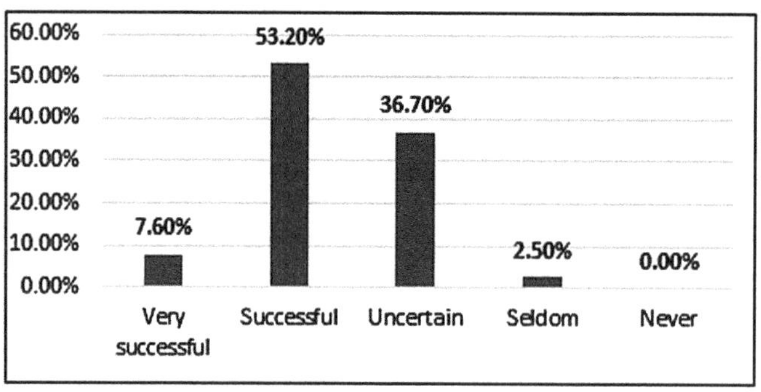

*Figure 4. Achieved business goal*

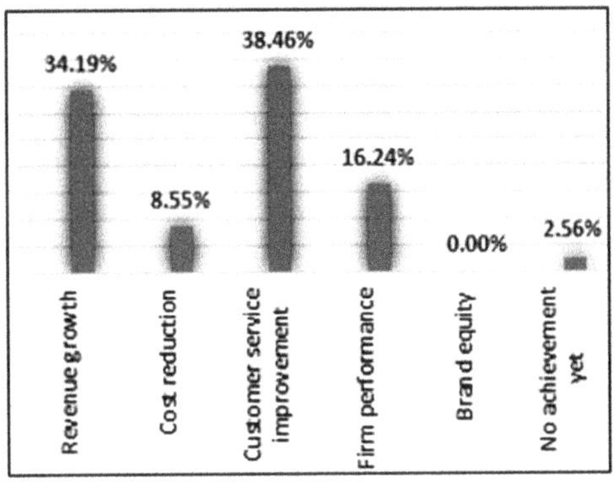


*Figure 5. Achieved marketing objective*

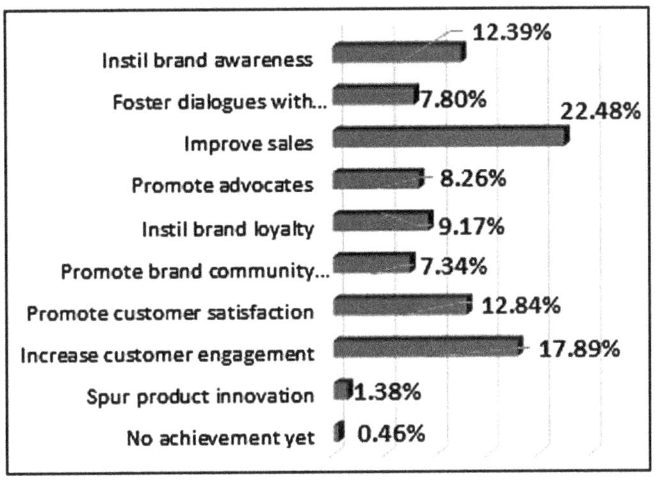


## CURRENT CHALLENGES OF THE CASE

In this study, a specific issue appeared to the SME profiles that drawn randomly from the database of the responsible agency, SME Corporation Malaysia. These profiles, which consist of general information such as the firm's name, contact number, office address, size, sector, and industry group, are antiquated and cannot fulfill the study requirements. According to the responsible officer, this data was recorded in 2015, resulting in the possibility of such profiles containing inactive businesses. When the invitation email as respondents has distributed, the progress of receiving the email is meager. This case is most likely due to a profile that receives an unwanted email (because the information mixed with SMEs that do not practice digital marketing), as well as an illegal and inactive email address (because the number of emails not reaching the recipient is high). Subsequently, the decision to change the approach was conducted to face-to-face with SME on the premises, and it has succeeded in achieving the goal of more than 100 respondents for descriptive analysis.

## SOLUTIONS AND RECOMMENDATIONS

This article emphasizes the evaluation of social media amidst the lack of literature and industry reports in Malaysia, with the justification that social media is a must in today's business world. In that regard, any issues that arise in the practice of valuation in social media must be properly addressed or continue to create major challenges to SME performance. The study is fortunate to receive 125 firm participation that really utilises social media for product marketing and enhancing interaction with customers. Various variants of social media use can be traced through the given scalable feedback. The relatively comfortable number of SMEs reveals the dominance of certain groups, namely size and sectors, from different firm backgrounds with the way social media is applied. Factors such as the age of the firm's establishment and the period of use also greatly affect the social media's pattern of use.

SME Annual Report 2016/2017 (SME Corporation Malaysia, 2018) demonstrated a discernible increase of social media use among SMEs to promote and market firms' products and services to 85.4% (1Q 2014:12.1%, 1Q 2016:78.2%). Previous studies supported the truth that marketing through social media is a good platform to increase SMEs' ability to access the wider reach of market and business opportunities (Meske and Stieglitz, 2013; McCann and Barlow, 2015). However, despite its widespread use, SMEs in Malaysia are far from being able to incorporate promising evaluation strategies. They are less able to accommodate the need to evaluate social media initiatives effectively. Among the reasons is SMEs often have limited resources to undertake in-house marketing activities or to engage a good marketing consultant (SME Master Plan 2012-2020, p.55). The findings in Table 2 prove the barrier of SMEs in funding marketing plans to utilize social media; they limit the workforce and run informal self-learning.

While most SMEs do not consider evaluation as routine on social media for some of the critical reasons such as cost issues mentioned in previous paragraphs, and their limited knowledge and guidance, these practices are still maintained by many owners. However, they implement it at a moderate pace (see Figure 1) may be due to the time constraints and employee confidence issues, expressed by the statistics in Figure 2.

SMEs in Malaysia are comfortable using intangible metrics found in the social media dashboard. For example, at Instagram, there are buttons, for example, *like*, *share*, and *follow* widely accepted as indicators for SMEs to evaluate and assume success levels in their performance. In reality, SMEs should consider the subsequent-stage metrics usually called Key Performance Indicators (KPIs) that are more specific and will be used to see the effectiveness of advertised products and services on the brand sites.

SMEs may be able to classify the level of customer feedback within the period of the ad received. In whatever indicators they perform, marketers need to record and review the progress of evaluation activities so that the advertising and technical content can be improved. With limited financial, managers can implement organic ads and pay alternately by targeting customers and prospective buyers during peak times such as festive seasons. The study revealed the SME is not taking advantage of the platform diversity when only Facebook is significantly monopolised advertising activities. Instagram, though growing, is likely to be undermined by conservative-minded SMEs management because in general, this channel is more accepted by a younger audience.

Two implications were obtained from this study based on the relevant survey analysis and following the relevant research literature exposures. From a practical point of view, SME knows the current level of evaluation in a wider homogeneous network. The only thing to know, firms often adopt social media with a minimum strategic plan. With the sharing of knowledge as revealed in this paper, more efforts can be shared and done to improve existing shortcomings and then outline a more structured guide to addressing virtual challenges. Professional digital trainers can use this information to enhance their expertise in assisting SME marketers who are interested in strategic techniques. Meanwhile, in terms of social implications, discussions in these manuscripts can be a borders to SME policy makers, for example, SME Corporation Malaysia in formulating policies that help SMEs in their digital marketing mix.

The endeavour will shade a light on assisting Malaysian SMEs upon the achievement of their business objectives and goals, in particular, through the social media platform. Building upon the entire discussed statements, the objectives of the study to mark the high potential new issue on social media evaluation has been achieved and bring remarkable points to literature and industries.

## REFERENCES

Aichner, T., & Jacob, F. (2015). Measuring the degree of corporate social media use. *International Journal of Market Research*, *57*(2), 257–275. doi:10.2501/IJMR-2015-018

Barger, V. A., & Labrecque, L. (2013). An integrated marketing communications perspective on social media metrics. *International Journal of Integrated Marketing Communications*, 64–76.

Bartholomew, D. (2013). *Social Media Measurement at a Crossroads, MetricsMan Blog*. Retrieved from https://metricsman.wordpress.com/2013/08/21/social-media-measurement-at-a-crossroads-2/

Beier, M., & Wagner, K. (2016). *Social Media Adoption: Barriers to the Strategic Use of Social Media in SMEs*. In *Twenty-Fourth European Conference on Information Systems (ECIS)*, İstanbul, Turkey.

Digital News Asia. (2016). *SMEs are laggards, but with good reason*. Retrieved from https://www.digitalnewsasia.com/content/dna-snap-week-may-9-2016

Doherty, J. (2015). Online Marketing for Small Businesses in Easy Steps: Includes Social Network Marketing. Warwickshire, UK: In Easy Steps Limited.

Drula, G. (2012). Social and online media research – data, metrics and methods. *Review of Applied Socio-Economic Research*, *3*(1), 77–86.

Effing, R., & Spil, T. A. (2016). The social strategy cone: Towards a framework for evaluating social media strategies. *International Journal of Information Management*, *36*(1), 1–8. doi:10.1016/j.ijinfomgt.2015.07.009

Felix, R., Rauschnabel, P. A., & Hinsch, C. (2017). Elements of strategic social media marketing: A holistic framework. *Journal of Business Research*, *70*, 118–126. doi:10.1016/j.jbusres.2016.05.001

Garcia-Madariaga, J., & Rodriguez-Rivera, F. (2017). Corporate social responsibility, customer satisfaction, corporate reputation, and firms' market value: Evidence from the automobile industry. *Spanish Journal of Marketing-ESIC*, *21*, 39–53. doi:10.1016/j.sjme.2017.05.003

Grosser, B. (2014). What do metrics want? How quantification prescribes social interaction on Facebook. *Computational Culture*, (4), 1–38. doi:10.100713398-014-0173-7

Hassan, S., Nadzim, S. Z. A., & Shiratuddin, N. (2015). Strategic use of social media for small business based on the AIDA model. *Procedia: Social and Behavioral Sciences*, *172*, 262–269. doi:10.1016/j.sbspro.2015.01.363

Hoffman, D. L., & Fodor, M. (2010). Can you measure the ROI of your social media marketing? *MIT Sloan Management Review*, *52*(1), 41.

Interactive Advertising Bureau (IAB). (2009). *Social Media Ad Metrics Definitions*. Retrieved from http://www.iab.net/media/file/Social-Media-Metrics-Definitions-0509.pdf

Jeffrey, A. (2013). *Social media measurement: A step-by-step approach using the AMEC valid metrics framework*. Institute for Public Relations. Retrieved from http://www.instituteforpr.org/wp-content/uploads/Social-Media-Measurement-Paper-Jeffrey-6-4-13.pdf

Karpinska-Krakowiak, M. (2014). Conceptualising and measuring consumer engagement in social media - implications for personal involvement. *International Journal of Contemporary Management*, 49-65.

Kaushik, A. (2011). *Best Social Media Metrics: Conversation, Amplification, Applause, Economic Value*. Retrieved from https://www.kaushik.net/avinash/best-social-media-metrics-conversation-amplification-applause-economic-value/

Keegan, B. J., & Rowley, J. (2017). Evaluation and Decision Making in Social Media Marketing. *Management Decision*, *55*(1), 15–31. doi:10.1108/MD-10-2015-0450

Lovett, J. (2011). *Social media metrics secrets*. Indianapolis, IN: Wiley Publishing, Inc.

Madison, I. (2012). *Why Your Social Media Metrics Are a Waste of Time*. Retrieved from https://hbr.org/2012/12/why-your-social-media-metrics

Marketing Science Institute. (2016). *2016-2018 Research Priorities*. Retrieved from http://www.msi.org/research/2016-2018-research-priorities/

Marzouk, W. G. (2016). Usage and Effectiveness of Social Media Marketing in Egypt: An Organization Perspective. *Jordan Journal of Business Administration*, *12*(1), 209–238. doi:10.12816/0030062

McCann, M., & Barlow, A. (2015). Use and measurement of social media for SMEs. *Journal of Small Business and Enterprise Development*, *22*(2), 273–287. doi:10.1108/JSBED-08-2012-0096

McLeod, B. (2017). *How Much Does It Cost to Advertise on Facebook?* Retrieved from https://www.bluecorona.com/blog/how-much-facebook-advertising-costs

Owyang, J. (2010). *Social Media Analytics*. Retrieved from https://www.slideshare.net/jeremiah_owyang/altimeter-report-social-marketing-analytics

Oztamur, D., & Karakadilar, I. S. (2014). Exploring the role of social media for SMEs: As a new marketing strategy tool for the firm performance perspective. *Procedia: Social and Behavioral Sciences*, *150*, 511–520. doi:10.1016/j.sbspro.2014.09.067

Pan, B., & You, Y. (2017). Conceptualizing and Measuring Online Behavior Through Social Media Metrics. In *Analytics in Smart Tourism Design* (pp. 203–211). Springer International Publishing. doi:10.1007/978-3-319-44263-1_12

Peters, K., Chen, Y., Kaplan, A. M., Ognibeni, B., & Pauwels, K. (2013). Social media metrics—A framework and guidelines for managing social media. *Journal of Interactive Marketing*, *27*(4), 281–298. doi:10.1016/j.intmar.2013.09.007

Praude, V., & Skulme, R. (2015). Social Media Campaign Metrics in Latvia. *Procedia: Social and Behavioral Sciences*, *213*, 628–634. doi:10.1016/j.sbspro.2015.11.462

Radhakrishna, R. B. (2007). Tips for developing and testing questionnaires/instruments. *Journal of Extension*. Retrieved from https://www.joe.org/joe/2007february/tt2.php

Rahman, N. A., & Hj Hassan, M. S. (2015). Social Media Research Trends in Malaysia: An Analysis of Three Major Communication Journals in Malaysia. *Communication. Mass Media*, 87-96.

Reuter, C., Ludwig, T., Kaufhold, M.-A., & Spielhofer, T. (2016). Emergency services' attitudes towards social media: A quantitative and qualitative survey across Europe. *International Journal of Human-Computer Studies*, *95*, 96–111. doi:10.1016/j.ijhcs.2016.03.005

Romaniuk, J. (2012). Are you ready for the next big thing? New media is dead! Long live new media! *Journal of Advertising Research*, *52*(4), 397–399. doi:10.2501/JAR-52-4-397-399

Ruhi, U. (2014). Social Media Analytics as a business intelligence practice: current landscape & future prospects. *Journal of Internet Social Networking & Virtual Communities*. doi:10.5171/2012.920553

Schultz, D. E., & Peltier, J. (2013). Social media's slippery slope: Challenges, opportunities and future research directions. *Journal of Research in Interactive Marketing*, *7*(2), 86–99. doi:10.1108/JRIM-12-2012-0054

Sime Darby. (2016). *Innovating For The Future, Sime Darby Annual Report 2016*. Retrieved from http://www.simedarby.com/sites/default/files/annualreport-pdf/sime-darby-annual-report-2016.pdf

Singh, A. S. (2017). Common procedures for development, validity and reliability of a questionnaire. *International Journal of Economics, Commerce and Management*, (5). Retrieved from http://ijecm.co.uk/wp-content/uploads/2017/05/5549.pdf

SME Corporation Malaysia. (2017). *2016/2017 SME Annual Report* (ISSN 1823-8734). Kuala Lumpur: National SME Development Council (NSDC).

SME Corporation Malaysia. (2018a). *SME Annual Report 2017/2018:SME Developments and Outlook*. Retrieved from http://www.smecorp.gov.my/index.php/en/laporan-tahunan/3342-laporan-tahunan-pks-2017-18

SME Corporation Malaysia. (2018b). *SME Masterplan (2012-2020)*. Retrieved from www.smeinfo.com.my/policy-regulation/sme-masterplan

Sterne, J. (2010). *Social media metrics: How to measure and optimize your marketing investment*. John Wiley & Sons.

Stockdale, R., Ahmed, A., & Scheepers, H. (2012). Identifying business value from the use of social media: An SME perspective. *PACIS 2012 Proceedings*. Retrieved from https://pdfs.semanticscholar.org/902b/4a16bb0f6364c77f1f7c06ed6e490cc1ed11.pdf

Vivek, S. D., Beatty, S. E., Dalela, V., & Morgan, R. M. (2014). A generalized multidimensional scale for measuring customer engagement. *Journal of Marketing Theory and Practice*, *22*(4), 401–420. doi:10.2753/MTP1069-6679220404

Wang, Z., & Kim, H. G. (2017). *Can Social Media Marketing Improve Customer Relationship Capabilities and Firm Performance*. Dynamic Capability Perspective. doi:10.1016/j.intmar.2017.02.004

Wiesel, T., Pauwels, K., & Arts, J. (2011). Practice prize paper—Marketing's profit impact: Quantifying online and off-line funnel progression. *Marketing Science*, *30*(4), 604–611. doi:10.1287/mksc.1100.0612

## KEY TERMS AND DEFINITIONS

**Brand:** Represents the face of the company, the recognizable logo, slogan, or mark that the public associates with the company.

**Customer Engagement:** A customer's behavioral manifestations that go beyond transactions and have a brand or firm focus, allow firms to involve specific interactive experiences in iterative processes within the online SNS community, to build long-term relationship, and create more meaningful and sustainable interactions between the focal object (in this study, a brand) and the other actor or actors (in this study, a customer, or among customers), resulting from motivational drivers.

**KPI:** Measures (a collective of metrics) that constitute indicators of success to help firms in understanding how they are doing by mapping and defining progress toward their target objectives.

**Metric:** A measuring system that quantifies a trend, dynamic, or characteristic, has been categorized to five different types: amounts (e.g., volumes, sales), percentages (e.g., fractions or decimals), counts (e.g., unit sales or number of competitors), ratings (e.g., scales), and indices (e.g., price index).

**Social Media:** A group of internet-based applications that build on the ideological and technological foundations of Web 2.0, and that allow the creation and exchange of user-generated content.

**Social Media Measurement:** A process to gauge, explain and report the success of social media initiatives begins with the identification of social media marketing objectives, then the selection of appropriate KPIs and metrics that involves the collection of quantitative and qualitative data, and to populate metrics and generate insights which are distilled into report format to ascertain the successful alignment against firms' business objectives, and therefore (corporate) goals.

## Appendix

*Table 5. Definition of SME in Malaysia*

| Category | Micro | Small | Medium |
|---|---|---|---|
| Manufacturing | Sales turnover of less than RM300,000 OR less than 5 full-time employees. | Sales turnover from RM300,000 to less than RM15 million OR full-time employees from 5 to less than 75 | Sales turnover from RM15 million to not exceeding RM50 million OR full-time employees from 75 to not exceeding 200 |
| Services and Other Sectors | | Sales turnover from RM300,000 to less than RM3 million OR full-time employees from 5 to less than 30 | Sales turnover from RM3 million to not exceeding RM20 million OR full-time employees from 30 to not exceeding 75 |

# Chapter 11
# Cybersecurity Governance on Social Engineering Awareness

**Vasaki Ponnusamy**
*Universiti Tunku Abdul Rahman, Malaysia*

**Leslie Mark Pannir Selvam**
*Universiti Tunku Abdul Rahman, Malaysia*

**Khalid Rafique**
*Azad Jammu and Kashmir Information Technology Board, Pakistan*

## ABSTRACT

*Raising awareness on social engineering is becoming more essential. When we are connected to the internet, we are at the risk of becoming victims of numerous attacks. It is very difficult to protect ourselves from all of these threats, but the least we could do is not jeopardize our own safety. In order to achieve this, the need to have substantial knowledge on social engineering techniques and how to safeguard one from being victimized are mandatory. This research presents the overview of social engineering attacks. This plays a crucial role in supporting victims of cyber assaults.*

## INTRODUCTION

Security threats, when users surf the internet is growing exponentially. Vulnerabilities that are present in the internet can be exploited by skilled hackers or performing well-crafted social engineering techniques. Breach in cyber security has had massive repercussions in the past in terms of business, liability, reputation staining, customer confidence and productivity (Bradshaw, 2015). Such catastrophic consequences have led to governments and corporations to invest in many resources to build a secured cyberspace thus protecting themselves as well as their users. In terms of social engineering, these resources in training has to be poured unto the employees of the organization especially key personnel as they may poses some vital information of an organization. Leaking of such information can be detrimental for the particular organization in question. Therefore, people with such sensitive information, has to be well versed of

DOI: 10.4018/978-1-7998-1851-9.ch011

the risks they may be of being victims of social engineering as well as other cyber threats. Understanding these risks and knowing how to protect themselves from such scenarios may save an organization secrets. Orhan Sari and Karyda (2017) found that in the first half of 2018, over more than 4.5 billion of records have been breached. Based on Figure 1, most of the breaches (about 56%) occurs in the social media platforms due to the overwhelming usage and also the sophistication of social media attacks. So Orhan Sari and Karyda (2017) stated that these data breaches mostly happen due to human factor and lack of security practices. In addition to that, employees should possess a distinct and clear mind set in order to be able to distinguish potential threats from legitimate contents and their source of originality (Karyda, 2017). Immature perception of employees may lead to massive breaches within the organization (Karlsson et al.,2013). It has to be said that despite stern efforts up to this point, it is quite eminent that knowledge on cyber security still lacks competence. That is the reason why basic security breaches using simple social engineering techniques are still happening to this very date. Adequate knowledge regarding this problem would have seen some improvements of late. However, evidence collected up to this point suggests otherwise. All these factors reflect on how important it is to foster cybersecurity governance in organizations and on social engineering threats.

## Problem Statement

Ever since the emergence of social engineering, about 272 criminal cases using social engineering attacks are reported from January to September 2017 in Malaysia (Yuen, 2017). People are becoming aware of social engineering attacks but the actual technique of social engineering attacks to get confidential data are still unknown. As seen in Figure 2, social engineering is the second common attack after malware attacks. The attackers exploit human vulnerabilities to take advantage from the victims. The major problem why people are easily falling into the trap of social engineering is due to lack of education on the various types of social engineering attacks as well as the techniques the attackers are using. Social engineering comes in diverse form of reaching the target victims and also purposes. The easier the people's emotion get manipulated, the higher the chances of falling into social engineering attacks. Majority of the social engineering victims are the elderly because the elderly community is less familiar with the technology knowledge. Another problem of social engineering that exists in the society is the oversharing of per-

*Figure 1. Number of records breached by industry first half of 2018*
*source: Karyda (2017)*

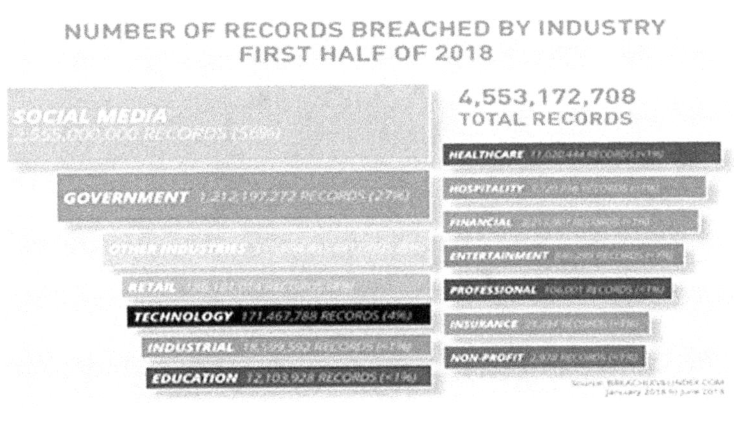

sonal data on online platforms. Online platform is now becoming the daily entertainment to majority of the people, thus online platform has also become a medium for the attacker to spot for their victims. People would share their personal information online by holding the mindset of sharing is caring, but oversharing is careless, and this habit has become the main reason to encounter social engineering, said CyberSecurity Malaysia. One of the renowned online social media platform – Facebook, would be a great medium for attacker to seek victim. Using the features in Facebook, people are able to share their name of school, the courses they are taking and even their workplace on their Facebook profile. For this feature, they are given the option to fill up the information, but in order to let other better understanding them, they choose to public their information online. Therefore, the attacker could use the information published to find them. Even the user does not publish anything online, the cybercriminal could still exploit the vulnerability of the online platform and get the user data. For example, from the workplace name the user has published on Facebook, attacker is able to get the workplace address and the office number. With further call to the office, the attackers will provide complete information of the user (acquired via Facebook) and start their social engineering plan. Eventually, because of the completeness of information provided by the attacker, the victim would easily provide their other confidential information to the attacker.

Besides that, the lack of social engineering countermeasure enforcement in an organization or institution is another reason for the social engineering exploitation. To many of the small and medium-sized enterprises (SMEs), they are less emphasized on practicing social engineering awareness. For example, no cyber security relevant policies and training are enforced within the organization. Although they have advanced technical support to protect the organization data, but the organization information could still be exposed and leaked due to human factor. Hence, the organization should come out with some social engineering countermeasures to prevent financial lost or customer information lost. Organization should strictly disallow their staff to use the organization computer for personal use.

*Figure 2. Attack methods*
*Source: (Frumento, 2019).*

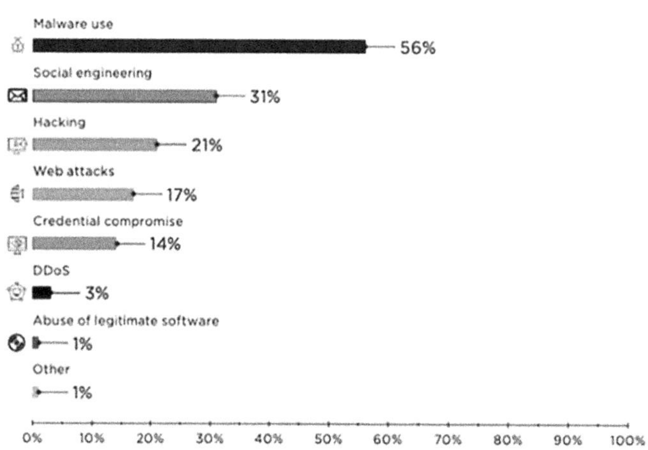

## Motivation

Normally, social engineering attacks is very common in small and medium-sized enterprises (SMEs). One reason could be that small and medium-size enterprise are unable to afford for intrusion detection system (IDS) and also intrusion prevention system (IPS). SMEs has limited funds to enhance their network security, and training their employees too. On the other hand, most of the small and medium-sized enterprises have shortage of skilled security workers to strengthen their network in order to counter against the attack. As compared to cybersecurity awareness, SMEs are more focused on financial gain. Thus, SMEs need to provide training on social engineering attacks in order to raise awareness of employees when facing similar attacks. People with pivotal information or sensitive information of an organization must be aware of all up to date threats when they are accessing the internet. This would require very efficient and up to date training so that these personnel can defend themselves from such threats. Organizations are not giving adequate importance towards social engineering techniques. This is because attackers these days have adequate skills to bypass an organizations security with ease and their reliance on employees working in the organization to disclose sensitive information is very less to none. This has prompted organizations to pump more resources and funds towards building a more highly secured system so that attackers will find it more difficult to gain access to the system and they often overlook the most basic and simplest way to attack an organization which is through social engineering. Regardless of how secured a system is designed, the employees within the organization will be able to access because they would need to. But if the employees themselves are the weakness within the organization, attackers could easily exploit that by using social engineering techniques. This is the essential problem that many organizations are not giving adequate importance. Figure 3 shows the four stages on how an attacker gain trust and collect information from a potential victim.

**Step 1:** Information Gathering

Most of the attackers spend most of their time and attention onto this step as this is one of the keys to success in using social engineering. The attacker would first identify their target and gather portion of background information and determine possible passwords, security numbers and more. On the other hand, attackers would identify several attack methods to achieve their targets.

**Step 2:** Establish Relationship and Rapport

*Figure 3. The cycle of social engineering*

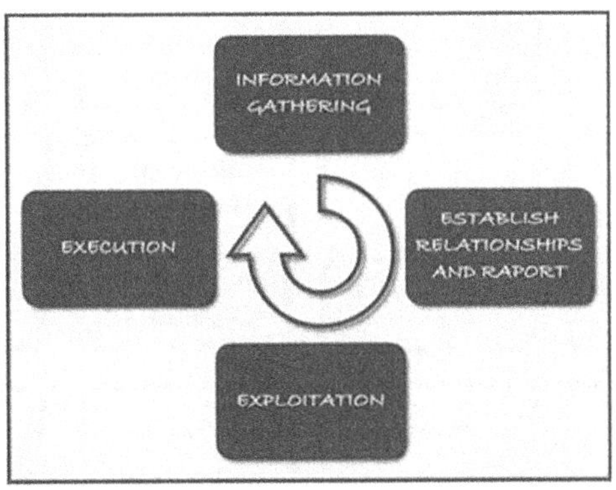

In this step, attackers will cheat on its target to obtain a foothold. Attackers will be taking control of their interaction with its target to establish a relationship. For example, creating a fake profile on social media or social networking site in order to have a connection with its target in order to accomplish its goals.

**Step 3:** Exploitation

When attackers had successfully achieved step 1 and 2, he/she will be focusing on maintaining the relationship with its target without raising any suspicion.

**Step 4:** Execution

This is the ultimate goal of social engineering attack where the attacker ended its attack without raising any suspicion. Generally, attackers will not leave their target immediately by just questioning while they will do something that will be benefit to both side in order to allow possible future interactions to be continued.

## Types of Social Engineering Attacks

Figure 4 shows that there are many different types of social engineering attacks in which phishing, baiting, pretexting and tailgating are the main categories of such attacks. The attacks with some examples will be presented in the following sections.

From Figure 5, it is understood that there is only 3% of malware are exploits through vulnerability of the systems. The remaining 97% of the attacks are through social engineering approaches through human interaction to find the vulnerability to hack into the systems. The statistics collected might vary but majority of the attacks are through human vulnerability instead of finding flaws of the hardware or software of the system. The biggest challenge of cybersecurity of SMEs nowadays are social engineering which are instigated through human error.

*Figure 4. Types of social engineering attacks*

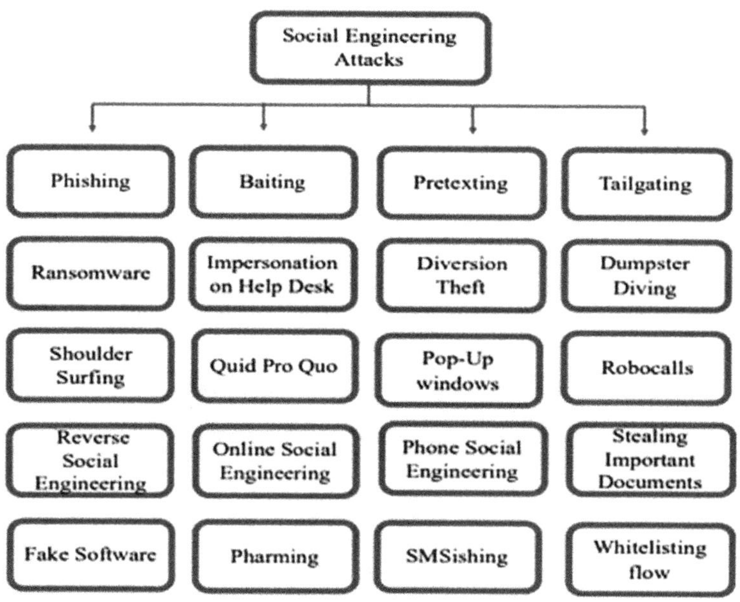

*Figure 5. statistics reported by APWG for Q2 and Q4 2016. CEFRIEL's elaboration of the APWG data reported in the periodic "Phishing activity trends report. Unifying the global response to Cybercrime" (APWG, 2016)*

| Malware Infections by Type | % of malware samples |
|---|---|
| Trojans | 67.01% |
| Virus | 1.54% |
| Worms | 3.33% |
| Adware / Spyware | 1.09% |
| PUPs | 27.03% |

Human enabled threats: 95,22%

Automatic attacks: 4,87%

| Malware Infections by Type | % of malware samples |
|---|---|
| Trojans | 74.99% |
| Virus | 1.55% |
| Worms | 1.50% |
| Adware / Spyware | 0.51% |
| PUPs | 21.45% |

Human enabled threats: 96,95%

Automatic attacks: 3,05%

From Figure 6, it is noted that social engineering attack had increased from 6% to 29% in Q1, from 15% to 25% in Q2 and in Q3 social engineering attacks had increased from 15% in year 2017 to 37% in year 2018 which had increased by 22%. Q4 of 2017 received 11% of social engineering attack and had increased to 31% in year 2018 . Overall, there is a high increasing trend of social engineering attacks from year 2017 to year 2018 (Salahdine and Kaabouch, 2019).

## Phishing

Phishing attacks (Figure 7 and Figure 8) are amongst the most common forms of social engineering. Phishing attacks idea is to obtain sensitive information let it be names, address, security pins, phone numbers, medical information and financial status. This method utilizes link shorting or embeded links that would redirect the user to a suspicious website. The uniform resource locator (URL) of this suspi-

*Figure 6. Percentage of social engineering attack on year 2017 and 2018 Source: (Salahdine and Kaabouch, 2019)*

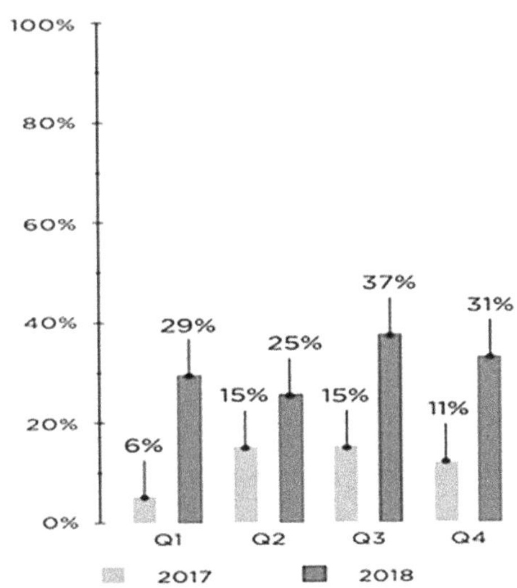

*Figure 7. Types of phishing attacks*
*Source: (Salahdine and Kaabouch 2019)*

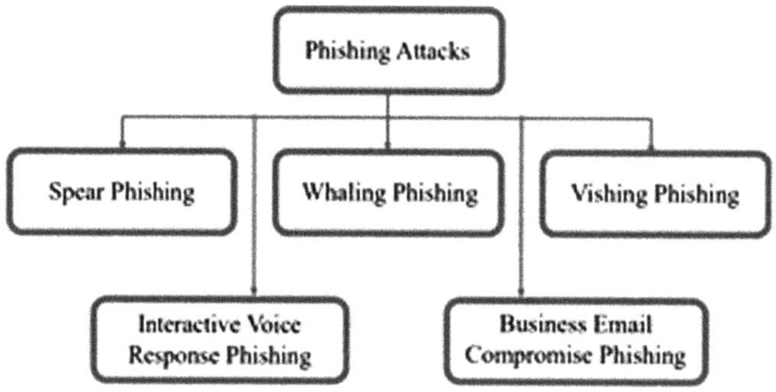


*Figure 8. An instance of a phishing attack targeted to a victim*

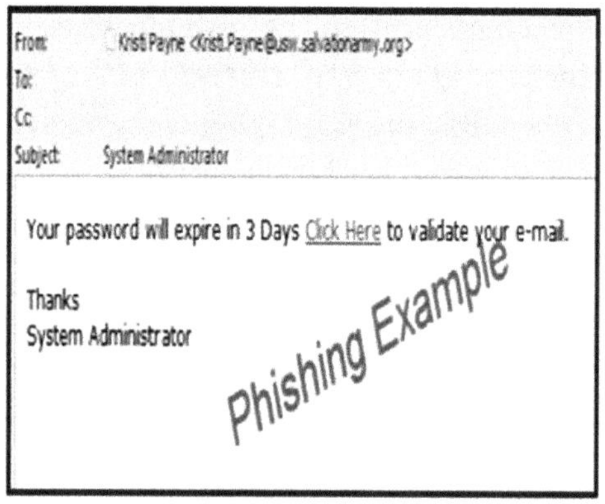


cious website would have a similar layout to that of the original website itself. This would make the victim enter these sensitive information into the website as he or she thinks that the website is actually legitimate (Walker, 2016). Other methods would include the incorporation of threats and fear in an attempt to manipulate the victim to react in a way that the attacker wants them to. Phishing attack can be generally classified into 5 major categories namely spear phishing, whaling phishing, vishing phishing, interactive voice response phishing, and business email compromise phishing (Walker, 2016).

Another example of phishing attack is by utilizing email links as shown in Figure 9. The attackers would mask themselves from a trusted organization and send a malicious link to the victims to gain further data. In a normal type of phishing email, the attackers would modify the email by changing the website URL with a different URL. The users being not aware on thinking that this is a legitimate

*Figure 9. Modify website URL*

and might reveal their confidential information. Most of the time phishing attacks comes with sense of urgencies, that would lead to giving out confidential information.

According to Phishing Activity Trend Report in May 2019 (Anon., 2019), phishing through Webmail and SAAS (Software-as-a-Service) was the highest at 36% as compared to other sectors (Figure 10). Figure 11 presents the statistics of phishing attack of different organizations in 2018 by Vergelis et al, (2019).

## Pretexting

This form of social engineering is when attackers emphasize on developing a good pretext, or a fabricated scenario in which they can steal their victim's information. For instance, an attacker would pretend that they require bits of information in order to verify their identity (Figure 12). This would trick the victim into entering the information as they feel that it is not significant enough to be a threat (Walker, 2016). These forms of attack can be more complicated as they can manipulate personnel within an organization to exploit certain structural weaknesses of a particular company. Comparatively, phishing attacks uses fear and urgency to exploit an individual, but pretexting rely on building a false sense of trust in order to exploit individual. The attacker in this scenario has to make himself or herself credible to the victim. The victim is this scenario does not experience a sense of fear or pressure to act in a way that is wanted by the attacker, however, the victim feels they can trust the attacker so that it helps them in some sort of way not knowing that they are attempting to exploit weaknesses within the victim's organization (Jain et al, 2016).

*Figure 10. Most-Targeted industry sectors 2019*
*Source: (Anon., 2019)*

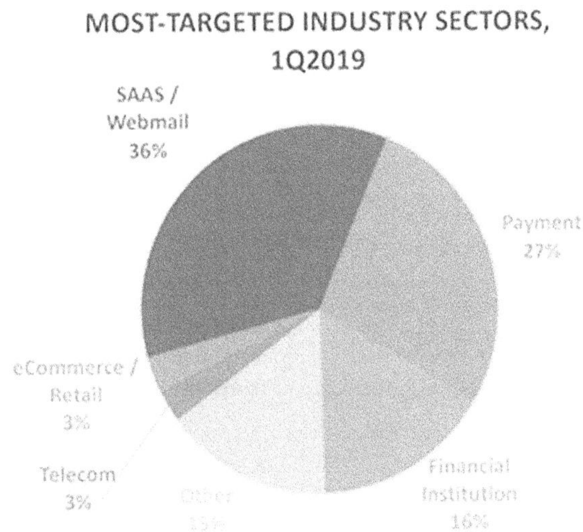

*Figure 11. Distribution of organisation subject to phishing attacks by categor*
*Source: (Vergelis et al., 2018)*

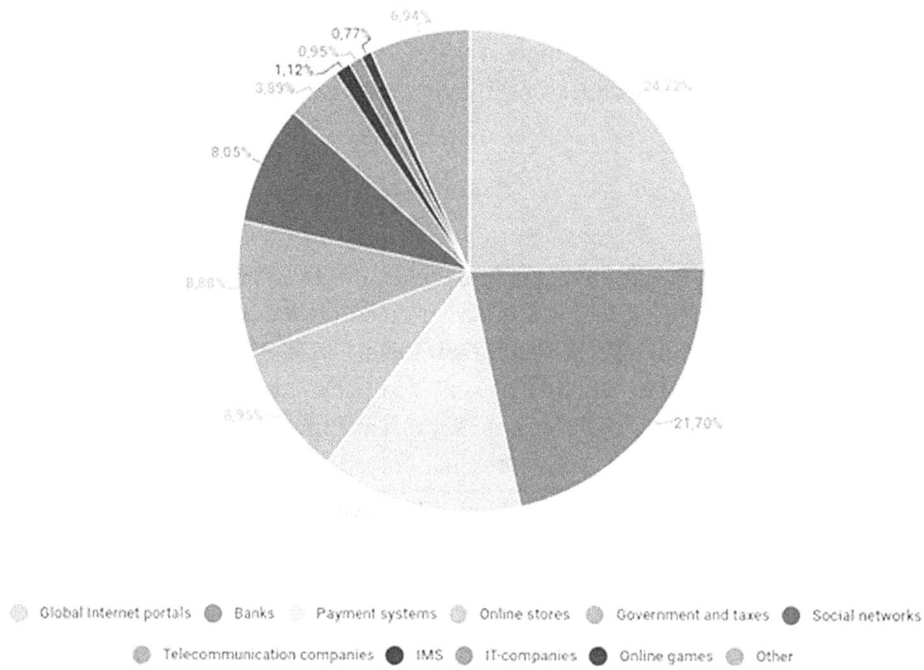


*Figure 12. A link appearing on a trustworthy email prompting a download*

## Baiting

Baiting (Figure 13) share very similar characteristics as phishing attacks. However, the difference between the two would be the promise that the victim would receive as a whole. This would then entice the victim to act in a way that would benefit the attackers as the victims thinks that they are also gaining from all of this (Jain et al, 2016). Baiting has a unique characteristic to it though. It is not restricted to online schemes; attackers may use it to exploit human curiosity as well by utilizing physical media to entice the victim. An example is shown in Figure 14 whereby the attacker might lend a thumb drive to the victim but indirectly collecting some information from the victim.

## Quid Pro Quo

This social engineering methods has very similar roots to baiting. This attack promises a benefit in an exchange for information. This is usually done in the form of a service. For example, if a user wants to

*Figure 13. An unofficial Google browser update. Enticed victim by promising better performance*

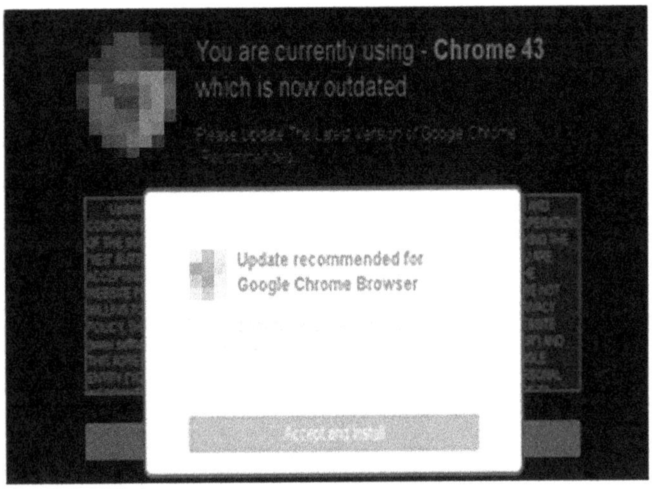


*Figure 14. Shows an example of a fake thumb drive (Kessier, 2017)*

enhance his software, the attacker would then ask for user information in order to validate the install. Therefore, the victim would then input the information that is requested by the attacker via the software (Koyun et al, 2017). It is to be noted though, using this social engineering attack there can be a few cases where it is much less sophisticated. Employees are willing to give out information for very cheap rewards as they lack of understanding of how important information is and can be misused due to one's negligence (Figure 15).

## Tail Gating

Another social engineering attack would be the tail gating attack. This attack is also known as the "piggybacking". This attack involves a person who is lack of proper authentication following a person who does into a restricted area which may contain sensitive information (Conteh and Schmick, 2016). This attack involves impersonation as well in which the attacker would impersonate as a delivery driver, maintenance staff or other viable positions which is not the true job of the attacker. By following people (Figure 16) with access to confidential areas they can exploit highly sensitive data. It has to be said that, this form of social engineering is not very effective nowadays (Ahmad, 2017) (Abas et al, 2017). This is

*Figure 15. A malicious software asking victim to sign up for pro subscription of a desired software*

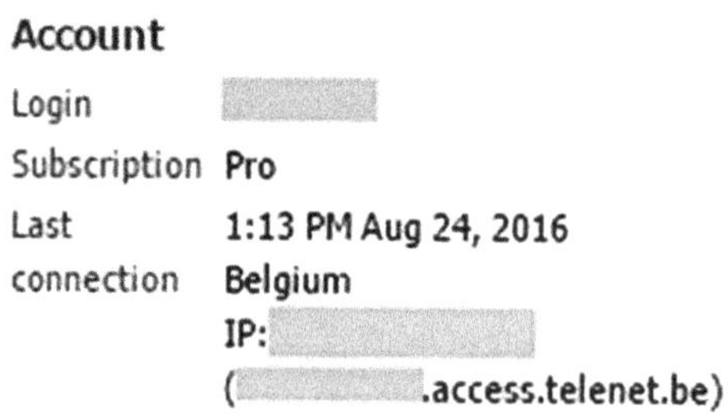


*Figure 16. Demonstration of tail gating*
*Source: (Tandon, 2018)*

because physical security these days are more effective, especially in large corporations as very limited personnel is even allowed to areas with sensitive information such as data centres.

## Watering Hole

Watering hole attack is an attack by injecting malicious code into public Web pages that the targets used to visit frequently. It is also a common social engineering attack that is used by cybercriminal and hackers. The main goal of this attack is to infect the target's computer in order to gain access to the network at the target's working place. In a watering hole attack, attackers would look for chances to infect the websites with malware to make the target vulnerable (Figure 17). Once the target visited the page, a backdoor Trojan is installed into his/her computer. Watering hole attack is an attack that focuses on popular websites. To gain access of the website, the attacker would first attack an employee profile who are some key personnel in the top management and they would identify the type of websites visited most frequently. The attacker then injects malicious HTML code to the target to separate site where the malware is hosted. Once the injected malware is accessed, the attacker is ready to infect the target.

# GUIDELINES AND RECOMMENDATIONS ON PROVIDING SECURITY AWARENESS

Creating awareness on social engineering techniques and how to safeguard is very crucial to avoid becoming a victim mainly employees. Informative handouts should be given to all employees of an organization and not only to those who are in positions that deals with sensitive data or information. In addition to that, organizations can conduct safety programs that trains employees to be able to detect any form of social engineering threats (Koyun,and Janabi, 2017). Furthermore, undergoing tests to see if one is well equipped to identify potential social engineering threats. This test would show how knowledgeable is an organization's staff towards social engineering techniques and how to safe guard themselves (Abas et al, 2017). Organizations should also provide online assistance to their staffs should they have

*Figure 17. An example of water hole attack*

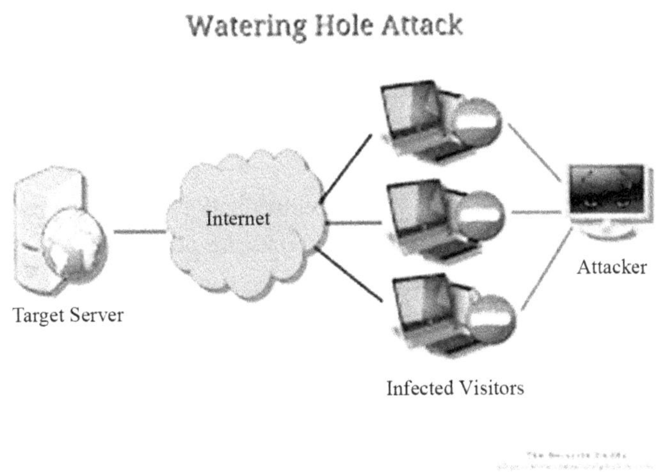

any queries associated with any form of social engineering threats that they may feel they are exposed to. For example, providing step-by-step guidance of how to deal with a scenario in which they feel they are being exploited (Aaron et al., 2018).

In order to prevent social engineering attacks, the very first thing one should do is to learn and understand how does these social engineering attacks work. Reading materials on social engineering attacks would improve one's knowledge. Moreover, attending working or training related to cybersecurity will improve our awareness regarding to social engineering.

1. Always think before you click on any message or email

Attackers are normally targeted on our sense of urgency to make sure we act first before thinking about it for example like phishing or baiting. When we get a highly urgent or suspicious message, do remember to double or triple check any messages we got. The best solution towards this problem is to fully utilize different methods of communication like texting the person to check whether they sent an urgent message to us or that was an act from an attacker or by setting up a spam filter in email.

2. Reject requests for help or offers to help

Legitimate companies and organizations will not contact us to provide help. But criminals may pretend as legitimate companies and organizations responding to our requests for help and on the same time, offering more help. They would normally pick companies or organizations that are well-known to avoid suspicion on us. It will be a good opportunity for the attackers because the victim would respond instead of getting stuck with the problem.

3. Install anti-virus software

Installing an anti-virus software is the best way to prevent computers form getting infected by viruses such as malware and spyware. Hackers and viruses go hand in hand, a good anti-virus software will guarantee a protection when surfing the Internet, preventing attackers to gain access to sensitive information like username, password, bank account and many more. Make sure to always keep the anti-virus up to date and also set automatic updates so that when a new patch is released, the software is patched immediately. Last but not least, remember to perform a daily routine anti-virus scanning on computers for prevention of malware and possible infections.

4. Using File Transfer Tools instead of email

File sharing program is a more secure way to electronically transfer sensitive information. A few different options for private sharing of documents and data are provided by applications like ShareFile. This is because they are included with Outlook add-on software that will encrypt an email message simply by just clicking an icon. Next, Dropbox is also one of a commonly used file-sharing tools for small business. Users can set up a Dropbox account and upload the files and Dropbox would share the file to the specified people. Those people would be notified by Dropbox. So, using a file transfer tool are more secure than using an email.

5. Use unique passwords

Password is a secret word that is used to allow access to a computer system. The strength of password would determine the security in protecting against any social engineering attacks. When creating a strong password, remember to use eight or more characters with a mix of numbers, symbols, capital letters and lower-case letters. Avoid using password that are easily guessable like 1234, abcd, date of birth or own name. Never use the same password more than once. If there are too many different passwords for different sites, use a tool like password managers to secure it in one place. Moreover, using a two-factor authentication will make password mush more difficult for hackers to get into it.

## Secure All Devices

Everyone should know how to prevent and protect their own assets. Secure all devices can simply mean to protect or keep our own pieces of equipment or assets well. We may begin from a simple step which is refusing to lend or let someone use our personal computer or any other devices, this may prevent our computer falls into a trojan-based attack or keylogger; because when you lend your computer to others, mean they have the full right to access and perform anything they wanted to at will and without your knowledge. No one knows if they are intended or accidentally plugged in a USB drive that could contain a trojan-based program; a trojan-based program will simply destroy or perform any malicious action towards your computer. Alternatively, you may enable the filelock feature on your system which requires the passcode before accessing the folders or files. It's used to prevent someone else from unauthorized access to your personal folders or files.

1. Do not expose everything on social media

As we are moving towards the 21st century, social media is one of the most famous trends and attractive things that everyone is doing. Posting ourselves to the internet is no longer a strange thing to us. Thus, from social media, a social engineer might find the simplest way to dig our personal details. We may not be aware that information, photos or personal details that were posted on the internet are in an invisible risk. For example, Facebook is a social media platform that more than millions of people

*Figure 18.*

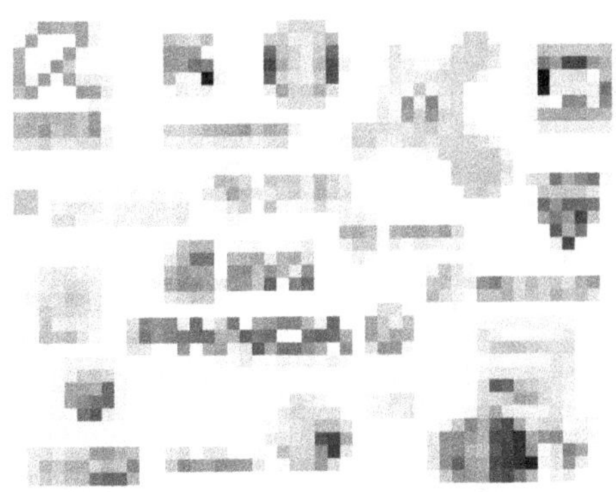

are using it every day. We often update our personal details including, life and whereabouts. But we do not know that behind it could benefit social engineer in collecting and gathering our personal details. Be conscious of what you posted on social media and knowing what should and shouldn't be posted

## TRAININGS AND POLICIES TOWARDS SOCIAL ENGINEERING AWARENESS

In terms of training an organization's staff, the training they receive must be more frequent and also educating them on how to deal with the latest threats (Aaron et al., 2018). The organization may not need to only focus on social engineering threats but also other forms of threats to organizations information safety (Safa et al., 2016). Up-to-date training would aid employees to deal with threats that are relevant in today's day and age. With regards to the policies, organizations would need to come up with information security policies compliance as this would protect organization assets (information) and also ensure employees are more careful when dealing with such information. Negligence should result in a heavier penalty. At the same time, attackers who are caught should be given greater punishments as that would put a higher risk for those who intend on performing social engineering attacks.

In data security, technical side of security such as firewalls, scanning and penetration testing are used to prevent social engineering. But did you know an attacker could bypass IT security during a social engineering attack? If employees are not being trained on how to overcome it, social engineers would make themselves as the owner of the company, walking in and out as they wish and also stealing data in a very short amount of time. Figure 21 shows a detailed overview of security awareness programs for personnel at all levels.

1. Training

When new employees are first hired, train them to have a training session once a year. The purpose of the training is to make sure that new employees are able to be aware of social engineering attacks. After completing the training, new employees will be able to fully ensure the safety of organization by preventing social engineering attacks. An example of training system is shown in Figure 20.

The organization should train every new employed staff in a one to one basis about social engineering awareness and have an unannounced, periodic test of the security framework. Besides, an organization should implement a policy that the employee is not allowed to use company devices for personal use. Only using personal computer to do personal matters is allowed. Moreover, organization should also encourage

*Figure 19. Shows the depth of security training program for all levels of personnel*

*Figure 20. Awareness training system educates employees to prevent cyber-attacks, social engineering*

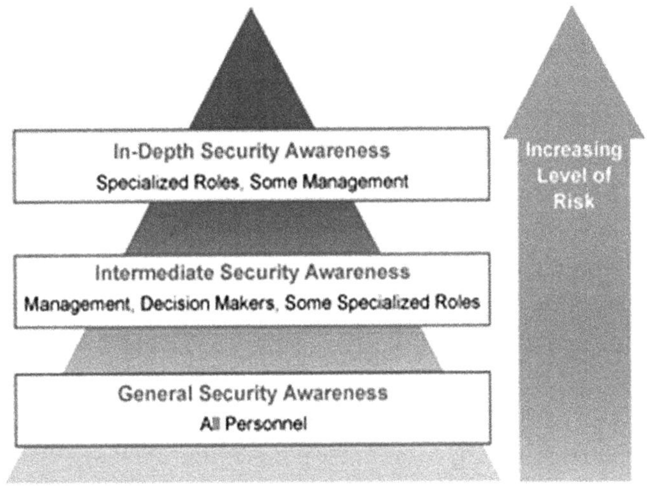


*Figure 21. Tips for avoiding a social engineering attack*

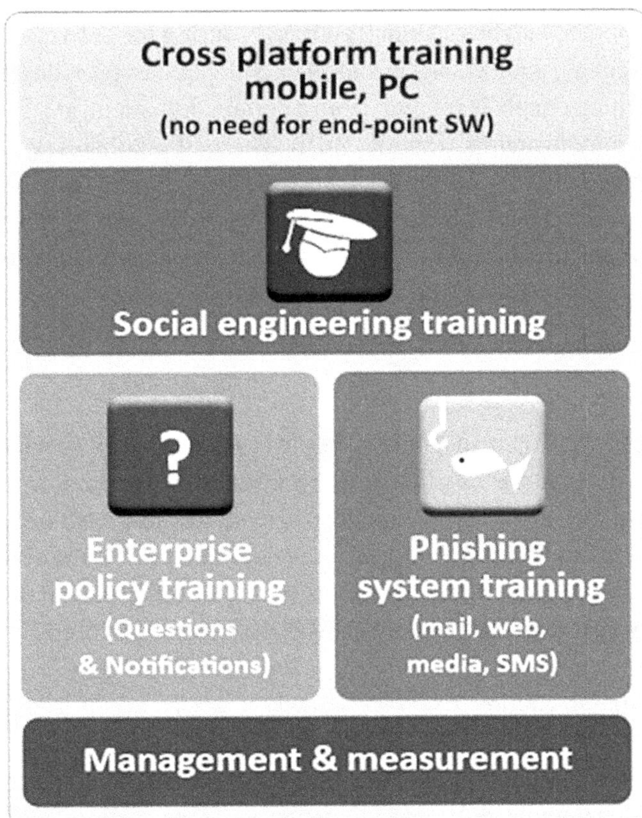

employees to get themselves involved in security education and training awareness (SETA) program. It helps to develop the employee's information security technique and awareness by running through a series of education and training. An organization or individual should always ensure that the firewalls is activated and an up-to-date and anti-virus software is installed. Both parties should be very careful to a suspicious physical media. Organization should implement a policy that disallow their employee to plug-in any physical media into the organization's devices. If such action is found, a punishment will be given to the employee. While for an individual, if a suspicious physical media is picked, should return it to the nearest lost-and-found centre. For human factor concern, before allowing authorized user to enter restricted area, a form is required to be filled up as an entry record for the organization.

2. Create policies

Most of the employees do not know how to solve the social engineering attack when they find themselves in this uncertain situation. Creating some of the policies like only use USB that are directly obtained from IT department, reporting lost of ID within 12 hours of discovery and immediately informing the top management when encountered a social engineering attack. By establish those policies, it will help the companies or organization to spot out suspicious activities quickly. Every company should strictly have a Sender Policy Framework (SPF). It is an email authentication or validation tool that is designed to detect and block the spoofed and forged email addresses during the delivery of email. The successful rate of using SPF to prevent phishing emails is high. The SPF policies provides the mail services that are able to detect if the incoming emails is sending from a certain domain inside the authorized list of hosts approved by the domain administrators (Raulot, 2019). These SPF policies will be implemented in two parts, that is SPF checking and SPF publishing. In the SPF publishing process, it is able to detect if the email server is authorized to send emails from the company. While in SPF checking process, the email provider can do a nslookup procedure to check whether the domain name is listed inside the domain name system (DNS). Else, it is a failed SPF validation.

3. Put staff to a test

People learn best by doing. Top management should conduct a secret social engineering situation to see how do the staff react to it and giving them an opportunity to practice how to solve social engineering attacks. By this way, it may also help the top management to know what needs is needed and what to be improved within a company's security. Some of the common social engineering like leaving a USB device around your site and obverse where it ends up, posing as an IT person that fixes the network and see how do those employees stop it before they can get into the server room.

4. Encourage a questioning culture

Working under a questioning culture will help the employees feel empowered to question why someone is visiting the company without an ID and also when someone needs to access into the server room. Moreover, it will also empower them with real-life skills to stop social engineers from getting into the server room. They would have a greater decision making and problem solving skills when the social engineering attack happens. Employees will work as a team to prevent social engineering attacks. For example, if they feel that the personnel is suspicious, they might need to ask the personnel about "Can I see your ID please? I need to verify your identity."

In order to stop attacks that are related to social engineering, the best defences are to have early detection programs and good defence programs that could fight all types of attacks. If the attacks are detected early, the sooner the action could be taken in order to overcome the attacks. There are several types of hints that show a social engineering attack (Figure. 23). For example, when someone created a massive sense of urgency it could be a type of attack. This urgency was created to trick someone into making mistakes. Second common clues that related to an attack is when someone asks for private information that they have no access to or they should have already known. Examples of the private information are account number, phone number, address and others. Thirdly, the most common attack would be some attackers asking for password. There are no legitimate companies or organizations that would be asking for password when accessing into websites or folders. Fourth, when someone asking to ignore or bypassing all the securities at work, it could be one type of social engineering attacks. By ignoring or bypassing the securities system, the attackers can get into the company's or organisation's systems much easily. Fifth, the most common attacks that could be found in the internet is fraud, where everything seems too good to be true. The most common example of fraud would be emails notifying someone that they have won a lottery or contest. To prevent falls into these types of fraud, a person should not simply share their private information to the attacker. The last clues that could be found is when a person receives an email from their close ones but using wording that does not sound like them. This could happen as the hacker could have hacked into their account and trying to fool someone. Using verification when reaching out the close ones using different communication methods could be a good self-protection for the users.

When there is suspicious person trying to attack or fool the system, the best way is to cut off all communication with the person. If the attacks are related to work, report to the information security team right away so that the information security team could come out with several prevention actions that could be taken so that the company's or organisation's information would not be leaked out to their competitors or outsiders.

## Social Engineering Attack Detection Methods

Social engineering detection methods can help to decrease the impact of social engineering attacks mainly in SMEs. Some kind of detection model can raise employee awareness towards social engineering attacks. Enhancing existing solution or developing new prevention and detection techniques are also important, but research efforts also need to focus on measuring the impact of social engineering attacks. Investing efforts on social engineering detection models are important to overcome the social engineering attacks. First and foremost, text-independent voice signature authentication system had been used as a social engineering detection model in an organization. Text-identification speaker verification is a process of verifying the identifying without enforcement on what content had been spoken. First the organization need to collect the voice of their employees and store their voiceprint in the database of the

Figure 22. Phishing e-mail detector "PILFER"

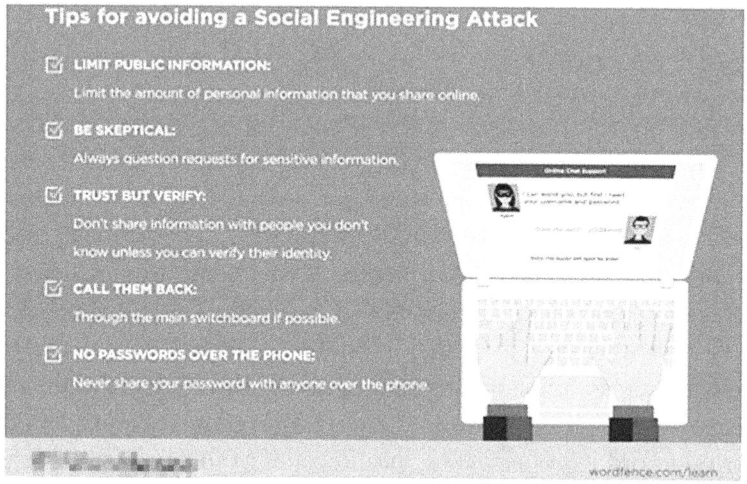


Figure 23. Social engineering attacks detection model (SEADM)

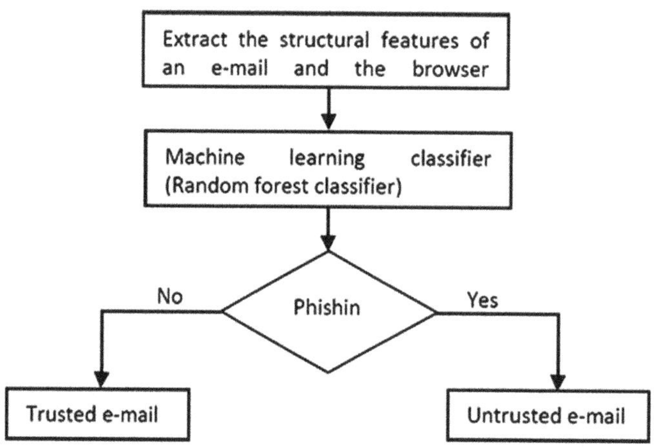


company. When the attackers need to masquerade themselves as company employees and gaining trust of other employees via voice call, they need to match their voice print before performing other activities. This is a very difficult method as the voice signature changes, the attackers need to modify their voice according to the new voice signature.

Another type of approaches is through Automated Individual White-List (AIWL). This approach would prepare a list of URLs that are legitimate websites and are of trusted source. The AIWL will be able to identify the pattern of the trusted websites and later when the user trying to access a different website, AIWL would compare the patterns of the websites. Any mismatch in pattern will be detected as untrusted site and warning would be sent to the user. Sometimes the hackers will slightly modify the domain name of the URL link, and most of the users are not aware of this change in the URL. The detection model is able to detect the suspicious email that had sent the suspicious URL and alert the end user to not to click into the email and to delete it immediately.

Another phishing e-mail detector "PILFER" had been used. As shown in Figure 24, PILFER extracts structural feature from the email, such as IP-based URLs, the domain names, performing WHOIS query to get the registration date and the domain name and comparing that with the of the e-mail information. The machine learning classifier learn how to examine the email pattern, such as number of links, number of domains, JavaScript. After that it would be able to classify which is the trusted email and which is not. The phishing email always comes with the urge of asking user to click on the link in the email, thus, if the user is not aware of the phishing email then alert message will be shown warning user to not to click on the suspicious URL link.

Emotional state of a person can be an advantage and disadvantage when it comes to social engineering attacks. When facing social engineering attacks, the person with positive emotion state will be more rationale when facing the social engineering attacks. Designation of SEADM model (Figure 25) are useful for detecting social engineering attacks when breaking down the decision-making process into manageable components. Before this model has been developed, people found it difficult to make rational decision in very limited time-framed especially regarding complex matters because human does not have only one ideal decision option (Monique Bezuidenhout, Francois Mouton, and H.S. Venter, 2010). Subjective expected utility theory model is an old method which has the decision analysis to improve the theory model before the author created SEADM. (N Braisby and A Gellatly, 2005) They did not have a set of pre-existing values and their decisions will be based the present context and demands of the decision.

This model would show each of these states individually before the full model is discussed with examples. The term individual is defined as the person dealing with the incoming call and the term requester is defined as the person whom is making the call and requesting the information. (Monique Bezuidenhout, Francois Mouton, and H.S. Venter, 2010) This model would individually analyse the emotional state of the individual by asking some questions pertaining to authentication and only then would provide access to the requester. The strength of this model is, it analyses the requester emotion from the perspective of phycology which is very hard for the attacker to hide from. One such example is requesting account balance. If a person requesting account balance, emotion state will be analysed and authorisation questions will then be asked. However, there are some limitations, for example, when a person has some knowledge about phycology and then may try to hide their emotion from others.

SEADM allow users to examine the social engineering attack by just following the steps given. However, no matter how strong the model is with a negative emotion state, the person might become the victims for social engineering attacks, the attacker is able to trigger the victim's emotion and causing them to follow the attacker's request. Another limitation found in this model is that it only works in bidirectional communication. Human based mitigation technique is based on human judgement whether is legitimate or malicious. Two approaches used which are auditing and policy in which policy is to defend and control on employees' action when facing social engineering attacks. Whereas education, training and awareness are used after applying auditing and policy approach. However, human judgement is a crucial part even having strong awareness towards social engineering attacks (Salahdine and Kaabouch, 2019). Technology-based mitigation used to increase the accuracy of human-based mitigation technique, but are not trusted by human at times.

Another method that had been used to solve social engineering attacks is the use of Social Engineering Defence Architecture (SEDA). SEDA is able to detect social engineering attacks through a phone call and records down for forensic investigation. SEDA system uses text-independent voice signature authentication systems which records down the employees' voice signature and will be store in SEDA system (Raskin et al., 2004). It receives any input secretly without notifying others to examine what had

*Figure 24. Social engineering attack framework*

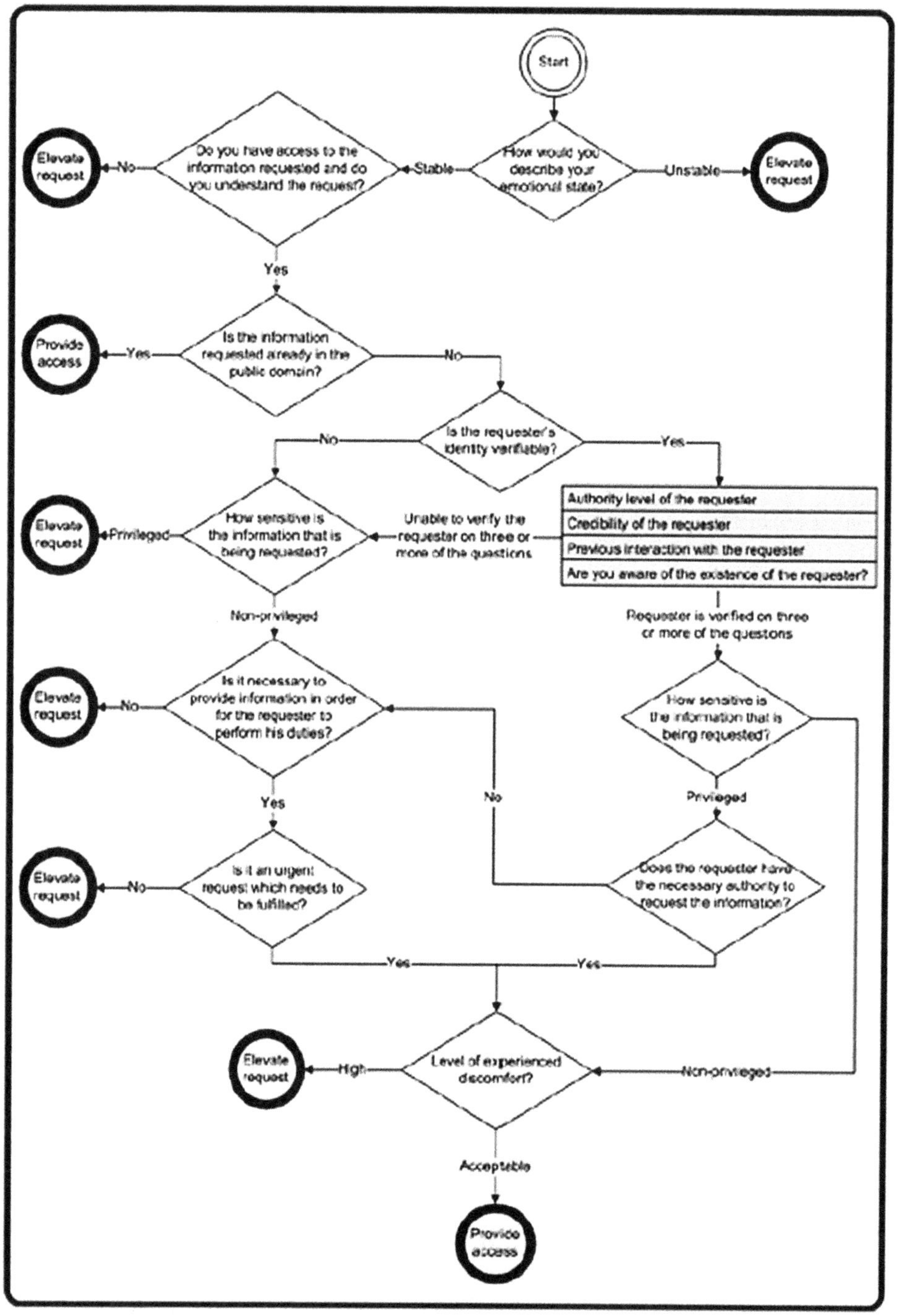


been spoken in the phone call so it would not easily detected by the attackers. This strategy successfully prevents the social engineers that masquerade as employee and attempt to collect all information through a phone call. However, the attackers are able to modify their voice signature and with another

*Figure 25. Social engineering defensive framework (SEDF)*
*Source: (Valerie Thomas (2014)*

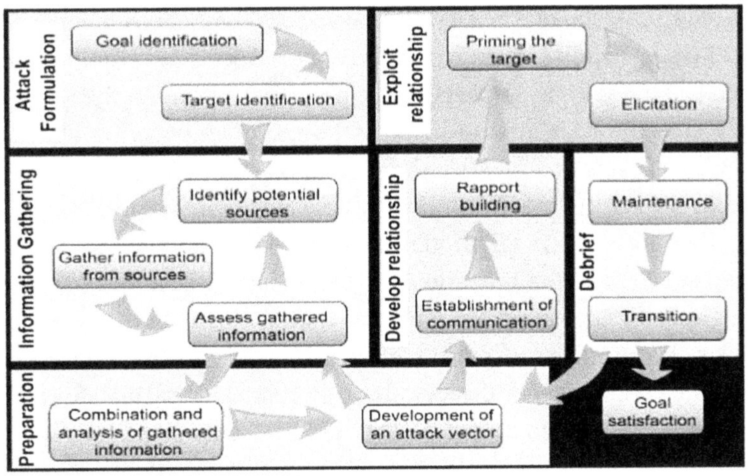


name and can change their voice by using voice modulation module. Although this model passes skimming for voice signature, SEDA also designed other powerful social engineering detection models. For instance, SEDA had designed a voice-to-text engine, which are able to convert voice conversation into text. However, this voice-to-text conversion engine need to be strong enough to deal with bad phone connection (Hoeschele and Rogers, 2006)

The authors (Mouten, F., Louise, L., Mercia, M. M., & Venter, H. S., 2014) proposed an attack framework based on Kevin Mitnick's social engineering attack cycle. The authors use a previously proposed social engineering attack ontological model which provides a formal definition for a social engineering attack. The Ontological model contains all the things of social engineering attacks and also the social engineering framework. The framework shows all the techniques of social engineering attack such as attack formulation, information gathering, preparation, developing relationship and debrief. The whole model of the attack framework can refered to Figure 24.

Another Social Engineering Defensive Framework (SEDF) was proposed by Valerie Thomas. SEDF has defined 4 phases (Maseno, 2017) as shown in Figure 25.

1. **Determining Exposure:** This phase focuses on seeing sites and other available resources as the attacker.
2. **Evaluating Defenses:** This phase can be used to evaluate employee resistance and reaction to simulated attacks.
3. **Educate Employees:** This phase involves teaching employees how attacks are executed and their impacts.
4. **Streamlining Existing Technology and Policy:** This is through improving effective defensive technologies, which are likely in your environment.

All of these methods can be performed according to the needs of company. This method can provide the training and awareness to the employees. It can also define the policies to the specific company. However, it more on physical technique and it did not have the computer technique. The research adopted the conceptual model (as shown in Figure 26) of the major aspects of social engineering-based attack by L. Janczewski and L. Fu (2010) as it meets all the objectives of the study.

Machine learning algorithm are able to identify the social engineering attacks before it happens by learning from a big data sets and identify the pattern of malicious activity. The machine learning algorithm also used in IP traffic classification (Nguyen and Armitage, 2008)(Zander et al., 2005) in helping of intrusion detection systems (IDS) identify DDOS attacks. Machine learning analyse historic cyber-attack dataset and calculate the probability of the attack happen in same network again. On the other hand, machine learning algorithm are launched automatically to analyse report without human intervention. However, the attacker uses machine learning as a tool to perform cybercriminal attacks such as use machine to gain unauthorized access for a system, and machine learning had been used to write malicious malware code that could not be detected by the machine learning based systems. Another limitation found in machine learning algorithm is the data set quality provided to machine learning algorithm to study have to be good, because the bad data and model will show bad result.

Machine learning had been used as an automatic tool to identify and observe the patterns of data and making predictions as outcome. By giving input data, machine learning algorithms are be able to learn themselves and coming out with most desirable results. Thus, when coming of new type of data, machine learning algorithms are able to observe the pattern of the input, with the updating outcome. There are few examples of machine learning which are fraud detection system, recommendation engine, face recognition in which these methods uses pattern recognition method to recognize pattern of data and coming out with the most similar result. The method that had been widely used in machine learning is supervised learning and unsupervised learning (Tagliaferri, 2017). Supervised learning provides the input and output for the machine learning algorithm to observe the pattern of both input and output. Another method is unsupervised learning, which is a learning algorithm which uses unlabelled data to find the similar part of the input.

Ying et al. 2006 presents a phishing detector as a plug-in for web browser that consists of: Identity extractor and page classifier. (i) The identity extractor, the website considered as set of words and the identity defined by objects or properties like organization name in the webpage. (ii) The page classifier depends on structural features that are identity related. Then SVM classifier is used to classify a website

*Figure 26. Conceptual framework*

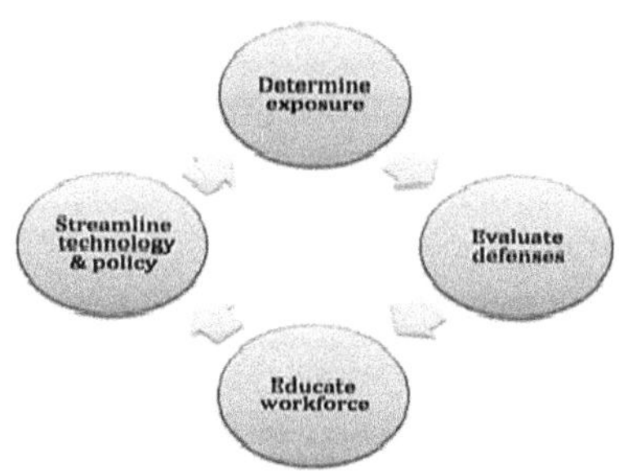

*Figure 27. Phishing attack detection*

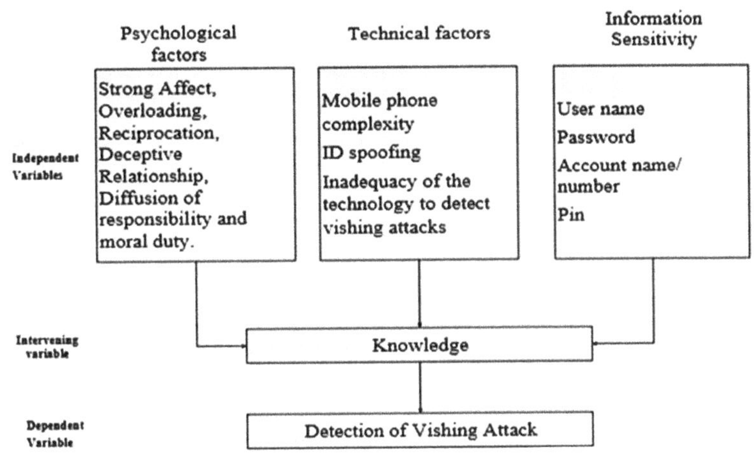


as legitimate or phishing website. The SVM is trained by large dataset of identities and features from both phishing and legitimate websites. Example on W3C Dom objects and properties: URI of links, title of elements, action of a form, text of body.

## DISCUSSION

This paper presents an overview of various types of social engineering attacks, the enforcement of training and policies against the social engineering attacks and some of the detection methods in the body of knowledge. The technique used in social engineering attack is believed to be improved and more damages over times. Although the social engineering attack will not be fully eliminated one day, but at least these attacks could be mitigated using technology and advanced security system. Therefore, a country, an organization or an individual must invest more in cyber security domain to prevent from any form of attacks and any form of damages. A human method could also be used to mitigate the social engineering attack. The main key to prevent from falling into the social engineering attack is to know self-protection and self-awareness. Self-protection and self-awareness are done by always stay alert to any suspicious incident and control emotion to not be easily altered by anyone else. Before sharing or storing any information online, users should ensure that the online platform provide user a strong information protection and will not easily leak the user information. When employee notice any unusual action perform within the organization computer, employee should report immediately before any damages occur.

## CONCLUSION

In conclusion, although social engineering is a very old way of gaining information, it is still has good means of doing so till this very day. Users being a little more pragmatic should improve security against social engineering attacks massively. Fostering cyber security governance and giving some sense of importance towards social engineering techniques should be bringing us much closer towards eradicating the problem as a whole. Organizations will also have to play a pivotal part to train the employees to be able to deal with scenarios in which they are exposed to social engineering attacks. Proper training will be amongst the solutions. Giving more respect and importance to cyber security laws will also help enhance security in organizations. More stern policies will have to be introduced to make people in organizations in particular more careful when dealing with sensitive data or information. If all the methods are implied, social engineering threats would become a thing of the past.

## REFERENCES

Abas, S.N., Bhimrajkar, S., & Raina, C.K. (2017). *Social Engineering*. Academic Press.

Ahmad, S. (2017). Social Engineering Techniques Contrast Study. *International Journal of Engineering*, *9*(1), 105–110.

Anon. (2019). *Phishing Activity Trends Report 1st Quarter 2019*. Brazil: APWG.

Anon. (n.d.). *Vircom*. Available at: https://www.vircom.com/moduscloud-cloud-email-security/

APWG. (2016a). Retrieved from http://docs.apwg.org/reports/apwg_trends_report_q2_2016.pdf

APWG. (2016b). Retrieved from http://docs.apwg.org/reports/apwg_trends_report_q4_2016.pdf

Bezuidenhout, M., Mouton, F., & Venter, H. S. (2010, August). Social engineering attack detection model: Seadm. In *2010 Information Security for South Africa* (pp. 1-8). IEEE.

Bradshaw, S. (2015). Combating Cyber Threats: CSIRTs and Fostering International Cooperation on Cybersecurity. *Global Commission on Internet Governance Paper Series*, (23)

Braisby, N., & Gellatly, A. (Eds.). (2012). *Cognitive psychology*. Oxford University Press.

Conteh, N. Y., & Schmick, P. J. (2016). Cybersecurity: Risks, vulnerabilities and countermeasures to prevent social engineering attacks. *International Journal of Advanced Computer Research*, *6*(23), 31–38. doi:10.19101/IJACR.2016.623006

Frumento, E. (2019). Social Engineering: an IT Security problem doomed to get worse. *Medium*. Available: https://medium.com/our-insights/social-engineering-an-it-security-problem-doomed-to-get-worst-c9429ccf3330

Hoeschele, M., & Rogers, M. (2006). Detecting Social Engineering. In IFIP — The International Federation for Information Processing, (vol. 194). Springer.

Information Supplement: Best Practices for Implementing a Security Awareness Program. (2014). *Security Awareness Program*. Special Interest Group PCI Security Standards Council. Available at: https://www.pcisecuritystandards.org/documents/PCI_DSS_V1.0_Best_Practices_for_Implementing_Security_Awareness_Program.pdf

Jain, A., Tailang, H., Goswami, H., Dutta, S., Sankhla, M. S., & Kumar, R. (2016). Social Engineering: Hacking a Human Being through Technology. *IOSR Journal of Computer Engineering*, *18*(5), 94–100. PMID:28540284

Karlsson, F., Åström, J., & Karlsson, M. (2014). *Information security culture: State-of-the-art review between 2000 and 2013*. Information and Computer Security.

Karyda, M. (2017). Fostering Information Security Culture In Organizations: A Research Agenda. *MCIS 2017 Proceedings*, 28.

Kessier, J. (2017). *Defending Public Safety from Social Engineering & Boxing*. Academic Press.

Koyun, A., & Al Janabi, E. (2017). *Social engineering attacks. Journal of Multidisciplinary Engineering Science and Technology*.

Mouton, F., Leenen, L., Malan, M. M., & Venter, H. S. (2014, July). Towards an ontological model defining the social engineering domain. In *IFIP International Conference on Human Choice and Computers* (pp. 266-279). Springer. 10.1007/978-3-662-44208-1_22

Nguyen, T. T., & Armitage, G. (2008). A survey of techniques for internet traffic classification using machine learning. *IEEE Communications Surveys and Tutorials*, *10*(4), 56–76. doi:10.1109/SURV.2008.080406

Pan, Y., & Ding, X. (2006, December). Anomaly based web phishing page detection. In *2006 22nd Annual Computer Security Applications Conference (ACSAC'06)* (pp. 381-392). IEEE. 10.1109/ACSAC.2006.13

People are the weakest links in cybersecurity, says CSM. (2018, Jul. 22). *The Star Online*. Available: https://www.thestar .com.my/news/ nation/2018/07/22/people-are-the-weakest-links-in-cybersecurity-says-csm/

Positive Technologies. (2019). *Cybersecurity threatscape 2018: trends and forecasts. Cyber Security Threat Trends 2018 and Forecasts for 2019: Attack Statistics and Facts*. Available: https://www.ptsecurity.com/ww-en/analytics/cybersecurity-threatscape-2018/

Price, R. (2018). *Artificial intelligence-powered malware is coming, and it's going to be terrifying*. Retrieved: September 19, 2018, from https://www.businessinsider.com/darktrace-dave-palmerartificial-intelligence-poweredmalware-hacks-interview-2016-10?r=UK&IR=T

Raskin, V., Hempelmann, C. F., & Triezenberg, K. E. (2004, July). Semantic forensics: An application of ontological semantics to information assurance. In *Proceedings of the 2nd Workshop on Text Meaning and Interpretation* (pp. 105-112). Association for Computational Linguistics. 10.3115/1628275.1628289

Raulot, A. (2019, Feb.). Bypassing phishing protections with email authentication. *Master Security and Network Engineering*, 5.

Safa, N. S., Von Solms, R., & Furnell, S. (2016). Information security policy compliance model in organizations. *Computers & Security, 56*, 70-82.

Salahdine, F., & Kaabouch, N. (2019). Social Engineering Attacks: A Survey. *Future Internet, 11*(4), 89. doi:10.3390/fi11040089

Sari, O. (2018, October 26). *Data Breach Record of 2018 in the First Half*. Retrieved from https://www.keepnetlabs.com/social-engineering-stats/

Smith, A., Papadaki, M., & Furnell, S. M. (2009, July). Improving awareness of social engineering attacks. In *IFIP World Conference on Information Security Education* (pp. 249-256). Springer.

Tagliaferri, L. (2017). *An Introduction to Machine Learning*. DigitalOcean.

Tandon, G. (2018). *Piggy Backing & Tail Gating*. Security.

Vergelis, M., Shcherbakova, T., & Sidorina, T. (2019, March 12). *Spam and phishing in 2018*. Retrieved from https://securelist.com/spam-and-phishing-in-2018/89701/

Walker, L.E. (2016). *Deception of Phishing: Studying the Techniques of Social Engineering by Analyzing Modern-day Phishing Attacks on Universities*. Academic Press.

Yuen, M. K. (2017, November 2). Don't fall prey to identity thieves. *The Star*, 12. Available https://www.thestar.com.my/news/nation /2017/11/02/dont-fall-prey-to-identity-thieves-rising -cases-of-criminals-stealing-personal-details-and-exploitin/

Zander, S., Nguyen, T., & Armitage, G. (2005, November). Automated traffic classification and application identification using machine learning. In *The IEEE Conference on Local Computer Networks 30th Anniversary (LCN'05) l* (pp. 250-257). IEEE.

# Chapter 12
# Fostering Public–Private Partnership:
## Between Governments and Technologists in Developing National Cybersecurity Framework

**Vasaki Ponnusamy**
*Universiti Tunku Abdul Rahman, Malaysia*

**N. Z. Jhanjhi**
https://orcid.org/0000-0001-8116-4733
*Taylor's University, Malaysia*

**Mamoona Humayun**
*College of Computer and Information Sciences, Jouf University, Saudi Arabia*

## ABSTRACT

*This chapter intends to provide a review of cooperation between public and private sectors towards cybersecurity governance. With the partnership, government can have confidence towards the safety and protection of their national critical digital infrastructure. The goal of this chapter is achieved by analyzing some of the cybersecurity frameworks adopted by the developed and developing nations. The analysis is further carried out by investigating the public-private policy initiatives in their national cybersecurity framework. The chapter also investigates the effectiveness of the National Institute of Standards and Technology Cybersecurity Framework (NIST) adopted by the US government.*

DOI: 10.4018/978-1-7998-1851-9.ch012

## INTRODUCTION

This Survey starts with brief discussion of some existing structures for Internet governance from the cyber security point of view. The survey will further embark on how different nations and states are responding to cyber security governance by studying the framework that they adopt and synergy between different nations. Among the many governing bodies of the Internet, Internet Corporation for Assigning Names and Number (ICANN) and Internet Engineering Task Force (IETF) play important roles in terms of setting up policies and protocols (Becker, 2019; Bradshaw & DeNardis, 2018). ICANN is dedicated to creating and distributing domains whereas the IETF is more on the technical aspects of the Internet protocols and computer codes (Handley, Bonaventure, & catholique de Louvain, 2013; Veillette, van der Stok, Pelov, & Bierman, 2018) . Although some pieces of the domain of the Internet could be owned by private sectors and the government owners, the Internet by itself is not owned by anyone. The IETF actually sets rules on how the Internet domain works, and in other words, this is an "open international community of network designers, operators, vendors and researchers concerned with the evolution of Internet architectures and smooth operation of the Internet" (Fidler, 2013; Rosenzweig, 2012).

As far as cyber security is concerned, one of the attempts made by IETF is the adoption of Domain Name System Security Extension (DNSSEC). The IETF has proposed some add on a suite of security functionalities that will become part of the Internet Protocol. Under this suite, the user should be able to authenticate the origin of the DNS data, to authenticate the domain name existence and to ensure the integrity of the DNS (Devarapalli & Joshi, 2013). This protocol suite is integrated to ensure that users actually access the legitimate domain name instead of some fake domain names. Without this effort, there is a high tendency for man-in-the-middle attack in which an intruder can intercept the communication between the sender and receiver to perform malicious activities. As far as DNSSEC is concerned, it acts as a synergy between IETF and ICANN. IETF provides the protocol whereas ICANN plays a role in terms of domain name authentication (Devarapalli & Joshi, 2013; Mundy, 2011; Satola & Judy, 2010). In order to authenticate any domain names, a chain of trust has to be established in which a root certification body can provide authentication to any domain names and websites. To trust these root certification bodies, a trust anchor has to be established into the chain of trusts. This trust anchor is provided by ICANN, although some people are not in favor of ICANN since it is an American company and they believe that it is of American interest (Lever et al., 2016).

Establishing trust and providing neutrality is of great challenge and currently, IETF sets the technical standards whereas ICANN is in charge of the creation of new domain names. These two nonprofit organizations are the current role players in international governance (O'Mahony, 2005; Shackelford & Craig, 2014). Although it has become standard practice, the trend is changing at the nation level whereby some countries are trying to alienate themselves from the Internet or even trying to sensor traffics coming to their borders, China a good example of this. While some countries have greater restrictions on Internet use, some countries have even much greater restrictions like the use of domestic Internet domains for all uses of entrepreneurial online services. Some countries have even proposed to impose restrictions on Internet contents as well as to have greater control over the Internet operation. A greater concern for countries like China and Russia is to have their own cyberspace control and to perform censorship on the Internet at the domestic level (Kalathil & Boas, 2010; Taubman, 1998). This is the new trend seen in the new initiative of state leaders to have their own cyberspace borders. This cyberspace territorial sovereignty is not only seen in major key players like China and the United States but also rampant in other countries that are developing in its cyberspace (Heintschel von Heinegg, 2013; von Heinegg, 2012).

The Convention on Cybercrime was one of a great effort in developing a unilateral approach to combat cybercrime. The intention of the convention was to create a unique set of cyber laws to keep cybercriminals away from a safe harbor. But this process was much slower than anticipated and it took nine years for thirty-eight countries to endorse the Agreement. Moreover, cultural and legal barriers hindered the process further. Besides combatting cybercrime, data sovereignty has become another great challenge to be tackled. The Internet generally is an interconnected network of devices and applications that comprises a huge collection of data. The big question to raise here is where this data is stored? And who owns the servers where the data reside? Although the physical location of the data center is of no great importance, the physical location of the data and information could be critical. The data has become so powerful in the Internet domain, and the sovereign nation that controls the data is very crucial in controlling the cyberspace. With the growth in cloud systems and the lack of standard rules shows the decline in international governance (Cavelty & Mauer, 2016). The ideal situation should be that at least all nations should come to a consensus on an international standard on the rules that govern. But such agreement is very unlikely since no specific organization or nation will take such effort in fostering a universal standard and law. There is a growing argument over whose law would be applied to conflicts on the Internet and this trend will further enlarge the gaps in a standard reference pertaining to cyber law. The geography or the storage location will become the deciding factor in deciding whose law is applicable. The nation that stores the data in its geography may become the control in whose law is applicable. One good example is China's Great Firewall and Iran's "halal" network shows the sovereign's effort of controlling its own data and content. The International Strategy for Cyberspace was a great effort in creating an "open, interoperable, secure and reliable" communications and information architecture (Deibert & Pauly, 2019).

While the role of ICANN and IETF has been clearly defined whereby IETF focuses on technical standards and ICANN deals with domain names, their roles have been continuously criticized. Instead, the ITU has been proposed by countries like China and partly take over the IP address allotment tasks from ICANN since ITU plays a much fairer game according to them (national government role). The ITU started gaining more attention as not only being active in Internet governance but also started pursuing more cyber security activities since the late 2000s (Akatyev & James, 2017; Dupont, 2018). With the "International Multilateral Partnership against Cyber Threats (IMPACT)", ITU was responsible for providing support and assistance in cyber security related activities. Some countries even proposed ITU to form a forum for cyber security. Countries like Russia and China even proposed to have stringent censorship on Internet traffic to overcome cybercrime and to combat national security. The role of ICANN in IP address allocation and domain name creation has been called for a major revision to have Internet traffic control at the national level. In 2012 about eighty-nine countries signed the WCIT resolution "Fostering an Enabling Environment for the Internet," that emphasizes Internet governance by stakeholders from all governments and to have equal roles and responsibilities to ensure the security and stability of the Internet (Price, 2017).

## AN ANALYSIS OF NATIONAL CYBER SECURITY INFRASTRUCTURES

These days, developed as well as developing countries are paying great attentions to rapidly growing cyber-attacks and preparing strategies to overcome from these malicious attacks. Currently, the trend is changed from government mandate towards public private partnership. Public sector knows the importance of involving private sectors in cyber decision making strategies. Therefore; countries are making various policies to protect their cyber environment at national and international level. Canada and the United States have come to bilateral cooperation to create a cross border agreement in terms of cyber infrastructure. They see the potential benefit of working cooperatively with the integrated infrastructure to protect their government and digital infrastructures which is the most critical information.

## United States

The United States, in particular, has been always concerned about cyber security and its first initiative began in 1988 with the Cyber Emergency Response Team (CERT) after observing an increasing trend in network intrusions. Based on their investigation, the intrusive activities had risen from six in 1988 to more than 106,000 in 2011 (Gray, 2013). In 2012 the SECURE IT Act emerged as a cyber-security act supported by the democrat's senate and the republican's senate to oversee the government cyber security related activities and to set requirements (Dillon, Belmar, & Sebastine, 2009). The SECURE IT Act was more of a voluntary approach and was geared towards public-private information sharing, The National Institute of Standards and Technology was engaged in developing a voluntary cyber security Framework to adopt the best practices of private sectors in enabling a "prioritized, flexible, repeatable, and cost-effective approach" to manage "manage cyber security related risk while protecting business confidentiality, individual privacy and civil liberties"(Purser, 2014; Scarfone, Benigni, & Grance, 2008).

In 2013 the public-private information sharing was expounded with the focus on the NIST framework (Figure 1) for securing critical infrastructures. The first version of NIST, Framework for Improving Critical Infrastructure cyber security was available in 2014. The framework does not have strict binding regulations and no strict regulations to enforce it. The success of this framework could be seen when Intel, Apple, and Walgreen implemented the NIST framework in the year 2015 into their cyber security initiatives (Shackelford & Bohm, 2016). With the private sector involvement, the NIST framework focuses on five components as identity, protect, detect, respond and recover. The structured approach of NIST framework allows organizations to assess their cyber security risks and to incorporate it into

*Figure 1. NIST cyber security framework*

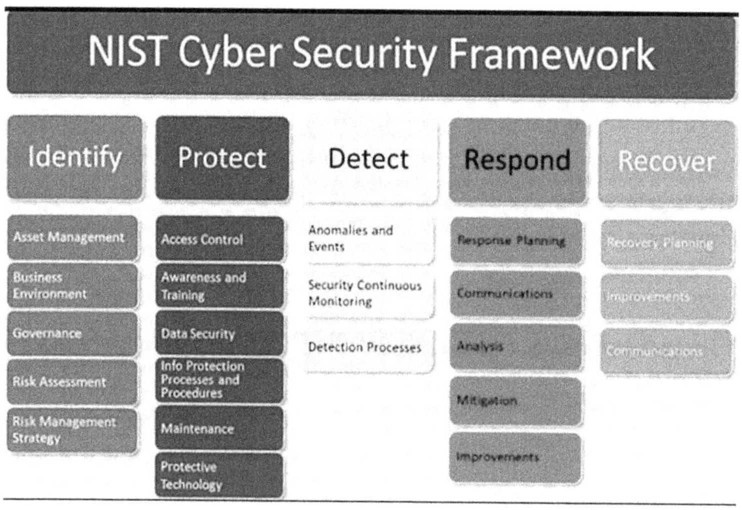

*Figure 2. Role of private public partnership in cyber security law enforcement*

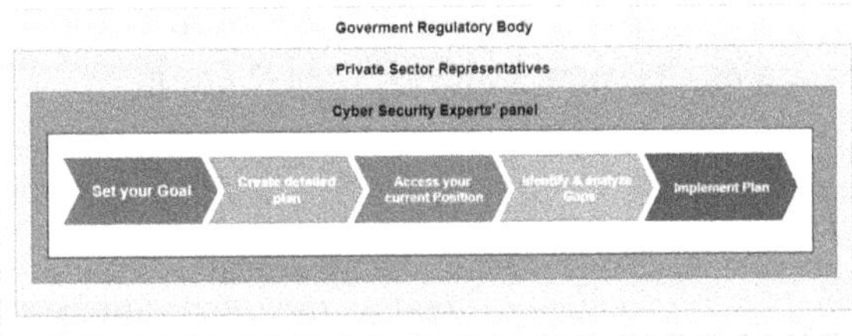


their business goals. The flexibility of the NIST framework allows CI operators and the policymakers to trust the framework and allows to standardize the cyber security operations. The NIST framework has even been well accepted in countries like United Kingdom, Japan, Korea, Estonia, Israel, Germany, and Australia. A detailed analysis is discussed in (Shackelford, 2017).

## Canada

Similar to the US, the Canadian government has set up various cyber security policies and frameworks to secure their CI. The major cyber security task force in designing cyber security policy exists in the Department of Public Safety and Emergency Preparedness Canada (PSEPC) (Safety, 2006). The organization is responsible for ensuring the security of the CI in the government and private industry networks. Under the PSEPC, the Canadian Cyber Incident Response Center (CCIRC) was formed in overseeing the public and private sector CI. There are three areas where this was seen, the first to advise both sectors on how to prepare and mitigate cyber-attacks, the second to provide technical skill and the last to create a platform for experts to collaborate and share information on how to tackle and safeguard CI in Canada. CCIRC works similarly to US-CERT in providing the private and government sectors with expertise in helping to mitigate cyber threats (Gallagher, McMahon, & Morrow, 2014; Gendron & Rudner, 2012). Information on best practices and threat information is shared. Under PSEPC again, the Cyber Security Cooperation Program (CSCP) was established in 2014. It is a grant project that to promote research projects that can improve the cyber security system in Canada in the aspects of "best practices, standards, operational methodologies, and cyber assessment tools" for CI. In line with that, in 2010 PSEPC published the National Strategy for Critical Infrastructure (National Strategy). The National Strategy identified ten CI areas that are more prone to cyber vulnerability and how the areas could be improved (Boyle & Speed, 2018).

In 2010, the PSEPC also published Canada's Cyber Security Strategy which focused on three main aspects, i) to secure all government systems, ii) to secure the non-government systems to secure the CI and iii) to promote public safe browsing on the Internet. Another version of the same Action Plan 2010-2015 was published to identify the different stakeholder's actions to achieve the goals identified. The same was updated again for 2014-2017 more specifically on protecting the infrastructures and roles of a public-private partnership to assess and to strengthen the CI. Most of the objectives stated in the action plan were similar to the NIST framework, especially in the critical risk identification and mitigation. Although the NIST framework does not explicitly identify the stakeholders responsible for each individual activity it has clear categorization of the activities to be carried out and this has caught the attention of some of the Canadian industries. The Information Technology Industry Council (ITI) has shared "the benefits of a public-private partnership-based approach to developing globally workable cyber security policies." The ITI also seeing the NIST framework as the best policy best practices and global standards (Bilodeau, Lari, & Uhrbach, 2019; Fowler & Quigley, 2019; Mussington et al., 2018).

## Africa

Africa cannot be negligent in its cyber security threat as the cyber-attack on Sony by North Korea was a reminder to strengthen its cyber security defense(Holm, 2017). With the modernization of infrastructures, cyber-attack is getting more sophisticated and Africa cannot be negligent on this issue although the threat may be very dormant for now. The lack of coordination between AU member states is quite worrisome as Africa is prone to "cyber-attacks such as hacktivism, hostile government action, espionage and cyber-terrorism" (Dutta, Geiger, & Lanvin, 2015). African Union (AU) Convention on cyber security and Personal Data Protection (AUCCPDP) focuses on the focus on AU in terms of "culture on cyber security, the role of governments, public-private partnerships followed by education and training to combat cybercrime (Dalton, van Vuuren, & Westcott, 2017) "Unfortunately the convention did not achieve the goal due to lack of progress among the AU member states. This has led to the need for international collaboration in terms of information exchange and legal assistance and this is well documented in Article 28 of the AUCCPDP. Although AUCCPDP was adopted in 2014 and received mixed feedback (NATO, 2015), the convention was generally criticized for being too elusive. With the attack sophistication and proliferation growth of mobile use, the need for more resilient cyber security regulations and standards were needed especially with the attacks such as Shamoon malware and DDOS attacks (Ahmad, Webb, Desouza, & Boorman, 2019). AUCCPDP encourages cyber security awareness amongst Internet users, schools and in the workplace. AUCCPDP and the Council of Europe Convention on Cybercrime were among the national and International legislation ensuring cyber security but the adoption and response were distinct among different African countries (Gallardo et al., 2017; Makinda, Okumu, & Mickler, 2015)

With that, a National cyber security Policy Framework (NCPF) was adopted in year 2012 by the South African government as an inspiration from the more advanced countries on the use of technology such as the EU and the US (Sutherland, 2017). Experiences from foreign government texts were used in order to identify the effectiveness of the framework in the political and cultural systems. A lot of government involvement and commitment is needed especially from the justice, crime prevention and ministers in the security cluster. But unfortunately, the 2010 NCPF draft lacked coordination within government and legal measures. So an improvement on the draft was aimed at fostering cooperation between government and private sectors, improve international cooperation and to reduce cybercrimes and vulnerabilities. One of the main objectives of NCPF was to ensure cyber security for the military to combat cybercrimes, to review and update regulations and to build confidence in the use of information and communication (Gcaza & Von Solms, 2017). Besides NCPF, South Africa also adopted the Computer Emergency Response Team (CERT) from the US and three CSIRTs including the First National Bank, government (ECS-CSIRT) and Standard Bank Group participated in the Forum for Incident Response and Security Teams (Gcaza & Von Solms, 2017; Sutherland, 2018) exchanging information and face-to-face meetings. South Africa also paid enough attention to personal data protection and surveillance. The Constitutional Court (SALRC, 2005) provided general guidelines on data protection with Constitution of 1996 paid attention that everyone has the rights on privacy and section 10 more generally focused on human dignity. As for the surveillance security, all the intelligence activities are monitored by National

Strategic Intelligence Act 1994, Intelligence Services Oversight Act 1994 and Intelligence Services Act 2002. Whereas the Regulation of Interception of Communications and Provision of Communication-related Information Act would oversee the domestic and international surveillance. South Africa is also trying to build cyber security skills through various initiatives such as cyber security awareness training, developing ICT skills in small-medium sized enterprises (SMEs). They also focus on improving ICT curriculum in universities and colleges by the initiative from The South African Cyber Security Alliance (SACSAA) which comprises of three groups from universities. Besides curriculum design their initiative was also on running awareness campaigns and competitions. Kenya and Nigeria also adopted similar internationally renowned standards by emulating countries leading in cyber security resilience (Sutherland, 2017).

## United Kingdom

The United Kingdom also realized that cybercrime attack as the second-highest threat to national security after terrorism in twenty-first century. British Military Intelligence, Section 5 (MI5) called for immediate action to curb the astounding increase in cyber-attacks. The UK has formed a Center for the Protection of National Infrastructure (CPNI) that involves in protection of critical infrastructure through a public-private partnership in promoting cyber security (Stoddart, 2016) Similar to the United States, United Kingdom also engages industries in voluntary cyber strategies and law enforcement. The 2011 U.K. Cyber Security Strategy, together with government contractors and industry partners intends to build cyber security standards. But the strategy does not have clear vision of how influential this voluntary strategy is going to help to act against cyber threats and in creating awareness among the public and the business entities. The strategy also revealed that the National Cyber Security Programme (NCSP) investments would focus on the government entities. In 2014, a report on the 2011 Strategy was released that comprises of the cyber security programs for the SMEs and also information sharing on cyber threat and best practices among businesses. The enhancement to this was the Advice Sheets that has prepared good practices in terms of actions and measures as well as information risk management. Many of the NIST framework categories and sub categories have been adopted into the Advice Sheet although the Active Sheet does not specifically highlight the best practices within each category or subcategory. As an update to the 2011 Strategy was released in 2015 with the theme 2010 to 2015 Government Policy: Cyber Security. The paper focused on Cyber Essentials Scheme while strengthening the cooperation between the UK and US on the cyber security practices and standards (Carr, 2016; Heitzenrater & Simpson, 2016)

## European Union

The European Union (EU) first initiated its critical infrastructure (CI) security after the terrorist bombing in Madrid in March 2004. The initiative was on how to prepare, prevent and respond to critical infrastructure attacks. The critical infrastructure referred by EU ranges from "telecommunication and energy networks, financial services, health services, transport systems, and the provision of safe drinking water and food." One of the greatest challenges of EU would be the universal cyber security approach for all interconnected UE countries without compromising the cyber security dysfunction in any countries within the connected borders. The EU is very concerned about engaging all the stakeholders from all the EU nations with some countries are far ahead in upgrading cyber security facilities while some are still lacking due to an ineffective approach in creating an international public-private partnership (Bendiek,

2012; Rughiniş & Rughiniş, 2014). The cyber security initiative in EU nations is not very welcoming as mostly depending on voluntary mechanisms or some requisite principles. But the effort took off in 2004 where the European Union comprising the heads of each EU member states was asked to come out with a strategy to protect the critical infrastructures. Within the same year the EU formed The European Network and Information Security Agency (ENISA) to provide the network and information security culture. ENISA plays a role in providing services to the member states as well as the private. Although ENISA suffered from some turmoil similar to the United States and China, the entity was given mandate to enhance cyber security achieving towards 2020 (Mitrakas, 2006).

The European Programme for Critical Infrastructure Protection (EPCIP) was another approach by EU in securing the CI. The framework focused on procedures to protect EU CI and also to support activities pertaining to CI protection. In October 2008, as a part of EPCIP, the Critical Infrastructure Warning Information Network was established in order to share information to enhance the IT systems pertaining to cyber security goals. In line with that Communication on Critical Information Infrastructure Protection (CIIP) was adopted to provide support to member states in responding to CI threat. In May 2012, The Digital Agenda for Europe (DAE) was proposed focusing on economic and cyber security development. Its main task is on the prevention, preparedness, and awareness of cybercrime and cyber-attacks (Marcus & Elixmann, 2012). In 2013, a new communication was formed to focus on e-commerce and to create a cyber-security culture. Communication is a recognized policy and framework to provide interactions between the public and private sectors in response to attacks. The EU cyber security focuses on five key points: i) to provide cyber resilience, ii) to reduce cybercrime, iii) to create a new policy on cyber defense, iv) to develop resources for cyber security and v) to establish international cyber security policy for EU. In order to achieve the first goal, public-private cooperation is much anticipated although it is much harder to be materialized. The second goal was focused more on reducing attacks on information systems by deploying the use of botnets. The third goal was focusing on "detection, response, and recovery from sophisticated cyber threats." A synergy between civilian and military in defending attacks against cyber assets was called for similar public and private partnerships. The commission was also focusing on the less dependence on the information and communication resources produced elsewhere yet calling for an in house industrial and technological resource creation. The last goal on achieving a coherent cyber security policy to promote EU values, closing the digital divide gaps and to promote freedom on the Internet and lastly to create international cyber security policy(Catteddu & Hogben, 2009)

One of the greatest efforts in EU cyber security is their 2013 cyber security Strategy which encourages member states to respond to the cyber incidents at the national level. In line with that, the Network and Information Security Directive (NIS) was proposed to ensure the standard security directives among all member states. The directive was also intended to "ensure a high common level of network and information security" standards and to "facilitate exchange of best practices" and to enhance "risk management practices and information sharing." One of their efforts was to authorize ENISA to work with the public and private to adopt the NIS guidelines. The NIS Public-Private Platform (NISP) was formed to ensure EU public and private sectors to adopt the "industry-led standards and the technical norms.' In May 2015 an enhanced version of the NIS Platform was introduced by adopting the NIST core components to ensure enterprise risk management although there is a mild difference between the NIST and the NISP (Carrapico & Barrinha, 2017; Shackelford, 2017)

## China

The critical infrastructure in China is seen from a different perspective as compared to other nations like the US, UK, and the EU. The cyber security policy in China started since 1994 which emphasized much on securing the "state affairs, economy, national defense and science, and technology" (Ventre, 2014). In 1999 a policy focusing on "multiple levels of security priorities and requirements" was proposed by ranking the industries on a scale from 1 to 5 with 5 being the most critical systems and contributing to national security. The emphasis was given on technical aspects like identification and authentication. According to (McGuire & Dowling, 2013) China Internet Network Information Center, 45% of Internet population in China were victims of malware attacks in the first half of year 2011 and about 8% were targeted by scammers. And according to the Ministry of Public Security's network security bureau about 80% of the computers were controlled by botnet. Chinese government officials said that most of these attacks were of foreign origin and the lack of interest from Western countries in curbing cyber-attacks has become a concern. According to Gu Jian of the Chinese Ministry of Public Security, more than 200 Chinese government websites experience cyber-attacks daily which are mostly initiated by foreign countries. And over one million Chinese IP addresses were taken control, the State Council Information Office says. Alternatively, a lot of attacks were also originated from China as it was ranked number two in the years 2005 and 2009 for instigating cyber-attacks (Harris, 2008) And the Anti-Phishing Working Group (APWG) stated that 70% of malicious domain names registered were actually originated from the Chinese to attack the businesses within the country itself.

China's approach to tackle cybercrime is different from the West in which China focuses on the broader aspects of information security rather than focusing on the cyber security alone. The Shanghai Cooperation Organization (SCO) Agreement in International Information Security feels the need to control their information from the western monopoly and also to inhibit other nations from disturbing their political, economic and social stability. A draft on International Code of Conduct for Information Security was submitted in 2011 to control the stability of information flow within the country. The Chinese government has always ensured a healthy Internet environment at domestic level that is free of pornography and crimes. This has shown to be quite promising in regulating the standards on the Internet. Another great effort was the Green Dam Youth Escort firewall software which was launched in 2018 and the government intended to make all new PCs in the country to have the firewall installed. The Green Dam firewall was intended to protect children from ill contents on the Internet and other cybercrimes but it has its limitations like some flaws in the system and also the increase in manufacturer cost. And due to strong resistance from the manufacturers and the public, the Green Dam project was finally halted (Frost, 2009; Kizekova, 2012).

Similar to the role of NASSCOM in India, The Internet Society of China (ISC) also played a significant role in China's information security although their approach was different. ISC controlled Internet companies from spreading information that could be a threat to state and social security. ISC also awarded some of the biggest search engine companies for keeping the Internet environment healthy and pleasant. The lack of western support (unlike the coalition between US and EU) to tackle cybercrime has hindered the cyber security initiatives in China. Besides that, some of the private companies were not in favor of imposing cyber security regulations and policies due to fear of losing customers (Lindsay, Cheung, & Reveron, 2015).

## India

The increase in cybercrime attacks in India has become the main concern for businesses and clients in India. In the 2014 reports submitted to KPMG's Cybercrime survey report, about 89% foresee cybercrime as a major threat and it is estimated that about 42 million Indians were victims of online attacks (Kumar, 2016). According to the National Crime Records Bureau 2016 (NCRB) report, about 11,592 cases of cyber-attacks were reported in 2015 which is a 300% increase from the year 2009 (Kshetri, 2016). Resource constraints and the relatively new trend of cybercrime has led to the poor defense and coordination of cybercrime attacks among government, business, and consumer agencies in India (Kundi, Nawaz, Akhtar, & MPhil Student, 2014) And according to a study, India is one of the most vulnerable countries in terms of cyber-attacks due to poor cyber security governance within the public and private sectors. Due to these trends, the government and the private sectors are emerging to strengthen cyber security in India. One of the main reasons for the increase in cyber security attacks in India was due to the shortage of skilled professional in cyber security domain (Kshetri, 2016; Kumar, 2016). This has hindered the effort to enforce cyber security laws especially due to shortage of experts in the police sectors and judicial systems. Moreover, huge number of IT auditors are needed in order to evaluate the business processes, IT infrastructures, project management controls, and websites. Most of the National Informatics Centre (NIC)-hosted websites were very vulnerable due to lack of auditors to identify security breaches on these websites. Lack of research (R&D) culture, facilities and lack of antivirus based companies were also some of the contributing factors to poor cyber security governance in India (Dua, 2017).

The first cyber security governance effort by India was by the IT Act 2000 although it did not cover some of the key offenses such as cyber stalking, phishing and cyber-harassment. In line with that, the IT Act 2008 added specific regulations to curb cyber offenses such as "cyber-terrorism, Wi-Fi hacking, child pornography, publication of sexually explicit materials, identity theft and spam (Kumar, 2016; Monga, 2008). " One of the main efforts by the Indian government to take countermeasures on the abuse of information privacy of the citizens was the introduction of biometric systems to get government benefits. The biometric system comprises of a 12-digit number which requires 10 fingerprints and iris scans and other information that could be used in the cloud-based E-Governance platform. Although there is no data breach disclosure laws in India, the central bank, RBI has issued all the banks and financial institutions in India to disclose information on data breach which can be good initiation to learn about cyber security and potential threats. This can be a good platform for banks to learn about potential malware and associated signatures to prevent such malware. Banks can co-operate in order to update on the latest malware and signatures to be updated and required firewall setup if needed. Not only in banks, even insurance companies started looking into cyber security guidelines and acts. The Insurance Regulatory and Development Authority of India (IRDAI) issued all insurance companies to appoint Chief Information Security Officer (CISO) oversee the cyber security matters. The cyber liability insurance industry started showing improvement in India, for example in order to buy coverage at a lower rate, a company is required to strengthen its cyber security standards. Moreover, coverage was given for the loss of first and third-party data (Kshetri & Kshetri, 2017).

In July 2013 the government of India has released the National Cyber Security Policy (NCSP) in order to protect its CI and to develop 500,000 professionals skilled in cyber security domain. Together with that, 12 other objectives had been identified to foster cyber security. Apart from that, private sectors such as the National Association of Software and Services Companies (NASSCOM) have also taken countermeasures to improve the cyber security governance in India (Kshetri, 2015; Raghavan, 2018).

They are seen as one of the main actors that manages India's cyber security risks in the Information Technology-Business Process Management (IT&BPM) of India. A lot of emphases has been put in cyber security domain whereby 30-40% of overall IT budgets were allotted to cyber security domain (Baporikar, 2017; Kshetri & Kshetri, 2017; Raghavan, 2018). NASSCOM targets to help IT&BPM to be a "trustworthy, respected, innovative and society friendly industry in the world". As far as data privacy and protection is concern, NASSCOM has set up a registry to keep IT employee records and to have a background inspection on their activities. One of the initiatives under this was the implementation of a database system that stores criminal and public records. In 2008, NASSCOM also instituted Data Security Council of India (DSCI) to focus on data protection act and to build trust among Indian companies in terms of economic development. Their task is to monitor member companies that they have self-regulation to secure the data by providing additional level of security. Both NASSCOM and DSCI work hand in hand to create cyber security awareness mainly in the cloud computing areas. Since India has very limited anti-virus companies, with the initiative from NASSCO and DSCI, they incubated 80 cyber security startups (Kshetri, 2015; Kshetri & Kshetri, 2017). Despite all the efforts in setting up policies, regulations, and organizations responsible in cyber security, India still has one major hindrance to the growth of cyber security domain. The lack of resources and the complexity in setting up provision of allocation for modern sectors compromising the traditional sectors is not welcomed by the majority population in India. For the data breach, data protection, and security are almost irrelevant. So in order to fill this gap, collaboration between the public sectors and the private sectors is needed and will be discussed in the next section.

## Public-Private Partnership (India)

PPP (public-private partnership) is an effort that requires expertise and knowledge in diverse types of domains and cyber security and is one of the main areas to be looked into. The PPP effort was initiated by the NCSP to create a cyber security related institutions in order to enhance cyber security background. And in the mid-2000s, NASSCOM urged the government to set up a special court to act on that breach the Information Technology Act of India (Kalia, Arora, & Law, 2017) . Besides, NASSCOM also initiated the CyberCop committee to be advisors to the CyberCrime Investigation Cell. It works with bar councils, police forces and industry bodies to enforce cyber security rules and to educate and train them in this area. DSCI also helped to set up cyber forensic labs and conducted training programs on cybercrime awareness and investigations. The government of India also encouraged cyber security by announcing that it would give financial incentives to local companies that would work with high-end cyber security enabled foreign companies. Self-regulatory bodies, NASSCOM together with DSCI has contributed a lot towards fostering cyber security awareness and regulations among IP&BPM in India. They have been the key players in PPP initiatives in India and also worked with government and law agencies to strengthen cyber security legislation. The model created by them helps to create a chain effect in which smaller firms try to mimic the cyber security initiatives exemplified by larger or rather more successful firms in cyber security domain. The transfer of knowledge among employees and trade associations further helps to promote cyber security awareness and also the enforcements. The NASSCOM had a very significant and influential role in putting India's cyber security at the right orientation (Mukhopadhyay, 2016).

## Italy

Similar to other countries, Italy too fell victim to cyber-attacks and one significant such attack was when its' cyber security Hacking Team's Twitter account was hijacked in 2015 and a link to a Torrent file revealed the company's confidential information such as employee emails, documents, and financial information. Besides this attack, Adobe was also been attacked with "zero-day" vulnerability whereas Microsoft was attacked with Windows kernel exploit. More "zero-day" attacks were found in Internet Explorer also on the Windows operating systems but were later being patched. Italy's cyber breach is comparatively lower due to the lower Internet usage (73%) as compared to the US (84%) and UK (90%) in the year 2014. Nevertheless, Italy started progressing on its cyber security governance along with its Internet infrastructure and connectivity enhancements. Their first cyber security approach was rather a law enforcement than setting up some standards of frameworks. In 2011 it formed the European Directive on Critical Infrastructure to classify the CI and to oversee their CI for possible vulnerabilities. And in 2012 and 2013 the government took steps in Italian Digital Agenda in-line with the Digital Agenda for Europe. In 2013 a cyber-legislation was enacted with the cooperation between the government and the private sector, the Intelligence and Security Department (DIS) and the Inter-Ministerial Committee for the Security of the republic. In line with that, The National Strategic Framework for Cyber security ("National Framework") and the National Plan for Cyberspace Protection and ICT Security ("National Plan") was introduced. The National Plan works towards the cooperation between the private and public sectors and also to identify the "international best practices" as a strategic development plan. The plan also works towards the coordination between the government, NATO and also the EU to work towards escalating the National CERT. Whereas the National framework works similar to the NIST framework in identifying key areas in the cyber security plan to set assessment strategies, mitigation techniques, planning training and awareness as well as implementing the management plans. Unlike the NIST framework, The National Framework pays more attention to the threat activities such as "hacktivism", "cyber terrorism", "cyber warfare" and the "computer crime market". The main motive of the National Framework was to protect and enhance the national cyber security infrastructure rather than just creating a standard framework for the personal adoption of an enterprise to strengthen its own cyber security. The main intention of the National Framework is to strengthen the public-private partnership but the outcome was not welcoming, only a handful of such efforts (Balta-Ozkan, Boteler, & Amerighi, 2014; Shackelford, Russell, & Haut, 2015).

## Japan

Japan faced cyber-attacks similar to other countries in which in the year 2014 alone Japan experienced about 12.8 billion cyber-attacks as compared to the year 2012 with 7.8 billion attacks. Japan's cyber security strategy have been improved a lot after its' data breach incident in year 2015. The strategy aimed at private sector adherence to their cyber security standards which works similar to NIST framework. Their cyber security strategy works similar to the US by focusing on the private sector-initiated cyber security standards. The First National Strategy on Information Security (FSIS) was one of the cyber security attempts made at the national level in 2006. FSIS focuses more on protecting the critical sectors and also to develop the cyber security infrastructure. The Second National Strategy on Information security was introduced in 2009 with more emphasis on the risk-management strategies followed by Cyber security Strategy in 2013 with resilience as the main strategy. Although there were regulations at the national level, Japan's cyber security emphasized the private sector participation with "autonomy" and "self-governance" as their main intention. The 2015 Cyber security Strategy placed more importance on private sector involvement whilst trying to form a standard like the NIST Framework and sharing of information between the public and private sectors. The strategy underlines the "Autonomy" and "Collaboration among Multi-stakeholders" as the two of five principles were they were focusing on. With the proliferation of cyber attacks the government role in cyber security strategy was seen as a necessity. The government started providing incentives for cyber security initiatives and also started promoting cyber security practices using security by design approaches for some new systems such as Internet of Things (IoT). Another example is the Intelligent Transportation Systems that consist of "manufacturers, government agencies and academics", whereby all these parties can come together to form a cyber security standard whereas the government still leads in some of the key roles while maintaining self-governance (Kallender & Hughes, 2017; Luiijf, Besseling, Spoelstra, & De Graaf, 2011; Shackelford et al., 2015; Tsuchiya, 2016).

## South Korea

South Korea is one of the countries with the best "connected" world whereby almost 80% of the population has Internet access. But South Korea is not an exception when comes to cyber-attacks with the government been a victim of 114,000 cases of cyber-attacks since 2011. Unlike the US and Japan, South Korea's approach towards cyber security is more on the pro-active side with its broad-spectrum legislation on safeguarding personal data while the sector-specific legislation on other aspects of cyber security. The Personal Information Protection Act (PIPA) was one of the most important legislation formed in 2011. PIPA ensures the security of personal information by data collectors and processors and special attention was given in securing the South Korean resident registration numbers. PIPA urges companies to take protections against these and to conform to South Korea's "general-purpose breach notification requirements." Initial cyber security legislation of South Korea was much outdated using obsolete encryption mechanisms for its online transactions. It needs to be understood that the rapid change of technology can outpace legislation very quickly. When it comes to South Korea's cyber security policy, it lies in between the economic and political ties with the US and the physical ties with China. South Korea has much stronger cyber security approaches with "multistakeholderism" as compared to the US and Japan although does not emphasize much on the state intervention as observed by China. South Korea is seen as the bridge between these regions with much diplomatic approach in international cyber security policy. South Korea works towards a bottom-up approach and also looking at the NIST model to enhance its cyber security regulations. South Korea also intends to reinforce its cyber security by having closer ties with Australia, a much "market-driven" authority (Luiijf et al., 2011; Shackelford et al., 2015; Thomas, 2009; Tsuchiya, 2016)

## Australia

Cyber attacks in Australia are termed as "undeniable, unrelenting and continue to grow" with a rise of 20% attacks during 2014. In 2010 the Australian government websites underwent two days distributed denial of service attacks putting the cyber security governance at the forefront. In 2011, their main telecommunications provider Telstra have exposed their customer's personal data to hackers. This was an alarm to strengthen Australia's cyber security regulations. Their regulation was a combination of that the US and the EU with small areas of data protection and sector-specific focus on cyber security. Their one of the most important law was the Privacy Act which focused on the federal government entities and private organizations. One of the most important principles of the act as the "data security principle" that whoever acquires personal information to adhere to the principles safeguarding the data. As for the sector-specific laws, Australia engages laws for Healthcare, Finance, and Internet service providers as practiced in the US. Some voluntary frameworks like ISO 27001/2, COBIT and Protective Security Policy Frameworks were proposed to oversee some critical infrastructures. Only in 2009 the Australian Cyber security Strategy was amended and started focusing on cyber security awareness, technologies and promoting public-private partnerships. Although the main lead on the strategy was by the government, more focus was given to private sectors on self-governance as practiced by Japan. The Cyber security Strategy underwent a rigorous amendment in 2015 by incorporating the NIST framework entities while giving mandate to private sectors in defining their own cyber security strategies for their business needs (Sabillon, Cavaller, & Cano, 2016; Shackelford et al., 2015).

## CONCLUSION

Cyber security is the main pillar with the current trend in attack vectors and the sophistication of attack technology. The article has reviewed the cyber security strategy and the framework adopted in each country. Based on the discussions public-private partnership is crucial in strengthening a nation's cyber security. The private sectors can be a technical advisory to the public sectors, whereas the public or the government should take measures to implement and provide financial support to the proposed strategies. Without the public-private partnership, the goal of cyber security governance is hard to be achieved. Developed countries like Japan, Australia, EU, and the US have fully embraced the NIST framework for its cyber security policymaking. The success of NIST framework can be seen in their sophistication of attack countermeasures and bottom-up approach in setting up CI protection. Italy and South Korea use more of a top-down approach by using some aspects of the NIST framework. Countries like India, China, and Africa can benefit from the NIST framework adoption by these countries. A much comprehensive effort is needed to identify the key success areas of the NIST framework in public-private partnerships. Unquestionably, the NIST Framework is not the ultimate solution, but it could be a guideline in setting up national cyber security framework tailored to their requirements. Besides the NIST framework, countries can also look into the Supervisory control and data acquisition (SCADA) cyber security framework as technology is driving towards Industry 4.0 whereby smart manufacturing of industrial control systems and SCADA system security is of utmost importance.

## REFERENCES

Ahmad, A., Webb, J., Desouza, K. C., & Boorman, J. (2019). Strategically-motivated advanced persistent threat: Definition, process, tactics and a disinformation model of counterattack. *Computers & Security*, *86*, 402–418. doi:10.1016/j.cose.2019.07.001

Akatyev, N., & James, J. I. (2017). Legislative Requirements for Cyber Peacekeeping. *Journal of Digital Forensics. Security and Law*, *12*(3), 4.

Balta-Ozkan, N., Boteler, B., & Amerighi, O. (2014). European smart home market development: Public views on technical and economic aspects across the United Kingdom, Germany and Italy. *Energy Research & Social Science*, *3*, 65–77. doi:10.1016/j.erss.2014.07.007

Baporikar, N. (2017). Significance of critical success factors for Indian eBusiness. In *Transcontinental Strategies for Industrial Development and Economic Growth* (pp. 184–206). IGI Global. doi:10.4018/978-1-5225-2160-0.ch011

Becker, M. (2019). When public principals give up control over private agents: The new independence of ICANN in internet governance. *Regulation & Governance*, *13*(4), 561–576. doi:10.1111/rego.12250

Bendiek, A. (2012). *European cyber security policy*. Stiftung Wissenschaft und Politik Berlin.

Bilodeau, H., Lari, M., & Uhrbach, M. (2019). Cyber security and cybercrime challenges of Canadian businesses, 2017. *Juristat*, *2*, 85–002.

Boyle, P. J., & Speed, S. T. (2018). From protection to coordinated preparedness: A genealogy of critical infrastructure in Canada. *Security Dialogue*, *49*(3), 217–231. doi:10.1177/0967010617748541

Bradshaw, S., & DeNardis, L. (2018). The politicization of the Internet's Domain Name System: Implications for Internet security, universality, and freedom. *New Media & Society*, *20*(1), 332–350. doi:10.1177/1461444816662932

Carr, M. (2016). Public–private partnerships in national cyber-security strategies. *International Affairs*, *92*(1), 43–62. doi:10.1111/1468-2346.12504

Carrapico, H., & Barrinha, A. (2017). The EU as a coherent (cyber) security actor? *Journal of Common Market Studies*, *55*(6), 1254–1272. doi:10.1111/jcms.12575

Catteddu, D., & Hogben, G. (2009). The European Network and Information Security Agency (ENISA) is an EU agency created to advance. Computing, 72(1), 2009-2013.

Cavelty, M. D., & Mauer, V. (2016). *Power and security in the information age: Investigating the role of the state in cyberspace*. Routledge. doi:10.4324/9781315601793

Dalton, W., van Vuuren, J. J., & Westcott, J. (2017). Building Cybersecurity Resilience in Africa. In *12th International Conference on Cyber Warfare and Security 2017 Proceedings*. Reading, MA: Academic Conferences and Publishing International Limited.

Deibert, R. J., & Pauly, L. W. (2019). *Mutual entanglement and complex sovereignty in cyberspace. In Data Politics: Worlds, Subjects*. Rights.

Devarapalli, S. J., & Joshi, P. S. (2013). Domain name system security extensions (DNSSEC) for global server load balancing. Google Patents.

Dillon, D., Belmar, R., & Sebastine, S. (2009). *Pre-fetching secure content using proxy architecture*. Google Patents.

Dua, S. (2017). Digital India: Opportunities and Challenges. *International Journal of Science Technology and Management*, *6*(3), 6.

Dupont, B. (2018). The global anti-cybercrime network: Mapping the polycentric regulation of online harms. In *Criminal Justice and Regulation Revisited* (pp. 163–185). Routledge. doi:10.4324/9781315174044-10

Dutta, S., Geiger, T., & Lanvin, B. (2015). *The global information technology report 2015*. Paper presented at the World Economic Forum.

Fidler, D. P. (2013). *Internet Governance and International Law: The Controversy Concerning Revision of the International Telecommunication Regulations*. Academic Press.

Fowler, T., & Quigley, K. (2019). Information, Innovation and the Boogeyman: Contextual Factors that Influence the Canadian Government's Response to Cyberspace Risk. In Cyber Law, Privacy, and Security: Concepts, Methodologies, Tools, and Applications (pp. 188-209). IGI Global.

Frost, A. (2009). *The Collective Security Treaty Organization, the Shanghai Cooperation Organization, and Russia's Strategic Goals in Central Asia*. Paper presented at the China & Eurasia Forum Quarterly.

Gallagher, H., McMahon, W., & Morrow, R. (2014). Cyber Security: Protecting the Resilience of Canada's Financial System. *Bank of Canada Financial System Review*, 47-53.

Gallardo, C., Soler, A., Nieto, R., Cano, C., Pelayo, V., Sánchez, M., ... Arias, M. (2017). Experimental infection of domestic pigs with African swine fever virus Lithuania 2014 genotype II field isolate. *Transboundary and Emerging Diseases*, *64*(1), 300–304. doi:10.1111/tbed.12346 PMID:25808027

Gcaza, N., & Von Solms, R. (2017). A strategy for a cybersecurity culture: A South African perspective. *The Electronic Journal on Information Systems in Developing Countries*, *80*(1), 1–17. doi:10.1002/j.1681-4835.2017.tb00590.x

Gendron, A., & Rudner, M. (2012). *Assessing cyber threats to Canadian infrastructure*. Canadian Security Intelligence Service.

Gray, C. P. (2013). Cyber utilities infrastructure and government contracting. *Nat'l Sec. & Armed Conflict L. Rev.*, *3*, 151.

Handley, M., Bonaventure, O., & catholique de Louvain, U. (2013). *Internet Engineering Task Force (IETF) A. Ford Request for Comments: 6824 Cisco Category: Experimental C. Raiciu*. Academic Press.

Harris, S. (2008). China's cyber-militia. *National Journal Magazine, 31*.

Heintschel von Heinegg, W. (2013). Territorial sovereignty and neutrality in cyberspace. *International Law Studies*, *89*(1), 17.

Heitzenrater, C. D., & Simpson, A. C. (2016). Policy, statistics and questions: Reflections on UK cyber security disclosures. *Journal of Cybersecurity*, *2*(1), 43–56. doi:10.1093/cybsec/tyw008

Holm, L. (2017). *Cyber attacks & coercion in the digital era.: A qualitative case analysis of the North Korean cyber attack on Sony Pictures*. Academic Press.

Kalathil, S., & Boas, T. C. (2010). *Open networks, closed regimes: The impact of the Internet on authoritarian rule*. Carnegie Endowment.

Kalia, P., Arora, R., & Law, P. (2016). Information Technology Act in India: e-Commerce value chain analysis. *NTUT Journal of Intellectual Property Law and Management*, *5*(2), 55–97.

Kalia, P., Arora, R., & Law, P. (2017). *Information technology act in India: E-commerce value chain analysis*. Academic Press.

Kallender, P., & Hughes, C. W. (2017). Japan's Emerging Trajectory as a 'Cyber Power': From Securitization to Militarization of Cyberspace. *The Journal of Strategic Studies*, *40*(1-2), 118–145. doi:10.1080/01402390.2016.1233493

Kizekova, A. (2012). *The Shanghai cooperation organisation: challenges in cyberspace*. Academic Press.

Kshetri, N. (2015). India's cybersecurity Landscape: The roles of the private sector and public-private partnership. *IEEE Security and Privacy*, *13*(3), 16–23. doi:10.1109/MSP.2015.61

Kshetri, N. (2016). Cybercrime and cybersecurity in India: Causes, consequences and implications for the future. *Crime, Law, and Social Change*, *66*(3), 313–338. doi:10.100710611-016-9629-3

Kshetri, N., & Kshetri, N. (2017). Cybersecurity in India: Regulations, governance, institutional capacity and market mechanisms. *Asian Research Policy*, *8*(1), 64–76.

Kumar, P. V. (2016). *Growing cyber crimes in India: A survey*. Paper presented at the 2016 International Conference on Data Mining and Advanced Computing (SAPIENCE). 10.1109/SAPIENCE.2016.7684146

Kundi, G. M., Nawaz, A., Akhtar, R., Phil, M., & Student, I. (2014). Digital revolution, cyber-crimes and cyber legislation: A challenge to governments in developing countries. *Journal of Information Engineering and Applications*, *4*(4), 61–71.

Lever, C., Walls, R., Nadji, Y., Dagon, D., McDaniel, P., & Antonakakis, M. (2016). *Domain-Z: 28 registrations later measuring the exploitation of residual trust in domains*. Paper presented at the 2016 IEEE Symposium on Security and Privacy (SP). 10.1109/SP.2016.47

Lindsay, J. R., Cheung, T. M., & Reveron, D. S. (2015). *China and cybersecurity: Espionage, strategy, and politics in the digital domain*. Oxford University Press. doi:10.1093/acprof:oso/9780190201265.001.0001

Luiijf, H., Besseling, K., Spoelstra, M., & De Graaf, P. (2011). *Ten national cyber security strategies: A comparison*. Paper presented at the International Workshop on Critical Information Infrastructures Security.

Makinda, S. M., Okumu, F. W., & Mickler, D. (2015). *The African Union: Addressing the challenges of peace, security, and governance*. Routledge. doi:10.4324/9781315688152

Marcus, J. S., & Elixmann, D. (2012, Sept.). Re-thinking the Digital Agenda for Europe (DAE): A richer choice of technologies. *Studie für Liberty Global*.

McGuire, M., & Dowling, S. (2013). *Cyber crime: A review of the evidence*. Summary of key findings and implications. Home Office Research Report, 75.

Mitrakas, A. (2006). Information security and law in Europe: Risks checked? *Information & Communications Technology Law*, *15*(01), 33–53. doi:10.1080/13600830600557984

Monga, A. (2008). E-government in India: Opportunities and challenges. *JOAAG*, *3*(2), 56.

Mukhopadhyay, C. (2016). A nested framework for transparency in Public Private Partnerships: Case studies in highway development projects in India. *Progress in Planning*, *107*, 1–36. doi:10.1016/j.progress.2015.02.001

Mundy, G. R. (2011). *Internet Deployment of Domain Name System (DNS) Security*. Academic Press.

Mussington, D., Arnold, B. J., Dupont, B., Hilts, S., Grayson, T., Leuprecht, C., . . . Tupler, J. (2018). *Governing Cyber Security in Canada*. doi:10.21236/ADA549761

O'Mahony, S. (2005). *20 Nonprofit Foundations and Their Role in Community-Firm Software Collaboration*. Academic Press.

Price, M. (2017). The global politics of Internet governance. *Technology and World Politics: An Introduction.*

Purser, S. (2014). *Standards for Cyber Security*. Academic Press.

Raghavan, R. (2018). Cyber Insurance–A Risk Mitigation Tool for Cyber Risk in India. *Bimaquest, 18*(1).

Rosenzweig, P. (2012). The international governance framework for cybersecurity. *Canada–United States Law Journal*, *37*, 405.

Rughiniş, C., & Rughiniş, R. (2014). Nothing ventured, nothing gained. Profiles of online activity, cyber-crime exposure, and security measures of end-users in European Union. *Computers & Security*, *43*, 111–125. doi:10.1016/j.cose.2014.03.008

Sabillon, R., Cavaller, V., & Cano, J. (2016). National cyber security strategies: Global trends in cyberspace. *International Journal of Computer Science and Software Engineering*, *5*(5), 67.

Safety, P. (2006). *Emergency Preparedness Canada (PSEPC)*. Canadian Disaster Database.

Satola, D., & Judy, H. L. (2010). Towards a dynamic approach to enhancing international cooperation and collaboration in cybersecurity legal frameworks: Reflections on the proceedings of the workshop on cybersecurity legal issues at the 2010 United Nations Internet Governance Forum. *Wm. Mitchell L. Rev.*, *37*, 1745.

Scarfone, K., Benigni, D., & Grance, T. (2008). Cyber security standards. Wiley Handbook of Science and Technology for Homeland Security, 1-10.

Shackelford, S. J. (2017). The Law of Cyber Peace. *Chi. J. Int'l L.*, *18*, 1.

Shackelford, S. J., & Bohm, Z. (2016). Securing North American critical infrastructure: A comparative case study in cybersecurity regulation. *Canada–United States Law Journal*, *40*, 61.

Shackelford, S. J., & Craig, A. N. (2014). Beyond the new digital divide: Analyzing the evolving role of national governments in internet governance and enhancing cybersecurity. *Stan. J. Int'l L.*, *50*, 119.

Shackelford, S. J., Russell, S., & Haut, J. (2015). Bottoms up: A comparison of voluntary cybersecurity frameworks. *UC Davis Bus. LJ*, *16*, 217.

Stoddart, K. (2016). UK cyber security and critical national infrastructure protection. *International Affairs*, *92*(5), 1079–1105. doi:10.1111/1468-2346.12706

Sutherland, E. (2017). Governance of cybersecurity-The case of South Africa. *African Journal of Information and Communication*, *20*(20), 83–112. doi:10.23962/10539/23574

Sutherland, E. (2018). *Digital Privacy in Africa: Cybersecurity*. Data Protection & Surveillance.

Taubman, G. (1998). A not-so World Wide Web: The Internet, China, and the challenges to nondemocratic rule. *Political Communication*, *15*(2), 255–272. doi:10.1080/10584609809342369

Thomas, N. (2009). Cyber security in East Asia: Governing anarchy. *Asian Security*, *5*(1), 3–23. doi:10.1080/14799850802611446

Tsuchiya, M. (2016). Cyber security of financial sectors in Japan, South Korea and China. In *Managing Cyber Risk in the Financial Sector* (pp. 112–127). Routledge. doi:10.4324/9781315675930-5

Veillette, M., van der Stok, P., Pelov, A., & Bierman, A. (2018). CoAP Management Interface. Internet Engineering Task Force.

Ventre, D. (2014). *Chinese cybersecurity and defense*. John Wiley & Sons.

von Heinegg, W. H. (2012). *Legal implications of territorial sovereignty in cyberspace*. Paper presented at the 2012 4th International Conference on Cyber Conflict (CYCON 2012).

# Chapter 13
# Democratic Governance:
## A Review of Secured Digital Electoral Service Infrastructure

**Leslie Mark Pannir Selvam**
*Universiti Tunku Abdul Rahman, Malaysia*

**Vasaki Ponnusamy**
*Universiti Tunku Abdul Rahman, Malaysia*

**Khalid Rafique**
*Azad Jammu and Kashmir, Information Technology Board, Pakistan*

## ABSTRACT

*In any democratic nation, electoral processes pave the way to democratic governance. In regards to this, a fully digitized online voting system should be introduced, and governments should invest efforts in studying the possibilities in adopting online voting system to curb the issues mentioned above. In that sense, any user should be allowed to cast his/her vote from any remote location, and the processes should be held in full confidentiality without any intervention or possible vulnerabilities. Therefore, this chapter investigates how governments can learn various initiatives and provide funding to further conduct research and implement such methodologies. The chapter begins by exploring some traditional and current practice of electoral system and further embarks into a review of secured online voting systems that has been proposed by the body of knowledge. The chapter then presents how some countries have adopted this secured online voting systems and giving an avenue to secured digital electoral governance.*

## INTRODUCTION

The need to develop a secured Digital Electoral Service Infrastructure is nowadays becoming more prominent than ever before. The benefits users and administrators get to experience is overwhelming and not to mention the ease of use when we compare the traditional electoral system compared to a digital electoral system. The concern for the digital electoral system lies in the security aspects of it. With

DOI: 10.4018/978-1-7998-1851-9.ch013

potential breaches, if a digital electoral system is exploited by attackers, it would lead to an enormous political unrest within the nation. A lack of trust in a digital electoral system majorly contributes to why traditional methods are still vastly used till this very day. With most governments thinking it is too much of a risk to take, they opt for the more reliable yet tedious approach that are profound in traditional electoral systems. This brings us to the next crucial element when implementing a digital electoral system, which is cybersecurity. Cybersecurity serves as a protective layer for internet-connected systems. This includes protecting hardware, software and data that makes up the system. Physical security is also required in protecting against unauthorized access to secured data centers. Among the elements we associate with cybersecurity includes application security which ensures that there are no vulnerabilities in the web page or application being used; information security which safeguards critical information of voters; network security in which channels used in the system has to be secured to access the application; data recovery strategies in situations where unforeseen problems arise; operational security where the systems are built without operations which could hinder the grade of security and end-user education whereby voters are given in-depth knowledge on how to operate and navigate within the system.

## TYPES OF ELECTORAL SYSTEMS

Electoral system is a distinguishable set of rules that determine procedures on how elections and referendums can be conducted and how the results are obtained. Generally, political electoral systems are organized by governments. But this does not mean it is not feasible in non-profit organizations as well as informal organizations. These are some examples in which we can associate electoral systems in non-political elections. Electoral systems inherit a standardized set of rules that governs pivotal aspects of the overall voting procedures. When elections occur, determining eligible voters, identifying eligible candidates, procedures in determining how the ballots are marked and casted, steps taken into consideration when the ballot is counted are all part of a fully automated electoral system. On that note, electoral systems can also be designed in a manner in which campaign spending is controlled. This is because excessive spending during the campaigning phase may lead to favoritism by the voters. This will directly affect the results of the elections. Political electoral systems are defined with respect to constitution and electoral laws. Currently, constitutional laws during the election phase is abided by the election commission. As long as it does not violate a particular country's constitution, the election commission has the authority to utilize all viable forms of election during the election phase. Traditional electoral systems are still used profoundly till this very date. That has somewhat stalled the evolution of digital electoral systems. Security concerns amid digital electoral systems has resulted in most governments globally to continue trusting on traditional electoral systems instead of reverting to digital means. Traditional electoral systems have many different ways in order to announce a winner. They are differentiated in the types of electoral system. For a country has particular type of electoral system need not abandon it as they can continue the same type of electoral system in a digital environment. The use of traditional electoral systems will continue to be in discussion until people fully trust the security measures implemented in a digital or online electoral system which should eventually be the case in our ever-developing technologies. Online electoral system can be a turning point for providing electronic services to governance similar to other systems that uses ICT for governance (Rura et al., 2011; Roy and Karforma 2011a;Roy, Banik and Karforma, 2011b); Learning (Queensland Government, 2013; Itmazi 2013); Shopping (Adeline et al., 2006); Medicine (Gartner (2013) and in democratic decision making (Olaniyan, Mapayi, and Adejumo

2011;, Olaniyi et al., 2011; Okediran et al., 2011a, Olaniyi et al., 2013). But their success rate in terms of secured communications is always determined by the security, integrity and authenticity of these electronic applications (Roy et al., 2011b). Digital voting system becomes the pillar for digital governance whereby electronic voting involves the entire process of registration, casting, counting, and recording to be performed using computerized system in a very secured manner. (Cetinkaya & Koc, 2009; Olaniyi et al.,2011). The following section provides a summary of electoral systems which consists of plurality systems, proposed systems and the mixed systems.

## Plurality Systems

Among the most common electoral system is the plurality systems (Figure 1). The plurality system methods work in a way in which the candidate with the highest number of votes wins. In this system, it is not mandatory for the candidate to get the majority of votes. There are scenarios in which there is only one place that requires filling, this is known as first-past-the-post and is the second most common electoral system for national legislatures (Blais, A. and Massicotte, L., 2016). Other scenarios include multiple positions to be elected. This scenario is most common in multi-member constituencies. Plurality voting can be referred to as bloc voting or plurality-at-large. This voting method has two main forms, one in which voters are eligible to multiple votes as there are seats and they have the luxury of voting any candidate disregarding the party in which the candidate is from. As shown in Figure 1, it is emminent that in order to win the election, a candidate is not reqired to win most of the available votes however, he or she is required to get more votes compared to the opposing candidates. In short, a candidate can win the elections even with lesser than 50% of the overall vote count.

## Proportional Systems

Another widely used electoral system is the majoritarian system. Majoritarian voting is a system whereby candidates has to win majority of the votes in order to be elected. Majoritarian voting can be broken into two major forms. One uses the single round of ranked voting and the other uses more than one round. If no candidate receives more than 50% of the vote, the process is repeated until that is achieved (Blais, A. and Massicotte, L., 2016). As shown in Figure 2.0, seat allocation is based on the number of votes won. To gain full power in the scenario above would require at least 50% of the overall seats available which would also mean that the candidate or party would have to also obtain at least 50% of the overall vote as well. If that is the case, only then would the individual or party would be eligible to form a government.

*Figure 1. Graphical representation of plurality system*

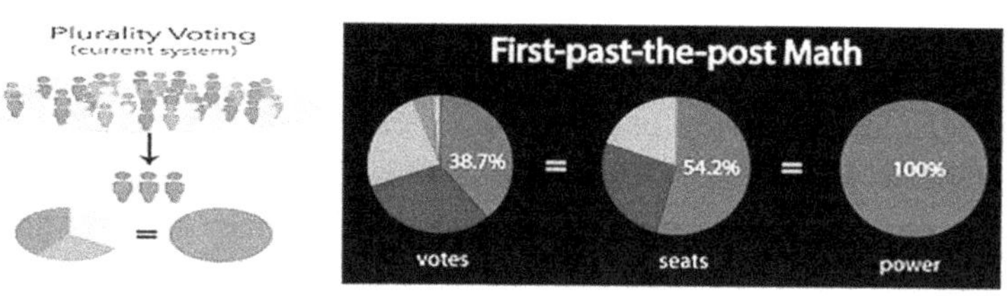

*Figure 2. Graphical representation of proportional systems*

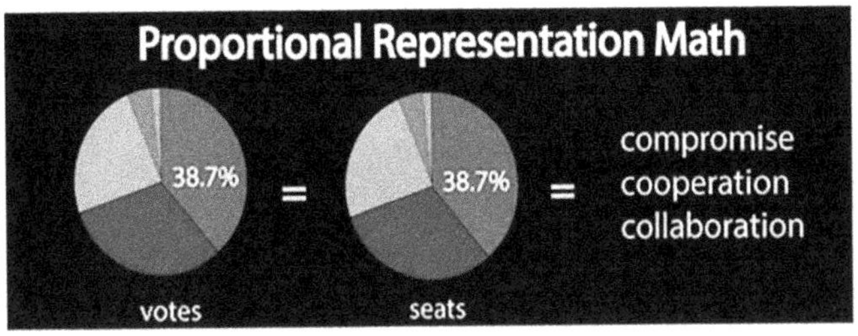


## Mixed Systems

In addition to the electoral systems that has been covered up to this point, the mixed system is also a recognizable electoral system as it uses a combination of different electoral system mostly parallel voting and mixed member proportional representation. Ideally, these are two methods in which members of a legislature can be elected. In this electoral system, the result of the constituency vote does not impact the outcome of the proportional vote. In terms of using a mixed electoral system, various combination can be used in order to draft an effective electoral system that best suits the needs of a particular nation. Among the types of electoral systems that exists in the mixed electoral systems include the alternative vote plus system, the Scorporo system, the majority bonus system, the dual member proportional and the already mentioned parallel voting and mixed member proportional. Different countries have implemented different combination of electoral systems in order to build a comprehensive mixed electoral system.

These are amongst the most common electoral systems that are globally used in a large scale as most countries still rely heavily on the traditional electoral systems. It should be known that by using a digital electoral system the concepts, methods, mathematics and algorithms used in determining the winner of the elections can be retained. None of these components needs to be sacrificed in order to facilitate an online / digital voting system. This further emphasizes on the fact the security may not just be a major concern in a digital electoral system rather it is the only concern that is hindering the development of online / digital electoral systems which would have been a reality by now had it not been an issue.

## WEAKNESSES IN TRADITIONAL ELECTORAL SYSTEMS

The major issue associated with online / digital electoral system is undeniably the security concerns. This does not mean the traditional electoral systems are flawless. Integrity of a particular election has been called into question on various occasions. With some of them being so controversial that a re-election had to be called. The political unrest in such instances clearly tarnishes the image of a nation in a global stage. This is evident that using traditional electoral systems may not just be very tedious which is the most apparent flaw of the traditional electoral systems but may also gravely lack integrity. When such cases surfaces into the media globally, we cannot help but question the integrity of every election that has happened in the past as to whether or not it was conducted in an ethical manner. Some of the weaknesses in traditional voting systems is clearly stated in Okediran et al., 2011a and Malkawi et al., 2009 for accusation of violence, coercion, ballot stuffing, under-age and multiple voting, counting error and many more such allegations.

*Figure 3. Calls for re-election amid fraud controversies in the 2017 Honduras Elections*

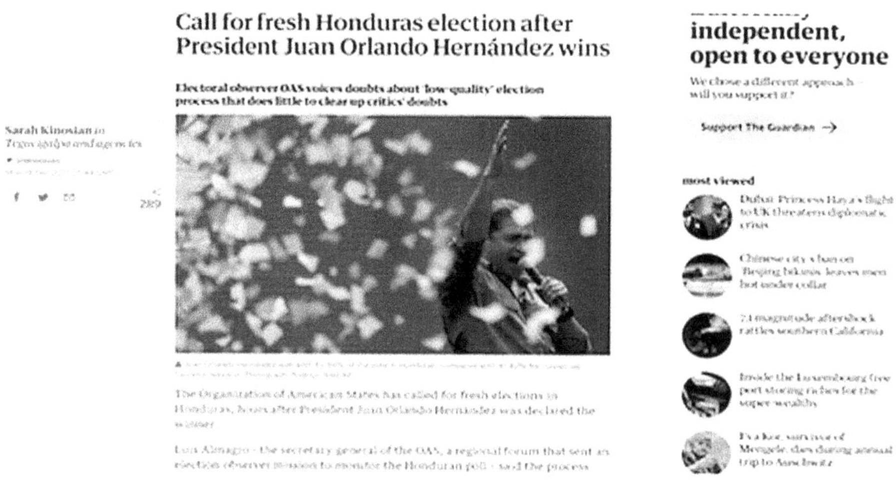


## Electoral Fraud

Among the security concerns highlighted using traditional electoral systems is electoral fraud also known as election fraud, election rigging or election manipulation. These are instances in which the vote count for the favored candidate is increased and the vote count for the opposing candidate is suppressed intentionally. Immense trust is put upon the election commission to carry out their tasks honestly. However, elements such as bribery, corruption and favoritism may prompt influential personnel in the election commission to act in an unethical manner. Elements such as intimidation, defamation and vote buying are all categorized in electoral fraud.

A reported case linking to electoral fraud was the 2017 Honduran general election which was held on the 26th of November. It was reported that there were electoral fraud in a very high scale to an extent in which the opposing candidate asked for justice and help from international bodies to help resolve the issue.

## Ballot Stuffing and Phantom Voters

Other concerns that is directly related to the traditional electoral systems is recognized as ballot stuffing or "ballot-box stuffing". This is where one eligible voter submits multiple ballots during an election where the permitted number of ballots submitted per voter is only one. This issue is strongly related to phantom voters because if all eligible voters turn up during the polling day and a single eligible voter votes more than once, the ballot count would exceed the maximum number of ballot paper that should be present. This makes it very apparent that there has been some foul play involved. This is hard to detect using traditional electoral systems because voter turn up during polling day is never reached 100% and since the ballot paper does not reveal information as to who casted the vote it makes it impossible to detect any duplicated ballot paper submissions. This is because phantom voters may have inadvertently casted a vote on behalf of another eligible voter that did not cast his or her vote on the polling day. This security concern can be averted using a digital electoral system.

A reported case involving ballot stuffing happened during the 2018 Republic of Congo general elections that was held on the 30th of December (Figure 4). The winning coalition won with a majority which was lesser than 2% of the overall votes. However, in the senatorial and legislative elections the margin of victory was huge, basically a landslide victory. This rose questions by the people as to whether ballot stuffing took place in the senatorial and legislative elections.

## INTRODUCTION TO DIGITAL ELECTORAL SYSTEMS

Digital electoral system is where most countries if not all aspires to reach. As mentioned, numerous numbers of times before, if the security concerns associated with online / digital electoral systems can be eradicated it would be beyond any form of doubt that it will be the most prominent form of voting when you consider how smooth the election proceedings may occur without any hindrance. There would be no need for millions of people to queue up in hundreds of polling centers for hours just to cast a single vote. People still do that these days because the political landscape is massive for the development of a nation. People would want their nation to be led by a governance that is not properly equipped to lead the nation forward. The importance people give to this factor is what prompts people to be patiently awaiting their turn to cast their vote. Online / Digital electoral systems significantly simplify all the electoral procedures as voters need not even leave the comforts of their own homes in order to cast their votes. Not to mention the efficiency of an online / digital electoral system. Millions of votes can be accounted for seamlessly in a very short span of time with the accuracy being much sharper than that of an average human being. Results of elections can be announced on a very short note after the end of the stipulated polling time-frame. Delays in announcing election results would become a thing of the past.

Digital / Online voting system can be classified in two distinctive categories. The first being the use of machine voting and the other touches on the use of online electoral systems. The level of automation can be differed based on the security concerns if they still somewhat exist to a certain extent. Among the well-known electoral systems include punched cards, specialized voting kiosks, optical scan voting and cryptographic electoral systems. The advantages of reverting to an online / digital electoral system clearly outperforms the traditional electoral system. It is no surprise that the need for online / digital electoral system is growing exponentially as the years pass. A lot of digital or electronic voting systems

*Figure 4. Immensely disputed 2018 general elections in Republic of Congo*

have been proposed previously, Kitcat 2010 proposes a FREE e-DEMOCRACY Project, using open code, on Java language, Qadah, and Taha 2007 proposed an e-Voting system to be accessed via Wireless Application Protocol (WAP), Short Message Service (SMS) and Hyper Text Transport Protocol (HTTP). Whereas Saleh et al, 2003 developed an e-voting system using a Universal Modeling Language (UML) which can be a platform for Foundation for Intelligent Physical Agents - FIPA 2010. This works in a very automatic manner whereby users cast their votes using a personal computer and mobile device within the proximity collects the data and passes to the server for further processing.

## REVIEW OF ONLINE / DIGITAL ELECTORAL SYSTEMS

Secured electronic voting system can be the ultimate solution for an improved digital governance in the aspects of authenticity and confidentiality. Therefore, a review of secured online voting systems is provided in the following sections followed by a summary provided in Table 1.

### Online Voting Systems Based on Image Steganography and Visual Cryptography

The idea of image steganography (Figure 5 and 6) is by replacing bits of useless or unused data in regular computer files (such as graphics, sound, text, HTML, or even floppy disks) with bits of different, invisible information. This information that is hidden can be in the form of plain text, cipher text, or even images. This algorithm densely involves visual cryptography which is a cryptographic technique which gathers visual information (pictures, text, etc.) to be encrypted in such a way that the resulting decrypted information appears as a visual image in the other end. The voting systems with such characteristics are categorized as end-to-end verifiable voting systems (E2E). Many E2E systems have been proposed and are widely used nowadays. These systems fare better compared to system that uses data Encryption Schemes like Data Encryption Standard (DES), and Advanced Encryption Standard (AES) have not only been adequate but are not very efficient in the encryption of large volume of digital data. Steganography technique is premised along transmission of confidential message on a channel. A steganographic system consists of the secret message, the cover data and the stego message. The purpose of the secret message is intended to be hidden at all times whereas the cover acts as a cover to the container for hiding the secret message. The traditional solution to these problems is achieved through Private Key Encryption

*Figure 5. General framework of steganographic system*
*Source: (Navaneet 2012)*

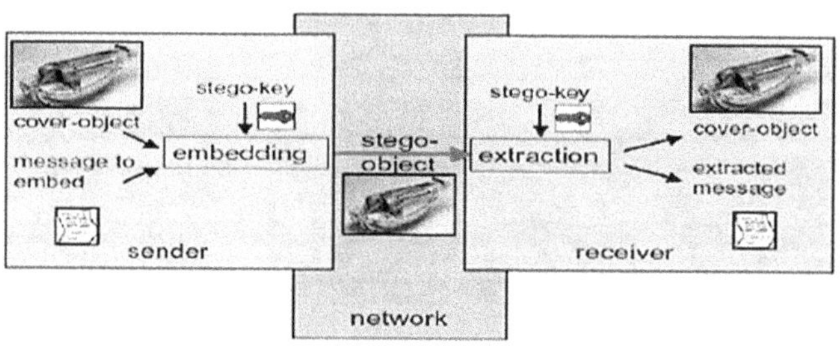

*Figure 6. Steganographic system transmission from sender to receiver*
*Source: (Navaneet 2012)*

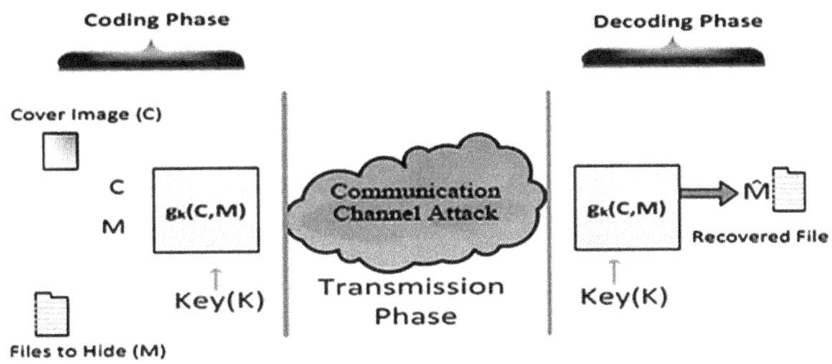


(PKE). However, it is noticeable that this proposed method by (Olaniyi, O.M., Tayo, A.O., Olusayo, O.E. and Olusola, O.O., 2013) to ensure security in online voting systems does not efficiently utilize steganography and also possess alignment problems during pixel expansion.

## Online Voting Systems Using a HYBRID Crypto Realm

This research reformed a protocol that integrates blind signature scheme, secret sharing technique and homomorphic encryption which ensures fair voting and eliminate criminal deception in voting. With the inclusion of public proxy server, we have successfully simulated the protocol which solves anonymity of the voters. The cryptographic approach securely transmits the vote of each voter in a high security lane. Homomorphic encryption is a special form cryptography in which the sum of two encrypted values is equal to the encrypted sum of values. The encryption algorithm E () is homomorphic if given E(x) and E(y), one can obtain E(x ¬ y) without decrypting x; y for some operation ¬. Homomorphism is an algebraic property useful in electronic voting schemes because it allows finding of the sum of the ballots without actually decrypting them. This ensures the security of the system with regards to its accuracy which is one of the major benefits of the system as a whole. This method as proposed by (K.N. Sandhya Sarma, S. Umamaheswari, 2012) (Figure 7) has weaknesses in which voters cannot complete the voting session until the tallying. In addition to that, voters are relied on to check whether their vote is counted correctly or not. Then again voter has to revisit the polling site after the announcement of the results to verify their votes

## A Secured Online Voting System Using Encryption and Digital Signature

This research (Figure 8) provide security from intruders both passive as well as active. The passive intruder can access the casted vote of a voter and create challenge to secrecy and privacy characteristics of voting system. The active intruder has the opportunity to tamper the casted vote and encounter problems when questioning integrity of the casted vote. In order to resolve this security concern, we are using the concept of cryptography and taking advantage of digital signature simultaneously. To provide security measures from passive intruders, the casted vote is encrypted on the client's system, and then

*Table 1. Summarizing the current existing systems*

| Description of Existing System | Strengths | Weaknesses |
|---|---|---|
| 1. Online Voting Systems Based on Image Steganography and Visual Cryptography | • Provides secret communication accessible by encoding a secret message to various types of cover data.<br>• Output data of steganography (stego-object) would still look the same as its input data thus it will be difficult to identify. | • Inefficient utilization of steganography. Unused as an additional layer of security.<br>• Alignment Problems. Pixel expansion is an important parameter for Visual Cryptography Schemes. |
| 2. Online Voting Systems using a HYBRID crypto Realm | • Homomorphism is an algebraic property that finds the sum of the ballot without decryption. The count accuracy will be secured and effective.<br>• The homomorphic algorithm used also ensure user verifiability.<br>• Blind signature in which a message is blinded before it is signed. | • Voter cannot complete voting session until the tallying.<br>• Voter cannot submit the decryption key until after the voting phase of the election is over.<br>• Voters are relied on to check whether their vote is counted correctly or not. |
| 3. A secured online voting system using encryption and digital signature | • The use of digital signature ensures the privacy of each voter.<br>• Simplicity makes the voter system easily adaptable as users need not spend too much time to understand how to operate the system. | • Security concerns referring to the level of security offered by using digital signature.<br>• Legal issues with regards to laws associating to voting in democratic countries. |
| 4. Online voting system powered by biometric security using steganography | • Anonymity as the user identification is not reviled throughout the process.<br>• Thwart a wide range of attacks, including "ballot stuffing" by voters and incorrect tallying by insiders.<br>• Usable by the entire voting population, regardless of age, infirmity or disability. | • Eavesdroppers can detect distortions in resulting stego image. This aids the attackers in some regards. |
| 5. In Okediran *et al.*,(2001a), shows a proposal in which the requirement, design and implementation of this e-voting system utilizes a variety of electronic devices which includes a private computer network, website and mobile phone. RSA encryption algorithm is used to encrypt end to end messages. | • As the system uses Public key encryption, it provides high flexibility through end to end key agreement protocols.<br>• Adding to that, the accuracy of biometric fingerprint authentication reduces security concerns dramatically. | • Using this system would require substantially high computing time as well as storage in any platform it may be used. This high resource requirements significantly reduces performance levels. |
| 6. In Sodiya *et al.*, 2011, the proposed e-voting system uses a combination of ECC and El-Gamal Cryptosystems in order to encrypt voters vote during the transmission phase and is decrypted during the tally phase. | • This system makes efficient use of probabilistic encryption and Elliptic encryption as it carves points (x,y) from it. This prevents anonymity, coercion and bribery in this particular voting system when compared to others. | • Among the major flaws in this system is that it easily compromises for the integrity of the elections as hackers are lured to exploit the system due to the use of authoritative cryptographic model. |
| 7. In Purushthomata and Alwyn (2009), the system is designed using Identity based encryption. This is used to ensure privacy, voter anonymity, eligibility and verifiability in this proposed e-voting system. National ID Number is used for unique identification and as the public key for encryption purposes. This system then generates its own private key for each user in which the only the owner would know. | • Possess effective enterprise key management system that would be used for data encryption and voter authentication<br>• Can be scaled as per requirements for future use. | • System easily compromises for the integrity of the elections as hackers are lured to exploit the system due to the use of authoritative cryptographic model (same as the limitations found in Sodiya *et al.*, 2011). |

*continued on the following page*

*Table 1. Continued*

| Description of Existing System | Strengths | Weaknesses |
|---|---|---|
| 8. In Sujata and Bandshidhar (2010), the proposed system utilizes multi-authority and blind signature to ensure security elements such as privacy, anonymity, eligibility and verifiability. It uses bitwise XOR operation in order to generate the blind signature in the system. | • A huge advantage in this proposed system lies in the denial of all relevant voting authorities in knowing any portion of the ballot content nor are able to associate a ballot to the particular voter. | • System easily compromises for the integrity of the elections as hackers are lured to exploit the system due to the use of authoritative cryptographic model (same as the limitations found in Sodiya *et al.*, 2011). |
| 9. In Tohari *et al*, (2009), the system is designed to facilitate mobile voting for better computing performance. This system ensures integrity, confidentiality and voter anonymity. The system uses a small key-size elliptic curve cryptographic algorithm for encryption. | • Substantially improves mobile computing performance with no compromise in other regards.<br>• Confidentiality of voters are well preserved (similar system as proposed by Light and David -2008). | • Factors relating to authentication was not put into consideration in the proposed system<br>• Potentially threats such as security of voters which defeats the purpose of the system. |
| 10. In Li and Hwang (2012), the writers proposed an e-voting system that compensates for the liabilities that are present in the voting system proposed by Chang and Lee (2006).<br>Introduced the use of RSA for voter registration. | • Developed a protocol that avoids insider attack. Administrator foul-play, denial of service attacks are all resolved in this system. | • Emphasizes on authoritative cryptographic model. Thus it is limited to the 2 above. |
| 11. Gina *et al.*, 2010 proposed an e–voting system that utilizes cryptography and protocol that require two bilinear pairing cryptography to ensure privacy, eligibility and transparency during elections are achieved. The system stresses on threshold encryption scheme and blind signature as the foundation for this encryption algorithm. | • A unique aspect of this proposed system is that it would not require public key storage nor public key binding management. This reduces computing time and the processing complexity thus making the system more efficient. | • System easily compromises for the integrity of the elections as hackers are lured to exploit the system due to the use of authoritative cryptographic model (same as the limitations found in Sodiya *et al.*, 2011). |
| 12. In Gupta *et al.*, (2011) authors had proposed an e-voting system in which it requires a blind signature based on a cryptographic scheme. This would ensure voter anonymity and ballot confidentiality which are essential requirements in a voting system. This system uses both blind signatures to ensure voter privacy and confidentiality. | • This system pays respect to confidentiality and authentication requirements that are needed in an e-voting system. | • This system has vulnerabilities when it concerns frauds as there are possibilities in which a voter may cast more than one vote which is what every voter should be limited to. |
| 13. In Patil (2010), the system proposed guarantees voter privacy, authenticity, integrity and verifiability. The system uses blind signature, pseudo code, and dynamic vote changing. This guarantees voter privacy. The digital signature is used to authenticate the voter who casts a ballot into the system. | • This online voting system ensures that one eligible voter casts only one vote and ensures voter confidentiality by utilizing the combination of blind signature, pseudo coding and dynamic change method of voting. | • System easily compromises for the integrity of the elections as hackers are lured to exploit the system due to the use of authoritative cryptographic model (same as the limitations found in Sodiya *et al.*, 2011). |

it would be sent to the voting server with the help of the internet and on server side decryption of a particular vote is done before taking them into account. To ensure a secured system and to maintain the privacy and authenticity of polled vote, two distinct cryptographic algorithms are used. The algorithms are Advanced Encryption Standard (AES) and Riverst Shamir Adleman (RSA) algorithm. In order to

*Figure 7. The flowchart for the counting phase in the proposed algorithm by by*
*Source: (K.N. Sandhya Sarma, S. Umamaheswari, 2012)*

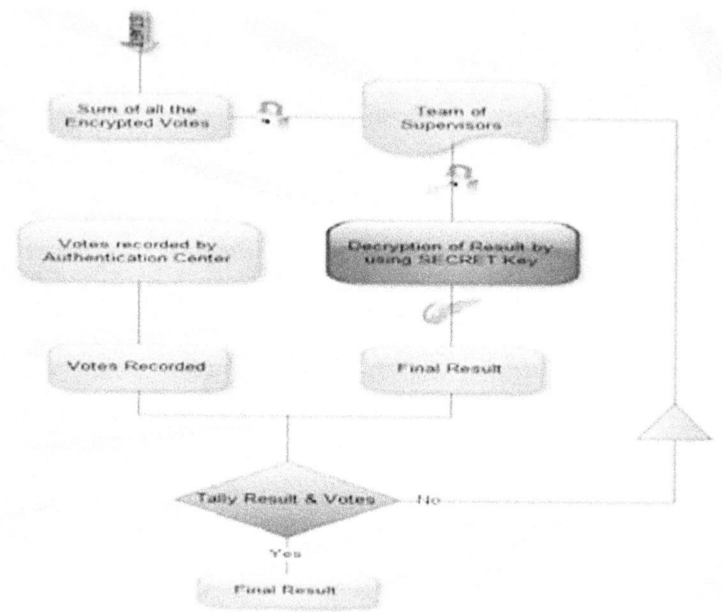


*Figure 8. The proposed method in which users will be able to cast their votes across a secured medium*

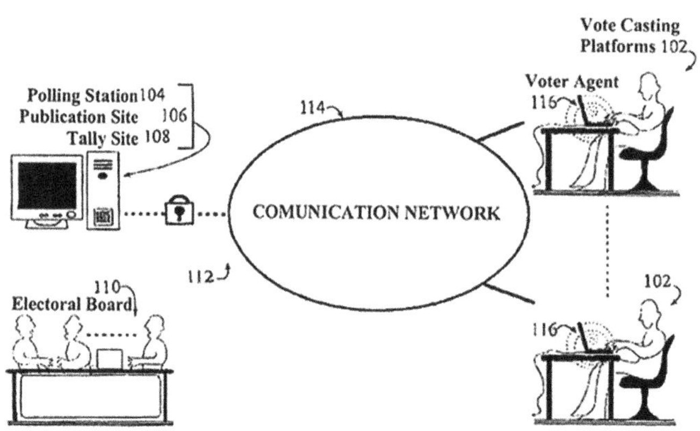


encrypt the vote, AES algorithm will be used and for the digital signature RSA algorithm is used. This algorithm proposed by (Jorba, A.R. and Roca, J.C., 2007) does lack technological compatibility and still does possess some security concerns as the algorithm used is not complex for the attacker to decrypt.

## Online Voting System Powered by Biometric Security Using Steganography

As shown in Figure 9, this project emphasizes on using cryptography and steganography at the same time. There will be attempts to provide biometric as well as password security to all eligible voters. The proposed scheme uses images as cover objects for Steganography and as keys for Cryptography.

*Figure 9. The basic mechanism used in the proposed online voting system*

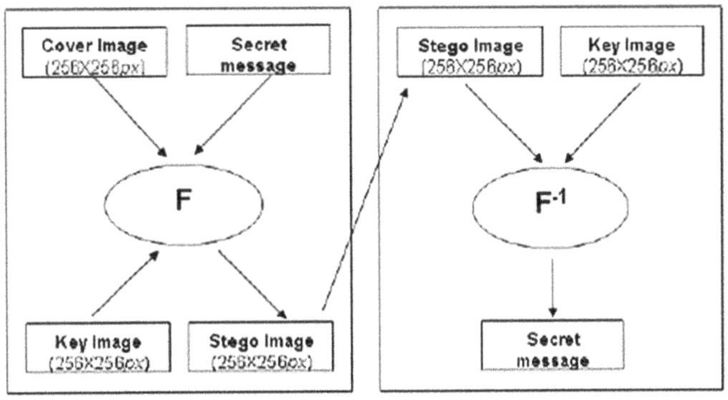


The key image is mainly a biometric measure, fingerprint image is a form of an instance. The idea in general is to merge the secret key together with the cover image on the basis of the key image. The end result of this process as a whole produces a stego image in which it would look very similar to the cover image but not detectable by the human eye. The generalized model of Steganography uses images as cover media because after digitalization images contain the quantization noise which provides space to embed data. The method easily works with still images as it yields random outputs, which in turn makes the steganalysis more difficult to be deciphered and the message is more secured. To cast a vote, a voter logs in to the system by entering the personal identification number and secret key along with his thumb impression as well. At the server side, it will use the Optical Character Recognition technique to read the personal identification number represented on the image. A clear cover image is then created so that the text written over it is machine readable. The secret key is then expanded using hashing. So, to further enhance the complexity of analysis, the two byte secret key will be expanded to a thirty two byte key by applying the SHA 256 hashing algorithm. So, even if eavesdroppers manage to identify the stego image, it would be more difficult for them to predict the embedded data. Though this proposed method by (Shivendra Katiyar, Kullai Reddy Meka, Ferdous A. Barbhuiya, Sukumar Nandi, 2011) is mostly secured, but due to the minor vulnerability caused by eavesdroppers, the security of this system is still exploitable as eavesdroppers weakens it.

## ADVANTAGES OF DIGITAL / ONLINE ELECTORAL SYSTEM

As mentioned before the advantages of using an online / digital electoral system vastly outnumbers the advantages that can be found in the currently used traditional electoral systems. Cybersecurity concerns contributing as a major factor as to why digital electoral systems are not as prominent as they should be at this day and age. If these issues or concerns are cleared or at the very least partially resolved, we would definitely get more governments internationally that would revert to using the digital electoral system in order to conduct elections. This is because it significantly improves the accuracy, efficiency and reliability of the electoral system.

Among the benefits one would get to experience by using an online / digital electoral system is the immense speed of counting ballot papers. Using the traditional electoral system, humans count each ballot paper manually. This is a very tedious procedure which significantly reduces the speed in which the election result may be announced. This effect is further compounded if the population of the country is enormous. Announcement of the results could take days after the closure of the polls. Disable voters will find it very useful as it is more troublesome for them to be moving around which is what that is required of them to cast their votes using the traditional electoral system. These ease of access when using a digital / electoral system will provide them with the much-needed aid in order for them to cast their votes.

Furthermore, the cost of having elections will be noticeably reduced. Preparing polling centers, hiring staffs to work for the election commission, preparing polling booths, hiring security personnel and preparing ballot papers would considerably increase the cost of the elections. The election commission gets to save all these direct and hidden costs by reverting to an online / digital electoral system. In addition to that, voters who are staying abroad need not travel back to their home country in order to cast their vote during the polling day. They are allowed to cast their vote irrespective of their geographic location. This would also improve the overall turnout during the general elections.

It becomes very clear that digital / online electoral systems have massive advantages that overshadows the use of traditional electoral systems. As mentioned from the beginning, the only drawback of the digital / online electoral system is the security concerns. When these issues and concerns are resolved, the use of digital / online electoral systems will definitely be used in an enormous scale.

## FUTURE RESEARCH DIRECTIONS

Many government and private organizations have spent millions on research in order to develop a secured and seamless experience when using a digital electoral system. Developing prototypes and conducting attacks onto it in order to ensure the system is not vulnerable to any form of attacks. The growing interest to develop a secured digital / online voting system has resulted in good funding on research. To come up with the ideal electoral system for a particular nation may differ. It was mentioned in the type of electoral system that are used by countries these days are different compared to others. The differences have to be highlighted when developing the digital / online electoral system for a particular nation by making sure the system abides by the practices of that particular nation. Digital / Online electoral systems have numerous advantages when compared to the traditional electoral systems. However, these systems are not flawless. There will be limitations when designing the system. In this scenario, it is important to wholly understand these limitations and find ways to enhance them in the future. Changes in trends and need for new features will come into the picture later on. Integrating these changes into the already existing digital electoral system is important in ensuring that they do not create any new vulnerabilities within the system.

Many organizations are putting efforts and resources to conduct research in developing an effective and reliable digital / online electoral system. With such efforts, it is only a matter of time when a reliable and effective digital electoral system will be developed completely. Researchers are putting a lot of effort in identifying key criteria for digital electoral systems (Cranor and Cytron, 1996); NSF (2001); Okediran et al., (2011a); Katiyar et al., (2011), Lambrinoudakis, Gritzallis, Tsoumas, Karyda and Ikonomopulos (2003).These requirements in Okediran et al., 2011a) Lambrinoudakis et al.,(2003) (Cranor and Cytron, 1996). And proper security measures in terms of integrity, authenticity and confidentiality should be considered (Ibrahim, Kamat, Salleh and Abdul Aziz (2003); Abo-Rizka and Ghounam (2007). Steganography could be another viable solution for secured digital electoral system besides encryption mechanism like Data Encryption Standard (DES) and Advanced Encryption Standard (AES) whereby steganography provides information hiding rather than just changing the information to a different form (Olaniyi, Arulogun, and Omidiora 2012). In steganography the information is hidden from intruders and draws lesser attention as compared to encrypted data that is still visible while in transit.

## CONCLUSION

In conclusion, although traditional electoral systems are still actively used to this very day, it is certain the eventually technology would have developed to an extent in which we would have eased off any security concerns associated with a digital / online electoral system. There are many benefits administrators as well as users may experience using a digital / online electoral system. There are simply far too many benefits to overlook. Security however, is crucial element in digital / online electoral systems. Therefore, instead of sacrificing security aspects of a digital / online voting system, finding solutions to this problem is more appropriate. Okediran et al., (2011a) provided a survey on online remote voting system vulnerabilities which includes the client-server communication as the primary vulnerability and social engineering and digital divide as the secondary vulnerabilities. These vulnerabilities need to be further studied in order to create a much-secured digital electoral system. Efforts to make a secured and reliable digital / online electoral system are well and truly on its way. Solutions will be discovered to finally put to rest any viable concerns with regards to a digital electoral system. Digital electoral systems would be built with respect to the constitutions and policies of a country which are different when compared with others. These systems will be designed in a manner in which it also facilitates future innovation and enhancement of the system. Digital / Online electoral system will be widely used in the near future.

## REFERENCES

Abo-Rizka, M., & Ghounam, H. R. (2007). A Novel E-voting in Egypt. *International Journal of Computer Science and Network Security*, *7*(11), 226234.

Adeline,, C.P., & Ali,, K., & Hishamnddin. (2006). E-Commerce: A study on Online Shopping in Malaysia. *Journal of Social Science*, *3*(3), 231–242.

Blais, A., & Massicotte, L. (2016). Electoral systems. Comparing Democracies: Elections and Voting in Global Perspective. *Sage (Atlanta, Ga.)*, 40–69.

Cetinkaya, O., & Koc, M. L. (2009). Practical Aspects of DynaVote E-voting Protocol. *Electronic Journal of E-Government*, *7*(4), 327–338.

Chang, C., & Lee, J. (2006). An Anonymous Voting mechanism based on the key exchange Protocol. *Elsevier Computer and Security Journal*, *25*(4), 307–314. doi:10.1016/j.cose.2006.02.004

Cranor, L.R. & Cytron, R.K. (1996). *Design and Implementation of a Practical Security-Conscious Electronic Polling System*. Washington University: Computer Science Technical Report.

FIPA - Foundation for Intelligent Physical Agents. (n.d.). Retrieved from http://www.fipa.org/

Gartner. (2013). *E-health for a Healthier Europe*. Retrieved online at www.calliopenetwork.eu/Link-click.aspx

Gina, G. G., Roberto, G., & Gonzalo, I. D. (2010). Identity based Threshold Cryptography and Blind signatures for Electronic Voting. *WSEAS Transactions on Computers*, *9*(1), 62–71.

Governnment, Q. (2013). *E-learning for Smart classrooms*. Retrieved online at http://gld.gov.au/smart-classroo ms/documents/strategy/pdf/scb yte-elearning.pdf

Gupta, N., Kumar, P., & Chokar, S. (2011). A Secure Blind Signature Application in E-voting. *Proceedings of the 5th National Conference, Computing for National Development*, 1-4.

Ibrahim, S., Kamat, M., Salleh, M., & Abdul Aziz, S. (2003). *Secure voting using blind signature*. available at http://eprints.utm.my/3262 /1/IEEE02EVS_full_paper_ver14Nov.pdf

Jorba, A. R., Roca, J. C., & Scytl Online World Security, S. A. (2007). *Secure remote electronic voting system and cryptographic protocols and computer programs employed*. U.S. Patent 7,260,552.

Katiyar, S., Meka, K. R., Barbhuiya, F. A., & Nandi, S. (2011). Online Voting System Powered By Biometric Security Using Steganography. *Second International Conference on Emerging Applications of Information Technology*, *1*(11). 10.1109/EAIT.2011.70

Kitcat, J. (n.d.). *Free E-Democracy Project*. Retrieved from http://www.gnu.org/software/free/

Li, C., & Hwang, M. (2012). Security Enhancement of Chang-Lee Anonymous E-Voting Scheme. *International Journal of Smart Home*, *6*(2), 45–52.

Light, J., & David, D. (2008). An efficient security algorithm in mobile computing for resource Constrained mobile devices. *Proceedings of the 4thACM symposium on QoS and security for wireless And mobile networks*. 10.1145/1454586.1454608

Malkawi, M., Khasawneh, M., & Al-Jarrah, O. (2009). Modeling and Simulation of a Robust E-voting System. *Communications of the IBIMA, 8*.

Navaneet, S. S. (2012). A model for Performance Enhancement of Steganography through Dynamic Key Cryptography. *International Journal of Advanced Networking and Applications*, *3*(6), 1395–1401.

Okediran, O. O., Olabiyisi, S. O., Omidiora, E. O., & Ganiyu, R. A. (2011). A Survey of Remote Internet Voting Vulnerabilities. *World of Computer Science and Information Technology Journal (WCSIT)*.

Okediran, O. O., Omidiora, E. O., Olabiyisi, S. O., & Ganiyu, R. A. (2011b). A Comparative Study of Generic Cryptographic models for Secure Electronic Voting. *British Journals of Science*, *1*(2), 135–142.

Okediran, O. O., Omidiora, E. O., Olabiyisi, S. O., Ganiyu, R. A., & Alo, O. O. (2011a). A framework for a Multifaceted Electronic Voting System. *International Journal of Applied Sciences*, *1*(4), 135–142.

Olaniyan, O. M., Mapayi, T., & Adejumo, S. A. (2011). A Proposed Multiple Scan Biometric-Based System for Electronic Voting. *African Journal of Computing and ICT*, *4*(2), 9–16.

Olaniyi, Arulogun, Omidiora, & Adeoye. (2013). Design of Electronic Voting System Using Multifactor Authentication and Cryptographic Hash Functions. *International Journal of Computer and Information Technology*, *2*(6), 1122–1130.

Olaniyi, O. M., Adewumi, D. O., Oluwatosin, E. A., Arulogun, O. T., & Bashorun, M. A. (2011). Framework for Multilingual Mobile E-Voting Service Infrastructure for Democratic Governance. *African Journal of Computing and ICT*, *4*(3), 23–32.

Olaniyi, O. M., Arulogun, O. T., & Omidiora, E. O. (2012). Towards an Improved Stegano-Cryptographic Model for Secured Electronic Voting. *African Journal of Computing and ICT*, *5*(6), 10–16.

Olaniyi, O.M., Tayo, A.O., Olusayo, O.E., & Olusola, O.O. (2013). A survey of cryptographic and stegano-cryptographic models for secure electronic voting system. *Covenant Journal of Informatics and Communication Technology, 2*(1).

Pardos, A.B.C., Eucins, A.H., White, S.H., Del Rey, A.M., & Sanchez, G.R. (2007). A Simple Protocol for Yes-No Electronic Voting. *International Journal of Computer Science and Network Security, 7*(7), 72-76.

Patil, V. M. (2010). Secure Electronic Voting System by Using Blind Signature and Cryptography for voter‟s privacy and Authentication. *Journal of Signal and Image Processing*, *1*(1), 1–6.

Purusthomata, B. R., & Alwyn, R. P. (2009). Design and Implementation of Secure Internet Based Voting System with User Anonymity using Identity Based Encryption System. *Proceedings of IEEE International Conference on Services Computing*, 474-481.

Qadah, G. Z., & Taha, R. (2007). Electronic Voting Systems: Requirements, Design, And Implementation. *Computer Standards & Interfaces, 29*(3), 376-386. Doi:10.1016/j.csi.2006.06.001

Roy, A., Banik, S., & Karforma, S. (2011). Object oriented modeling of RSA digital signature in E-governance. *International Journal of Computer Engineering and Information Technology*, *26*(1), 2433.

Roy, A., & Karforma, S. (2011). Risk and Remedies of E-governance Systems. *Oriental Journal of Computer Systems and Technology*, *4*(2), 329–339.

Roy, A., & Karforma, S. (2012). A survey of digital Signatures and Its Applications. *CIT. Journal of Computing and Information Technology*, *3*(1), 45–46.

Saleh, K., El-Morr, C., Mourtada, A., & Morad, Y. (2003). *Specifications Of A Mobile Electronic Voting System And A Mobile Agent Platform*. IADIS e-Society 2003, Lisbon, Portugal.

Sandhya Sarma, K. N., & Umamaheswari, S. (2012). A Powerful Protocol for Electronic - Voting using a HYBRID Crypto Realm. Journal of computing and information technology. *International Journal of Clothing Science and Technology*, *3*(3).

Sodiya, A., Onashoga, S., & Adelani, D. I. (2011). Secure E-Voting Architecture. *Proceedings of Eighth International Conference on Information Technology: New Generations*, 342-347.

Sujata, M., & Banshidhar, M. (2010). A Secure Multi Authority Electronic Voting Protocol based on Blind Signature. *Proceedings of International Conference on Advances in Computer Engineering*, 271-273.

Tohari, A., Jainkun, H., & Song, H. (2009). An Efficient Mobile Voting System Security Scheme based on Elliptic Curve Cryptography. *Proceedings of Third International Conference on Network and System Security*, 474-479.

# Chapter 14
# Economic African Development in the Context of FinTech

**Youssra Ben Romdhane**
*University of Sfax, Tunisia*

**Sahar Loukil**
*University of Sfax, Tunisia*

**Souhaila Kammoun**
*IHEC, CODECI, University of Sfax, Tunisia*

## ABSTRACT

*The purpose of this chapter is to analyze the effect of FinTech and political incertitude on economic growth through a multiple regression. Thus, the authors employ the method of generalized least square (GLS) with panel data. The sample concerns 21 African countries during (2001-2014-2017). The authors use a wide range of measures from Global Findex Database 2017, the World Bank platform, the World Bank national accounts data, and the OECD National Accounts data files base in the context of Africa. Empirical results show that FinTech is a driver of economic growth unless it is actively used in a developed digital infrastructure. In fact, the authors prove that, when financial technologies are used in both transactions (receive and made digital payment), they significantly contribute to the economic cycle. Passive use like simple consumption actions are not a significant lever for the economy. The principal contribution is to highlight that the active use of financial innovations and not passive one and the developed digital infrastructure do promote economic growth in African countries.*

## INTRODUCTION

Despite numerous constraints that hinder economic development, Africa does not escape to the digital revolution through technological innovations. Concerning financial and banking services, Fintech appears as an innovative start-up allowing bankers and traders rethink the model of finance and consequently economic solutions.

DOI: 10.4018/978-1-7998-1851-9.ch014

Since 2010, Fintech has been presented as an important factor in improving the growth rate of African countries (Chakravorti et al., 2017). In this context, several studies have analysed the role of Fintech in the development of countries such as Nambisan (2017), Autio et al. (2017), O'Connor and Allan (2016) and Puschmann (2017). Boot (2016) and Puschmann (2017) have defined Fintech as a new type of financial service (remittances, asset management, and payment) associated with computer technology. In fact, technologies have made remarkable progress in the telecommunications, IT and audio-visual sectors. This will lead to an increasing improvement in service productivity, which meets consumer expectations. Alt et al. (2018) confirmed the obligation to stimulate the expectations of consumers who accept new technological products to seize market opportunities. As a result, these technologies appear to be an important element of economic development, particularly for emerging economies (Guzy, 2016).

In the African context, Chishti and Barberis (2016) concluded that there are more mobile money accounts than bank accounts. According to the authors, financial technology makes it possible to create a new infrastructure capable of ensuring a better inclusion of millions of people in the real economy. These results are confirmed by Claessens et al. (2018) who stated that financial technologies promote investment growth in the region by improving the efficiency of financial market infrastructure, including regulation and clearing in sub-Saharan Africa, compared to other regions of the world. In this sense, Lenzhofer et al. (2013) confirmed that citizens do not need banks only for banking services. On the other hand, they found that some customers around the world remain sceptical despite all the advantages of Fintech innovations. Triki and Faye (2013) noted that Africa is in the phase of transformation and technological and entrepreneurial revolution. These results are also confirmed with the Kanza (2016) and Ndemo and Weiss (2017) studies but are invalidated by the results of other studies, including Malecki and Moris (2007), Foster et al. (2018) and Murphy and Carmody (2015). This explains why this region lags behind Europe or Asia in terms of Internet connectivity or the amount of digital production (OECD, 2016; World Bank, 2016, Pejkovska, 2018; Bahrini, 2019 and Graham and Mann, 2018). Furthermore, Watanabe et al. (2018) found that ICTs are striving to transform themselves into a new economic model generating GDP; they assume that the Internet is a source of happiness for Internet users through its consumption, but cannot be captured by GDP data, which measures economic values. Therefore, the above-mentioned transformation of global ICT companies can be seen as a spontaneous creation not captured by GDP. Moreover, technological entrepreneurship is now the main driver of economic growth and cultural progress, which involves both the application and production of technology (Siyanbola et al. 2011, Goutam and Rishiraj, 2015, Nambisan, 2017 and Puschmann, 2017). According to Rob Nichols (2017), the success of profitable innovation is linked to customers since it is their desires and needs that must guide solutions. Therefore, FSB (2019) and European Economy (2017) point out that technology facilitates the unbundling of services offered by banks and reduces several barriers to entry into the financial market.

Similarly, recent studies have noted that mobile payments are already popular in Africa and that central banks often encourage faster systems and have found that financial technologies favor retail transactions (Philippon, 2017; Chishti and Barberis, 2016). In this sense, the European Economy (2017) found in a recent survey that one in four people have a bank account. Therefore, it has been confirmed that even a low account rate in banks has a positive and very significant effect on mobile payment systems in the African context. Furthermore, financial technologies provide an opportunity to promote the financial inclusion of low-income households and can be a catalyst for poverty eradication, business development and cost reduction. Deloitte (2019) noted that Fintech companies are characterized by stricter regulation than regulations in other sectors of activity, which could limit financial market risks and protect players and consumers in this sector. The study assesses the opportunities that could be seized in West Africa using Fintech's financial system and highlights its impact on economic growth. It describes the current West African financial sector, the settlement of digitization and Fintech's financial market segments. According to this study, the financial sector creates a huge amount of data that Fintech companies use to segment customer populations, identify opportunities for new products and services and optimize prices. In each segment, products can use data and analysis to computerize decision-making processes.

In this light, the underlying aim of this paper is to analyse the impact of digital finance (Fintech) on economic growth and household spending in African countries, a problem that has not yet been explored or is under-explored in the literature. The structure of the paper is organized as follows. Following the introductory section, we provide an overview of the theoretical and empirical literature on the effect of digitalization on economic growth in African countries and a brief summary of the digitization strategies of some African countries. After that, we present the data set and methodology of the research. And lastly, the study highlights the empirical findings, states the main conclusions of the research, and provides some practical implications.

## BACKGROUND

The technological revolution achieved in the world during these years has attracted the attention of many researchers who have focused on studying the impact of Fintech on the development of African countries. The literature review will provide a backdrop to the issues addressed in these research areas and will highlight the effect of Fintech on economic growth and household consumption spending.

The first category of studies provides an economic explanation of the effect of Fintech on economic growth. Several studies have analyzed this effect such as Copenhagen Economics, 2015; Coyle, 2016; Ahmad and Schreyer, 2016; Watanabe et al., 2018; Naveed et al., 2018. Kasekende (2014) and Chakraborty (2018) found that the inclusion of financial and economic systems in Africa has been improved over the period 1995-2013. Syverson (2017) treated the issue of digitalization as a determinant of economic growth and found that technology provides economic stability and improves the economic growth index. These results are confirmed by more recent studies, including McDonagh, 2008, Feldstein, 2017 and Groshen et al., 2017. In this sense, Watanabe et al. (2017) confirmed the existence of an economic evolution measured by GDP not captured and explained by the change in consumer preferences via the Internet.

Using a dynamic and static panel model for a period between 1990- 2014, Aghaei and Rezagholizadeh (2017) analyze the direction of causality between growth and ICT. Empirical results show that the 1% increase in ICT investment leads to an economic growth of 0.52% in the Organisation for Islamic Co-operation (OIC) countries over time. On the other hand, Pradhan et al. (2018) analysed the relationship between ICT infrastructure and economic growth for the G-20 countries over the period 2001-2012. By analyzing the Granger causality test, the results confirm the existence of a positive relationship between ICT and economic growth. Sepehrdoust (2018) also analyzed the same hypothesis but for oil-exporting countries (OPEC) using the generalized current method (GMM) over the period 2002-2015 and confirmed that the 1% increase in the financial development index and ICT variables increased economic growth by 0.048%.

The second category of studies provides an economic explanation of the effect of Fintech on household spending. Bernanke (2009) noted the existence of two favorable factors for household finances, namely reducing complexity and improving transparency. The existence of these two factors in the household environment influences economic activity, Thaler (2016). These results are confirmed by studies by Chishti and Barberis (2016), who pointed out that Fintech facilitates the transparent management of household incomes and thereby improves financial transparency for consumers. Certainly, financial technologies could provide a new dynamic for household consumption and the response to income shocks and savings behavior of individuals, Cocco et al. (2016). In the African context, Chakravorti et al. (2015) noted that the Harvard Business Review has included 3 countries in its global list of the most rapidly moving digital economies (Kenya, Nigeria and South Africa). On the other hand, Mutegi (2016) argues that Senegal is the leading African country in terms of technology and some Maghreb countries such as Tunisia have exploited the growth of information and communication technologies.

However, other research has focused on the risks of technology such as Lee and Shin (2018), Temelkov (2018) and Ramsay (2017). These authors found that Fintech could have negative effects on the country's development and economic systems would be more exposed to systemic risk, Chui, (2017). It should also be noted that the technology used in consumer goods and financial services sector is developing at an increasing and remarkable pace in the African economy. Based on the ADF study (2017), we propose to highlight the digitization strategies of different African countries, as summarized in the following Table 1:

*Table 1. Digitization strategies in African countries*

| African countries | Digital Government Strategy |
|---|---|
| Kenya | The Kenyan government is trying to present itself as a regional leader in different sectors of the digital economy until 2030 through the creation of a smart city ''Konza Technology City'', planning to create 200,000 jobs. |
| Mali | Mali has attempted to provide capacity building programmes through project financing and has set up a one-stop shop to facilitate business creation. |
| Senegal | The Senegalese government created a platform for interoperability of digital payment methods in 2015 that facilitates payment on websites and mobile applications. |
| Tunisia | The National Strategic Plan "Digital Tunisia 2020" was drawn up in order to position Tunisia as an international reference for digital development considered as an important lever for socio-economic development. This plan will enable Tunisia to build a technological infrastructure and thus a more modern economy. |
| Nigeria | The State created the payment company Pay-stack in 2018, which attracts very significant investments to finance the government's budget. In addition, Cowry Wise Corporation helps citizens save and invest in government bonds. Nigerian economists argue that Fintech companies could rapidly improve their country's economic growth. |
| Madagascar | This country has seen an improvement in the rate of access to financial services for the population of Madagascar through an increase in the number of adults with access to formal financial services from 29% in 2016 to 45% by 2022. |
| South Africa | In terms of Fintech, South Africa ranks first compared to the whole African continent with the highest amount of funding. It is therefore not surprising that there are the most start-ups in South Africa. |

Source: Authors' development

## DATA AND METHODOLOGY

In this chapter, we use countries level data concerning 21 African countries during (2001-2014-2017). We gather data from the Global Findex Database 2017, the World Bank platform, the World Bank national accounts data and the OECD National Accounts data files concerning 21 African countries as representative of Fintech during 2011, 2014 and 2017[1].We explain our choice by the fact that African countries are considered as a pilot experience for Fintech development since it represents a Greenfield opportunity for the expansion of financial technologies. We approximate Fintech through several proxies. We consider the possession of debit card, Mobile money, digital transactions as to make or receive digital payment, the use internet to pay and the use mobile to check account. We study Fintech's impact on economic development measured by annual GDP then on household consumption. We also measure the potential effect of political stability that captures the absence of violence and democracy approximated by the variable Voice and accountability that captures perceptions of the extent to which a country's citizens are able to participate in selecting their government, as well as freedom of expression, etc. Finally, we approximate the effect of corruption in the country measured by perceptions of the extent to which public power is exercised for private gain, including both petty and grand forms of corruption. The table below presents all variables measures used in our study.

Based on a multiple regression to estimate, our study uses a panel generalized least square regression (GLS).The multiple regression models for estimating the main effects of the Fintech have the following form:

$$Yi = \beta_0 + \beta_{i1} FINTECHi + \beta_{i2} CVi + \varepsilon$$

where:

$\beta_0$ – intercept or constant amount;

$\beta_{i1}$ – coefficients of the explanatory variables;

$\beta_{i2}$ – coefficients of the control variables

$\varepsilon$ – error term.

The following table reports the variables indications and measurements used in this research.

## DESCRIPTIVE ANALYSIS:

Next, we intend to analyze the importance of Fintech and economic development in the 21 chosen African countries during 2011-2014-2017. Then, we study the impact of Fintech proxies on both GDP and Household final consumption expenditure. We notice that on average 10% of Africans own debit cards whit a maximum of 40% in some African countries and that on average 17.76% have got mobile money and that this percentage could reach up 73%. These weak mean values explain the low development of

*Table 2. Variable definitions and measures*

| Variables | Indication | Measure |
|---|---|---|
| Debit card ownership | DEBIT CARD | The percentage of respondents who report having a debit card. |
| Voice and Accountability | VOICE | Voice and Accountability captures perceptions of the extent to which a country's citizens are able to participate in selecting their government, as well as freedom of expression, freedom of association, and a free media. Estimate gives the country's score on the aggregate indicator, in units of a standard normal distribution, i.e. ranging from approximately -2.5 to 2.5. |
| POLIICAL STABILITY | STABILITY | Political Stability and Absence of Violence/Terrorism measures perceptions of the likelihood of political instability and/or politically-motivated violence, including terrorism. Percentile rank indicates the country's rank among all countries covered by the aggregate indicator, with 0 corresponding to lowest rank, and 100 to highest rank. Percentile ranks have been adjusted to correct for changes over time in the composition of the countries covered by the WGI. |
| Gross domestic production | GDP | The logarithm of gross domestic production per capital. |
| Make or receive digital payment | MADE/RECEIVE DP | The percentage of respondents who report using mobile money, a debit or credit card, or a mobile phone to make a payment from an account, or report using the internet to pay bills or to buy something online, in the past 12 months. It also includes respondents who report paying bills, sending or receiving remittances, receiving payments for agricultural products, receiving government transfers, receiving wages, or receiving a public sector pension directly from or into a financial institution account or through a mobile money account in the past 12 months. |
| Make digital payment | MADE DP | The percentage of respondents who report using mobile money, a debit or credit card, or a mobile phone to make a payment from an account, or report using the internet to pay bills or to buy something online, in the past 12 months. It also includes respondents who report paying bills or sending remittances directly from a financial institution account or through a mobile money account in the past 12 months, out of labor force. |
| Mobile money users | MOBILE MONEY | The percentage of respondents who report personally using a mobile money service in the past 12 months |
| Evolution of time | TIME | The evolution of time during the three past years (2011-2014-2017). |
| Users of internet to pay | INTERNET TO PAY | The percentage of respondents who report using the internet to pay bills in the past 12 months. |
| Users of mobile to check account | MOBILE CHECK ACCOUNT | The percentage of respondents who report using a mobile phone or the internet to check their balance for a financial institution account in the past 12 months. |
| Household final consumption expenditure | CONSUMPTION | Household final consumption expenditure (formerly private consumption) is the market value of all goods and services, including durable products (such as cars, washing machines, and home computers), purchased by households. It excludes purchases of dwellings but includes imputed rent for owner-occupied dwellings. It also includes payments and fees to governments to obtain permits and licenses. Here, household consumption expenditure includes the expenditures of nonprofit institutions serving households, even when reported separately by the country. Data are converted to constant 2011 international dollars using purchasing power parity rates. |
| Perception of corruption | Corruption | Control of Corruption captures perceptions of the extent to which public power is exercised for private gain, including both petty and grand forms of corruption, as well as "capture" of the state by elites and private interests. Estimate gives the country's score on the aggregate indicator, in units of a standard normal distribution, i.e. ranging from approximately -2.5 to 2.5. |

Source: Authors' development

*Table 3. Descriptive variables*

| VARIABLE | Mean | Std. Dev. | Min | Max |
|---|---|---|---|---|
| VOICE | -0.4339394 | 0.629338 | -1.594412 | 0.5919073 |
| DEBIT CARD | 0.1037975 | 0.1109667 | 0 | 0.41 |
| MADE/RECEIVE DP | 0.287125 | 0.1844883 | 0.04 | 0.79 |
| MADE DP | 0.237875 | 0.1757998 | 0.02 | 0.76 |
| MOBILE MONEY | 0.177625 | 0.1684422 | 0 | 0.73 |
| GDP | 3.550542 | 10.01452 | -62.07592 | 26.67587 |
| INTERNET TO PAY | 0.0805 | 0.0720917 | 0 | 0.33 |
| Political Stability | -0.5679542 | 0.8597314 | -2.353407 | 1.064447 |
| CORRUPTION | -0.6182547 | 0.5631981 | -1.590927 | 0.995537 |

Source: Authors' development

digital transactions like the use of internet to pay with an average value of 8.05% and the making and reception of digital payment with an average of 10.3% and 28.7%. On the social level, we notice that African countries are characterized by a negative mean value (-0.5679542) reflecting the high likelihood of political instability and/or politically-motivated violence, including terrorism. Moreover, in terms of the ability to participate in selecting their government, as well as freedom of expression, freedom of association, and a free media, African countries are characterized by a low level of democracy with a mean value of (-0.4339394). To make it worse, we notice that the spread of the corruption phenomenon (with a mean value of -0.6182547) measured by the perceptions of the extent to which public power is exercised for private gain, including both petty and grand forms of corruption. These results are confirmed with the studies (Arner et al., 2016; Pousttchi and Dehnert, 2018; Philippon, 2018, Loukil et al., 2019 and Kammoun et al., 2019). These situations could affect the impact of Fintech on economic development and household consumption. We will present details in the following section through econometric regressions.

The disparity of values between African countries could be explained by the fact that the adoption of financial technology vary in intensity and in nature from one country to another. The study made up by the AFD[2] on digital innovation in Africa and emerging countries classify African countries into four groups according to the degree of Fintech development as follows:

- The leading countries of English-speaking Africa, like Kenya, the regional hub of innovation
- The emerging countries of North Africa, including Morocco, which has a relatively developed ICT sector
- New actors in West Africa such as Côte d'Ivoire, experiencing a rapid development of its ecosystem
- And the rest of the continent, like Mali, whose startup scene remains embryonic, facing structural obstacles.

*Figure 1. The percentage of users of mobile to check account in Africa*

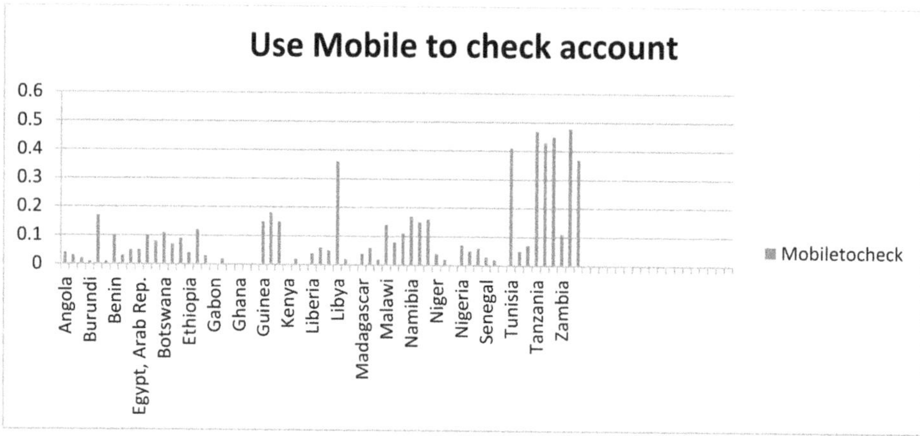


*Figure 2. The percentage of users of internet to pay in Africa*

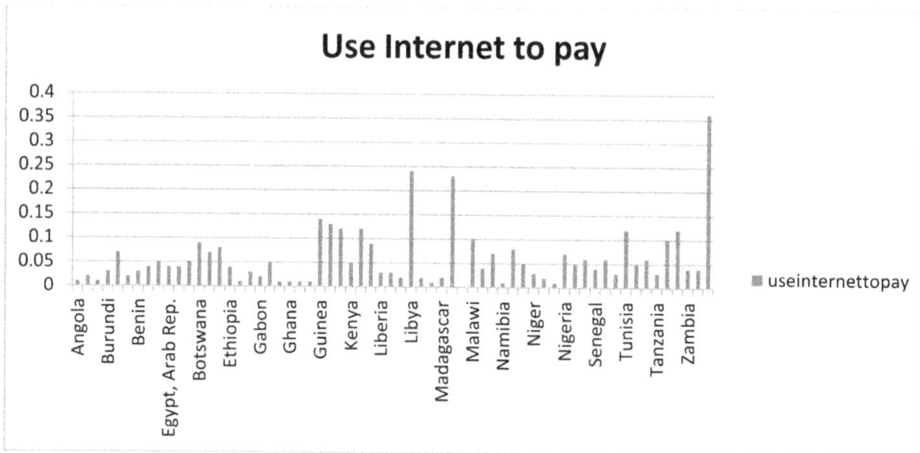


## EMPIRICAL RESULTS

Before estimating of the panel multiple regressions, it is essential to test the absence of correlation between the explanatory variables. We test for the existence of bi-variate and multivariate multi-colinearity. To do so, we apply the Pearson test and measure the Variance inflation factor. We calculated the Pearson correlation coefficients between these variables as well as the VIF "Variance Inflation Factor" which measures how much the variance of an estimated regression coefficient is increased because of co-linearity and the index of tolerance of variables measured by the expression (1-R2j) of the regression of each variable on the (P-1) other independent variables. The closer this value is to zero, the more there is a problem of multi-collinearity. Examination of the correlation matrix shows that all correlation coefficients are less than 0.8. According to Kennedy (1985), there is no serious problem of multi-collinearity. Moreover, we can note that all our explanatory variables have a value of the Variable Inflation Factor (VIF) less than 10, the limit recommended by Chatterjee et al. (1991).

*Figure 3. The percentage of Africans making or receiving digital payment*

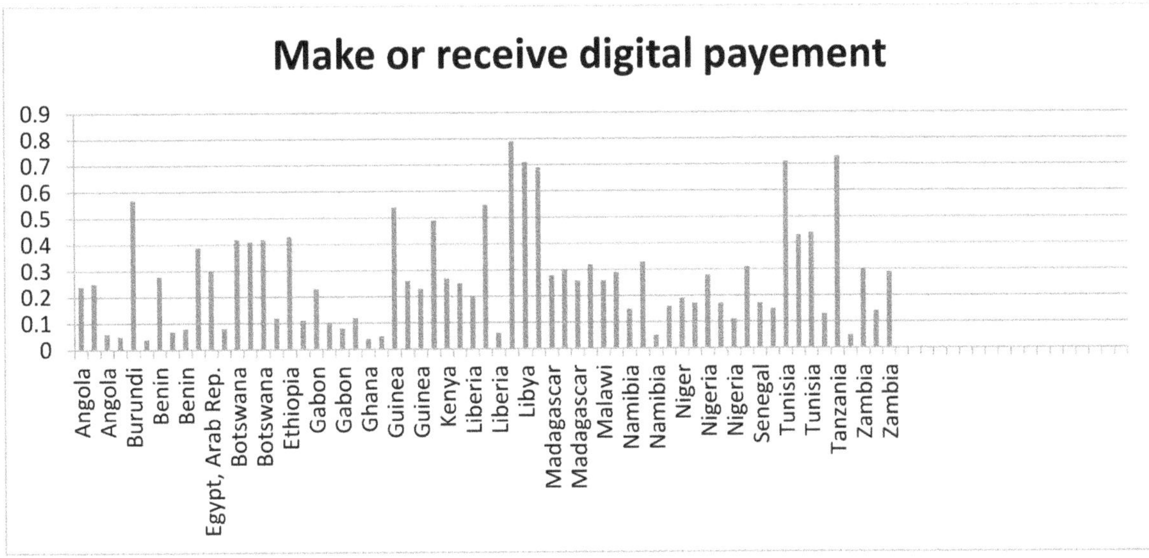


*Figure 4. The percentage of owners of mobile money account*

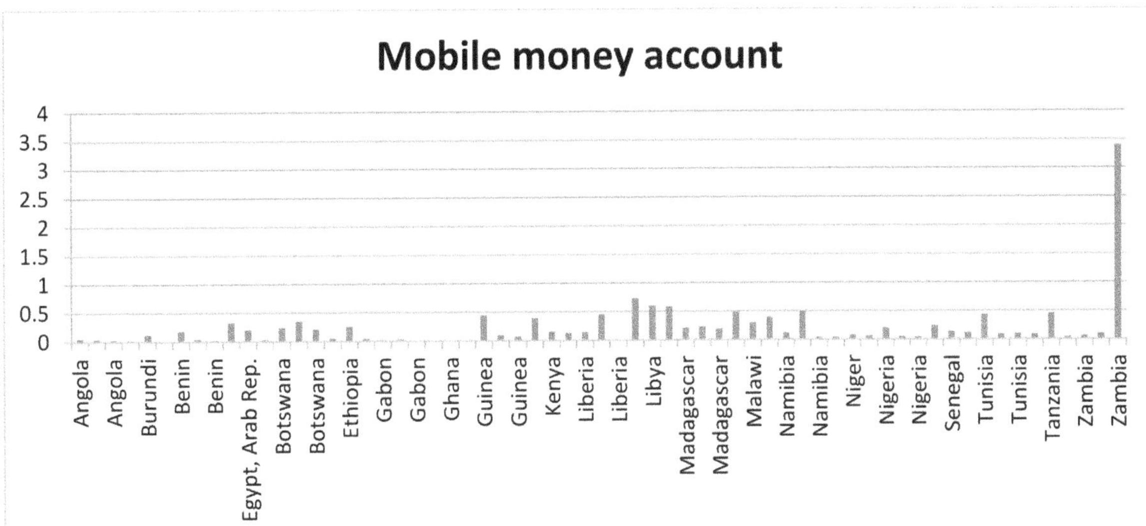


## Impact of Fintech on Economic Development

The results of the econometric regression of the dependent variables measuring economic development and Fintech proxies and control variables through the GLS regression based on panel data are presented in the next table.

Before the implementation of a linear regression it is relevant to determine the relevance of the global model. We specify the representation quality of the model and the global significance of the coefficients associated to the explanatory variables. The Wald chi2 test of the overall significance of the regression

*Table 4. Pairwise correlation matrix*

| | GDP | VOICE | MOBILE MONEY | MADE DP | DEBIT CARD | USE INTERNET TO PAY | POLITICAL STABILITY |
|---|---|---|---|---|---|---|---|
| GDP | 1 | | | | | | |
| VOICE | 0.204 | 1 | | | | | |
| | 0.1087 | | | | | | |
| MOBILE MONEY | -0.3993* | -0.089 | 1 | | | | |
| | 0.0012 | 0.4877 | | | | | |
| MADE DP | -0.3878* | -0.172 | 0.8070* | 1 | | | |
| | 0.0017 | 0.1776 | 0 | | | | |
| DEBIT CARD | -0.2519* | -0.0594 | 0.2760* | 0.6168* | 1 | | |
| | 0.0464 | 0.6438 | 0.0132 | 0 | | | |
| USE INTERNET TO PAY | -0.4240* | -0.0126 | 0.3723* | 0.3942* | 0.2786 | 1 | |
| | 0.0005 | 0.922 | 0.0007 | 0.0003 | 0.0123 | | |
| POLITICAL STABILITY | 0.1124 | 0.6771 | * -0.1562 | -0.2253 | 0.0063 | -0.0558 | 1 |
| | 0.3802 | 0 | 0.2215 | 0.0758 | 0.9609 | 0.6638 | |

* Significant at 10% level. ** Significant at5% level. *** Significant at 1% level.
Source: Authors' development

*Table 5. Tolerance value and variance inflation factor*

| Variable | VIF | 1/VIF |
|---|---|---|
| MADE DP | 5.57 | 0.17945 |
| MOBILE MONEY | 3.83 | 0.261407 |
| DEBIT CARD | 1.98 | 0.50469 |
| POLITICAL STABILITY | 1.97 | 0.506366 |
| VOICE | 1.87 | 0.53487 |
| USE INTERNET TO PAY | 1.24 | 0.808475 |
| Mean VIF | 2.74 | |

Source: Authors' development

thus represented in Table 6 is significant at the 1% level (Prob> chi2 = 0.000). Thus, the explanatory power of the model is satisfactory. Then, we evaluate the linear adjustment quality of the regression equation between the dependent variable and the independent variables through the coefficient of determination R2. It represents the percentage of the explained variance of the dependent variable by the model, Evrard et al. (2000)[3].

*Table 6. Panel regression analyze results (GLS regression)*

| GDP | Coef. | Std. Err. | Z | P>\|z\| | [95% Conf. Interval] | |
|---|---|---|---|---|---|---|
| VOICE | 4.882799** | 2.318982 | 2.11 | 0.035 | 0.3376785 | 9.42792 |
| MOBILE MONEY | -28.18274** | 12.67332 | -2.22 | 0.026 | -53.02199 | -3.34349 |
| MADE DP | -113.6976*** | 40.32356 | -2.82 | 0.005 | -192.7304 | -34.66491 |
| MADE/RCEIVE DP | 126.3004*** | 41.93541 | 3.01 | 0.003 | 44.10848 | 208.4923 |
| DEBIT CARD | -18.97119* | 11.03674 | -1.72 | 0.086 | -40.6028 | 2.66042 |
| USE INTERNET TO PAY | -60.42388** | 24.59397 | -2.46 | 0.014 | -108.6272 | -12.22057 |
| POLITICAL STABILITY | -2.501691 | 1.77784 | -1.41 | 0.159 | -5.986194 | 0.9828121 |
| R2 | 0.3866 | | | | | |
| Wald chi2 (P-value) | 34.67***(0.000) | | | | | |
| Notes: *p< 0.1, ** p < 0.05, *** p < 0.01 | | | | | | |

Source: Authors' development

According to the table, the coefficient of determination is equal to 38.66%. The results of the estimation of our model show that the some features of Fintech have got significant negative impact while other positively affects GDP. We find that making digital payment, owning mobile money account and a debit card have negative and significant effect respectively at 1%, 5% and 10%. In addition, the use of internet as a proxy measure for Fintech affect negatively and significantly GDP at 5%. But, we find evidence for the significant effect of digital transaction (made and receive digital payment) at 1% level. In order to analyze these results, we should first evoke that the benefits of FinTech Mobile depends on access to a phone subscription and possession of a mobile. In fact, by the end of 2015, only 46% of the African population had a mobile subscription, compared to 63% worldwide. This figure is expected to increase to 54% by 2020, compared with 72% worldwide. In this context, Fintech proxies do not have a leverage effect on economic development in African countries. In fact, while mobile money transfers services have had a major impact on developing countries it still has not yet prove its impact in African countries. The expansion of this service allows everyone to avoid going through a bank account, and transfer money to each other and merchants via mobile. The most famous example is the Kenyan M-Pesa, which allows users to deposit money into an account stored on their mobile, to transfer it to other users using PIN-secured SMS, including sellers of goods and services, and exchange deposits for money. Another possible explanation to the divergent effect of Fintech on the economic development is that the nature of use of these financial technologies. So, we can conclude that the Fintech has got a leverage effect on economic development in a country unless it is actively used in a well-developed digital infrastructure. These results are confirmed with the studies of McKinsey, 2011, Langenegger, 2015, Nitcheu (2016) and Mutegi (2017). Another interesting result is the positive and significant impact of democracy on economic development at 5% level. This empirical finding is expected since a democratic political environment (characterized by the ability to participate in selecting their government, as well as freedom of expression, freedom of association, and a free media) is opportune for the success of new technological investment and start-ups. Because of the luck of empirical results on the potential effect of political and social environment, more investigation should be implemented on this research focus.

*Table 7. Panel regression analyze results (GLS regression)*

| CONSUMPTION | Coef. | Std. Err. | z | P>\|z\| | [95% Conf. Interval] |
|---|---|---|---|---|---|
| MOBILE TO CHECK | 4.98*** | 1.76 | 2.82 | 0.005 | 1.528.43 |
| TIME | 9.70* | 5.81 | 1.67 | 0.095 | -1.682.11 |
| CORRUPTION | -1.32* | 7.19E+09 | -1.83 | 0.067 | -2.739.43 |
| MOBILE MONEY | -1.57 | 1.32E+10 | -1.19 | 0.235 | -4.161.02 |
| Cons. | -1.96* | 1.17E+12 | -1.67 | 0.095 | -4.253.38 |
| R-sq | 0.2736 | | | | |
| Wald chi2(6) (P-value) | 14.89*** (0.004) | | | | |

Source: Authors' development

## Impact of Fintech on Household Final Consumption Expenditure

In the previous sub-section we empirically prove that the positive impact of Fintech on GDP depends on the use of these technologies and the level of digital infrastructure development. In order to offer more detailed explanation to the relationship between Fintech and a country's economic development, we turn to study the effect of these financial technologies on household consumption. We notice that to our knowledge, our study is among the few studies to dress this issue.

According to our empirical results, we find that the ability to use a mobile to check account affect significantly and positively household consumption at 1% level. This is due to the fact that the more it is easy to get access to a baking account the more it is easy to make an economic transaction through a digital payment. We note that the household consumption is positively correlated with time. This could be an evidence for the gradually effect of the adoption of technologies especially in finance on household consumption behavior. These results are confirmed with the studies of Consumers International (2017). On the other hand, we interestingly find that the perception of corruption in the country has a negative and a significant impact on consumption. This finding is logically accepted since the perception of the extent to which public power is exercised for private gain, including both petty and grand forms of corruption, does influence the conditions of supply and demand. In fact, in such circumstances, the household behavior could be influenced in impure and imperfect competitions. Thus, the luck of confidence in one country hinders the household consumption. Further researches should be set up to offer more details on the effect of Fintech on economic development and household consumption with respect to differences at countries level.

## CONCLUDING REMARKS AND RECOMMENDATIONS

The adoption of financial technology is an aspect of the economic environment and thus plays a crucial role in economic development through ease of payment, digital currencies and improved customer relations (Peer to Peer). In this context, Africa constitutes a favorable field for the adoption and development of Fintech. Indeed, the existence of mobile payments, the use of mobile phones, the decrease in the number of bank accounts are factors stimulating economic growth, especially in sub-Saharan Africa, which is the fastest growing region in the world for Fintech.

In this study, we examined the effect of the digital revolution on the economic development of 21 African countries over the three years 2011-2014-2017 through two multivariate econometric regression models. The first less square regression model analyses the effect of Fintech on economic growth. The second model explores a new avenue of research to study the impact of Finteh on household consumption.

The study findings show that the benefits of Fintech in terms of a country's economic development depend on two main factors. The first determinant of this positive relationship is the nature of the use of these new technologies, whether passive or active (in the sense of establishing an economic transaction). The second factor is the existence of a well-developed digital infrastructure. However, a detailed analysis of African countries shows significant differences in terms of infrastructure. Moreover, we have found that democracy has a positive impact on economic development through the provision of a timely environment for the recovery of start-ups. Furthermore, we found that household consumption is positively correlated with Fintech since these technologies facilitate access to finance.

In this respect, several recommendations can be given. Indeed, to ensure sustainable economic growth, authorities and decision-makers in African countries should invest more in technology infrastructure, especially in Fintech companies. Fintech can indirectly ensure the country's growth through the adoption of innovation processes and by promoting more transparent competitive conditions. In this context, an e-government will be a powerful tool to help to promote good governance and improve the efficiency and transparency of public administration. However, even though Fintech's services are revolutionizing the consumer financial services market, they have a dual positive and negative impact on consumer behaviour (Deloitte, 2019). That being said, all the necessary measures should be taken to protect customers and operators against the risks of piracy and data breaches, which could increase tariff discrimination and financial exclusion.

## REFERENCES

ADF. (2017). *Etude sur l'innovation numérique en Afrique et dans les pays émergents. Etat des lieux de l'innovation numérique et éléments de benchmark.* Roland Berger.

Aghaei, M., & Rezagholizadeh, M. (2017). The Impact of Information and Communication Technology (ICT) on Economic Growth in the OIC Countries. *Environmental & Socio-economic Studies*, *17*(22), 255–276. doi:10.25167/ees.2017.42.7

Ahmad, N., & Schreyer, P. (2016). Measuring GDP in a Digitalised Economy. *OECD Statistics Working Papers*. doi:10.1787/5jlwqd81d09

Alt, R., Beck, R., & Smits, M. (2018). FinTech and the transformation of the financial industry. *Electronic Markets*, *28*(3), 235–243. doi:10.100712525-018-0310-9

Bahrini, R., & Qaffas, A. (2019). Impact of Information and Communication Technology on Economic Growth: Evidence from Developing Countries. *Economies*, *7*(1), 21. doi:10.3390/economies7010021

Bernanke, B. S. (2009). Financial Innovation and Consumer Protection, speech at the *Federal Reserve System's Sixth Biennial Community Affairs Research Conference.*

Boot, A. (2016). Understanding the Future of Banking: Scale & Scopes Economics and Fintech. In The Future of Large, Internationally Active Banks, (pp. 431-449). World Scientific Studies in International Economics. doi:10.1142/9789813141391_0025

Boot, A. (2017). *The Future of Banking: From Scale and Scope Economies to Fintech, European Economy banks, regulation, and the real sector*. Retrieved from: https://european-economy.eu/2017-2/the-future-of-banking-from-scale-scope-economies-to-fintech/

Chakraborty, S. (2018). *Fintech: Evolution or Revolution*. Business Analytics Research Lab India.

Chishti, S., & Barberis, J. (2016). *The FinTech Book*. New York: Wiley. doi:10.1002/9781119218906

Claessens, S., Frost, J., Turner, G., & Zhu, F. (2018). Les marchés du crédit Fintech à travers le monde: taille, moteurs et enjeux de politique publique. BRI Quarterly Review.

Cocco, F. (2016). Most US Manufacturing Jobs Lost to Technology, Not Trade. *Financial Times*. Retrieved from https://www.ft.com/content/dec677c0-b7e6-11e6-ba85-95d1533d9a62

Coyle, D. (2016). *The Political Economy of National Statistics*. Economics Discussion Paper Series EDP-1603. Available at SSRN: https://ssrn.com/abstract=2850061

Deloitte. (2019). *Beyond Fintech: A Pragmatic Assessment of Disruptive Potential in Financial Services*. Author.

European Union. (2017). *Economics of Industrial Research and Innovation*. EU.

Feldstein, M. (2017). Underestimating the Real Growth of GDP, Personal Income and Productivity. *The Journal of Economic Perspectives*, *31*(2), 145–164. doi:10.1257/jep.31.2.145

Foster, C., Graham, M., Mann, L., Waema, T. & Friederici, N. (2017). *Digital Control in Value Chains: Challenges of Connectivity for East African Firms*. . doi:10.1080/00130095.2017.1350104

FSB. (2019). *Fintech and market structure in financial services: Market developments and potential financial stability implications*. Financial Stability Board.

Graham, M., & Mann, L. (2018). Digital Control in Value Chains: Challenges of Connectivity for East African Firms. *Economic Geography*, *94*(1), 68-86.

Guzy, M. (2016). *Fintech Book*. Chichester, UK: Wiley.

International Monetary Fund. (2019). *FinTech in sub-Saharan African countries: a game changer? Author.*

Jung, I., & Zawacki-Richter, O. (2019). Open and Distance Education in Asia, Africa and the Middle East National Perspectives in a Digital Age? *Springer Briefs in Open and Distance Education.*

Kammoun, S., Loukil, S., & Ben Romdhane, Y. (2019). The Impact of FinTech on Economic Performance and Financial Stability in MENA Zone. In Impact of Financial Technology (FinTech) on Islamic Finance and Financial Stability. IGI Global. Doi:10.4018/978-1-7998-0039-2.ch013

Kasekende, L. (2014). What Role Does Financial Inclusion Play in the Policy Agenda for Inclusive Growth in Sub-Saharan Africa? *Development*, *57*(3-4), 481–487. doi:10.1057/dev.2015.32

Lenzhofer, A., Reber, C., Diemers, D., & Kramer, S. (2013). Taking Wealth Management Digital. Retrieved from: http://www.strategyand.pwc.com/reports/taking-wealth-management digital.

Loukil, S., Ben Romdhane, Y., Kammoun, S., & Ibenrissoul, A. (2019). Empirical Determinants of Fintech: Panel Data Analysis in the MENA Zone. *International Journal of Science and Engineering Invention*, *05*(05), 90–97. doi:10.23958/ijsei/vol05-i05/159

Malecki, E., & Moriset, B. (2008). *The digital economy: business organization, production processes and regional developments.* Abingdon, UK: Routledge.

McDonagh, D. (2008). Satisfying Needs beyond the Functional: The Changing Needs of the Silver Market Consumer. Presented at *the International Symposium on the Silver Market.*

McKinsey Global Institute. (2011). *Internet Matters: The Net's Sweeping Impact on Growth, Jobs, and Prosperity.* San Francisco: McKinsey & Company.

Murphy, J., & Carmody, P. (2015). *Africa's information revolution: Technical regimes and production networks in South Africa and Tanzania.* Chichester, UK: Wiley.

Mutegi, L. (2017). *ICT sector to contribute up to 8% of Kenya's GDP, CIO/East Afric.* Retrieved from: http://www.cio.co.ke/news/ main-stories/ict-sector-to-contribute-upto-8-of-kenya%27s-gdp-by-2017

Mutegi, M. (2016). *Vodafone Launches MPesa Ghana in Africa Expansion Bid.* IPSOS 7.

Nambisan, S., & Majchrzak, A. (2017). Digital Innovation Management: Reinventing Innovation Management Research in a Digital World. *Management Information Systems Quarterly*, *41*(1), 223–238. doi:10.25300/MISQ/2017/41:1.03

Naveed, K., Watanabe, C., & Neittaanmäki, P. (2018). The Transformative Direction of Innovation toward an IoT-based Society: Increasing Dependency on Uncaptured GDP in Global ICT Firms. *Technology in Society*, *53*, 23–46. doi:10.1016/j.techsoc.2017.11.003

Ndemo, B., & Weiss, T. (2017). Digital Kenya, Une révolution entrepreneuriale en devenir, Palgrave Studies of Entrepreneurship in Africa.

Nitcheu, P. (2016). *Apport de l'économie numérique sur le développement économique au Cameroun.* halshs-01970291f.

O'Connor, E., Allan, S., & Audretsch, B. (2016). *Entrepreneurial Ecosystems.* ISEN.

OECD. (2016). *The Digital Economy.* Paris: OECD.

OECD. (2016). *Tax Challenges in the Digital Economy.* Paris: OECD.

OECD. (2016). *Measuring GPD in a digitalized economy.* STD/CSSP.

Pejkovska, M. (2018). *Potential negative effects of Fintech on the financial services sector*. Examples from the European Union.

Puschmann, Th. (2017). Fintech. *Business & Information Systems Engineering: The International Journal of Wirtschaftsinformatik*, *59*(1), 69-76. Doi:10.100712599-017-0464-6

Sepehrdoust, H. (2018). Impact of information and communication technology and financial development on economic growth of OPEC developing economies. *Kasetsart Journal of Social Sciences*. doi:10.1016/j.kjss.2018.01.008

Thaler, R. (2016). Behavioral Economics: Past, Present, and Future. *American Economic Association*, *106*(7), 1577–1600.

The World Bank. (2016). *Digital Dividends*. Washington, DC: The World Bank.

Triki, T., & Faye, I. (2013). Financial inclusion in Africa: An overview. African Development Bank, 148.

Watanabe, C., Moriya, K., Tou, Y., & Neittaanmäki, P. (2018a). Consequences of the digital economy: Transformation of the growth concept. *International Journal of Managing Information Technology*, *10*(2), 21–39. doi:10.5121/ijmit.2018.10202

Watanabe, C., Moriya, K., Tou, Y., & Neittaanmäki, P. (2018b). Structural Sources of a Productivity Decline in the Digital Economy. *International Journal of Managing Information Technology*, *10*(1), 1–20. doi:10.5121/ijmit.2018.10101

Watanabe, C., Naveed, K., Tou, Y., & Neittaanmäki, P. (2018c). Measuring GDP in the digital economy: Increasing dependence on uncaptured GDP. *Technological Forecasting and Social Change*, *137*, 226–240. doi:10.1016/j.techfore.2018.07.053

## KEY TERMS AND DEFINITIONS

**Corruption:** Illegal behaviour by powerful people or people in positions of authority.

**Digitalization:** A trend phenomenon that moves the economy from the era of the physical world to a virtual world based on digitization technologies (mainly internet, "Big Data," and mobile devices).

**FinTech:** Financial technology used to help companies, business owners and consumers in order to better manage their financial operations, processes, and lives.

**Household Final Consumption Expenditure:** The market value of all goods and services purchased by households.

**Macroeconomic Stability:** A term which describes a policy prescription that a national economy has to make to foster growth by buffering it against external shocks.

**Mobile Money:** A technology that allows people to transfer funds between banks or accounts, deposit or withdraw funds, or pay bills by using a mobile phone.

**Voice and Accountability:** Perceptions of the extent to which a country's citizens are able to participate in selecting their government, as well as freedom of expression and a free media.

## ENDNOTES

1 Demirgüç-Kunt, Asli, Leora Klapper, Dorothe Singer, Saniya Ansar, and Jake Hess, 2018. The Global Findex Database 2017: Measuring Financial Inclusion and the Fintech Revolution. World Bank: Washington, DC.

2 https://www.afd.fr/sites/afd/files/2018-05-05-57-55/etude-innovation-numerique-afrique-pays-emergents.pdf

3 The proportion that remains unexplained (1- R2) is due either to the omission of explanatory variables that could contribute to the explanation of the cost of credit or to experimental error.

# Chapter 15
# Tactile Internet and the Remote Surgeon

**Muhammad Imran Majid**
*Institute of Business Management, Pakistan*

**Tahniyat Aslam**
*Institute of Business Management, Pakistan*

**Ahmed Mujtaba Hashmi**
*Institute of Business Management, Pakistan*

**Danyal Subzwari**
https://orcid.org/0000-0002-5584-0555
*Institute of Business Management, Pakistan*

**Babar Siddiqui**
*Institute of Business Management, Pakistan*

## ABSTRACT

*The delivery of healthcare in remote rural areas is a global phenomenon. Using digital governance, a discretized cluster of node relaying time-sensitive, and reliable data, life-critical health services can be ensured. Here tactile internet-based internet of things (IoT) and its variant internet of medical things (IoMT), as well as tactile internet, is introduced. Its relationship with 5G is explored. A model for e-health implementation using unmanned autonomous vehicles (UAVs) in the backhaul is presented. The link is analyzed over fiber optics, satellite, and 5G network in terms of latency and reliability: the crucial metrics for robotic surgery. It is also shown that 5G implemented in Italy, Zimbabwe, and Cooks Island have supported mobile hospitals in rural regions. Hence, induction of 5G networks is a feasible approach. Finally, a critical review of expected future technologies including telematics is presented. Hence, tactile internet-based robotic surgery can reduce the digital divide.*

DOI: 10.4018/978-1-7998-1851-9.ch015

## INTRODUCTION

Digital governance for health infrastructure requires a context for patient-doctor interaction.

Traditional medicine requires diagnosis by a doctor who examines a patient and diagnoses the cure. Initially, the doctor relies on his / her previous training and user experience as well as communication of the diagnosed region of the human body.

Since the 2000s, an estimated 60% of physicians make use of electronic and Personal Health Records. Eventually, there was a need to increase mobility hence the introduction "m-health" or mobile health. Using sensors in an intelligent ambient environment (a-health) have helped in providing in the home health interventions. This in-the-home or district hospital (as compared to city hospital) health care service is the aim of this book chapter. It aims to provide services to a group of users mostly in rural or limited connectivity setting to health provisions which are crucial using state of the art communication infrastructure.

## INTERNET OF THINGS

IoT is a novel technology for accessing machine to machine interaction capable of transmitting from any place via the internet to any connected machine or interface or object in a network at any time. IoT will convert the globe and introduce smartness in all aspects. Here smartness implies efficiency, access to data and easy of working without unnecessary human interface. Some examples include smart home and cities, smart health, smart energy and grid, transportation (Kaur & Singh, 2016). In this chapter we focus mostly on the aspect of life saving surgeries or medical provision in remote locations where traditional medical services are not available readily. We designate this variant of the machine to machine interface in terms of medical devices and ecosystems and call it Internet of Medical Things or IoMT.

### Internet of Medical Things

IoT allows smart devices to be operated distantly all over the network by uninterrupted incorporation between the real and digital worlds resulting in enhanced efficiency and accuracy. With the ultra-fast connectivity, smart administration, and data competencies, the 5G network supports novel prospects in terms of medical care including imaging, diagnostics, and treatment. IoMT is an amalgamation of smart medical equipment such as clinical wearables and remote sensors that transmit health records such as vital signs and patients 'safety. Smart clinical devices provide diagnosis and cure facilities at affordable prices. Various clinical complexities expect high dependability and accessibility with latency intervals that are dependent on human interaction down to a few milliseconds. Currently, several healthcare applications will profit a range of medical businesses that need high bandwidth and consistent connectivity, and these applications are a function of the evolving 5G test circumstances

## Digital Health Imaging and IoT

The benefit of telemedicine is to not only access the images in distantly areas but rapidly share it across the geographic area especially the rural and undeveloped areas where the patients do not have access to the modern clinical expertise. X-rays or CT scans aids the patients to rapidly attain other opinions across a global network of medical professionals. This technique improves healthcare proficiency to small societies with no thorough connectivity without established healthcare infrastructure.

## Wearable Medical Device

The diagnostic is an important capability of observing wearable medical devices. It helps the patient's experience from serious or prolonged wellbeing issues such as cardiovascular disease, diabetes or cancer, distant observing equipment can monitor vital signs and transmit this information to doctors. This medical device provides an initial warning that aids doctors to detect possible complications and give clinical care to patients in a preventive manner.

The Michael J. Fox Foundation has established equipment's that detect the tremors linked with Parkinson's disease. Instead of relying on patients' self-complaining, physicians are installing wearable motion-based sensors that deliver consistent information in real-time for several different sides of the disease. These kinds of observing tools are particularly beneficial to older citizens. Doctors and nurses can detect essential signs, motion, falls, and speech slurring, among further effects, to deliver real-time identification of patient's healthcare problems (Looney, 2016). Wearable devices can transmit medical alerts if the individual suffers from health emergencies such as victims of heart attack or individuals who are weakened by a stroke obtain rapidly aid from doctors with smartphones or wearable devices. Distant surgery will be possible once latency levels are decreased to a few intervals. (Crommenlaan, 2015).

Recent advances in information and communication technology have enabled media content sharing and feedback between devices and nodes in an unprecedented manner. This together with seamless integration of devices in a network to enhance activation of multiple devices including sensors and actuators give rise to IoT (Gubbi, Buyya, Marusic, & Palaniswami, 2013). However, the data throughput has increased manifold as the lowered latency over long-haul links to less than 10ms. This breakthrough has enabled to develop networks of various platforms whereby feedback and control are driven faster than our perceptual speed of interaction with our environment, thus giving rise to an era of "Tactile Internet" (Fettweis, 2014). Haptic interaction is any kind of mode requiring the perception of touch.

## Tactile Internet

Tactile Internet aims to advance interaction between humans and machines in virtual, real and remote environments. It democratizes access to skills and expertise in the same way as current internet infrastructure has helped evolve and enable access to information. The near consistent, fast-responsive, and versatile connection helps realize the real-time control and physical haptic experiences instantly. In terms of digital governance, it has implications for controlling industrial systems, Internet of Things, wearable sensors, health monitoring and robotics which aid in the delivery of essential medical services.

The Tactile Internet will improve what is called as a human in the loop. The order of ms in latency is essential to enable the human sensory and reaction method to gain meaningfully when an end-user is a machine which helps prevent errors especially when humans are the client to the haptic interaction.

Tactile refers to the perception of information using gesture and mechanical features. Tactile Internet ensures internet connectivity and remote physical interaction networked control highly dynamic processes and contour of touch experience. Its features include not only latency but reliability, security, and availability.

The evolution towards the Internet of Medical Things (IoMT) is making a new model of controlling devices or equipment through communications. Concept of Tactile Internet (Fettweis, 2014) is evolving because it will enable the control remotely in real-time and experience physical haptic. The Tactile Internet support ultra-reliability, very low latency, provides sufficient capacity for communicating devices and intelligent connectivity within the network. Tactile Internet is defined as a network of networks for retrieving, observing, directing real or virtual devices or processes observed in real-time by human or machine distantly.

## Critical Review of Emerging Communications Technologies

For digital governance, internet links various societies across the world for sharing of information and resources. In this environment, smart equipment is connected through several communication standards. These facilities need an effective, consistent, low latency, and end on information security which influences the spectral efficiency of the network. But for effective utilization of spectrum, multiple access (MA) techniques are employed that allocates the existing bandwidth between the users. All of the techniques improve the capability of the network by delivering an improved quality of services (QoS) to the users. (Budhiraja, Tyagi, Tanwar, Kumar, & Rodrigues, 2018) The various communication developments are presented in Table 1.

*Table 1. Comparison of communication developments*

| Generation Type | 1G | 2G | 2.5G | 3G | 3.5G | 4G | 4.5G | 5G |
|---|---|---|---|---|---|---|---|---|
| **Standards** | AMPS | GSM | GPRS | UMTS | HSDPA/ HSUPA | LTE-A, Wi-Max (IMT-A) | LTE-A Pro (IMT-A) | IMT-2020 |
| **Techniques** | SISO | SISO | SISO | SISO | SISO | MIMO | MIMO | mMIMO |
| **Bandwidth** | 30KHz | 200KHz | 200 KHz | 5-20 MHz | 5-20 MHz | 100 MHz | 100 MHz | 1-2 GHz |
| **Frequency** | 800-900MHz | 850-1900 MHz | 850-1900 MHz | 1.6-2.5 GHz | 1.6-2.5 GHz | 2.0-8GHz | 2.0-8GHz | 3-300 GHz |
| **Data Rate Downlink/uplink** | 2Kbps | 14 Kbps | 115/20 Kbps | 7/2 Mbps | 56/22 Mbps | 128/56 Mbps | 3/2 Gbps | 20/10 Gbps |
| **Throughput** | < 20Kbps | <20Kbps | <200Kbps | <4Mbps | <15Mbps | <300 Mbps | <1Gbps | <10 Gbps |
| **Spectral Efficiency** | 0.0015 bps/Hz | 0.17 bps/Hz | 0.33 bps/Hz | 1.3 bps/Hz | 4.22 bps/Hz | 30 bps/Hz | 30 bps/Hz | 120 bps/Hz |
| **Latency** | >1000ms | ≅300ms-1000ms | ≅300ms-1000ms | ≅100ms-500ms | <200ms | <15ms | <5ms | <1ms |
| **Connection density** | Limited | Limited | Limited | 100/km$^2$ | 100/km$^2$ | 1000/km$^2$ | 10,000/km$^2$ | 10,00000/km$^2$ |
| **Mobility** | - | - | - | - | - | 350Kmph | 350Kmph | 500Kmph |

Source: (adapted from (Budhiraja, Tyagi, Tanwar, Kumar, & Rodrigues, 2018)

This circumstance is a novel technology where entirely every IoT based devices can share information to the Internet, hence creating a smart environment. (Mohammed & Ahmed, 2017). The 5G technology as seen in Table 1, highlights that the technology is based on massive MIMO technique implying a high connection density and mobility reaching at least 500 Kmph. This is what enables the adoption of IoT for future standards.

## LITERATURE REVIEW

The state of the art related to the development of IoT applications and latency-sensitive but reliable tactile internet applications. The impact of TeleHealth is finally explored.

## APPLICATIONS OF INTERNET OF THINGS

Internet of things has numerous uses in daily routine, making simple, secure and smart life. Following are the applications of the internet of things:

### Smart Cities

Cities are observed as hubs of smart life. It requires careful designing in all stages, with assistance from governments and people to employ the IoT technology in each aspect. Currently, many cities like Seoul, New York, Tokyo, Shanghai, Singapore, Amsterdam, and Dubai were followed by these projects.

Cities will become smarter by improving the transportation system, health care system, security system, weather observing systems along with the support by providing the internet facility everywhere to retrieving the database of airports, railways, and other transportation. (Nuaimi, Neyadi, Mohamed, & Al-Jaroodi, 2015)

Managing a Smart City based on IoT is a complex task due to the wide variety of interfaces available at physical and data link layers. (Zanella, Bui, Castellani, Vangelista, & Zorzi, 2014). Combining various end-user systems and equipment into a seamless and unified network of communication devices generally referred to as the Internet of Things. For Smart Cities to openly exploit the underlying data and digital devices for optimized services, efficient employment of IoT is considered paramount. This can cover the backhaul as well as the major public utilities. However, to provide a better Digital Governance maximum usage of underlying communication technologies is to be utilized.

### Smart Home and Buildings

A smart home provides online streaming facilities to operate the equipment's functionality through the network. Here cell phones guarantee that users have an approach to a transportable 'controller' for the connected electronics technology. Both sorts of equipment can behave as gateways for application of IoT. (Amrita Sajja, 2016)

Several companies are currently creating standards that assimilate the home automation with entertainment, healthcare observing, noting of energy and wireless sensor observing in the house. Through the perception of the IoT, the smart house might control different equipment's smartly like electricity, environments, air controlling and central heating, and security. (Tiago D. P. Mendes et al., 2015)

## Smart Energy and Smart Grid

A smart grid assimilates the information and communications technologies (ICTs) to the energy system will allow continuous communication between providers and customers, making an additional interface on the flow of energy and help in delivering electricity more proficiently and sustainably. Main features will comprise of detecting and observing technologies for power flows; digital infrastructure to transfer information through the grid; smart meters to notify the usage of energy; management and computerization systems to combined and process several information and to generate an extremely interactive electricity (Mohassel, Fung, Raahemifar, & Mohammadi, 2014). Various uses like industrial, solar power, automobiles and cities power control can be handled due to the smart grid.

## Smart Health

IoT a based technology provides continuous real-time monitoring to the admitted patients whose physiological condition would be examined constantly. Health sensors are used to gather patient's health-related information and after examine and saving it transmit the analyzed information to caregivers for additional analysis. It replaces regular interval checkup of the patient's by providing a real-time automated sensor that gathers patient's vital signs. By this technology, it enhances the quality of healthcare. (Chamberlin, 2016)

Peoples around the globes especially in the remote areas are experiencing from the bad health for the reason that they don't have an approach to efficient health observing and monitoring system. By the effective wireless solutions linked with the IoT, are now making feasible to examine these patients. They can be used to gather health information of patients from various sensors, direct different algorithms to examine the information and then communicate with medical specialists for suitable health recommendations (Ahmed, Björkman, Čaušsević, Fotouhi, & Lindén, 2015).

## Smart Transportation

The main perception of smart transportation is crowd tracking and monitoring. The procedure initiated with customer acknowledged their route by indicating some points in the smart phone's application (Talasila, Curtmola, & Borcea). The smart transportation is deal with some concepts; such as transportation analytic, controlling vehicles and automobile connectivity. The transportation analytics signifies the analysis of prediction of demand and irregularity detection. The routing of automobiles and speed controlling as well as management of heavy traffic are all known as controlling of vehicles.

Various environments have been integrated into complex networks due to the advent of recent hardware and software technological advancements. An aspect of such networks widely being implemented and researched upon is the concept of VANET, that is, Vehicular Ad-Hoc Networks. Such types of networks have the perspective of being developed into a standardized system of road safety, traffic optimization and vehicle safety for drivers as well as passengers. VANET research work has focused on specific areas including routing, broadcasting, Quality of Service (QoS), and security. (Zeadally, Hunt, Chen, Irwin, & Hassan, 2012). The driverless car is now being considered a reality in the sense of giving autonomy to the human driver to only prompt in various complex adventures. This has implications for the elderly and vulnerable people who will not be dependent on humanistic interference.

## APPLICATION OF TACTILE INTERNET

Tactile Internet has a wide range of applications in the fields of industrial automation, transport systems to telemedicine and virtual and augmented reality.

### Self-Driving Vehicles

Technological gap is reduced between traditional human-driven cars and self-directed motors that are partially and fully drive themselves. Self-directed automobile technology with self-driving capabilities will transform the driving especially, self-parking and emergency braking. Several firms have now distributed self-driving automobile. Moreover partially self-driving cars, make parking easier, reduces accident rates, and saves time. Self-directed car drivers might be capable to employ traveling period working, enjoyable, or retrieving leisure/virtual media on the way that has a great impact on economics. By entirely autonomous driving traffic congestion and road accidents can be reduced as the time required for automobile safety to avoid a collision is less than 10 ms. The latency of a millisecond will be required in two way communication of fully automated driving (Mer, 2015).

### Industrial Automation

The traditional industry of manufacturing is now transformed by a digital revolution that is enhanced by developing technologies like robots, sensors, etc. Following three revolutions of industries are observed by many companies, that might be defined as disrupting leaps in the procedures resultant in greater production (Industry 4.0, 2015). The initial industrial revolution originated in the 1780s, where manufacture amenities by the aid of water and steam power. In the 1870s, the next industrial riot facilitated bulk manufacture with the aid of electrical energy. The third industrial riot is started in 1969, mutually connected with the programmable logic controller.

Industry 4.0 is established on cyber-physical methods in real-time. It changes the manufacture, engineering, material and provides supply chain and other management, resulting in self-systematized smart factories. The industry 4.0 includes smart machinery, storing and manufacturing amenities able to originally communicate each other autonomously. (Weiner, Jorgovanovic, & Sahai, 2014). Latency requirements are in the range of few milliseconds. Therefore, smart factories simultaneously with the robotics industry are a main application in the Tactile Internet.

## Tele-Medicine

E-health and healthcare facilities are used to improve patient health. So, the Information Technology (IT) and wireless standards are needed to be enhanced to provide effective, consistent, and continuous health facilities in telediagnosis, telesurgery and telerehabilitation specifically telemedicine. Doctors will control a telerobot at the patient's site not only through video/audio information but it will provide haptic feedback. Similar technology is implemented for telesurgery purposes. E-health based on the requirements of the reliable wireless connection. Because inconsistent connections may lead to delayed images with bad image resolution.

An exact telemedicine cure can only be done with haptic feedback if the doctor/ human-to-device interface can be enabled in real-time. A few milliseconds latency with connection links and information transmission is necessary for e-health that could be achieved through the Tactile Internet. (Simsek, Aijaz, Dohler, Sachs, & Fettweis, 2016)

## Virtual and Augmented Reality

Currently virtual and augmented reality uses are due to the accessibility of the Tactile Internet. In the haptic virtual environment, various users are together physically through simulation software to mutually accomplish tasks by observing not only through audio-video but also from touch sense. In the augmented reality the actual and computerized content is envisioned in the field of user's view.

The virtual reality through the touch sense practices numerous applications depending on the high level of accuracy which is a few milliseconds latency between the virtual reality and the users. The augmented reality can be shifted from immobile to dynamic in Tactile Internet. It allows a real-time virtual view for the user, to identified and avoided possible dangerous events (Simsek, Aijaz, Dohler, Sachs, & Fettweis, 2016)

## Telehealth

"Telehealth" is regarded as a revolution in providing centralized access to latency-sensitive network domain as well as IoT becoming the stepping stone for favorable outcomes in providing swift and prompt medical assistance and controlling a large number of vulnerable and critical medical situations where conditions are worsened only due to delay in necessary healthcare or correct diagnosis.

Owing to the recent advances in various telehealth methodologies, advancements have been achieved in many fields such as radiology coverage, urgent services including stroke or burn case and provisioning of healthcare where the patient is unable to physically reach the medical facility, such as inmates of a prison. (Weinstein et al., 2014).

One of the breakthrough dividend achieved as a result of telehealth is the concept of remote surgery, where a physician performs surgery on a patient positioned miles away but connected through a very high-speed low latency communication link. Medical practice around the world is mostly limited to the region where medical care and doctors are physically present. However, extremely low latency feedback and control through tactile internet have enabled the medical care to be available anywhere and anytime, without the limitations of geographical boundaries, with complete audio-visual and haptic feedback.

## Case Study

The innovative idea of bringing 5G technology into telehealth provision has long been debated. This provides an economical sample in rural regions and low-income areas of countries like Zimbabwe, Cook Island, and Italy. The major idea is to provide health facilities using UAVs (Unmanned Aerial Vehicle) by cutting off the delay. The researchers provided the preliminary results by approaching two methods, one through RRH (Remote Radio Head) mounted on UAVs and other through Large Cells by providing coverage in the region. In RRH over UAV (so-called RRH-UAV), the radio cells are implemented on a rotary-wing flying on low altitudes. The radio cells provide basic radio functionalities like digital processing, analog-digital conversion, filtering and amplifications which are coded in the hardware of UAV.

Solar Power Node (SP Node) has been used to get RRH-UAV charged as it contains batteries and solar panels. The RRH is integrated with Baseband Unit (BBU) at SP node via microwave link to perform baseband signal processing operation. The link between RRH and BBU is highly reliable and cost-efficient results in flexible coverage in the region. The provided scenario by the UAVs is only applicable for emergency calls.

Large Cells (LCs), as on the other hand, provide wireless access connectivity covering dozens of kilometers under a specified radius. The scenario requires base stations build up with a large array of antennas for better spatial use of the beamforming technique for transmission. The technique is energy efficient as it exploits to increase cell coverage. The ultra-lean technique has also been utilized for a reduction in power consumption as it uses spectral resources via 5G transmission schemes. LCs are also integrated with RRH and BBU to provide baseband operations under coverage radius. The two scenarios are mentioned in the figure to observe the preliminary results (Birke et al., 2017).

As in Figure 1, optical connections and radio links were established to provide link more efficiently. Solar Power (SP) nodes are hosting RRH-UAV and LCs to provide flexible, low-cost and less power-consumption in networking and computing resources. Power batteries and solar panels are controlled by

*Figure 1. Proposed 5G architecture for two techniques RRH-UAV and large cells*
*Source: (adapted from Birke et al., 2017)*

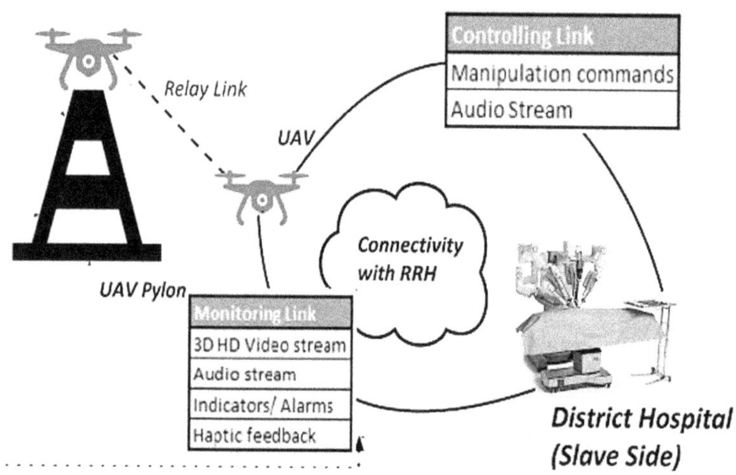

SP nodes. The centralized network core or orchestrator has been built up as brain of whole architecture to provide telehealth facilities in low-income regions of Zimbabwe, Cook Island, and Italy. The orchestrator monitors and controls the trajectories of UAVs while facing the challenges regarding power constraints, channel modelling, and interface management. The exploitation of optical provides the connectivity between LCs and UAVs. In Zimbabwe and Italy, a certain amount of grids are installed to provide electrical power as a backup if in case of battery and solar crises. On the other hand, in Cook Island, the case is quite different as it is perfectly manageable batteries equipped to provide the energy without having any backup of electrical power. Economically, LCs consumes less monthly user subscription charges, which is € 0.61, than UAVs, as it consumes € 1, in these low-income regions. This has implications for rural area communication and can be a good starting point for exploring backhaul.

## CRITICAL REVIEW

### Enabling Tactile Internet

Tactile Internet aims to advance interaction between humans and machines in virtual, real and remote environments. The Tactile Internet is identified where various stringent design challenges for the Tactile Internet have been described by (A. Aijaz, 2015). The interaction of human being with their environment is critical. The requirement for real-time interaction is based on human senses. The audio, visual and touch-based interaction needs reaction times around 100ms, 10ms, and 1ms respectively. By realizing these reaction times humans can seamlessly interact with machines. Tactile Internet aims to accurately transmit across the internet of skills (Dohler, 2013). Following are the requirements of Tactile Internet:

### Ultra-Responsive Connectivity

For ultra-responsive connectivity, the Tactile Internet requires 1ms end-to-end latency. (Z. Shi, 2009), (M. Di Luca, 2009). Because of conflicts among visual, vestibular, and proprioceptive sensory systems, the users will undergo cyber-sickness (K.M. Stanney, 2002). If eyes observe a slightly delayed contrasted to what is witnessed by the vestibular system even though the human body stays static, this delay directs to cyber-sickness. The delay or end-to-end latency comprises of the time utilized in transmitting information from an actuator (or haptic interaction) through communication setup to a main server and retransmission of the information back to the sensor. Figure 2, shows a typical haptic communication scenario.

The phenomenal success of cellular networks has been based on providing ubiquitous and reliable wide-area coverage for voice and text communications. Extension to provide rural coverage to emancipate population is a requirement. With 4G and the success of mobile computing devices, the industry is targeting to bring the same reliability and ubiquity of access to mobile internet applications such as web browsing and audio and video streaming.

Long Term Evolution (LTE) is providing effective data rates of around 50 Mb/s. However, considering technology and market forces in the last few years, cellular speeds of 10 Gb/s or more are achievable by introducing new applications (E. Dahlman, 2014), (Osseiran, 2013). While high data rates will be a key feature of 5G networks, another key challenge is to be able to provide carrier-grade access reliability.

*Figure 2. A meter controlled haptic communication system*
*Source: (adapted from (Dohler,2017))*

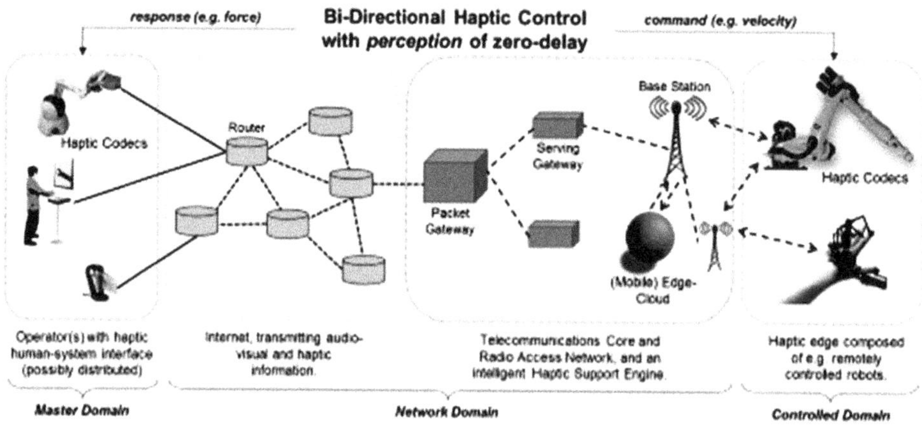


Beyond single-digit end-to-end latency, ultra-reliable network connectivity is an important requirement for the Tactile Internet. Reliability refers here to the probability to guarantee a required function/performance under stated conditions for a given time interval. The specific reliability requirements differ for various types of services and applications. Demands for the highest possible reliability are associated with requirements for real-time response.

## Security and Privacy

The fundamental requirement of Tactile Internet is safety and privacy. New techniques are required for tactile applications that allow only the designated receivers to route secure communication. Appropriate and novel measures such as biometric fingerprinting are one such method used for such a secure link.

## Tactile Data Encoding

The Tactile Data including audio/visual information is handled over a dedicated link known as tactile encoding mechanisms working over Tactile Internet. The tactile encoding mechanisms requires the transmission of the data over packet-switched networks. Further, there must be providing sensory feedback in terms of audio and visual, to the human for tactile perception (R. Chaudhari, 2015).

## Evaluating Tactile Internet and Robotic Telesurgery

Communication quality of services (QoS) requirement varies from application to application. The amount for robotic telesurgery is given Table 2 and the details are discussed below:

Real-time multimedia stream requires the latency rates below 150 ms to ensure better performance (M. Perez, 2015). Video conferencing involve jitter less than 30 ms and packet loss rate less than 1% (M.Eid, 2011). Generally, the sampling rate of an uncompressed haptic signal is 1 kHz. For typical teleoperation method (B. Cizmeci, 2017), the data rate for haptic data contrasts from 128 to 400 kbps, the latency and jitter must be less than 50 ms and 2 ms, respectively, and the expected loss of data is around 0.01%.

*Table 2. QoS requirements summary for sensor data in robotics telesurgery*

| | | Data Types | Latency | Jitter | Packet Loss Rate | Data Rate |
|---|---|---|---|---|---|---|
| Network Domain | Real-time Multimedia streams | 2D Camera flow | <150ms | 3-30ms | $<10^{-3}$ | <10 Mbps |
| | | 3D Camera flow | <150ms | 3-30ms | $<10^{-3}$ | 137-1600 Mbps |
| | | Audio flow | <150ms | <30ms | $<10^{-2}$ | 0.022-0.2 Mbps |
| Control Domain | Physical Vital Signs | Temperature | <250ms | - | $<10^{-3}$ | <0.01Mbps |
| | | Blood Pressure | <250ms | - | $<10^{-3}$ | <0.01Mbps |
| | | Heart Rate | <250ms | - | $<10^{-3}$ | <0.01Mbps |
| | | Respiration Rate | <250ms | - | $<10^{-3}$ | <0.01Mbps |
| | Cardiac Electrograms | ECG | <250ms | - | $<10^{-3}$ | 0.072 Mbps |
| | | EEG | <250ms | - | $<10^{-3}$ | 0.0864 Mbps |
| | | EMG | <250ms | - | $<10^{-3}$ | 1.536 Mbps |
| Master Domain | Haptic Feedback | Force | 3-10ms | < 2ms | $<10^{-4}$ | 0.128-0.4 Mbps |
| | | Vibration | <5.5ms | < 2ms | $<10^{-4}$ | 0.128-0.4Mbps |

Source: (adapted from (Zhang, 2018))

The research case study (Kajimoto, 2017) examined the result of latency feedback and found that the threshold is at 5.5 ms. Tele-surgery is a serious scenario, therefore more stringent latency, jitter and packet loss are required. The physiological data can allow 250 ms latency with average data rates (Wang, 2010). The distant surgeon primarily depends on the 3D video stream and haptic feedback. For audio conversation, the data rate varies from 22 to 220kbps. The delay and jitter must be lesser than 150ms and 30 ms, respectively. Only then can it accept a 1% packet loss.

## Critical Review of Tactile Internet in Remote Surgery

Medical profession around the world is mostly limited to the region where medical care and doctors are physically present. However, extremely low latency feedback and control through tactile internet have enabled the medical care to be available anywhere and anytime, without the limitations of geographical boundaries, with complete audio-visual and haptic feedback. Robot-assisted surgery has been around for quite some time (Bodner et al., 2004), however, due to the limitation of underlying communication networks, the concept of remote telesurgery or ‘Remote Surgeon’ could not flourish. With the advent of 5G, certain features could help to flourish the concept of Robotic Surgeon.

## Characteristics of M-Health Solutions Using Tactile Internet

5G network offers a wide range of connectivity to multiple devices across the network. Machine-to-machine devices (M2M) and Internet of Things (IoT) are the novel development of mobile technology. Upcoming 5G competences (5G-PPP White Paper on E-Health Vertical Sector, 2015) might generate major improvements in healthcare sectors, such as the administration and monitoring of hospital resources, robotic telesurgery, and distant smart medication.

### Low Latency

Latency in an m-health application has a direct effect on the procedure of controlling the robotic equipment's. Future predicted end-to-end latency of telesurgery has been accepted lesser than 200ms with a restriction of 100ms (Q. Han, 2015). 5G network will decrease the communication latency to 1ms. Shortly, a patient could be virtually operated by a doctor from a remote location.

### Bandwidth

The 3G and 4G network does not support real-time monitoring and observing in healthcare applications due to limited bandwidth. The major advantage of using the 5G network is the ability to use higher frequencies (high transmission rates of Gbps). The ultra-high-speed 5G network will provide high-resolution images to the doctor at a remote location. Besides, the 5G network will provide an accessible and flexible technique to assign bandwidth in the medical field.

### Network Capacity and Massive Numbers of Devices

Novel network infrastructure can help support several devices and biomedical sensors with IoT competencies. This generates a massive amount of data, thus exceeding the network capacity (Shariatmadari, 2015). Here network capacity is considered in terms of capacity and rate of information flow.

### Long Battery Lifetime

In the m-health scenario, billions of biomedical sensors and actuators are connected remotely for monitoring and observing the medical treatment. Therefore, the need for extended battery life is essential. The condition to avoid charging and replacement of the battery in 5G must be a few years.

### Security

As a result of the increase in the number of tenants and stakeholders, security threat will increase. 5G network model has an inbuilt network slicing concept. This aims to enhance privacy and security protection.

### High Speed Audio-Visual Feedback

A pre-requisite for a remote surgeon is a real-time, reliable, synchronous, hi-speed, audio, visual and haptic feedback. Authors in this chapter explore the challenges and possibilities of utilizing recently introduced reliable and low latency communication and networking technologies as underlying platforms for tactile internet particularly for enabling a stable and consistent remote surgeon in areas where the physical presence of a surgeon is deemed difficult.

### Critical Analysis of 5G and Digital Health Governance

The era of 5G improvises the health sector including mobile hospitals to facilitate the rural areas where the small ratio of doctor to patient and little income per capita exists. A surgeon in China has done the world's first distant operation using 5G technology. Doctor Liu Rong was able to use the next-generation network to direct robotic arms in a distant location. The 5G technology allowed the doctor to see a clear, real-time image of the operation, as well as communicate easily with his assistants. This link was established with a latency of just 0.1 seconds.

A telemedical device can simply complete haptic feedback if the doctor-to-patient interaction is enabled in real-time. The physician will control a telerobot at patients' location with the help of high-speed audio/video and haptic feedback. This scenario requires real-time deterministic performance in communication systems. E-health required few milliseconds end-to-end latency with ultra-high consistency which can be attained by Tactile Internet (M. Simsek, 2016). Tele-medical tools increase access to health care services by reducing the distance.

Tele-robots can be classified according to the technique used to direct commands and receive haptic feedback. These three categories are tele-programmed, tele-simulated and teleoperated. Tele-programmed robot includes only visual feedback in the form of video. The most popular telerobot employed by Tarn and Brady, where the operator delivers a path for the robot to execute (K. Brady and T. J. Tarn, 1998). Similarly, Simmons presented the autonomous mobile robot Xavier that can be ordered to go to different sites and perform different tasks (Simmons, 1998). Tele-simulated system was invented by Chong, where the operator controls in real-time, and feedback are supplied by overlapping real and predicted display (N. Y. Chong, 2000).

Tele operated or remote operated devices such as cargo-carrying drone service are used to aid remote areas with medical deliveries and assist in testing instruments and therapeutic agents containing medications, blood, sera, and vaccines. For efficient health care in distant areas is still dependent on effective logistic support to transporting samples to the nearest laboratory and to transfer back the result to the hospital.

## MODEL FOR E-HEALTH SURGERY IN REMOTE AREAS

IoT provides an advanced application related to telehealth surgery, especially in remote areas. Rural areas or low-income areas, where health facility is lacking due to poor infrastructure, the application of telehealth model is designed to provide emergency medical assistance. As shown in Figure 3, the model depicts a district hospital (DH) in a rural area where a critical patient is being monitored by a surgeon in City Hospital (CH) located far away from the operation theatre. The connectivity is provided via 3

*Figure 3. Proposed architecture of Mobile Hospital in rural areas*

| | Data Types | Latency | Jitter | Packet Loss Rate | Data Rate |
|---|---|---|---|---|---|
| **Real-time multimedia streams** | 2D Camera flow | < 150 ms | 3-30 ms | $< 10^{-3}$ | < 10 Mbps |
| | 3D Camera flow | < 150 ms | 3-30 ms | $< 10^{-3}$ | 137 -1600 Mbps |
| | Audio flow | < 150 ms | < 30 ms | $< 10^{-2}$ | 0.022-0.2 Mbps |
| **Physical Vital Signs** | Temperature | < 250 ms | - | $< 10^{-3}$ | < 0.01 Mbps |
| | Blood Pressure | < 250 ms | - | $< 10^{-3}$ | < 0.01 Mbps |
| | Heart Rate | < 250 ms | - | $< 10^{-3}$ | < 0.01 Mbps |
| | Respiration Rate | < 250 ms | - | $< 10^{-3}$ | < 0.01 Mbps |
| | ECG | < 250 ms | - | $< 10^{-3}$ | 0.072 Mbps |
| | EEG | < 250 ms | - | $< 10^{-3}$ | 0.0864 Mbps |
| | EMG | < 250 ms | - | $< 10^{-3}$ | 1.536 Mbps |
| **Haptic Feedback** | Force | 3-10 ms | < 2 ms | $< 10^{-4}$ | 0.128-0.4 Mbps |
| | Vibration | < 5.5 ms | < 2 ms | $< 10^{-4}$ | 0.128- 0.4Mbps |

different links namely satellite link, 5G intrasite towers, and optical link. The overview configuration is based on the da Vinci surgical system including the master side and slave side (Alkatout, Mettler, Maass, & Ackermann, 2016). City hospital act as a master side to control and monitor the patient located in the district hospital as a slave side. The implementation is based on the pre-requisites of features including high speed, high reliability, low latency, low jitter, and low packet loss. Three different links are based on priority wise during failover conditions. Normally, it works on 5G using Multiple Input and Multiple Output (MIMO) towers. The towers are situated 10 km away apart from each other. The dual Unmanned Aerial Vehicle (UAV) is implemented and travels a distance along with 5G MIMO towers to stay in a Line of Sight (LoS). Here both UAVs are mounted nearby and flown from CH to the electricity pylon established near and within 500meter range of DH. The pylon is required for a docking station and charging dock for Master UAV. The slave UAV is then connected to a docking terminal located via DH. The master UAV performs as a relay station between slave UAV and CH base station provider. During surgery, Master UAV stays on top of Pylon, to fetch control data from CH and forwarded it as per its build-in features to its dual connected Slave UAV via a relay. Relay act as a communication channel not only between two UAVs but also transmits control from and feedback to CH (Koyuncu, 2018). Master-Slave stays at the pylon and communicates with its Slave under LoS in the prevention of any obstacles and hindrances which might occur in between two UAVs. Forward and feedback links transports real-time commands to control the motions and rotation of robotic tactile arms while feedback link takes care of real-time multi-modal sensory feedback (Haptic feedback) which include but not limited to 3D video stream, force feedback e.g. pressure and tactile feedback like patients' blood pressure and heart rate. In District Hospital, the tactile glove has been implemented for providing an operational end for remote surgery. The concept of a tactile glove is taken from da Vinci system as the robotic hand act as a slave as it operates the patient according to given control command by the surgeon through UAV connected to operation theatre and acting as master during surgery.

The telesurgery operates according to a flowchart of data and control plane described in Figure 4 and Figure 5 respectively. Data plane consists of commands and instructions provided by the control plane. Control plane checks the priority of cheapest and reliable among three links and passes the instructions to the data plane to deliver data. Both control planes consist of closed-loop for monitoring and feedback.

## Ultra-Low Latency Use Cases

The healthcare sector enhances the availability of high-resolution clinical data, improvement in transmission of real-time video, supporting the number of wearable connected devices, active mobility aid, and ultra-low latency communications (BERG, GLANS, KONING, & KUIPERS, 2017). There are two main healthcare applications of 5G (5G and e-Health, 2015). The first application is distant healthcare monitoring and accurate medicine with the usage of bio-connectivity. In this perspective, hospitals have become decentralized, where healthcare can be provided either in hospitals/home or in remote areas.

Distant medication using robotic surgery in far off areas irrespective of geographical boundaries can be implemented by providing quality healthcare in numerous clinical interventions and surgeries. In a telemonitoring set-up, the level of communication is different from vocal instructions by observing a real-time video stream. This is to take charge of the supporter via a robotic arm. In the telesurgery scenario, real-time cases have determined that the extreme tolerable delay is 150ms (B. Challacombe, 2003).

Offering the doctor exploratory skills related to the touch senses is referred to as haptic feedback. This can enhance the accuracy in revealing certain tumors and nodules especially in distant surgeries and monitoring scenarios. Therefore, the robotic societies are aligned towards the addition of such sensors which track an efficient usage of the medical palpation (J. Konstantinova, 2014). By using haptic feedback, still, constrict the requirements of latency. The reported facts show that telesurgery in the existence of haptic feedback entails an end to end round trip times (RTTs) of lower than 10ms.

*Figure 4. Data plane flowchart representing telesurgery*

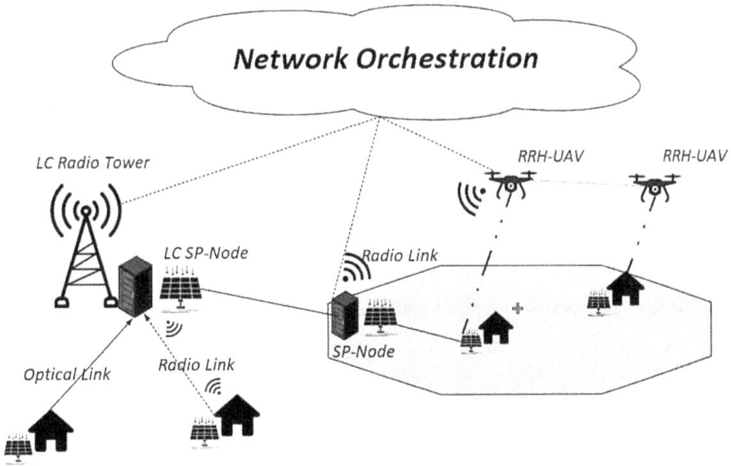

*Figure 5. Control plane flowchart representing telesurgery*

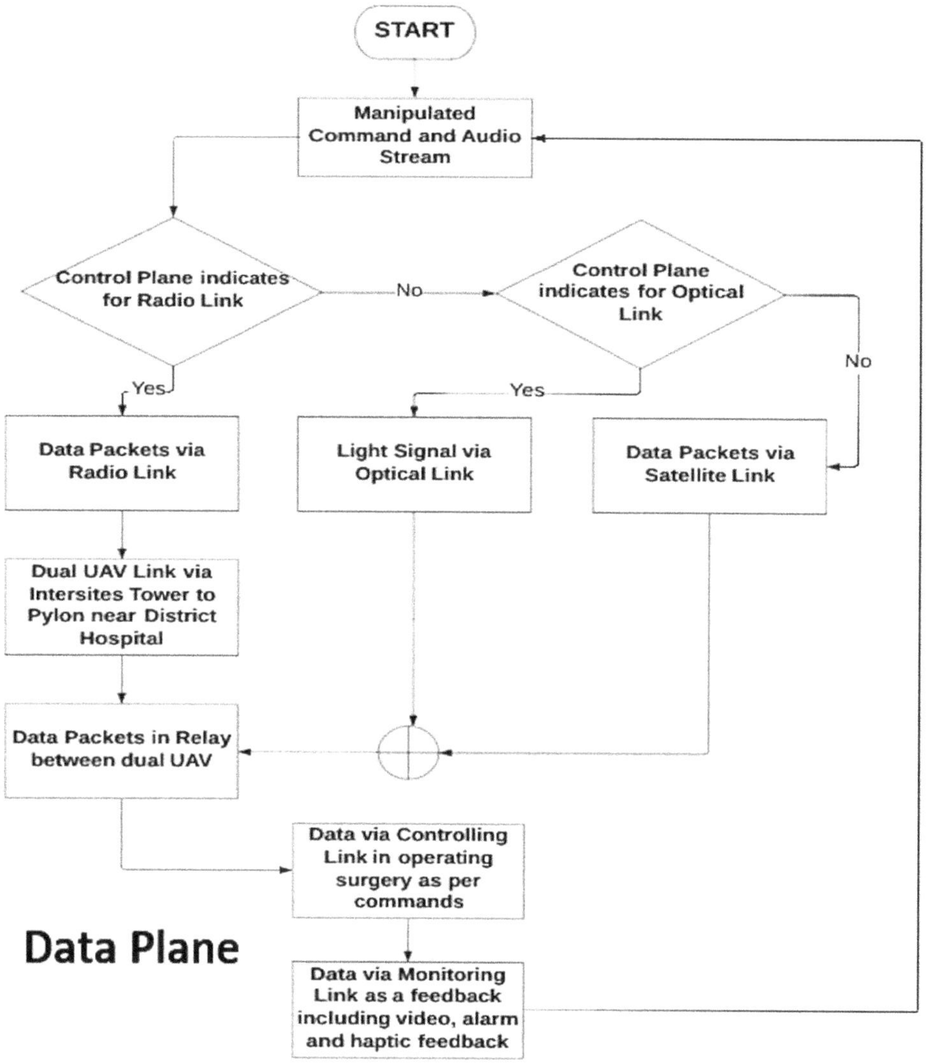


## BACKHAUL LINK ANALYSIS

The following discusses link analysis in terms of cost maintenance latency and durability for the backend for various solutions available including Optical, Satellite, and 5G enabled relay network.

### Backhaul Link: Satellite Communications

The Internet of things (IoT) found to be lethal evolution for monitoring and sensing devices that results in a smart city. Although, the throughput is much reliable in urban regions where the short-range connections like WiFi, Zigbee, Power line connections (PLC) and cellular towers can be easily deployed. However, the problem arises in remote regions like forests, deserts, and oceans, where the engineering difficulties arise which results in the impossibility of deploying range connections. To minimize the limitations, Low Earth Orbit (LEO) satellite creates compatibility with IoT devices to deliver smart services. SATCOM provides cost-efficient solutions rather than implying other terrestrial technologies (Qu, Zhang, Cao, & Xie, 2017). LEO satellite provides high speed and low latency due to its low altitude. Normally, the altitude of the LEO satellite was below 2000 km, and the minimum is recorded at 160 km. The round trip time (RTT) is approximately 100 ms of LEO satellite while comparing with GEO satellite having RTT equals to 600 ms (Christopher S.Allen, 2018). The penetrations of LEO satellite can bear and passes through any obstacles, occurring before reaching smart devices, in line of sight (LOS). This feature is not applicable in case of GEO. Due to shorter propagation distance, loss signals occur small in rate. Thus, LEO founds to be more compatible with IoT devices and applications than GEO satellite.

The constellation based scenario of LEO satellite with IoT is categorized into two groups: 1) Delay Tolerant Application (DTA) and 2) Delay Sensitive Application (DSA). DTA provides a store and forwards data communication services in networks. It features in water monitoring aims to secure the quality of water from natural disasters. The characteristics include frequent disconnections and long propagation delay. On the other hand, DSA provides smart grid and Internet of Battle Things (IoBT) which is essential in military applications. Wired or wireless communication can be utilized in the urban region using a smart grid. In the case of the remote region, offshore wind farms and solar energies can manage the power solution for the smart grid.

In telehealth, while constellation with IoT, LEO satellite link configuration must be focused on routing scheme, constellation design, layer design, interface mitigation, and MAC address mapping. Tele-health is one of the smart application in IoT, that can be facilitated by the LEO satellite benefiting in high speed, reliable monitoring and low latency (low RTT). Tele-Health is based on short range cellular as a connection type and compatibles with health monitoring devices. The model here opted for Low Earth Orbit (LEO) satellite as the alternate communication channel for telesurgery as, comparing with Geographical Earth Orbit (GEO) and Medium Earth Orbit (MEO). LEO is much closer to the earth around 1500 km which is approx. 25 times than GEO and 5 times than MEO. The latency of LEO is 25 ms (Qu et al., 2017). Like RLD and OLD, the model has also proposed for Satellite Link Deficiency (SLD), including the integration of Satellite Link Latency (SLL) in RTT and RTD and these are described in equation 1. After calculating with critical RTD, the SLD results as 3.8.

$$SLD = SLL \times RTD \tag{1}$$

## Backhaul Latency: Physical Medium and Latency Links

The backhaul (also referred as back-net or backbone network) is the cellular network that connects the evolved Node Base Stations (eNBs) to the core network using fibre, microwave, and satellite links (Jaber, Imran, Tafazolli, & Tukmanov, 2016). Enabled Cellular UAV communicates with RRH via 5G network architecture under URLLC and enhanced Mobile Broadband (eMBB) as mentioned in (Simsek, Aijaz, Dohler, Sachs, & Fettweis, 2016). URLLC and eMBB have been used for haptic feedback and 3D High Definition (HD) respectively.

Haptic feedback is the way of communication that provides a sense of touch to a user through motion, vibration and applied forces (Thiem et al., 2017). The closed control loop of tactile surgery end between Slave UAV and Tele-operator through connectivity with RRH. Controlling link indicates the passes of manipulation commands and audio stream that is coming to Slave UAV at the backend of its

5G infrastructure controlled from CH. In return, haptic feedback is being provided in a monitoring link along with video streaming, indicators, alarms and audio stream to the surgeon at CH via UAVs. Surgery measures such as an electrocardiogram (ECG), electroencephalogram (EEG), Electromyography (EMG), blood pressure, heart rate, respiration rate, and temperature have an average latency of 250 ms each with less than 10-3 packet loss rate. All these are concluded as alarm and indicators in monitoring link. Overall maximum latency of Controlling Link, with the inclusion of all factors defined, is 150 ms. In generating feedback, the monitoring link has a maximum latency of 450 ms (Koyuncu, 2018). Jitter and packet loss is due to the factor of physical medium established among nodes in the network. After receiving signals from all channels, the cumulative jitter and packet loss in the proposed loop occurs with 90ms and 2.21 x 10-4 respectively.

After the collection of data from references, this chapter proposes a formula of the rate of tactile deficiency (RTD) based on the value described in Equation 1. The rate of tactile deficiency describes the total time lacking in the response of tactile movement including factors of physical medium and latency rates. In the equation, Controller Latency (CL) and Monitor Latency (ML) represents the total latency of Controlling link and Monitoring Link respectively. Both of these are inclusively summed up with Loop Jitter (LJ) and Loop Packet Loss (LPL) of tactile surgery. The RTD is forwarded to Slave UAV where it sends the data to all the three proposed links for telesurgery. The RTD has been calculated as 0.152 which is donated as Critical RTD which signifies the maximum bearable latency utilized in Equation 2.

$$RTD = (CL+ML+LJ) \times LPL \qquad (2)$$

## Backhaul Link: Optical Link Deficiency

The proposed model focused on an alternate second link of communication which is based on optical fiber using Dense Wavelength Division Multiplexing (DWDM) which is known to have less attenuation and dispersion (Jazayerifar et al., 2013). The latency of optical fiber in Round Trip Time (RTT) is 0.66 µsec per 100 m which means it is 330 µsec (.33 ms) RTT for 50 km which is very low as compared to Radio Link (Tan, 2015). The issue arises with its wired connectivity as it is not secured with jamming, physical cutting, and an interception. However, for failover situation as described in the flowchart Figure 2, it can be implemented according to the International Telecommunication Union Telecommunication Standardization Sector (ITU-T) which provides G.652 standards using multimode optical fibre having feature of Optical Fibre Sensor (OFS) (Correia, James, Lee, Morgan, & Korposh, 2018).

To calculate the rate of deficiency of this link, Equation 3 has been proposed as Optical Link Deficiency (OLD), including Optical Latency in multiplication with RTD. The critical resultant of OLD will be 0.05 that is far low and seems to be more reliable and faster than the radio link.

$$OLD=OLL \times RTD \qquad (3)$$

As shown in Table 3 there is an obvious advantage of preferring the optical backhaul. For a low altitude LEO satellite link the latency is on the order of tens of ms and for multi-hop line of sight radio link the latency goes a little better. However, the installation cost is highest for satellite followed by Radio and then Optical link. However, this is the only option for disaster-prone locations or hard to reach places due to difficult terrain.

*Table 3. Comparison of three backhaul links for implementing Remote Surgery using 5G*

| Link | Latency/Rate of Deficiency | Installation Cost | Maintenance Cost | Durability |
|---|---|---|---|---|
| Radio Link (5G) | 208ms/ 31.616 | Normal | Normal | 5 years |
| DWDM Optical Link | 0.33ms/ 0.05 | High | Low | 25 years |
| LEO Satellite Link | 25ms/ 3.8 | Very High | High | 10-15 years |

## Challenges: Building Tactile Internet on 5G Communication

The important condition for supporting haptic communication over the Tactile Internet is a latency that is at most 1 ms. Another condition for haptic communication is reliability, that the probability of data packet loss does not exceed to 0.001%. Security is yet another important challenge in haptic communication. Quality of experience is further defined as a major challenge for a user at a remote location. 5G technologies proposed solutions to some but not all of the challenges. These are dictated as follows:

### Haptic Devices

Presently devices incapable of providing efficient dependability and transparency. Novel devices are needed so that one can send equally accurate and kinesthetic feedback.

### Compression Methods

Compression algorithm must not increase distortion beyond the thresholds. Selecting the correct factors for this procedure and attaining a trustworthy reduction of data for each arrangement of data samples still need an appropriate solution.

### Multiplexing

This challenge includes a reliable multiplexing technique that combines haptics modalities but they are still restricted and hence this is an open challenge.

### Software-Defined Networking (SDN) and Network Functions Virtualization (NFV)

The purpose of these approaches is to optimize the network domain in teleoperation model system. Existing networking algorithms are prominently efficient in implementing quality of services (QoS) control. The improvement of novel SDN algorithms linked with NFV can aid to resolve various challenges.

## Network coding schemes

Network coding depends on the latency requirement. The current methods are still not in the range of 1ms end-to-end delay. Therefore, more schemes need to be developed for improving latency.

## Haptic Data Prediction

Present haptic data prediction technique requires real-time testing because the propagation delay is larger than 1ms in many cases. Artificial Intelligence is hence required to be combined with the present technique.

## Edge Computing

Edge computing challenge resulting from the usage of computational prediction algorithms. A secure technique is required that does not exceed the latency.

## Haptic System Evaluation

At present, analysis of a haptic communication link varies by different techniques. An objective metric (HPWPSNR) has been established, but it requires additional specific testing before it is used in a practice.

## Haptic System Testbed

A generic testbed to test an end-to-end haptic communication system essentially required. Currently, haptic testbeds are aimed also for specific practice-case situations or testing dedicated haptic system models.

The key entity in provisioning of instantaneous medical and healthcare facilities is a reliable, very high speed and low latency communication link between the service provider and the beneficiary. However, currently, widely available communication links such as 3G or 4G cellular networks do not offer data rates and latency rates effective enough to conduct telesurgery involving risk of human life in case feedback or action is failed to be delivered at a specific time due to a delay in the network. Table 4 shows data rates of existing cellular networks and their respective latencies as per (Grigorik, 2013).

Medical practices around the world are mostly limited to the region where medical care and doctors are physically present. However, extremely low latency feedback and control through tactile internet have enabled the medical care to be available anywhere and anytime, without the limitations of geographical boundaries, with complete audio-visual and haptic feedback. The following section builds up the proposed architecture and compares the backhaul link.

*Table 4. Data rate and latency of cellular networks*

| Generation | Data rate | Latency |
|---|---|---|
| 2G | 100–400 Kbit/s | 300–1000ms |
| 3G | 0.5–5 Mbit/s | 100–500ms |
| 4G | 1–50 Mbit/s | < 100ms |

## CRITICAL REVIEW OF FUTURE TACTILE INTERNET APPLICATIONS: TELEMATICS

The implementation of IoT requires transportation of a huge number of data miles away to provide features and updates to the application users. Handling the huge amount of data requires a strong concept of protocol and network infrastructure. Telematics enhances the means of communications of IoT applications. Relation of Telematics and IoT is significant for transferring several amounts of data miles far away to provide interactions among all the IoT users. Telematic provides solutions for road safety as well as it considered to be the major concern all over the world. Telematic holds data and information about road safety including monitoring, controlling and e-challan. Identifying the speed of vehicles is getting more accurate and provides a path to avoid accidents by digitizing the traffic rules and regulations (Kirushanth & Kabaso, 2018). Telematics is also significant for the e-Health as it collects the data of remote telesurgery. One of the major challenges of achieving a seamless and flawless haptic feedback based tactile internet is to achieve latency of less than "1 ms" (Van Den Berg et al., 2017). One of the possibilities being explored in this chapter is to use 5G cellular technology as an underlying layer for tactile internet as 5G networking promising very low latency rates. By reducing the latency to near-instantaneous, 5G opens up new opportunities for prevailing technologies like augmented and virtual reality.

## CONCLUSION

The concept of tactile internet is introduced in the context of UN SDG driven telemedicine system. It is applied to resolve health governance issues within sparsely dense communities. Case studies from Cook Island, Italy and New Zealand were introduced for benchmarking. Tactile internet application and communication QoS requirements helped understand the context in telehealth. The concept of latencies, jitter and RTT was brought forward as metrics for measuring the quality of remote surgery's backbone system in the context of Smart City. A telesurgery concept was proposed and a comparison made between three different types of backhaul link including optical, satellite and upcoming 5G systems. The material provided forms the basis of evaluating feasibility in installation and commission of tactile internet for various latency, jitter and delay specific applications and to make informed decisions for future telehealth planners to use haptic communication for increasing the reach of specialist surgeries to places with limited infrastructures.

## REFERENCES

Ahmed, M. U., Björkman, M., Čaušsević, A., Fotouhi, H., & Lindén, M. (2015). An Overview on the Internet of Things for Health Monitoring Systems. *EAI International Conference on IoT Technologies for HealthCare*.

Ahvenniemi, H., Huovila, A., Pinto-Seppä, I., & Airaksinen, M. (2017). What are the differences between sustainable and smart cities? *Cities (London, England)*, *60*, 234–245. doi:10.1016/j.cities.2016.09.009

Aijaz, A. (2015). Realizing the Tactile Internet: Haptic Communications over Next Generation 5G Cellular Networks. In A. Aijaz, M. Dohler, A. H. Aghvami, V. Friderikos, & M. Frodigh (Eds.), *Realizing the Tactile Internet*. IEEE Wireless Communication.

Alkatout, I., Mettler, L., Maass, N., & Ackermann, J. (2016). Robotic surgery in gynecology. *Journal of the Turkish German Gynecological Association*, *17*(4), 224–232. doi:10.5152/jtgga.2016.16187 PMID:27990092

Amrita Sajja, D. K. (2016). *A Survey on efficient way to Live: Smart Home - It's an Internet of Things*. Academic Press.

5G. and e-Health. (2019). Retrieved 18 July 2019, from https://5g-ppp.eu/wp-content/uploads/2014/02/5G-PPP-White-Paper-on-eHealth-Vertical-Sector.pdf

Berg, D. V., Glans, R., Koning, D. D., & Kuipers, F. A. (2017). Challenges in Haptic Communications Over the Tactile Internet. *IEEE Access: Practical Innovations, Open Solutions*.

Birke, R., Blefari-Melazzi, N., Chen, L., Bagula, A., Idzikowski, F., Wu, J., & Liu, W. (2017). *Bringing 5G into Rural and Low-Income Areas: Is It Feasible?* IEEE Communications Standards Magazine. doi:10.1109/mcomstd.2017.1700023

Bodner, J., Wykypiel, H., Greiner, A., Kirchmayr, W., Freund, M. C., Margreiter, R., & Schmid, T. (2004). Early experience with robot-assisted surgery for mediastinal masses. *The Annals of Thoracic Surgery*, *78*(1), 259–265. doi:10.1016/j.athoracsur.2004.02.006 PMID:15223439

Brady, K., & Tarn, T. J. (1998, May). Internet-based remote teleoperation. In *Proceedings. 1998 IEEE International Conference on Robotics and Automation (Cat. No. 98CH36146)* (Vol. 1, pp. 65-70). 10.1109/ROBOT.1998.676261

Budhiraja, I., Tyagi, S., Tanwar, S., Kumar, N., & Rodrigues, J. (2018). Tactile Internet for Smart Communities in 5G: An Insight for NOMA-based Solutions. *IEEE Transactions on Industrial Informatics*.

Cellular Technologies enabling the Internet of Things. (2015). Retrieved 18 July 2019, from http://www.5gamericas.org/files/6014/4683/4670/4G_Americas_Cellular_Technologies_Enabling_the_IoT_White_Paper_-_November_2015.pdf

Challacombe, B., Kavoussi, L. R., & Dasgupta, P. (2003). Trans-oceanic telerobotic surgery. *BJU International*, *92*(7), 678–680. doi:10.1046/j.1464-410X.2003.04475.x PMID:14616444

Chamberlin, B. (2016). *Healthcare Internet of Things*. IBM Center for Applied Insights.

Chaudhari, C. S. (2015). Perceptual and bitrate-scalable coding of haptic surface texture signals. *IEEE Journal of Selected Topics in Signal Processing, 9*.

Chong, N. Y. T. K. (2000). Remote coordinated controls in multiple telerobot cooperation. *Proc. IEEE Int. Conf. Robotics and Automation*.

Cizmeci, B. X. X. (2017). A multiplexing scheme for multimodal teleoperation. ACM Trans. Multimedia Comput. Commun. Appl., 13.

Correia, R., James, S., Lee, S. W., Morgan, S. P., & Korposh, S. (2018). Biomedical application of optical fibre sensors. *Journal of Optics*. doi:10.1088/2040-8986/aac68d

Crommenlaan, G. (2015). *5G and e-Health*. 5G Infrastructure Association.

Dahlman, G. M. (2014). 5G Wireless Access: Requirements and Realization. *IEEE Comm. Mag., 52*(12), 42-45.

Di Luca, M., Machulla, T. K., & Ernst, M. O. (2009). Recalibration of multisensory simultaneity: Cross-modal transfer coincides with a change in perceptual latency. *Journal of Vision (Charlottesville, Va.)*, *9*(12), 7–7. doi:10.1167/9.12.7 PMID:20053098

Dohler, M. (2013). *Machine-to-Machine in Smart Cities & Smart Grids: Vision, Technologies & Applications*. WiFlex.

Dohler, M., Mahmoodi, T., Lema, M. A., Condoluci, M., Sardis, F., Antonakoglou, K., & Aghvami, H. (2017, June). Internet of skills, where robotics meets AI, 5G and the Tactile Internet. In *2017 European Conference on Networks and Communications (EuCNC)* (pp. 1-5). IEEE. 10.1109/EuCNC.2017.7980645

Eid, M. J. C. (2011). Admux: An adaptive multiplexer for haptic - audio -visual data communication. IEEE Transactions on Instrumentation and Measurement, 60(1).

Farhoudi, M., Palantas, P., Abrishamchi, B., Mihailovic, A., & Aghvami, A. H. (2017). A novel reliable routing scheme for Tactile-oriented Internet traffic. *Proceedings of the 24th International Conference on Telecommunications: Intelligence in Every Form, ICT 2017*. 10.1109/ICT.2017.7998248

Fettweis, G. P. (2014). The tactile internet: Applications and challenges. *IEEE Vehicular Technology Magazine*, *9*(1), 64–70. doi:10.1109/MVT.2013.2295069

Ghori, M. R., Zamli, K. Z., Quosthoni, N., Hisyam, M., & Montaser, M. (2018). Vehicular ad-hoc network (VANET): Review. *2018 IEEE International Conference on Innovative Research and Development, ICIRD 2018*. 10.1109/ICIRD.2018.8376311

Gubbi, J., Buyya, R., Marusic, S., & Palaniswami, M. (2013). Internet of Things (IoT): A vision, architectural elements, and future directions. *Future Generation Computer Systems*, *29*(7), 1645–1660. doi:10.1016/j.future.2013.01.010

Han, Q., Liang, S., & Zhang, H. (2015). Mobile Cloud Sensing, Big Data, and 5G Networks Make an Intelligent and Smart World. *IEEE Network*, *29*(2), 40–45. doi:10.1109/MNET.2015.7064901

Hopkins, J., & Hawking, P. (2018). Big Data Analytics and IoT in logistics: A case study. *International Journal of Logistics Management*, *29*(2), 575–591. doi:10.1108/IJLM-05-2017-0109

Hume, R., & Looney, J. (2016). Telemedicine and facility design. *Health Facilities Management*, *29*(2), 30. PMID:27017802

Jaber, M., Imran, M. A., Tafazolli, R., & Tukmanov, A. (2016). 5G Backhaul Challenges and Emerging Research Directions: A Survey. *IEEE Access: Practical Innovations, Open Solutions*, *4*, 1743–1766. doi:10.1109/ACCESS.2016.2556011

Kajimoto, T. H. (2017). Vibration feedback latency affects material perception during rod tapping interactions. *IEEE Transactions on Haptics*, *10*(2), 288–295. doi:10.1109/TOH.2016.2628900 PMID:28113957

Kamal, Z., Mohammed, A., Sayed, E., & Ahmed, A. (2017). *Internet of Things Applications, Challenges and Related Future Technologies.* World Scientific News. doi:10.1016/j.respol.2006.05.004

Katz, J. E., & Haddon, L. (2019). Domestication and Mobile Telephony. In Machines that become Us. doi:10.4324/9780203786826-4

Kaur, S., & Singh, I. (2016). A Survey Report on Internet of Things Applications. *International Journal of Computer Science Trends and Technology, 4*(2).

Khodashenas, P. S., Aznar, J., Legarrea, A., Ruiz, C., Siddiqui, M. S., Escalona, E., & Figuerola, S. (2016). 5G network challenges and realization insights. *International Conference on Transparent Optical Networks.* 10.1109/ICTON.2016.7550539

Kirushanth, S., & Kabaso, B. (2018). Telematics and Road Safety. *2018 2nd International Conference on Telematics and Future Generation Networks, TAFGEN 2018.* 10.1109/TAFGEN.2018.8580482

Konstantinova, J., Jiang, A., Althoefer, K., Dasgupta, P., & Nanayakkara, T. (2014). Implementation of tactile sensing for palpation in robot-assisted minimally invasive surgery: A review. *IEEE Sensors Journal, 14*(8), 2490–2501. doi:10.1109/JSEN.2014.2325794

Koyuncu, E. (2018). Power-Efficient Deployment of UAVs as Relays. *IEEE Workshop on Signal Processing Advances in Wireless Communications, SPAWC.* 10.1109/SPAWC.2018.8445865

Leath, B., Dunn, L. W., Alsobrook, A., & Darden, M. L. (2018). Enhancing Rural Population Health Care Access and Outcomes through the Telehealth EcoSystemTM Model. *Online Journal of Public Health Informatics*, *10*(2). doi:10.5210/ojphi.v10i2.9311 PMID:30349636

Li, Y., Cheng, X., Cao, Y., Wang, D., & Yang, L. (2018). *Smart choice for the smart grid: Narrowband internet of things (NB-IoT).* IEEE Internet of Things Journal. doi:10.1109/JIOT.2017.2781251

Maier, M., Chowdhury, M., Rimal, B. P., & Van, D. P. (2016). The tactile internet: Vision, recent progress, and open challenges. *IEEE Communications Magazine*, *54*(5), 138–145. doi:10.1109/MCOM.2016.7470948

Malche, T., & Maheshwary, P. (2017). Internet of Things (IoT) for building smart home system. *Proceedings of the International Conference on IoT in Social, Mobile, Analytics and Cloud, I-SMAC 2017.* 10.1109/I-SMAC.2017.8058258

Mendes, T. D. P., Godina, R., Rodrigues, E., Matias, J., & Catalão, J. (2015). Smart Home Communication Technologies and Applications: Wireless Protocol Assessment for Home Area Network Resources. *Energies*, *8*(7), 7279–7311. doi:10.3390/en8077279

Mer, S. B. (2015). Smart Vehicle-to-Vehicle Communication with 5G Technology. International Journal on Recent Innovation Trends in Computing and Communication, 3(5), 3241–3244.

Mohammed, Z. K., & Ahmed, E. S. (2017). Internet of Things Applications, Challenges and Related Future Technologies. *World Scientific News*, *67*(2), 126–148.

Mohassel, R. R., Fung, A., Raahemifar, K., & Mohammadi, F. A. (2014). A survey on Advanced Metering Infrastructure. *International Journal of Electrical Power & Energy Systems*, *63*, 473–484. doi:10.1016/j.ijepes.2014.06.025

Nuaimi, E. A., Neyadi, H., Mohamed, N., & Al-Jaroodi, J. (2015). Applications of big data to smart cities. *Journal of Internet Services and Applications.*

Osseiran. (2013). The Foundation of the Mobile and Wireless Communications System for 2020 and Beyond Challenges, Enablers and Technology Solutions. *Proc. IEEE Vehic. Technol. Conf. (VTC Spring'15).*

Patel, M., & Wang, J. (2010). Applications, challenges, and prospective in emerging body area networking technologies. *IEEE Wireless Communications*, *17*(1), 80–88. doi:10.1109/MWC.2010.5416354

Perez, S. X. (2015). Impact of delay on telesurgical performance: Study on the robotic simulator dv-trainer. *International Journal of Computer Assisted Radiology and Surgery*, *11*. PMID:26450105

Qu, Z., Zhang, G., Cao, H., & Xie, J. (2017). LEO Satellite Constellation for Internet of Things. *IEEE Access: Practical Innovations, Open Solutions*, *5*, 18391–18401. doi:10.1109/ACCESS.2017.2735988

Risteska Stojkoska, B. L., & Trivodaliev, K. V. (2017). A review of Internet of Things for smart home: Challenges and solutions. *Journal of Cleaner Production*, *140*, 1454–1464. doi:10.1016/j.jclepro.2016.10.006

Shahid, N., & Aneja, S. (2017). Internet of Things: Vision, application areas and research challenges. *Proceedings of the International Conference on IoT in Social, Mobile, Analytics and Cloud, I-SMAC 2017*. 10.1109/I-SMAC.2017.8058246

Shariatmadari, H., Ratasuk, R., Iraji, S., Laya, A., Taleb, T., Jäntti, R., & Ghosh, A. (2015). Machine-Type Communications: Current Status and Future Perspectives toward 5G Systems. *IEEE Comm.*, *53*(9), 10–17. doi:10.1109/MCOM.2015.7263367

Shi, Z. H. Z. (2009). Effects of Packet Loss and Latency on Temporal Discrimination of VisualHaptic Events. IEEE Trans. on Haptics, 1(1).

Simmons, R. (1998). Xavier: An autonomous mobile robot on the web. *Proc. IROS Workshop Robots on the Web.*

Simsek, A. A. (2016). The 5G-Enabled Tactile Internet: Applications, Requirements, and Architecture. *IEEE Wireless Communications and Networking Conference WS 6: IEEE WCNC.*

Simsek, M., Aijaz, A., Dohler, M., Sachs, J., & Fettweis, G. (2016). The 5G-Enabled Tactile Internet: Applications, Requirements, and Architecture. *IEEE Wireless Communications and Networking Conference WS 6: IEEE WCNC'2016 Workshop on 5G & Vertical Industry.*

Stanney, K. M., Graeber, D. A., & Kennedy, R. S. (2006). Virtual environment usage protocols. Handbook of standards and guidelines in ergonomics and human factors, 381-397.

Talasila, M., Curtmola, R., & Borcea, C. (n.d.). *Mobile Crowd Sensing*. New Jersey Institute of Technology. Retrieved from https://web.njit.edu/~mt57/publications/Chapter4.pdf

Tan, T. C. (2015). *Building Information Modeling and Intelligent Green Buidings*. Commscope White Paper.

Thiem, D. B., Neupert, C., Bilz, J., Matich, S., Polzin, J., Werthschutzky, R., & Hatzfeld, C. (2017). User-interface for teleoperation with mixed-signal haptic feedback. *IEEE International Conference on Intelligent Robots and Systems*. 10.1109/IROS.2017.8202251

Van Den Berg, D., Glans, R., De Koning, D., Kuipers, F. A., Lugtenburg, J., & Polachan, K. … Van Wijk, B. (2017). Challenges in haptic communications over the tactile internet. *IEEE Access: Practical Innovations, Open Solutions*. doi:10.1109/ACCESS.2017.2764181

Wahlström, J., Skog, I., & Handel, P. (2017). Smartphone-based vehicle. *IEEE Transactions on Intelligent Transportation Systems, 18*(10), 2802–2825. doi:10.1109/TITS.2017.2680468

Weiner, M., Jorgovanovic, M., & Sahai, A. (2014). Design of a Low-Latency, High-Reliability Wireless Communication System for Control Applications. *2014 IEEE International Conference on Communications (ICC)*. 10.1109/ICC.2014.6883918

Willner, A. (2018). The Industrial Internet of Things. Internet of Things A to Z. doi:10.1002/9781119456735.ch11

Zhang, Q., Liu, J., & Zhao, G. (2018). *Towards 5G enabled tactile robotic telesurgery*. arXiv preprint arXiv:1803.03586

# Chapter 16
# UAV-Based Smart Environmental Monitoring

**Muhammad Imran Majid**
*Institute of Business Management, Pakistan*

**Yunfei Chen**
*Warwick University, UK*

**Osama Mahfooz**
*Institute of Business Management, Pakistan*

**Wajahat Ahmed**
*Institute of Business Management, Pakistan*

## ABSTRACT

*In the context of digital governance, manmade activities will have the most impact on the environment. The concept of internet of things (IoT) is much more profound, supposedly having an impact on the way environmental sampling of data takes place. The cost benefits of unmanned aerial vehicles (UAVs) as alternate to satellite or high-altitude platform system (HAPS) is established as these are more flexible, mission specific, and versatile. UAV in sensor network context and communication relaying as well as data harvesting application is explored. Proposal to upgrade environmental laws and implementing network infrastructure security and cyber insurance for HetNets involving UAVs and environmental monitoring will be covered as well.*

## INTRODUCTION

### IoT Data Sampling and Smart Environment

The Industrial Revolution 4.0 is revolutionizing our world and its future. The future holds the key for connecting the home, work place, industry and the rest of the concept of a Smart City. In the next few years, an estimated 500 billion devices will be connected to the internet. There is the obvious energy

DOI: 10.4018/978-1-7998-1851-9.ch016

usage, but also the impact on natural IoT connects physical world to cyber world or computer world where, objects can have smart capabilities, and can make their own decisions. Design of technology infrastructures using fast and efficient delivery of services and data aims to cover the Digital Divide that is a need of the day. Here the natural consequence of evolution of architectures in the digital landscape requires interactivity and increased efficiency for betterment of the quality of life of residents. Climate change and environmental sustainability has assumed importance in the lives of the digital citizens as it impacts health, work life patterns, city operation, communication and socio-cultural contexts. The most profound effect of man-made activity is on environment.

It is aimed to explore the impact of transportation including man-made activities on environment using fixed sensors spread in an urban environment. Air pollution monitoring systems have low three-dimensional and time-based resolution using only a few measuring sensors with small footprint. This is because of limited resources using lower time resolution hence the huge sampling intervals.

UAVs aid in spatially efficient handling of environmental data from smart sensors. These can also house various sensors as well as collect data from the same. It is proficient as it can be in close proximity to data collection hardware in a variety of terrain. The data can be relayed remotely to a centralized location and processed using power efficient UAV to UAV and UAV to Ground links. The UAV platform and payload are discussed in following two configurations 1) fixed sensor on ground and 2) air borne sensor on board the UAV.

In the former context, the idea of information relaying is explored where data is stored and forwarded or processed and forwarded for various sensor configuration. It is explored with request to the data relaying portion and tests are discussed in detail to enable COTS development in the context of Technology Adoption for developing countries. The most important part is the communication payload because that is where most of the energy is utilized and where most of the operational problems like control, connectivity and application specific issues occur. For communication sensors, information is relayed to neighbor nodes to transfer the data but for transferring data. However, for other nodes it needs to be in the transmission radius referred to as the transmitting range. Finally, the relayed data is transferred to sink node which is the end node in wireless sensor network. Here all the traffic from different nodes gather at a point and from there, it is transmitted to the server.

In the later context, the architecture and present scenario for onboard scientific testing is normally taking place on board the Data Acquisition (DAQ) and Ground Control System (GCS). As in the wireless sensor networks, UAVs can also sense and monitor different parameters such as temperature, pressure, humidity, pollution indexes etc.

A single UAV is good but with limited coverage, it cannot cover the whole area of interest. UAV monitoring is usually associated with UAV swarms because swarms can provide larger coverage area. Hence the idea of multiple aerial platform deployment is considered. However, this concept is still in its infancy. There are issues concerning the autonomous operation and control of devices in 3D space. The tracking and accurate alignment is required for guidance as well as communication. Where the communication system is itself a payload this becomes more crucial.

Finally, the idea of merging risk transfer mechanism for upgradation of the scheme is attractive for future agriculturists and environmentalist is explored in some detail.

## ENERGY SCAVENGING AND WIRELESS SENSOR NETWORK

Wireless sensor networks are backbone of any IoT network. Main operations taking place are data sensing, processing the data, transmitting or relaying the data through communication with different nodes. When sensor nodes sense the data, it collects some information which needs to be processed by the processor to convert it to a useful form.

The central part of a network is communication such system which relays information to neighbor nodes within the transmission radius referred to as the transmission range. Finally, the relayed data is transferred to sink node which is the end node in wireless sensor network, where all the traffic from different nodes, it is transmitted to the server.

The main problem in wireless sensor network is limited power source. Currently there are two ways to tackle this issue. First is to minimize the energy consumption of the network by altering some parameters or adjusting duty cycle of the node. Second is to use ambient power source from surroundings referred to as energy harvesting.

Energy harvesting is a trending research topic which requires scavenging ambient energy from the environment. This ambient energy can be from sun as solar energy. It can be from vibration using piezoelectric devices, from wind as wind energy, or from waves as tidal energy etc. This ambient energy can be utilized in wireless sensor networks to supplement the power source.

The main problem in harvesting energy from the environment is that it is unpredictable. This ambient energy is variable in nature and thus it is very hard to design conditioning circuits for them. Second, the ambient energy sources also have their disadvantages such as that solar energy cannot be used at night. RF energy fades away at long distances because of the path loss function. Mechanical energy from vibration using piezoelectric devices produces very low output power. Thermal energy only works if there is a huge difference in temperature. Wind energy only works efficiently when there is a good steady wind pressure. Due to the very nature of these ambient energy sources, the application and trends are very specific and hence limited.

### Aerial Sampling

Unlike HAPs and satellites, UAVs have the capability of real time monitoring and in close proximity to the detection site. Exploiting their three-dimensional sampling or mobility, the target can be reached in time. A single UAV is good but with limited coverage, the resolution is high. It can be used to control the region of interest in different sorts. UAV monitoring is usually associated with UAV swarms because swarms can provide a large coverage area.

### UAV Sampling for Smart Environment

It is inevitable that growing pollution in cities motivates the development of smart environment so that air quality can be monitored and improved (Gu, Q., & Jia, C. 2019). In large cities, many people living in are subjected to bad air quality, which is a basic cause for disease burden. To eliminate these problems causing air pollution, many large cities have initiated to combine smart cities technology to observe air qualities and pollutants within.

Wireless sensor networks are manually calibrated. They can only acquire data in low spatial resolution and thus, cannot trace the change in air pollution within a spatial dimension. Aerial relay network is getting considerable attention for high-spatial-quality and near-to-surface vertical gradation of atmospheric pollution. For effective air quality monitoring, UAV should carry air pollutant sensors and must operate beyond simple platform.

### Air Pollution Sampling and Collection of Big Data

The use of UAVs has potential because it is flexible, mobile and is economical. People are now more aware of the hazards caused by air pollution and existing monitoring equipment cannot provide solution to requirement of modern cities UAVs are important assets for collecting data using sensor equipped UAVs embedded with existing Wireless Sensor Networks. It has been proven that air pollution has sudden changes at lesser distances horizontally and vertically.

## Air Pollution Substances

Pollution in air is very difficult to detect and analyze. Contamination is commonly caused by human activities, so it is mostly measured in urban areas. There are six common pollutants

**Carbon monoxide** is produced when fuels for example oil, gas, wood and coal do not burn fully.

Fired up charcoal, fumes emitting cars and people smoking cigarettes also produce carbon monoxide gas.

Wood, oil, Gas and coal are the source of fuel for many commonly used household appliances that may include:

1. Boilers
2. Cookers
3. Water heaters
4. Gas fires
5. Central heating systems
6. Open fires

Management of the above is crucial in terms of digital governance as it indirectly impacts the general public and domestic household

*Table 1. Description of six common air pollutants along with their significant characteristics*

| Symbol of Substance | Component Details and Characteristics |
|---|---|
| CO | Carbon Monoxide without color is a gas without odor. |
| NO2 | Nitrogen Dioxide is a highly reactive gas. |
| O3 | Ozone found abundantly in upper atmosphere and is a pale blue gas; responsible for GHE. |
| SO2 | Sulphur Dioxide is a gas with irritating smell with no color trace. |
| PM2.5 / PM10 | This matter contains particles which negatively impacts on health when inhaled. |

## Sensors for Pollution Sampling

Low-cost sensors are installed using readily available drivers and have better prospects for more sampling leading to better accuracy. They can be used in direct contact with air pollution sources. The sensor's data should be analyzed in detail: it can be affected by many parameters like temperature, humidity, wind and contamination of air-based pollutants in the region. As they are not bulky, only low-cost sensors can be mounted on UAVs. Following are the basic components that maybe used to serve the purpose of pollution sampling for UAVs:

1. A control device as in Raspberry Pi with the latest version i.e. version 3
2. An analogue processing device or Analogue Front End (AFE) from makers as in Alphasense
3. Analogue to digital solution as in by South Coast Science
4. Various reactive gas sensors as in one by Alphasense
5. Specific particulate size including Volatile Organic Compounds (VOCs) sensors by Alphasense

Here a comparative history of the development of Rasberry Pi 2 is discussed for context sake. It is proposed that Rasberry Pi be used due to its easy installation features. In February 2016, Raspberry Pi2 Model 2 was replaced by Raspberry pi 3 Model B, which is a third generation of Raspberry Pi. In the range of Raspberry Pi 3 Model B there is another model which is referred to as Raspberry 3 Model B+. This is shown in Figure 1 which is a photograph of a Rasberry Pi3 Module.

The Analog Front End (AFE) group of circuits is intended for use with the Alphasense A4 group of four-electrode gas sensors is as shown in Figure 2 where space is provided, for example, with portable air quality screens. Alternatives incorporate a 2-sensor AFE, 3-sensor AFE and 3-sensor +PID AFE. All circuits incorporate a low commotion bandgap to give 200 mV inclination voltage to NO sensors. Requiring from 3.5 to 6.4 VDC supply, the 3-sensor board for instance just needs 2 mA power.

*Figure 1. Rasberry Pi 3 module*

*Figure 2. Alphasense electrochemical gas sensors*
*Source: (Alphasense CO-A4 Carbon Monoxide Sensor, 2019)*

For software temperature corrections, each board includes a pt1000, provides the signal of 1mv/K. The group of circuits i.e. AFE can measure the both reducing (NO, O3) and oxidizing(CO, H2S,NO, SO2).

South Coast Science can deliver on all of the gasses supported by Alpha sense electrochemical sensors: CO, H2S, NO, O3, and SO2. In addition, an Alpha sense photoionization detector (PID) is available for measurement of volatile organic compounds (VOCs) to part-per- billion resolution. Electrochemical sensors and PIDs can be deployed in almost any combination of two, three or four sensors per device. Measurement of CO2 can be performed by devices with the Alphasense non-dispersive infrared (NDIR) sensor. South Coast Science boards are provided with an interface for the forthcoming Alpha sense digital NDIR product.

## Particulates

PM1.5 PM2.5 and PM10 – can be sampled using the Alpha sense optical particle counter OPC N2, and future OPCs. This has implication for monitoring particulate pollution of different sizes. A summary is provided in Table 2.

*Table 2. Summary of the range of air pollutant effect*

| *Substance* | **Range** | **Data Sheet** |
|---|---|---|
| CO | 500 ppm | Alphasense CO-A4 |
| NO | 20 ppm | Alphasense NO-A4 |
| NO2 | 20 ppm | Alphasense NO2-A43F |
| VOCs | 50 ppm | Alphasense PID-AH2 |
| PM | 0.38 to 17 um spheq size | Alphasense OPC-N2 |

## UAV System Design for Environmental Sampling Delivery

### Application and Types of UAVs

In recent years, the UAVs platforms are becoming increasingly popular in a wide range of domestic applications (photography, surveillance, environment monitoring, search-and-rescue, etc.) having reduced weight and size and a wide range of speeds. As the market develops various applications expand, more requirements are then imposed to extend the flight distance and increase the flexibility to adapt to the complex missions (Li, B, Jiang, Y, Sun, J, Cai, L. and Wen, C. 2016).

UAV works like a Wi-Fi hotspot, for example a switch or a router. There are two different sub networks, one for the sensors to the UAV and another for the connection among UAV and destination. Mobile Hotspots program, which aims to transform small UAVs into flying wireless hubs are also very popular.

A novel spatial domain data sampling system has been recently proposed, which can abruptly reduce the quantity of data sampled at sensors and get rid of the bandwidth requirement between cloud and UAV. The principle followed behind this proposed scheme is the temporal and spatial correlation between the sensor data. In complex environment, sensors data correlation is not as simple as linearity as the data is heterogeneous. Therefore, a new approach has been introduced, basically a neural network model; denoising auto encoder is utilized (Yu, T., Wang, X., & Shami, A, 2018), which has the ability of compressing both linear and nonlinear correlated data. The recently proposed sampling scheme basically comprised of three phases; system initialization, model training and data sampling. In the first stage, UAV flies above the target served by the IoT system and the cloud. All the sensor nodes will keep active and upload the data to cloud through UAV. Based on this collected data, sensor nodes will be clustered by bounded-size K-means clustering algorithm. In second phase, various sensor nodes within their clusters are selected as data sampling bodies. In the third phase of data sampling, data is sampled from selected bodies and encoded by UAV before it gets forwarded to the cloud. The final dataset is finally decoded and recreated in the cloud. With the complete support of UAV and cluster formation, the efficiency of data sampling can be improved.

### UAV Subsystem

UAV subsystem includes various parts mentioned below:

1. Sensors
2. Communication unit
3. Navigation unit
4. Ground equipment
5. Propulsion

The subsystems can be added or modified based on the type of application:

## Sensors

UAV sensors are of wide variety of types. Different sensors are for different types of applications. Below are the list of different types of sensors.

### Distance Sensors

A Laser Range Finder (LRF) type of sensor shoots beams of certain intensity in specific intervals. When those beams reflect from objects, the sensor finds out the time taken for the beams to reflect back to the UAV according to the reflected distance hence the range.

Light Detection and Ranging (LiDAR): are best for making 3D constructions and for surveillance activities needed for navigation purposes and aerial surveys.

### Thermal Sensors

Thermal sensors senses the heat that radiates from materials and objects as to transform this gathered data into videos and images

Thermal Sensors Cameras can capture a wide variety of organisms and objects, as to almost all source on our planet, as well as in the Universe, basically radiates some thermal energy. These includes, but is not limited to, buildings, living species, electrical sources, planes, land, machinery, rocks, gas, liquid, etc.

### Air Pollutant Sensors

Air pollutant sensors are used for smart environment purposes. These sensors monitor the quality of air and pollutants present in it at different altitudes. There are wide variety of sensors for different air pollutants

### Orientation Sensors

Accelerometers basically sense movement of the UAV in real time which is highly dependent on type of sensor used for the purpose. Magnetometers detects changes in UAV's direction and translates the data to processor that includes the data for speed, direction and orientation.

### Communication Unit

This is the main unit that serves communication purpose. RF unit contains radars, antennas, transmitter and receiver circuits, RF frequency converters, RF filters, RF power amplifiers etc. All the subtypes in communication unit can be added based on the type of application used.

### Navigation Unit

The navigation unit consists of inertial navigation unit along with GPS receiver.

## Ground Equipment

For remote operations, ground equipment is necessary, where the sensors sometimes directly sends data to ground station, where there video receivers for video processing. Communication unit also communicates with the ground unit about flight control etc.

## Propulsion

Engines, motors, batteries and propellers which helps put UAV in motion are all part of propulsion subsystem.

### UAV Payload: Wireless Sensor Network

UAV payload designing is an important part in design of any UAV. UAV payload is the amount of load that it can carry during its flight operation. Modern UAV payload consists of three applications; Data collection (using sensors), communication with other UAV'S, base station, and navigation support (Borky, J. M. (n.d.).1997). All these functions; sensors, communication and navigation are centralized using a controller, which is also a part of UAV payload where batteries are crucial. The application of UAV payload is for military purposes such as weapon delivery and warfare.

Communication payload often includes radio frequency equipment such as antennas. These antennas basically include radar and other spectral surveillance functions such as signal monitoring and RF radiometry for accurate estimation for signal of arrival. Behind these payload antennas, are associated transmitting and receive electronics for communication purposes. Typically, fiber optic serves that purpose.

Navigation payload basically consists of inertial navigation unit with a GPS. Automatic landing of UAV will require a differential GPS. UAV position and other data are passed to sensors, processing unit or flight controls.

Finally, sensors for data collection are major part of any UAV payload. There are wide variety of sensors categorized as per application requirement. IR sensors includes optical or IR cameras for image processing. Pressure sensors, thermal sensors, magnetometers, gyro meters, accelerometers are all part of sensors used as payload for UAV. These sensors collect data which is processed on board.

Finally, the heart of the UAV payload is the controller or processor. A processor controls the operations of the sensors and processed data gathered by the sensors. The UAV payload subsystem description is summarized in *Figure 3* which gives a description of different components.

The problem with free lancing UAVs is that it can fly up to 30 minutes on single fully charged battery. Carrying extra devices does reduce the UAV's flight time, because these additional devices contribute weights to the UAV and these devices themselves consumes electricity. A conventional microcontroller powered by usb consumes power in the range from 1W to 10W.\

*Figure 3. UAV payload components*

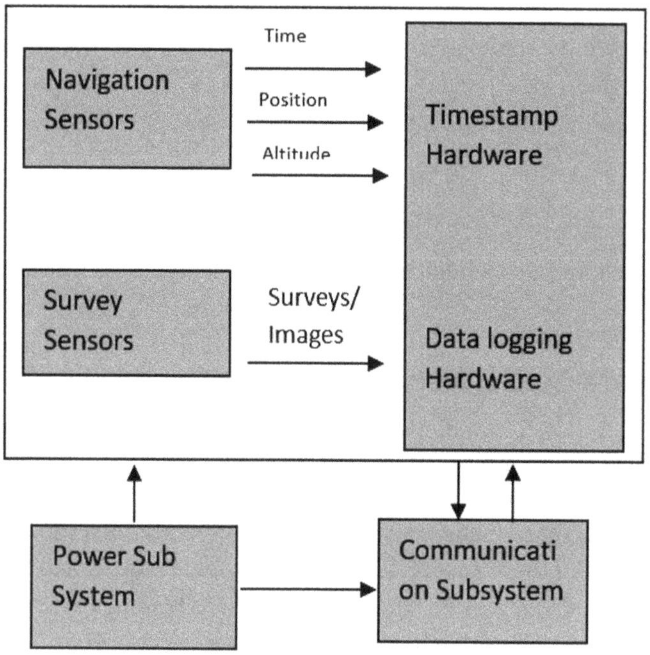


## TYPES OF UAVS

UAV are categorized into different types. Most are classified in terms of size, payload, flight range, flight endurance, altitude and its capabilities. The characteristics of these UAV are correlated. Usually, large UAVs carry more payload, move a longer distance, fly at higher altitudes, and thus have good flight timing and capability as compared to mini-ones. Based on the wing types, UAV are classified into five major types.

The first type of UAV's is the rotary-wing, which uses batteries and thus have very shorter flight times (Paneque-Gálvez, J., McCall, M.K., Napoletano, B.M., Wich, S.A.& Koh, L.P. 2014). This particular type of UAV basically includes the helicopter, Quad copter, octocopter and other types of multicomputer. This type of UAV has an advantage over fixed-wing UAV that it needs less space for landing. Second type of UAV is the fixed-wing UAV, which are suited for applications that demands longer flight time. They require very large space for landing and takeoff and as a result, they face problems in areas that are crowded with tall objects like mountains, buildings, trees, wind farms and other infrastructures. Regardless, they have simple structure which does not have maintenance problems (Klimkowska, A., Lee, I., & Choi, K. 2016).

Third type of UAVs are lightweight aerial systems basically called blimps such as airships and balloons, which are very light and thus have longer flight time, flies at low speeds and usually are very large (Gupta, S.G., Ghonge, M.M., & Jawandhiya, P.M. 2013). The fourth type of UAV is the flapping-wing UAV,hich has an advantage of flexibility and have morphing small wings, just like birds and flying insects. Finally, the Vertical takeoff & Landing (VTOL) UAV does not require takeoff and landing run (Watts, A.C., Ambrosia, V.G., & Hinkley, E.A. 2012)

The above shows that there is a vast choice to select an appropriate technology that is best suited for required application so that it conforms to the constantly prevailing environmental settings and purpose.

## UAV Characteristic Applications

Characteristics of UAVs in environmental monitoring can be segmented according to the specific applications and time to perform certain mission. These are detailed as follows:

**Hefty Operations:** Operations that need more time like 30 plus hours in a mission are most suitably carried out by using UAV. Here the time consuming workload with low intensity operations are suited to unmanned air vehicle. These missions can be automated by programming the UAV to navigate autonomously and requires human attention rather than continuous and direct control. Examples may include: change monitoring of life surveillance over fixed locations, acting as a communications relay for example wifi extender or an air-to-air refueling systems.

**Self-Trap:** Unmanned aerial vehicles are best suited when operation needs to be carried out in harmful environment such as nuclear clouds or radioactive region or highly polluted regions where human being has risk of life. Sensors may be installed for sampling data in such kind of atmosphere. Small UAVs can be used to detect fires in inaccessible areas or where smoke or fire is injurious to human being.

**Risk Taking:** Missions for example monitoring of enemy activities may put many lives in danger so UAS serve as valuable assets in such scenario. The intensity of risk maybe too high in case of using air support while having high alert for ground to air combat or even more when using ground troops. In this case UAS may be used to suppress such threats. Cheap UAS maybe used intended to be sacrificed so that enemy may expend its arsenal so that their command and control equipment is exposed (Suraj G. Gupta, Mangesh M. Ghonge, Dr. P. M. Jawandhiya 2013)

## UAV Environmental Applications

**Oil Refinery and UAVS:** Just like diesel fuel demand for hydrogen have increased according to United States Energy Information Administration (EIA; Washington, DC United states; www.eia.gov). It is used to lower the sulfur contents in diesel. Diesel price is rising globally and so is the regulation to decrease sulfur contents is becoming strict. A process of steam-methane reforming is conducted in which methane is brought into steam at high temperatures in catalyst pipes. As a result the combination of hydrogen, carbon monoxide and carbon dioxide is obtained. Pressure swing absorption is then used to remove carbon monoxide for purification of hydrogen.

**Dangerous Environment in Oil Refineries:** Environment in Oil refineries is very dangerous for example the hydrogen production is also known as the radiant box as temperature of furnace is uptill 4000 degree F containing high contents of chromium designed to withstand high temperature. Even a minor error in the radiant box is not accepted. It consists of burners that heat 143, 40 ft tubes which then heat the steam-methane mix. These tubes are kept 1 ft apart for uniform heat delivery.

If a minor error is even detected in manufacturing all 143 pipes are inspected that require scaffolds that have to be taken inside piece by piece as there is not much space in such areas.

**Use of UAVs for Inspection Instead of Scaffoldings:** The solution to this problem is UAVs. Industrial Skyworks proposed to use UAVs having 1920x1080, 30 fps camera. As a result, instead of using scaffolds for inspection UAVs can capture video of all pipes safely. They may even stream pictures at low resolution. Both video and pictures can be recorded in an SD card and may be transmitted from UAV to its controller. The UAVs can be equipped with carbon fiver structure to avoid damage to both pipes and UAV itself in case of collision. Additional software can then be used to analyze the pictures and videos for detailed inspection.

**UAVs and Wildlife Management:** In order to detect sea turtles Eric Becker of the world wild Fund (WWF; Woking, Surrey, UK; www.wwf.org.uk) has used a UAV that has 4k color, 336 x 256 micro bolometer based thermal image detector. Pictures were taken from the UAV week before hatching of their nests and then data was analyzed with thermal images finding temperature difference between hatching and non-hatching days. Water resources can also be mapped. UAVs are used to track species of algaes or marine animals using a relay station on an in-situ boat. The linear paths are scanned and fed back to a central hub where real time environmental monitoring of species can be implemented as in for tracking humpback whales, Indus River Dolphins of South Asia or other endangered species.

## UAV Relaying of Environmental Sensor Data

The UAV pollutant sensors vary significantly in their mechanisms. Each UAV sensor is for one type of air pollutant, like nitrogen dioxide or particles, and emitted different forms of signals. These signals are in response to the concentration of the pollutants and therefore, should be calibrated before flight (Yu, T., Wang, X., &Shami, A. 2018). Therefore, an individual data acquisition system is usually attached to each sensor to acquire the signal from sensor and gives sensor-type data to process in the data fusion system later. Each type of data acquisition system run its own source program designed for each sensor. In modular approach design, the sensors do not send data directly to the decision making entity but rather, make use of the data fusion system to aggregate the data gathered from the sensors. The aggregated data can be send to the base or ground station.

The data fusion system is the core and most important element in modular design approach. It is basically responsible for processing UAV location data, sensor data and time data. The UAV's flight controller basically provides this location data to the data fusion system. The location data is usually gathered from a GPS that the flight controller is connected to but, because GPS data also carries time information, the flight controller gives time data as well. Moreover, the flight controller can also provide the UAV attitude data, such as its velocity, orientation, acceleration to the data fusion system. This information, accompanies the air pollutant data from air pollutant sensors and thus, can show that how the data was collected by the UAV and this helps in understanding the context of the data measurements. All sensors, including the flight controller provides data in many formats having different rates, so the data fusion system can put on these heterogeneous data types. Moreover, multiple sensors can also be added or removed from the system.

Many researchers have reviewed UAV platform performance. Gupta (Gupta, Jain, & Vaszkun, 2016)] detailed issues regarding wireless transmission in UAV communication networks and emphasized on providing stable and reliable communication. Hayat (Hayat, Yanmaz, & Muzaffar, 2016) studied the civil applications of unmanned aerial vehicle from a perspective of communication. Author also reviewed their characteristics and showed experimental results from different projects. Author Motlagh (Motlagh, Taleb, & Arouk, 2016) presented a detailed survey and discussed the potentials of UAV for low-altitude delivery services from the sky. Cybersecurity related issues were reviewed in (Krishna & Murphy, 2017) where simulated and actual attacks were discussed. J. jiang (Jiang & Han, 2018) discussed most relevant routing protocols for unmanned aerial vehicle and compared different major routing protocols. Khawaja (Khawaja, Guvenc, Matolak, Feibig, & Schneckenberger, 2019) merely focused on air-to-ground channel modelling and measurement along with channel characterization for UAV propagation. Lu (Lu, Bagheri, James, & Phung, 2018) introduces various wireless charging techniques designed for UAV flight time improvement. Author classified wireless charging techniques based on electromagnetic and non-electromagnetic methods. Cao (Cao, et al., 2018) was concerned with protocols and mechanisms for designing of air-borne networks by merely focusing on HAP-based networks, LAP-based networks and integrated networks for air-borne communication. The above literature is outlined in table-1 which help reader to grasp the main study from the literature.

## Risk Transfer Mechanism for Climate Change Resilience using UAV based Digital Governance

Places which are hit by calamity for example by causes relate to Climate Change are to be governed by Paris Climate Change Conference of Parties (COP 21, December 2015). This amongst others call for a comprehensive risk transfer mechanism strategy in the context of climate risk management.

For this purpose, various micro level financial insurances have been promoted to support small farmers and local dwellers in dealing with environmental degradations. Here small premiums are used to not only insure life death but also insure personal assets both tangible and livestock, agriculture and fisheries related equipment.

HetNets are heterogeneous networks of varying cell size and coverage overlap. As crops fishers and various man-made environmental artefacts are assessed, the importance of HetNets cannot be overemphasized. Addressing malicious attacks in HetNets, the following difficulties are encountered:

1. Pinpointing attack sources, because traces can be disguised easily
2. Finding attackers' identities due to anonymity of some Internet protocols
3. Avoiding attacks due to the open accessibility of the wireless medium

*Table 3. Summary of UAV Application case studies and literature review*

| Publication | Summary |
|---|---|
| Gupta (Gupta, Jain, & Vaszkun, 2016) | Issues regarding wireless transmission in UAVs |
| Hayat (Hayat, Yanmaz, & Muzaffar, 2016) | Civil applications of UAVs from a communication perspective |
| Motlagh (Motlagh, Taleb, & Arouk, 2016) | Potentials for Low-altitude delivery services in UAVs |
| Krishna (Krishna & Murphy, 2017) | Cybersecurity issues in UAVs |
| J. Jiang (Jiang & Han, 2018) | Routing protocols comparison for UAVs |
| Khawaja (Khawaja, Guvenc, Matolak, Feibig, & Schneckenberger, 2019) | Air-to-ground measurement and channel modeling for propagation of UAVs |
| Lu (Lu, Bagheri, James, & Phung, 2018) | Wireless charging techniques for improving UAV flight time |
| Cao (Cao, et al., 2018) | Mechanisms and protocols for the design of air-borne networks for UAVs |

To deal with cyber risk damage, the global expenditure on cyber security is expected to reach US$1 trillion from 2017 to 2021. This potentially opens up a vast market for developing anti-risk. Nowadays, the mainstream of security research in HetNets focuses on developing system-based risk mitigation solutions in various enabling wireless technologies, such as massive multiple-input multiple-output (MIMO) technology, cloud radio access networks (C-RANs), and wireless caching. Most of these technologies are compatible with the concept of Internet of Mobile Things (IoMT) of which UAVs form an integral part.

By adding UAVs the MFIs (Micro financial Institutes) can further reduce the risk loss by determining in real time the impact of damage on personal assets as well as commonly used infrastructure reducing the different in actual loss and index as defined by various financial instruments including weather derivatives, catastrophic bonds and climate bonds.

### Recommendation for Enhancing Digital Governance in UAV aided Environmental Monitoring

It is advisable to pursue a policy of enhancing outreach using environmental data sampling. The philosophy differs if one has an agrarian base or industrial base. The risk assessment of the sector needs to be carried out. MFIs need to be involved in determining the environmental monitoring format- sampling interval as well as whether it is air borne or terrestrial. Also this can give a better outlook on the updated situation for a community's economic situation. The data gathered reflects the community's GDP and in turn reflects on the level of investment for the future. UAVs will see if the growth is variable. The Environmental variables will be defined.

Here Catastrophic bonds need to handle extreme effects of climate like hurricane and earth quake. These can be considered for capital market as UAV and Smart Sensor Technology fusion. Similarly, derivatives can be introduced based on weather indices as identified through temperature, humidity soil conditions Such a network connects GSM to LoRaWAN and intern to affixed wing UAV based relay network.

Such a work will help protect the most vulnerable people whose livelihood is dependent on variables linked to the environmental variables.

## FUTURE RESEARCH

The following summarizes the different trends happening in the field of UAV and its applications:

1. **Relaying Multiple Aerial Platforms**

Unmanned aerial vehicles come in different shapes and sizes with each having advantage and disadvantages. A large aerial vehicle has long range of connectivity as compared to small aerial vehicle because of the fact that the large aerial vehicle can carry big transceivers while small aerial vehicle carries Wi-Fi radios for communication. Small UAV'S although cheap, are constrained in their short flight time and lesser payload capacity. Flight time and payload are inversely related to each other. However, these problems can be countered by reliable communication network formation. For example, a single UAV that is taking images of an area, will do many trips for a complete coverage task and thus consuming battery. In order to avoid this back-and-forth activity, a chain of nodes can be placed for relaying between the UAV and ground-base station for transferring of data in a real-time scenario.

2. **Device autonomy**

Device autonomy is related to the control of UAVs and it specifies that whether the UAV can operate autonomously or need remote-controlled assistance. Whatever, the scenario, it is desirable for aerial vehicles to stay in range of a remote controller so, that in case of an emergency it can be controlled. For autonomous flights, UAV can fly based on adaptive and pre-computed waypoints. These waypoints are decided by the central processing unit such as a base station and sent over a link to the UAV. Moreover, the UAVs can also decide the path of flying by using information collected by the sensors.

3. **Connectivity**

It is the most important aspect in any UAV communication network. For simplicity, it can be divided into three categories. Firstly, connectivity with the decision-making entity such as a base station. It is necessary for the UAV to be connected with DME all the time so that its path can be tracked for security and safety. Secondly, UAVs connectivity with ground-personnel is necessary in case any disaster happens. Thirdly, the connectivity between different devices is an essential part of UAV communication networks.

4. **Traffic demands**

Understanding the network demands in UAV communication plays an important role in designing of the network protocols and system. Traffic demands can be sub-categorized into types of sensors for different type of data traffic, type of traffic that is being exchanged in the network, and the frequency of the traffic.

5. **Adaptability**

Robustness in UAV communication networks demands adaptability of such network. For aerial communication networks, the adaptability includes network adaptability and adaptability of different mission demands. A robust network must be adaptable to different nodes leaving and joining the network and thus, topology changes. These topology changes are more frequent in networks having high devices mobility. In large networks, covering a large area may cause variant terrains and channel conditions and thus, adaptability to this topology changes due to the mobility of devices, will be a basic requirement in UAV communication networks. The dynamism is not only caused by high mobility, channel conditions, terrain changes or node failure but due to the network mission demands as well. In simpler words, the objectives of UAV mission may change and therefore, the tasks assigned by DME to UAVs will also change for successful completion of the mission

6. **Synchronization**

UAV communication requires time synchronization and legitimate time-stamping for the traffic. It is important for the devices to be synchronized with each other. For example, if multiple UAVs are collecting data from different points using different sensors, the data collected from all devices will need to be synchronized in time and space and it basically refers to synchronous activity of the devices performing a joined coordinated task. Coordinating UAVs or multiple UAVs synchronizes with each other so that they do not perform the duplicate task at the same location and time.

## FUTURE RESEARCH DOMAIN WITHIN UAV NETWORKS

1. mmWave UAV networks

It is of no doubt that UAV communication requires high bandwidth because it has to deal with different types of data such as video, voice, huge data files. This spectrum crowding and growth encourages transferring to modern frequency allocations. In our context, Mm or millimeter wave communication are emerging as a new candidate that can make use of a huge amount of spectral resource at millimeter wave frequency band so that high bandwidth requirements can be dealt. A basic concern for millimeter wave communication is the high propagation loss as Friis law states that free space path loss grows at the rate of the square of the frequency. Fortunately, short wavelengths of millimeter-wave signals allow multiple rectennas to be wrapped in small UAV. This has application in smart agriculture and in urban cities and short distance channel measurements.

2. UAV NOMA (Non-Orthogonal Multiple Access) transmission

NOMA has drawn much attention as key enabling technology for 5G communications. NOMA basically integrates super position code at the transmitter along with successive interference cancellation at the receiver's side. In comparison to orthogonal schemes, NOMA serves variety of users with varied traffic patterns in non-orthogonal pattern by focusing on power domain. The large variety of wildlife and environmental groups can be connected using this technology.

3. Energy harvesting UAV networks

UAV's have limited power supply as it is powered with capacity-limited battery. UAV has to perform multiple operations such as flight control, data sensing, transmission of data and running applications. All of these operations consume battery, which creates energy deletion problem in UAV communication networks. Energy harvesting is a trending research in the field of communication networks. The idea is to scavenge energy from the environment such as solar energy, wind energy, vibrational energy, thermal energy etc.

## SUMMARY

The recent advances in the field of UAV payload development has lead to a vast variety of applications which were not considered for aerial remote sensing and communication due to the large cost associated with deployment logistics and legal issues with respect to HAPS, Satellites and LAPs. The concept of small and localized deployment of sensors conducting data collection and processing in situ and transmitting to a central server is evaluated. It has been suggested that such a method has its own benefits and advantages and can be readily deployed especially in developing countries where there is a lack of appropriate infrastructure and /or geographical restrictions in place which can only be circumvented with use of appropriate technology. The trends are expected to extrapolate with variety of users (NOMA), energy efficiency (UAV based Energy scavenging) and mmWave networks (varied connectivity).

## REFERENCES

Bolla, G. M., Casagrande, M., Comazzetto, A., Dal Moro, R., Destro, M., Fantin, E., & Lorenzini, E. C. (2018, June). ARIA: Air Pollutants Monitoring Using UAVs. In *2018 5th IEEE International Workshop on Metrology for AeroSpace (MetroAeroSpace)* (pp. 225-229). IEEE.

Borky, J. M. (1997). *Payload technologies and applications for uninhabited air vehicles (UAVs). I* EEE Aerospace Conference.

Cao, X., Yang, P., Alzenad, M., Xi, X., Wu, D., & Yanikomeroglu, H. (2018). Airborne Communication Networks: A Survey. *IEEE Journal on Selected Areas in Communications*, *36*(9), 1907–1926. doi:10.1109/JSAC.2018.2864423

CO-A4 Carbon Monoxide Sensor datasheet. (2017). Retrieved 17 July 2019, from http://www.alphasense.com/WEB1213/wp-content/uploads/2017/01/COA4.pdf

Gu, Q., & Jia, C. (2019). A Consumer UAV-based Air Quality Monitoring System for Smart Cities. *2019 IEEE International Conference on Consumer Electronics (ICCE).* 10.1109/ICCE.2019.8662050

Gupta, L., Jain, R., & Vaszkun, G. (2016). Survey of Important Issues in UAV Communication Networks. *IEEE Communications Surveys and Tutorials*, *18*(2), 1123–1152. doi:10.1109/COMST.2015.2495297

Gupta, S. G., Ghonge, M. M., & Jawandhiya, P. M. (2013). Review of unmanned aircraft system (UAS). *International Journal of Advanced Research in Computer Engineering & Technology*, *2*(4), 1646–1658.

Gupta, Ghonge, & Jawandhiya. (2013). Review of Unmanned Aircraft System (UAS). *International Journal of Advanced Research in Computer Engineering & Technology, 2*(4).

Hayat, S., Yanmaz, E., & Muzaffar, R. (2016). Survey on Unmanned Aerial Vehicle Networks for Civil Applications: A Communications Viewpoint. *IEEE Communications Surveys and Tutorials*, *18*(4), 2624–2661. doi:10.1109/COMST.2016.2560343

Jiang, J., & Han, G. (2018). Routing Protocols for Unmanned Aerial Vehicles. *IEEE Communications Magazine*, *56*(1), 58–63. doi:10.1109/MCOM.2017.1700326

Khawaja, W., Guvenc, I., Matolak, D. W., Feibig, U. C., & Schneckenberger, N. (2019). A Survey of Air-to-Ground Propagation Channel Modeling for Unmanned Aerial Vehicles. *IEEE Communications Surveys and Tutorials*, 1–1.

Klimkowska, A., Lee, I., & Choi, K. (2016). Possibilities of UAS for maritime monitoring. *The International Archives of the Photogrammetry, Remote Sensing and Spatial Information Sciences*, XLI-*B1*.

Krishna, C. L., & Murphy, R. R. (2017). A review on cybersecurity vulnerabilities for unmanned aerial vehicles. In *2017 IEEE International Symposium on Safety, Security and Rescue Robotics (SSRR)*. Shanghai: IEEE. 10.1109/SSRR.2017.8088163

Li, B., Jiang, Y., Sun, J., Cai, L., & Wen, C. Y. (2016). Development and testing of a two-UAV communication relay system. *Sensors (Basel)*, *16*(10), 1696. doi:10.339016101696 PMID:27754369

Lu, M., Bagheri, M., James, A. P., & Phung, T. (2018). Wireless Charging Techniques for UAVs: A Review, Reconceptualization, and Extension. *IEEE Access: Practical Innovations, Open Solutions*, *6*, 29865–29884. doi:10.1109/ACCESS.2018.2841376

Motlagh, N. H., Taleb, T., & Arouk, O. (2016). Low-Altitude Unmanned Aerial Vehicles-Based Internet of Things Services: Comprehensive Survey and Future Perspectives. *IEEE Internet of Things Journal*, 899 - 922.

NO2-A43F Nitrogen Dioxide Sensor datasheet. (2018). Retrieved 17 July 2019, from http://www.alphasense.com/WEB1213/wp-content/uploads/2018/12/NO2A43F.pdf

NO-A4 Nitric Oxide Sensor data sheet. (2016). Retrieved 17 July 2019, from http://www.alphasense.com/WEB1213/wp-content/uploads/2016/03/NO-A4.pdf

OPC-N3 Particle Monitor datasheet. (2019). Retrieved 17 July 2019, from http://www.alphasense.com/WEB1213/wp-content/uploads/2019/03/OPC-N3.pdf

Paneque-Gálvez, J., McCall, M., Napoletano, B., Wich, S., & Koh, L. (2014). Small drones for community-based forest monitoring: An assessment of their feasibility and potential in tropical areas. *Forests*, *5*(6), 1481–1507. doi:10.3390/f5061481

Photo Ionisation Detector datasheet. (2016). Retrieved 17 July 2019, from http://www.alphasense.com/WEB1213/wp-content/uploads/2016/09/PID-AH2.pdf

Stefan, A., Stefan, A., Constantin, D., Mateescu, C., & Cartal, L. A. (2015). Aspects of kinematics and dynamics for payload UAVs. In *2015 7th International Conference on Electronics, Computers and Artificial Intelligence (ECAI)* (pp. WR-1). 10.1109/ECAI.2015.7301228

U.S. Energy Information Administration (EIA). (2019). Retrieved 18 July 2019, from https://www.eia.gov/

Watts, A. C., Ambrosia, V. G., & Hinkley, E. A. (2012). Unmanned aircraft systems in remote sensing and scientific research: Classification and considerations of use. *Remote Sensing*, *4*(6), 1671–1692. doi:10.3390/rs4061671

We are WWF Woking, Surrey. (2019). Retrieved 18 July 2019, from https://www.wwf.org.uk/

Yu, T., Wang, X., & Shami, A. (2018). UAV-Enabled Spatial Data Sampling in Large-Scale IoT Systems Using Denoising Autoencoder Neural Network. *IEEE Internet of Things Journal*, *6*(2), 1856–1865. doi:10.1109/JIOT.2018.2876695

# Compilation of References


5G . and e-Health. (2019). Retrieved 18 July 2019, from https://5g-ppp.eu/wp-content/uploads/2014/02/5G-PPP-White-Paper-on-eHealth-Vertical-Sector.pdf

Abas, S.N., Bhimrajkar, S., & Raina, C.K. (2017). *Social Engineering*. Academic Press.

Abdullah, M. (2009). Review and comparison study of hybrid diesel/solar/hydro/fuel cell energy schemes for a rural ICT Telecenter. *Energy*, *35*, 639–646.

Abo-Rizka, M., & Ghounam, H. R. (2007). A Novel E-voting in Egypt. *International Journal of Computer Science and Network Security*, *7*(11), 226234.

Absher Official Webpage. (n.d.). *Ministry of the Interior of Saudi Arabia*. Retrieved 28 October 2019 from https://bit.ly/2r2Gt9g

Accordino, F. (2013). The futurium—A foresight platform for evidence-based and participatory policymaking. *Philosophy & Technology*, *26*(3), 321–332. doi:10.100713347-013-0108-9

Adeline,, C.P., & Ali,, K., & Hishamnddin. (2006). E-Commerce: A study on Online Shopping in Malaysia. *Journal of Social Science*, *3*(3), 231–242.

ADF. (2017). *Etude sur l'innovation numérique en Afrique et dans les pays émergents. Etat des lieux de l'innovation numérique et éléments de benchmark*. Roland Berger.

African Development Bank. (2017). *Support project for the implementation of the "digital Tunisia 2020" National Strategic Plan*. Retrieved from https://www.afdb.org/fileadmin/uploads/afdb/Documents/Project-and-Operations/Tunisia_-_Support_Project_for_the_Implementation_of_the_%E2%80%9CDigital_Tunisia_2020%E2%80%9D_National_Strategic_Plan.pdf

Agarwal, P., & Alam, M. A. (2018). Use of ICT for sustainable transportation. *28th International Conference on Future Environment and Energy (ICFEE 2018) IOP Publishing IOP Conf. Series: Earth and Environmental Science, 150*(2018). doi:10.1088/1755-1315/150/1/012032

Aghaei, M., & Rezagholizadeh, M. (2017). The Impact of Information and Communication Technology (ICT) on Economic Growth in the OIC Countries. *Environmental & Socio-economic Studies*, *17*(22), 255–276. doi:10.25167/ees.2017.42.7

Ahmad, M. O., Markkula, J., & Oivo, M. (2013). *Factors affecting e-government adoption in Pakistan : a citizen' s perspective*. doi:10.1108/17506161311325378

Ahmad, N., & Schreyer, P. (2016). Measuring GDP in a Digitalised Economy. *OECD Statistics Working Papers*. doi:10.1787/5jlwqd81d09

Ahmad, A., Webb, J., Desouza, K. C., & Boorman, J. (2019). Strategically-motivated advanced persistent threat: Definition, process, tactics and a disinformation model of counterattack. *Computers & Security*, *86*, 402–418. doi:10.1016/j.cose.2019.07.001

Ahmad, S. (2017). Social Engineering Techniques Contrast Study. *International Journal of Engineering*, *9*(1), 105–110.

Ahmed, M. U., Björkman, M., Čaušsević, A., Fotouhi, H., & Lindén, M. (2015). An Overview on the Internet of Things for Health Monitoring Systems. *EAI International Conference on IoT Technologies for HealthCare*.

Ahuja, K., & Khosla, A. (2019). *Driving the Development, Management, and Sustainability of Cognitive Cities*. IGI Global. doi:10.4018/978-1-5225-8085-0

Ahvenniemi, H., Huovila, A., Pinto-Seppä, I., & Airaksinen, M. (2017). What are the differences between sustainable and smart cities? *Cities (London, England)*, *60*, 234–245. doi:10.1016/j.cities.2016.09.009

Aichner, T., & Jacob, F. (2015). Measuring the degree of corporate social media use. *International Journal of Market Research*, *57*(2), 257–275. doi:10.2501/IJMR-2015-018

Aijaz, A. (2015). Realizing the Tactile Internet: Haptic Communications over Next Generation 5G Cellular Networks. In A. Aijaz, M. Dohler, A. H. Aghvami, V. Friderikos, & M. Frodigh (Eds.), *Realizing the Tactile Internet*. IEEE Wireless Communication.

Akatyev, N., & James, J. I. (2017). Legislative Requirements for Cyber Peacekeeping. *Journal of Digital Forensics. Security and Law*, *12*(3), 4.

Aktouf, O. (1987). *Méthodologie des sciences sociales et approche qualitative des organisations*. Québec: PUQ.

Alaba, F. A., Othman, M., Hashem, I. A. T., & Alotaibi, F. (2017). Internet of Things security: A survey. *Journal of Network and Computer Applications*, *88*, 10–28. doi:10.1016/j.jnca.2017.04.002

Alamri, M., Jhanjhi, N., & Humayun, M. (2019). Blockchain for Internet of Things (IoT) Research Issues Challenges & Future Directions: A Review. *Int. J. Comput. Sci. Netw. Secur*, *19*, 244–258.

Alavi, A. H., Jiao, P., Buttlar, W. G., & Lajnef, N. (2018). Internet of Things-enabled smart cities: State-of-the-art and future trends. *Measurement*, *129*, 589–606. doi:10.1016/j.measurement.2018.07.067

Al-Awidi, H., & Aldhafeeri, F. (2017). Teachers Readiness to implement Digital Curriculum In Kuwaite Schools. *Journal of Informationtechnology Education: Research, 16*.

Albino, V., Berardi, U., & Dangelico, R. M. (2015). Smart cities: Definitions, dimensions, performance, and initiatives. *Journal of Urban Technology*, *22*(1), 3–21. doi:10.1080/10630732.2014.942092

Alchemy Learning. (2014). *The Evolution of Digital Curriculum and the Role of Cultural Institutions*. Author.

Alexander, C. (2017). *Digital learning education and skills in the digital age*. Corsham Institute and RAND Europe.

Al-Fuqaha, A., Guizani, M., Mohammadi, M., Aledhari, M., & Ayyash, M. (2015, 4). Internet of things: A survey on enabling technologies, protocols, and applications. *IEEE Communications Surveys and Tutorials*, *17*(4), 2347–2376. doi:10.1109/COMST.2015.2444095

Alkatout, I., Mettler, L., Maass, N., & Ackermann, J. (2016). Robotic surgery in gynecology. *Journal of the Turkish German Gynecological Association*, *17*(4), 224–232. doi:10.5152/jtgga.2016.16187 PMID:27990092

Almulhim, M., & Zaman, N. (2018). *Proposing secure and lightweight authentication scheme for IoT based E-health applications*. Paper presented at the 2018 20th International Conference on Advanced Communication Technology (ICACT).

Almunawar, M. N., Low, P. K. C., Rahman, M. H., & Mohiddin, F. (2012). From e-government to e-governance: winning people's trust. In *E-Governance and civic engagement: Factors and determinants of e-democracy* (pp. 613–636). IGI Global. doi:10.4018/978-1-61350-083-5.ch030

Almusaylim, Z. A., & Jhanjhi, N. (2019). Comprehensive Review: Privacy Protection of User in Location-Aware Services of Mobile Cloud Computing. *Wireless Personal Communications*, 1–24.

Almusaylim, Z. A., & Zaman, N. (2019). A review on smart home present state and challenges: Linked to context-awareness internet of things (IoT). *Wireless Networks*, *25*(6), 3193–3204. doi:10.100711276-018-1712-5

Al-Mushayt, O., Haq, K., & Perwej, Y. (2009). Electronic government in Saudi Arabia: A positive revolution in the peninsula. *International Transactions in Applied Sciences*, *1*(1), 87–98.

Alt, R., Beck, R., & Smits, M. (2018). FinTech and the transformation of the financial industry. *Electronic Markets*, *28*(3), 235–243. doi:10.100712525-018-0310-9

Ammar, M., Russello, G., & Crispo, B. (2018). Internet of Things: A survey on the security of IoT frameworks. *Journal of Information Security and Applications*, *38*, 8–27. doi:10.1016/j.jisa.2017.11.002

Amrita Sajja, D. K. (2016). *A Survey on efficient way to Live: Smart Home - It's an Internet of Things*. Academic Press.

Anderson, J. M., Kalra, N., Stanley, K. D., Sorensen, P., Samaras, C., & Oluwatola, T. A. (2016). *Autonomous Vehicle Technology: A Guide for Policymakers*. Santa Monica, CA: RAND Corporation. Retrieved from https://www.rand.org/pubs/research_reports/RR443-2.html

Anon. (2019). *Phishing Activity Trends Report 1st Quarter 2019*. Brazil: APWG.

Anon. (n.d.). *Vircom*. Available at: https://www.vircom.com/moduscloud-cloud-email-security/

Anshari, M., & Alas, Y. (2015). Smartphones habits, necessities, and big data challenges. *The Journal of High Technology Management Research*, *26*(2), 177–185. doi:10.1016/j.hitech.2015.09.005

Anthopoulos, L. (2015). Defining smart city architecture for sustainability. *Proceedings of 14th electronic government and 7th electronic participation conference (IFIP2015),* 140-147.

Anthopoulos, L. (2017a). Smart utopia VS smart reality: Learning by experience from 10 smart city cases. *Cities (London, England)*, *63*, 128–148. doi:10.1016/j.cities.2016.10.005

Anthopoulos, L. G. (2017b). *Understanding Smart Cities: A tool for smart government or an industrial trick?* Springer. doi:10.1007/978-3-319-57015-0

APWG. (2016a). Retrieved from http://docs.apwg.org/reports/apwg_trends_report_q2_2016.pdf

APWG. (2016b). Retrieved from http://docs.apwg.org/reports/apwg_trends_report_q4_2016.pdf

Arfeen, M. I., & Khan, N. (2009). Public sector innovation: Case study of e-government projects in Pakistan. *The Pakistan Development Review*, *48*(4-II), 439.

Arfeen, M., & Kamal, M. (2014). *Future of e-Government in Pakistan: A Case Study Approach*. Academic Press.

Artmann, M., Kohler, M., Meinel, G., Gan, J., & Ioja, I.-C. (2019). How smart growth and green infrastructure can mutually support each other—A conceptual framework for compact and green cities. *Ecological Indicators*, *96*, 10–22. doi:10.1016/j.ecolind.2017.07.001

Asgarkhani, M. (2005). Digital government and its effectiveness in public management reform: A local government perspective. *Public Management Review*, *7*(3), 465–487. doi:10.1080/14719030500181227

Ashraf, M., Hanisch, J., & Swatman, P. (2009). ICT intervention in the "Chandanbari" village of Bangladesh: Results from a field study. *Information Systems Frontiers*, *11*(2), 155–166. doi:10.100710796-008-9133-0

Asian Development Bank. (2009). *Sri Lanka: Secondary Education Modernization Project.* Available at: http://www.adb.org/documents/sri-lanka-secondary-education-modernization-project

Awan, J., & Memon, S. (2016). Threats of cyber security and challenges for Pakistan. In *International Conference on Cyber Warfare and Security* (p. 425). Academic Conferences International Limited.

Azam, A., Qiang, F., & Abdullah, M. I. (2013). Determinants of e-government services adoption in Pakistan: an integrated model. *Electronic Government, an International Journal, 10*(2), 105-124.

Bahrini, R., & Qaffas, A. (2019). Impact of Information and Communication Technology on Economic Growth: Evidence from Developing Countries. *Economies*, *7*(1), 21. doi:10.3390/economies7010021

Bakıcı, T., Almirall, E. & Wareham, J. (2013). A smart city initiative: the case of Barcelona. *Journal of the Knowledge Economy, 4*, 135-148.

Balakrishna, C. (2012). Enabling technologies for smart city services and applications. In *2012 sixth international conference on next generation mobile applications, services and technologies*. IEEE.

Balta-Ozkan, N., Boteler, B., & Amerighi, O. (2014). European smart home market development: Public views on technical and economic aspects across the United Kingdom, Germany and Italy. *Energy Research & Social Science*, *3*, 65–77. doi:10.1016/j.erss.2014.07.007

Banafa, A. (2017). *Three Major Challenges Facing IoT*. Retrieved from https://iot.ieee.org/newsletter/march-2017/three-major-challenges-facing-iot.html

Bandauko, E., Bobo, T., & Mandisvika, G. (2018). Towards Smart Urban Transportation System in Harare, Zimbabwe. In Intelligent Transportation and Planning: Breakthroughs in Research and Practice (pp. 962-978). IGI Global.

Banerjee, M., Lee, J., & Choo, K.-K. R. (2018). A blockchain future for internet of things security: A position paper. *Digital Communications and Networks*, *4*(3), 149–160. doi:10.1016/j.dcan.2017.10.006

Baporikar, N. (2017). Significance of critical success factors for Indian eBusiness. In *Transcontinental Strategies for Industrial Development and Economic Growth* (pp. 184–206). IGI Global. doi:10.4018/978-1-5225-2160-0.ch011

Barger, V. A., & Labrecque, L. (2013). An integrated marketing communications perspective on social media metrics. *International Journal of Integrated Marketing Communications*, 64–76.

Barns, S. (2018). Smart cities and urban data platforms: Designing interfaces for smart governance. *City, Culture and Society, 12*, 5-12.

Baron, M. (2012). Do we need smart cities for resilience. *Journal of Economics & Management*, *10*, 32–46.

Bartholomew, D. (2013). *Social Media Measurement at a Crossroads, MetricsMan Blog*. Retrieved from https://metricsman.wordpress.com/2013/08/21/social-media-measurement-at-a-crossroads-2/

Batabyal, A. A., & Nijkamp, P. (2019). Creative capital, information and communication technologies, and economic growth in smart cities. *Economics of Innovation and New Technology*, *28*(2), 142–155. doi:10.1080/10438599.2018.1433587

Bătăgan, L. (2011). Smart cities and sustainability models. *Informações Econômicas*, *15*, 80–87.

Batty, M. (2018). *Inventing future cities*. MIT Press. doi:10.7551/mitpress/11923.001.0001

Beardsley, S. C., Enriquez, L., Bonini, S., Sandoval, S., & Brun, N. (2010). *Fostering the economic and social benefits of ICT.* The Global Information Technology Report 2009–2010: ICT for Sustainability

Becker, M. (2019). When public principals give up control over private agents: The new independence of ICANN in internet governance. *Regulation & Governance, 13*(4), 561–576. doi:10.1111/rego.12250

Beier, M., & Wagner, K. (2016). *Social Media Adoption: Barriers to the Strategic Use of Social Media in SMEs.* In *Twenty-Fourth European Conference on Information Systems (ECIS)*, İstanbul, Turkey.

Bekara, C. (2014). Security Issues and Challenges for the IoT-based Smart Grid. *Procedia Computer Science, 34*, 532–537. doi:10.1016/j.procs.2014.07.064

Bélanger, F., & Carter, L. (2012). Digitizing government interactions with constituents: An historical review of e-government research in information systems. *Journal of the Association for Information Systems, 13*(5), 363–394. doi:10.17705/1jais.00295

Belshaw, D. A. (2011). *What is 'digital literacy'? A Pragmatic investigation. Degree of Doctor of Education, Durham university.* Department of Education.

Ben Khalifa, M., Redondo, R. P. D., Vilas, A. F., & Rodríguez, S. S. (2017). Identifying urban crowds using geo-located Social media data: A Twitter experiment in New York City. *Journal of Intelligent Information Systems, 48*(2), 287–308. doi:10.100710844-016-0411-x

Bendiek, A. (2012). *European cyber security policy.* Stiftung Wissenschaft und Politik Berlin.

Benevolo, C., Dameri, R. P., & D'Auria, B. (2016). *Smart mobility in smart city. Empowering Organizations.* Springer.

Bently, D. (2019, July). Hajj Hackers Are Bringing Tech to the Mecca Pilgrimage. *Fortune Magazine.* Retrieved from https://fortune.com/2019/07/27/modern-pilgrim-hajj-mecca/

Berg, D. V., Glans, R., Koning, D. D., & Kuipers, F. A. (2017). Challenges in Haptic Communications Over the Tactile Internet. *IEEE Access: Practical Innovations, Open Solutions.*

Bernanke, B. S. (2009). Financial Innovation and Consumer Protection, speech at the *Federal Reserve System's Sixth Biennial Community Affairs Research Conference.*

Bertot, J. C., Jaeger, P. T., & Hansen, D. (2012). The impact of polices on government social media usage: Issues, challenges, and recommendations. *Government Information Quarterly, 29*(1), 30–40. doi:10.1016/j.giq.2011.04.004

Bertot, J., Jaeger, P., Munson, S., & Glaisyer, T. (2010). Engaging the public in open government: Social media technology and policy for government transparency. *Computer, 43*(11), 53–59. doi:10.1109/MC.2010.325

Beslay, L., & Hakala, H. (2007). *Digital territory: bubbles. European visions for the knowledge age: A quest for new horizons in the information society.* Academic Press.

Bezuidenhout, M., Mouton, F., & Venter, H. S. (2010, August). Social engineering attack detection model: Seadm. In *2010 Information Security for South Africa* (pp. 1-8). IEEE.

Bhatnagar, S. (2004). *E-government: from vision to implementation: a practical guide with case studies.* New Delhi: Sage Publications.

Bhowmick, P. (n.d.). *Samvidha: A ICT System for Personalized Offline Internet Access for Rural Schools.* Academic Press.

Bhoyar, P., Sahare, P., Dhok, S., & Deshmukh, R. (2018). Communication Technologies and Security Challenges for Internet of Things: A Comprehensive Review. *AEÜ. International Journal of Electronics and Communications.*

Bibri, S. E. (2018). A foundational framework for smart sustainable city development: Theoretical, disciplinary, and discursive dimensions and their synergies. *Sustainable Cities and Society*, *38*, 758–794. doi:10.1016/j.scs.2017.12.032

Bifulco, F., Tregua, M., Amitrano, C. C., & D'Auria, A. (2016). ICT and sustainability in smart cities management. *International Journal of Public Sector Management*, *29*(2), 132–147. doi:10.1108/IJPSM-07-2015-0132

Bilodeau, H., Lari, M., & Uhrbach, M. (2019). Cyber security and cybercrime challenges of Canadian businesses, 2017. *Juristat*, *2*, 85–002.

Birke, R., Blefari-Melazzi, N., Chen, L., Bagula, A., Idzikowski, F., Wu, J., & Liu, W. (2017). *Bringing 5G into Rural and Low-Income Areas: Is It Feasible?* IEEE Communications Standards Magazine. doi:10.1109/mcomstd.2017.1700023

Blair, T. (2005). *Prime Minister's Mission for E-Government Unit*. London: Cabinet Office.

Blais, A., & Massicotte, L. (2016). Electoral systems. Comparing Democracies: Elections and Voting in Global Perspective. *Sage (Atlanta, Ga.)*, 40–69.

Blurton, C. (2000). *New Directions of ICT-Use in Education. United National Education Science and Culture Organization*. UNESCO.

Bodner, J., Wykypiel, H., Greiner, A., Kirchmayr, W., Freund, M. C., Margreiter, R., & Schmid, T. (2004). Early experience with robot-assisted surgery for mediastinal masses. *The Annals of Thoracic Surgery*, *78*(1), 259–265. doi:10.1016/j.athoracsur.2004.02.006 PMID:15223439

Bolla, G. M., Casagrande, M., Comazzetto, A., Dal Moro, R., Destro, M., Fantin, E., & Lorenzini, E. C. (2018, June). ARIA: Air Pollutants Monitoring Using UAVs. In *2018 5th IEEE International Workshop on Metrology for AeroSpace (MetroAeroSpace)* (pp. 225-229). IEEE.

Bonanno, P. (2011). Developing an instrument to assess teachers' readiness for technology-enhanced learning. *14th International Conference on Interactive Computer Aided Learning*. 10.1109/ICL.2011.6059622

Boot, A. (2016). Understanding the Future of Banking: Scale & Scopes Economics and Fintech. In The Future of Large, Internationally Active Banks, (pp. 431-449). World Scientific Studies in International Economics. doi:10.1142/9789813141391_0025

Boot, A. (2017). *The Future of Banking: From Scale and Scope Economies to Fintech, European Economy banks, regulation, and the real sector*. Retrieved from: https://european-economy.eu/2017-2/the-future-of-banking-from-scale-scope-economies-to-fintech/

Borky, J. M. (1997). *Payload technologies and applications for uninhabited air vehicles (UAVs). I* EEE Aerospace Conference.

Borneo Bulletin. (2018a). *Revision in cover price for Borneo Bulletin and Media Permata*. Retrieved from https://borneobulletin.com.bn/revision-in-cover-price-for-borneo-bulletin-and-media-permata/

Borneo Bulletin. (2018b). *Brunei leapfrogs to 59th place in the United Nations E-Government Survey*. Retrieved from https://borneobulletin.com.bn/brunei-leapfrogs-to-59th-place-in-un-e-government-survey/

Borneo Bulletin. (2018c). *Fake news can destabilize the nation*. Retrieved from https://borneobulletin.com.bn/fake-news-can-destablise-the-nation/

Borneo Bulletin. (2019a). *'Informed Community' programme held at Information Department*. Retrieved from https://borneobulletin.com.bn/informed-community-programme-held-at-information-department/

Borneo Bulletin. (2019b). *Sultan says Brunei to see better growth in 2019*. Retrieved from https://borneobulletin.com.bn/sultan-says-brunei-to-see-better-growth-in-2019/

Bouaziz, F., & Chaabouni, J. (2013). Criteria for assessing the success of e-government projects. In J. R. Cil-Carcia (Eds.), E-Government Success Factors and Measures: Theories, Concepts, and Methodologies, (pp. 273-288). IGI Global. doi:10.4018/978-1-4666-4058-0.ch015

Boudersa, N. (2016). *The Importance of Teachers' Training Programs and Professional Development in the Algerian Educational Context: Toward Informed and Effective Teaching Practices*. Ministere de Lenseignementt Superieur et de la Recherche Scientifique.

Boulos, M. N. K., & Al-Shorbaji, N. M. (2014). *On the Internet of Things, smart cities and the WHO Healthy Cities*. BioMed Central.

Boyle, P. J., & Speed, S. T. (2018). From protection to coordinated preparedness: A genealogy of critical infrastructure in Canada. *Security Dialogue*, *49*(3), 217–231. doi:10.1177/0967010617748541

Bradshaw, S. (2015). Combating Cyber Threats: CSIRTs and Fostering International Cooperation on Cybersecurity. *Global Commission on Internet Governance Paper Series*, (23)

Bradshaw, S., & DeNardis, L. (2018). The politicization of the Internet's Domain Name System: Implications for Internet security, universality, and freedom. *New Media & Society*, *20*(1), 332–350. doi:10.1177/1461444816662932

Brady, K., & Tarn, T. J. (1998, May). Internet-based remote teleoperation. In *Proceedings. 1998 IEEE International Conference on Robotics and Automation (Cat. No. 98CH36146)* (Vol. 1, pp. 65-70). 10.1109/ROBOT.1998.676261

Braisby, N., & Gellatly, A. (Eds.). (2012). *Cognitive psychology*. Oxford University Press.

Briscoe, G., & Mulligan, C. (2014). *Digital Innovation: The Hackathon Phenomenon*. London: Creativeworks London/QMUL.

Brous, P., & Janssen, M. (2015). *Advancing e-Government Using the Internet of Things: A Systematic Review of Benefits*. doi:10.1007/978-3-319-22479-4_12

Bryson, J. M., Crosby, B. C., & Bloomberg, L. (2014). Public Value Governance: Moving Beyond Traditional Public Administration and the New Public Management. *Public Administration Review*, *74*(4), 445–456. doi:10.1111/puar.12238

Budhiraja, I., Tyagi, S., Tanwar, S., Kumar, N., & Rodrigues, J. (2018). Tactile Internet for Smart Communities in 5G: An Insight for NOMA-based Solutions. *IEEE Transactions on Industrial Informatics*.

Bu, L., Isakov, M., & Kinsy, M. A. (2018). A secure and robust scheme for sharing confidential information in IoT systems. *Ad Hoc Networks*. doi:10.1016/j.adhoc.2018.09.007

Camillo, E. (2013). *A brief history of the evolution of transportation over time*. Academic Press.

Cao, X., Yang, P., Alzenad, M., Xi, X., Wu, D., & Yanikomeroglu, H. (2018). Airborne Communication Networks: A Survey. *IEEE Journal on Selected Areas in Communications*, *36*(9), 1907–1926. doi:10.1109/JSAC.2018.2864423

Caragliu, A., Del Bo, C., & Nijkamp, P. (2011). Smart cities in Europe. *Journal of Urban Technology*, *18*(2).

Cardoso, R. M., Mastelari, N., & Bassora, M. F. (2013). Internet of things architecture in the context of intelligent transportation systems—a case study towards a web-based application deployment. In *22nd International Congress of Mechanical Engineering (COBEM 2013)* (pp. 7751-7760). Academic Press.

Carrapico, H., & Barrinha, A. (2017). The EU as a coherent (cyber) security actor? *Journal of Common Market Studies*, *55*(6), 1254–1272. doi:10.1111/jcms.12575

Carrizales, T., Holzer, M., Kim, S. T., & Kim, C. G. (2006). Digital governance worldwide: A longitudinal assessment of municipal web sites. *International Journal of Electronic Government Research*, *2*(4), 1–23. doi:10.4018/jegr.2006100101

Carr, M. (2016). Public–private partnerships in national cyber-security strategies. *International Affairs*, *92*(1), 43–62. doi:10.1111/1468-2346.12504

Carter, L., & Bélanger, F. (2005). The utilization of e-government services: Citizen trust, innovation and acceptance factors. *Information Systems Journal*, *15*(1), 5–25. doi:10.1111/j.1365-2575.2005.00183.x

Cats, O., & Loutos, G. (2016). Real-Time Bus Arrival Information System: An Empirical Evaluation. *Journal of Intelligent Transport Systems*, *20*(2), 138–151. doi:10.1080/15472450.2015.1011638

Catteddu, D., & Hogben, G. (2009). The European Network and Information Security Agency (ENISA) is an EU agency created to advance. Computing, 72(1), 2009-2013.

Causton, J., Theoharis, M. C., & Orsati, F. (n.d.). *Does self-contained special education deliver on its pronises? a critical inquiry in to research and practice.* Chaprnan University.

Cavelty, M. D., & Mauer, V. (2016). *Power and security in the information age: Investigating the role of the state in cyberspace*. Routledge. doi:10.4324/9781315601793

CBSL. (2011). *Annual Report 2011*. Available at: http://www.cbsl.gov.lk/pics_n_docs/10_pub/_docs/efr/annual_report/AR2011/English/content.htm

Cebrail, C., Tuncer, A., & Ozsashin, A. (2009). *A New Future Technology in Rural and Remote Education: IEEE 802.22 Cognitive Radio Based Wireless Regional Access Network (WRAN)*. IEEE.

Cellular Technologies enabling the Internet of Things. (2015). Retrieved 18 July 2019, from http://www.5gamericas.org/files/6014/4683/4670/4G_Americas_Cellular_Technologies_Enabling_the_IoT_White_Paper_-_November_2015.pdf

Cetinkaya, O., & Koc, M. L. (2009). Practical Aspects of DynaVote E-voting Protocol. *Electronic Journal of E-Government*, *7*(4), 327–338.

Chaabouni, J., & Kamoun, F. (2018). *Usage des applications en ligne dans les municipalités & perspectives de villes intelligentes*. Retrieved from http://www.aegid-foundation.org/events/archives/3/telechargements/rapport-integre-usages-des-applications-developpees-pour-les-municipalites.pdf

Chakraborty, S. (2018). *Fintech: Evolution or Revolution*. Business Analytics Research Lab India.

Challacombe, B., Kavoussi, L. R., & Dasgupta, P. (2003). Trans-oceanic telerobotic surgery. *BJU International*, *92*(7), 678–680. doi:10.1046/j.1464-410X.2003.04475.x PMID:14616444

Chamberlin, B. (2016). *Healthcare Internet of Things*. IBM Center for Applied Insights.

Chandio, A., Haider, Z., Ahmed, S., Ali, M., & Ameen, I. (2019). E-government in Pakistan: Framework of opportunities and challenges. *GSJ, 6*(12).

Chang, C., & Lee, J. (2006). An Anonymous Voting mechanism based on the key exchange Protocol. *Elsevier Computer and Security Journal*, *25*(4), 307–314. doi:10.1016/j.cose.2006.02.004

Changgen & Xuemei. (2017). The impact of virtual reality technology on digital media teaching. *3rd International Conference on Economics, Social Science, Arts, Education and Management Engineering*, 119.

Chatfield, A. T., & Alanazi, J. (2015). Collaborative governance matters to e-government interoperability: An analysis of citizen-centric integrated interoperable e-government implementation in Saudi Arabia. *International Journal of Public Administration in the Digital Age*, *2*(3), 24–44. doi:10.4018/ijpada.2015070102

Chaudhari, C. S. (2015). Perceptual and bitrate-scalable coding of haptic surface texture signals. *IEEE Journal of Selected Topics in Signal Processing, 9*.

Cheng, X., & Deak, J. G. (2019). *United States Patent No. US10240961B2*. Retrieved from https://patents.google.com/patent/US10240961B2/en

Chen, M., Li, G., & Feng, Z. (2009). Web-based Education Technology and Application. In *The Rural Basic Education*. IEEE.

Chen, Y., Ardila-Gomez, A., & Frame, G. (2017). Achieving energy savings by intelligent transportation systems investments in the context of smart cities. *Transportation Research Part D, Transport and Environment*, *54*, 381–396. doi:10.1016/j.trd.2017.06.008

Chen, Y.-C. (2010). Citizen-centric e-government services: Understanding integrated citizen service information systems. *Social Science Computer Review*, *28*(4), 427–442. doi:10.1177/0894439309359050

Chien, J. (2012). *How digital media and Internet transforming education*. University of Denver.

Chifor, B.-C., Bica, I., Patriciu, V.-V., & Pop, F. (2018). A security authorization scheme for smart home Internet of Things devices. *Future Generation Computer Systems*, *86*, 740–749. doi:10.1016/j.future.2017.05.048

Chinedu, P. U., & Hussaini, Y. (2018). ICT Deployment in Governance: A Potent Enabler of Accountability, Reliability and Transparency (ART) of Government Business. *Engineering Mathematics.*, *2*(2), 68–78.

Chishti, S., & Barberis, J. (2016). *The FinTech Book*. New York: Wiley. doi:10.1002/9781119218906

Chong, N. Y. T. K. (2000). Remote coordinated controls in multiple telerobot cooperation. *Proc. IEEE Int. Conf. Robotics and Automation*.

Chourabi, H., Nam, T., Walker, S., Gil-Garcia, J. R., Mellouli, S., Nahon, K., . . . Scholl, H. J. Understanding smart cities: An integrative framework. *2012 45th Hawaii international conference on system sciences*, 2289-2297.

Chowdhury, M., & Sadek, A. (2003). *Fundamentals of Intelligent Transportation Systems Planning*. Norwood, MA: Artech House.

Cizmeci, B. X. X. (2017). A multiplexing scheme for multimodal teleoperation. ACM Trans. Multimedia Comput. Commun. Appl., 13.

Claessens, S., Frost, J., Turner, G., & Zhu, F. (2018). Les marchés du crédit Fintech à travers le monde: taille, moteurs et enjeux de politique publique. BRI Quarterly Review.

CO-A4 Carbon Monoxide Sensor datasheet. (2017). Retrieved 17 July 2019, from http://www.alphasense.com/WEB1213/wp-content/uploads/2017/01/COA4.pdf

Cocco, F. (2016). Most US Manufacturing Jobs Lost to Technology, Not Trade. *Financial Times*. Retrieved from https://www.ft.com/content/dec677c0-b7e6-11e6-ba85-95d1533d9a62

Čolaković, A., & Hadžialić, M. (2018). Internet of Things (IoT): A review of enabling technologies, challenges, and open research issues. *Computer Networks*.

ColomboPage. (2011). *Sri Lanka national exam results blunder due to a computer error - Education Minister.* Available at: http://www.colombopage.com/archive_11B/Dec29_1325183041JR.php

Conteh, N. Y., & Schmick, P. J. (2016). Cybersecurity: Risks, vulnerabilities and countermeasures to prevent social engineering attacks. *International Journal of Advanced Computer Research*, *6*(23), 31–38. doi:10.19101/IJACR.2016.623006

Cordella, A., & Bonina, C. M. (2012). A public value perspective for ICT enabled public sector reforms: A theoretical reflection. *Government Information Quarterly*, *29*(4), 512–520. doi:10.1016/j.giq.2012.03.004

Correia, R., James, S., Lee, S. W., Morgan, S. P., & Korposh, S. (2018). Biomedical application of optical fibre sensors. *Journal of Optics*. doi:10.1088/2040-8986/aac68d

Cowley, R., Joss, S., & Dayot, Y. (2018). The smart city and its publics: Insights from across six UK cities. *Urban Research & Practice*, *11*(1), 53–77. doi:10.1080/17535069.2017.1293150

Coyle, D. (2016). *The Political Economy of National Statistics.* Economics Discussion Paper Series EDP-1603. Available at SSRN: https://ssrn.com/abstract=2850061

Cranor, L.R. & Cytron, R.K. (1996). *Design and Implementation of a Practical Security-Conscious Electronic Polling System.* Washington University: Computer Science Technical Report.

Criado, J. I., Sandoval-Almazan, R., & Gil-Garcia, J. R. (2013). *Government innovation through social media*. Academic Press.

Crommenlaan, G. (2015). *5G and e-Health.* 5G Infrastructure Association.

Cunningham, P., & Cunningham, M. (2006). eParticipation in policy-making: the research and the challenges. *Exploiting the knowledge economy: Issues, applications and case studies, 3*, 364.

Curzon, J., Almehmadi, A., & El-Khatib, K. (2019). A survey of privacy enhancing technologies for smart cities. *Pervasive and Mobile Computing*, *55*, 76–95. doi:10.1016/j.pmcj.2019.03.001

Dahlman, G. M. (2014). 5G Wireless Access: Requirements and Realization. *IEEE Comm. Mag.*, *52*(12), 42-45.

Dalton, W., van Vuuren, J. J., & Westcott, J. (2017). Building Cybersecurity Resilience in Africa. In *12th International Conference on Cyber Warfare and Security 2017 Proceedings*. Reading, MA: Academic Conferences and Publishing International Limited.

Das, A. K., Zeadally, S., & He, D. (2018). Taxonomy and analysis of security protocols for Internet of Things. *Future Generation Computer Systems*, *89*, 110–125. doi:10.1016/j.future.2018.06.027

David, N., Mcnutt, J. G., & Justice, J. B. (2018). *Smart cities, transparency, civic technology and reinventing government. In Smart Technologies for Smart Governments*. Springer.

Dawes, S. S. (2008). The evolution and continuing challenges of e-governance. *Public Administration Review*, *68*, S86–S102. doi:10.1111/j.1540-6210.2008.00981.x

De Jong, M., Joss, S., Schraven, D., Zhan, C., & Weijnen, M. (2015). Sustainable–smart–resilient–low carbon–eco–knowledge cities; making sense of a multitude of concepts promoting sustainable urbanization. *Journal of Cleaner Production*, *109*, 25–38. doi:10.1016/j.jclepro.2015.02.004

Deibert, R. J., & Pauly, L. W. (2019). *Mutual entanglement and complex sovereignty in cyberspace. In Data Politics: Worlds, Subjects*. Rights.

Deloitte. (2019). *Beyond Fintech: A Pragmatic Assessment of Disruptive Potential in Financial Services*. Author.

Deng, H., Karunasena, K., & Xu, W. (2018). Evaluating the Performance of e-Government in Developing Countries: A Public Value Perspective. *Internet Research*, *28*(1), 169–190. doi:10.1108/IntR-10-2016-0296

Department of Census and Statistics. (2012). *Census of Population and Housing*. Available at: http://www.statistics.gov.lk/PopHouSat/CPH2011/Pages/Activities/Reports/CPH_2012_5Per_Rpt.pdf

Department of National Planning Ministry of Finance and Planning. (2006). *Mahinda Chintana: Vision For A New Sri Lanka*. Colombo: State Printing Corporation.

Devarapalli, S. J., & Joshi, P. S. (2013). Domain name system security extensions (DNSSEC) for global server load balancing. Google Patents.

Di Luca, M., Machulla, T. K., & Ernst, M. O. (2009). Recalibration of multisensory simultaneity: Cross-modal transfer coincides with a change in perceptual latency. *Journal of Vision (Charlottesville, Va.)*, *9*(12), 7–7. doi:10.1167/9.12.7 PMID:20053098

Di Martino, B., Rak, M., Ficco, M., Esposito, A., Maisto, S., & Nacchia, S. (2018). Internet of things reference architectures, security and interoperability: A survey. *Internet of Things*, *1*, 99–112. doi:10.1016/j.iot.2018.08.008

Di Pasquale, G., dos Santos, A. S., Leal, A. G., & Tozzi, M. (2016). Innovative public transport in Europe, Asia and Latin America: A survey of recent implementations. *Transportation Research Procedia*, *14*, 3284–3293. doi:10.1016/j.trpro.2016.05.276

Digital News Asia. (2016). *SMEs are laggards, but with good reason*. Retrieved from https://www.digitalnewsasia.com/content/dna-snap-week-may-9-2016

Dillon, D., Belmar, R., & Sebastine, S. (2009). *Pre-fetching secure content using proxy architecture*. Google Patents.

Ding, H., Zhang, C., Cai, Y., & Fang, Y. (2017). Smart cities on wheels: A newly emerging vehicular cognitive capability harvesting network for data transportation. *IEEE Wireless Communications*, *25*(2), 160–169. doi:10.1109/MWC.2017.1700151

Dlodlo, N. (2009). Access to ICT education for girls and women in rural South Africa: A case study. *Technology in Society*, *31*, 168–175. doi:10.1016/j.techsoc.2009.03.003

Doherty, J. (2015). Online Marketing for Small Businesses in Easy Steps: Includes Social Network Marketing. Warwickshire, UK: In Easy Steps Limited.

Dohler, M. (2013). *Machine-to-Machine in Smart Cities & Smart Grids: Vision, Technologies & Applications*. WiFlex.

Dohler, M., Mahmoodi, T., Lema, M. A., Condoluci, M., Sardis, F., Antonakoglou, K., & Aghvami, H. (2017, June). Internet of skills, where robotics meets AI, 5G and the Tactile Internet. In *2017 European Conference on Networks and Communications (EuCNC)* (pp. 1-5). IEEE. 10.1109/EuCNC.2017.7980645

DPEPA. (2001). *Benchmarking E-government: A Global Perspective*. United Nations Division for Public Economics and Public Administration – American Society for Public Administration. Retrieved from http://unpan1.un.org/intradoc/groups/public/documents/un/ unpan021547.pdf/

Drula, G. (2012). Social and online media research – data, metrics and methods. *Review of Applied Socio-Economic Research*, *3*(1), 77–86.

Dua, S. (2017). Digital India: Opportunities and Challenges. *International Journal of Science Technology and Management*, *6*(3), 6.

Dunleavy, P., Margetts, H., Bastow, S., & Tickler, J. (2006). New public administration is dead — Long live digital-era governance —. *Journal of Public Administration: Research and Theory*, *16*(3), 467–494. doi:10.1093/jopart/mui057

Dupont, B. (2018). The global anti-cybercrime network: Mapping the polycentric regulation of online harms. In *Criminal Justice and Regulation Revisited* (pp. 163–185). Routledge. doi:10.4324/9781315174044-10

Dutta, S., Geiger, T., & Lanvin, B. (2015). *The global information technology report 2015.* Paper presented at the World Economic Forum.

Effing, R., & Spil, T. A. (2016). The social strategy cone: Towards a framework for evaluating social media strategies. *International Journal of Information Management*, *36*(1), 1–8. doi:10.1016/j.ijinfomgt.2015.07.009

e-Government mobile services (Ma'ak). (n.d.). Retrieved September 30, 2019 from https://www.yesser.gov.sa/EN/BuildingBlocks/Pages/M-gov.aspx

e-Government Program "Yesser". (n.d.). Retrieved September 30, 2019 from https://www.yesser.gov.sa/EN/Pages/default.aspx

Eid, M. J. C. (2011). Admux: An adaptive multiplexer for haptic - audio -visual data communication. IEEE Transactions on Instrumentation and Measurement, 60(1).

Elazhary, H. (2018). Internet of Things (IoT), mobile cloud, cloudlet, mobile IoT, IoT cloud, fog, mobile edge, and edge emerging computing paradigms: Disambiguation and research directions. *Journal of Network and Computer Applications.*

Enyedy, N. (2014). *Personalized Instruction: New Interest, Old Rhetoric, Limited Results, and the Need for a New Direction for Computer-Mediated Learning*. Boulder, CO: National Education Policy Center.

ETSI. (2013). Retrieved from https://www.etsi.org/newsroom/news/728-2013-12-press-release-its-interop-event

European Union. (2017). *Economics of Industrial Research and Innovation*. EU.

Express Mail Service. (EMS). (n.d.). *Mureeh*. Retrieved 25 October 2019 from https://www.ems.com.sa/en/services/mureeh

Ezell, S. (2010) Explaining International IT Application Leadership: Intelligent Transport Systems. *The Information Technology and Innovation Forum*, 1-45.

Ezell, S. (2010). *Explaining international IT application leadership: Intelligent transportation systems*. Washington, DC: ITIF—The Information Technology & Innovation Foundation.

facts.co. (n.d.). *Sri Lanka Map - blank Political Sri Lanka map with cities.* Available at: http://srilankamap.facts.co/srilankamapof/srilankamap.php

Falconer, G., & Mitchell, S. (2012). Smart city framework. *Cisco Internet Business Solutions Group*, *12*, 2–10.

Fang, Z. (2002). E-government in digital era: concept, practice, and development. *International Journal of the Computer, the Internet and Management, 10*, 1-22.

Fang, Z. (2002). E-government in the digital era: concept, practice, and development. *International Journal of the Computer, the Internet and Management, 10*(2), 1-22.

Farhoudi, M., Palantas, P., Abrishamchi, B., Mihailovic, A., & Aghvami, A. H. (2017). A novel reliable routing scheme for Tactile-oriented Internet traffic. *Proceedings of the 24th International Conference on Telecommunications: Intelligence in Every Form, ICT 2017*. 10.1109/ICT.2017.7998248

Feldstein, M. (2017). Underestimating the Real Growth of GDP, Personal Income and Productivity. *The Journal of Economic Perspectives*, *31*(2), 145–164. doi:10.1257/jep.31.2.145

Felix, R., Rauschnabel, P. A., & Hinsch, C. (2017). Elements of strategic social media marketing: A holistic framework. *Journal of Business Research, 70*, 118–126. doi:10.1016/j.jbusres.2016.05.001

Ferro, E., Caroleo, B., Leo, M., Osella, M., & Pautasso, E. The role of ICT in smart cities governance. *Proceedings of 13th international conference for E-democracy and open government. Donau-Universität Krems*, 133-145.

Fettweis, G. P. (2014). The tactile internet: Applications and challenges. *IEEE Vehicular Technology Magazine, 9*(1), 64–70. doi:10.1109/MVT.2013.2295069

Fidler, D. P. (2013). *Internet Governance and International Law: The Controversy Concerning Revision of the International Telecommunication Regulations*. Academic Press.

FIPA - Foundation for Intelligent Physical Agents. (n.d.). Retrieved from http://www.fipa.org/

Fletcher, G. (2014). *The accessibility of learning content for all students, including students with disabilities, must be addressed in the shift to digital instructional materials*. United States Leadership Technology Innovation Learning.

Foster, C., Graham, M., Mann, L., Waema, T. & Friederici, N. (2017). *Digital Control in Value Chains: Challenges of Connectivity for East African Firms*. . doi:10.1080/00130095.2017.1350104

Fowler, T., & Quigley, K. (2019). Information, Innovation and the Boogeyman: Contextual Factors that Influence the Canadian Government's Response to Cyberspace Risk. In Cyber Law, Privacy, and Security: Concepts, Methodologies, Tools, and Applications (pp. 188-209). IGI Global.

Frost, A. (2009). *The Collective Security Treaty Organization, the Shanghai Cooperation Organization, and Russia's Strategic Goals in Central Asia*. Paper presented at the China & Eurasia Forum Quarterly.

Frumento, E. (2019). Social Engineering: an IT Security problem doomed to get worse. *Medium*. Available: https://medium.com/our-insights/social-engineering-an-it-security-problem-doomed-to-get-worst-c9429ccf3330

Frustaci, M., Pace, P., Aloi, G., & Fortino, G. (2018). Evaluating critical security issues of the IoT world: Present and Future challenges. *IEEE Internet of Things Journal, 5*(4), 2483–2495. doi:10.1109/JIOT.2017.2767291

FSB. (2019). *Fintech and market structure in financial services: Market developments and potential financial stability implications*. Financial Stability Board.

FTA. (2000). US Department of Transport: Federal Transit Agency. Staff, CAAT. Automated and Connected Vehicles. autocaat.org.

Fund, H. L., & Britain, G. (1997). *Strategic Plan*. Environmental Defense Fund.

Furber, S. (2012). *Shut down or restart? The way forward for computing in UK schools*. The Royal Academy of Engeneering.

Gallagher, H., McMahon, W., & Morrow, R. (2014). Cyber Security: Protecting the Resilience of Canada's Financial System. *Bank of Canada Financial System Review*, 47-53.

Gallardo, C., Soler, A., Nieto, R., Cano, C., Pelayo, V., Sánchez, M., ... Arias, M. (2017). Experimental infection of domestic pigs with African swine fever virus Lithuania 2014 genotype II field isolate. *Transboundary and Emerging Diseases, 64*(1), 300–304. doi:10.1111/tbed.12346 PMID:25808027

Garcia-Madariaga, J., & Rodriguez-Rivera, F. (2017). Corporate social responsibility, customer satisfaction, corporate reputation, and firms' market value: Evidence from the automobile industry. *Spanish Journal of Marketing-ESIC, 21*, 39–53. doi:10.1016/j.sjme.2017.05.003

Garson, D. (2004). *What's a good value for Cronbach's Alpha*. Available at: http://www.pmean.com/04/CronbachAlpha.html

Gartner. (2013). *E-health for a Healthier Europe*. Retrieved online at www.calliopenetwork.eu/Linkclick.aspx

Gauld, R. (2006). E-government: What is it, and will it transform government? *Policy Quarterly, 2*(2).

Gauld, R., Goldfinch, S., & Horsburgh, S. (2010). Do they want it? Do they use it? The 'Demand-Side' of e-Government in Australia and New Zealand. *Government Information Quarterly*, *27*(2), 177–186. doi:10.1016/j.giq.2009.12.002

Gayon, C. (1997). De la managérialisation du secteur public à un véritable métissage public-privé. *Revue Française de Gestion*, 115-120.

Gcaza, N., & Von Solms, R. (2017). A strategy for a cybersecurity culture: A South African perspective. *The Electronic Journal on Information Systems in Developing Countries*, *80*(1), 1–17. doi:10.1002/j.1681-4835.2017.tb00590.x

Gendron, A., & Rudner, M. (2012). *Assessing cyber threats to Canadian infrastructure*. Canadian Security Intelligence Service.

Ghayur, A. (2006). Towards good governance: Developing an e-government. *Pakistan Development Review*, *45*(4II), 1011–1025. doi:10.30541/v45i4IIpp.1011-1025

Ghori, M. R., Zamli, K. Z., Quosthoni, N., Hisyam, M., & Montaser, M. (2018). Vehicular ad-hoc network (VANET): Review. *2018 IEEE International Conference on Innovative Research and Development, ICIRD 2018*. 10.1109/ICIRD.2018.8376311

Giffinger, R. & Gudrun, H. (2010). Smart cities ranking: an effective instrument for the positioning of the cities? *ACE: Architecture, City and Environment, 4*, 7-26.

Gil-Garcia, J. R., Dawes, S. S., & Pardo, T. A. (2018). Digital government and public management research: Finding the crossroads. *Public Management Review*, *20*(5), 633–646. doi:10.1080/14719037.2017.1327181

Gil-García, J. R., & Pardo, T. A. (2005). E-government success factors: Mapping practical tools to theoretical foundations. *Government Information Quarterly*, *22*(2), 187–216. doi:10.1016/j.giq.2005.02.001

Gina, G. G., Roberto, G., & Gonzalo, I. D. (2010). Identity based Threshold Cryptography and Blind signatures for Electronic Voting. *WSEAS Transactions on Computers*, *9*(1), 62–71.

Gliem, J., & Gliem, R. (2003). *Calculating, Interpreting, And Reporting Cronbach's Alpha Reliability Coefficient For Likert-Type Scales.* Available at: https://scholarworks.iupui.edu/handle/1805/344

Global Finance Magazine. (2019). *Richest Countries in the World 2019*. Retrieved from https://www.gfmag.com/global-data/economic-data/richest-countries-in-the-world

Glybovets, A. & Mohammad, A. (2017). E-Government Versus Smart Government: Jordan Versus The United States. *EUREKA: Social and Humanities*, 3-11.

Goodwin, K. (2012). *Use of Tablet Technology in the Classroom*. Strathfield, Australia: NSW Curriculum and Learning Innovation Centre.

Gottfried, J., DeLancey, L., Watwood, C., & Hardin, A. (2015). Virtual conferencing and meeting systems: Resources for online connections. *College & Research Libraries News*, *76*(2), 98–101. doi:10.5860/crln.76.2.9265

Gottwalt, F., & Karduck, A. P. (2018). *Securing smart cities—a big data challenge. In Information Innovation Technology in Smart Cities*. Springer.

Govemnment, Q. (2013). *E-learning for Smart classrooms*. Retrieved online at http://gld.gov.au/smartclassroo ms/documents/strategy/pdf/scb yte-elearning.pdf

Gow, J., & Dufour, C. (2000). Le nouveau management public est-il un paradigme ? cela a-t-il de l'importance? *Revue Internationale des Sciences Administratives, 4.*

Graham, M., & Mann, L. (2018). Digital Control in Value Chains: Challenges of Connectivity for East African Firms. *Economic Geography*, *94*(1), 68-86.

Gray, C. P. (2013). Cyber utilities infrastructure and government contracting. *Nat'l Sec. & Armed Conflict L. Rev.*, *3*, 151.

Gross Lopes, K. M., Macadar, M.A., & Luciano, E.M. (2019). Key drivers for public value creation enhancing the adoption of electronic public services by citizens. *International Journal of Public Sector Management*. doi:10.1108/IJPSM-03-2018-0081

Grosser, B. (2014). What do metrics want? How quantification prescribes social interaction on Facebook. *Computational Culture*, (4), 1–38. doi:10.100713398-014-0173-7

Gruening, G. (2001). Origin and theoretical basis of New Public Management. *International Public Management Journal*, *4*(1), 1–25. doi:10.1016/S1096-7494(01)00041-1

Guardian. (2019, Jan. 8). Universal internet access unlikely until at least 2050, experts say. *The Guardian.*

Gubbi, J., Buyya, R., Marusic, S., & Palaniswami, M. (2013). Internet of Things (IoT): A vision, architectural elements, and future directions. *Future Generation Computer Systems*, *29*(7), 1645–1660. doi:10.1016/j.future.2013.01.010

Guerrero-Ibáñez, J., Zeadally, S., & Contreras-Castillo, J. (2018). Sensor Technologies for Intelligent Transportation Systems. *Sensors*, *18*(4).

Gunasekaran, A., McGaughey, R. E., Ngai, E. W. T., & Rai, B. K. (2009). E-Procurement adoption in the Southcoast SMEs. *International Journal of Production Economics*, *122*(1), 161–175. doi:10.1016/j.ijpe.2009.05.013

Gupta, Ghonge, & Jawandhiya. (2013). Review of Unmanned Aircraft System (UAS). *International Journal of Advanced Research in Computer Engineering & Technology, 2*(4).

Gupta, L., Jain, R., & Vaszkun, G. (2016). Survey of Important Issues in UAV Communication Networks. *IEEE Communications Surveys and Tutorials*, *18*(2), 1123–1152. doi:10.1109/COMST.2015.2495297

Gupta, N., Kumar, P., & Chokar, S. (2011). A Secure Blind Signature Application in E-voting. *Proceedings of the 5th National Conference, Computing for National Development*, 1-4.

Gupta, S. G., Ghonge, M. M., & Jawandhiya, P. M. (2013). Review of unmanned aircraft system (UAS). *International Journal of Advanced Research in Computer Engineering & Technology*, *2*(4), 1646–1658.

Gu, Q., & Jia, C. (2019). A Consumer UAV-based Air Quality Monitoring System for Smart Cities. *2019 IEEE International Conference on Consumer Electronics (ICCE)*. 10.1109/ICCE.2019.8662050

Guzy, M. (2016). *Fintech Book*. Chichester, UK: Wiley.

Hague, C., & Payton, S. (2011). Digital literacy across the curriculum. *The Curriculum & Leadership Journal, 9*(10).

Hai, C., & Jeong, I. (2007). *Fundamental of development administration*. Scholar.

Hall, R. E., Bowerman, B., Braverman, J., Taylor, J., Todosow, H., & Von Wimmersperg, U. (2000). *The vision of a smart city*. Upton, NY: Brookhaven National Lab.

Hameed, T. (2007). *ICT as an enabler of socio-economic development*. Academic Press.

Handley, M., Bonaventure, O., & catholique de Louvain, U. (2013). *Internet Engineering Task Force (IETF) A. Ford Request for Comments: 6824 Cisco Category: Experimental C. Raiciu*. Academic Press.

Han, Q., Liang, S., & Zhang, H. (2015). Mobile Cloud Sensing, Big Data, and 5G Networks Make an Intelligent and Smart World. *IEEE Network*, *29*(2), 40–45. doi:10.1109/MNET.2015.7064901

Hanushek, E. A., & Wößmann, L. (2017). *The Role of Education Quality in Economic Growth*. Stanford University.

Harfouche, A., & Robbin, A. (2015). E-Government implementation in developing countries. In From Information to Smart Society (pp. 315-327). Springer.

Harris, S. (2008). China's cyber-militia. *National Journal Magazine, 31*.

Harsh, A., & Ichalkaranje, N. (2015). *Transforming e-government to smart government: A South Australian perspective. In Intelligent Computing, Communication and Devices*. Springer.

Hassan, S., Nadzim, S. Z. A., & Shiratuddin, N. (2015). Strategic use of social media for small business based on the AIDA model. *Procedia: Social and Behavioral Sciences*, *172*, 262–269. doi:10.1016/j.sbspro.2015.01.363

Hassan, W. H. (2018). Current research on Internet of Things (IoT) security: A survey. *Computer Networks*.

Hayat, S., Yanmaz, E., & Muzaffar, R. (2016). Survey on Unmanned Aerial Vehicle Networks for Civil Applications: A Communications Viewpoint. *IEEE Communications Surveys and Tutorials*, *18*(4), 2624–2661. doi:10.1109/COMST.2016.2560343

Heeks, R. (2001). *Understanding e-governance for development*. Manchester, UK: Institute for Development Policy and Management.

Heintschel von Heinegg, W. (2013). Territorial sovereignty and neutrality in cyberspace. *International Law Studies*, *89*(1), 17.

Heitzenrater, C. D., & Simpson, A. C. (2016). Policy, statistics and questions: Reflections on UK cyber security disclosures. *Journal of Cybersecurity*, *2*(1), 43–56. doi:10.1093/cybsec/tyw008

Hellaoui, H., Koudil, M., & Bouabdallah, A. (2017). Energy-efficient mechanisms in security of the internet of things: A survey. *Computer Networks*, *127*, 173–189. doi:10.1016/j.comnet.2017.08.006

Hobbs, R. (2010). *DIGITAL and media literacy: A Plan of Action*. The Aspen Institute.

Hoeschele, M., & Rogers, M. (2006). Detecting Social Engineering. In IFIP — The International Federation for Information Processing, (vol. 194). Springer.

Hoffman, D. L., & Fodor, M. (2010). Can you measure the ROI of your social media marketing? *MIT Sloan Management Review*, *52*(1), 41.

Holm, L. (2017). *Cyber attacks & coercion in the digital era.: A qualitative case analysis of the North Korean cyber attack on Sony Pictures*. Academic Press.

Hopkins, J., & Hawking, P. (2018). Big Data Analytics and IoT in logistics: A case study. *International Journal of Logistics Management*, *29*(2), 575–591. doi:10.1108/IJLM-05-2017-0109

Hu-Au, E., & Lee, J. (2018). *Virtual reality in education: a tool for learning in the experience age*. Columbia University, Department of Communications.

Hughes, O. (2003). *Public management and administration: an introduction* (3rd ed.). Creative Print and Design.

Hume, R., & Looney, J. (2016). Telemedicine and facility design. *Health Facilities Management*, *29*(2), 30. PMID:27017802

Hussain, K., Hussain, S. J., Jhanjhi, N., & Humayun, M. (2019). *SYN Flood Attack Detection based on Bayes Estimator (SFADBE) For MANET.* Paper presented at the 2019 International Conference on Computer and Information Sciences (ICCIS). 10.1109/ICCISci.2019.8716416

Hussain, K., Jhanjhi, N. Z., Mati-ur-Rahman, H., Hussain, J., & Islam, M. H. (2019). Using a systematic framework to critically analyze proposed smart card based two factor authentication schemes. *Journal of King Saud University - Computer and Information Sciences*. doi:10.1016/j.jksuci.2019.01.015

Hussain, K., Jhanjhi, N., Mati-ur-Rahman, H., Hussain, J., & Islam, M. H. (2019). Using a systematic framework to critically analyze proposed smart card based two factor authentication schemes. *Journal of King Saud University-Computer and Information Sciences*.

Hussain, S. J., Ahmed, U., Liaquat, H., Mir, S., Jhanjhi, N., & Humayun, M. (2019). *IMIAD: Intelligent Malware Identification for Android Platform.* Paper presented at the 2019 International Conference on Computer and Information Sciences (ICCIS). 10.1109/ICCISci.2019.8716471

Hussain, A., Wenbi, R., Da Silva, A. L., Nadher, M., & Mudhish, M. (2015). Health and emergency-care platform for the elderly and disabled people in the Smart City. *Journal of Systems and Software*, *110*, 253–263. doi:10.1016/j.jss.2015.08.041

Ibrahim, S., Kamat, M., Salleh, M., & Abdul Aziz, S. (2003). *Secure voting using blind signature*. available at http://eprints.utm.my/3262 /1/IEEE02EVS_full_paper_ver14Nov.pdf

ICTA. (2009). *eSri Lanka.* Available at: http://www.icta.lk/en/e-sri-lanka.html

Ilyas, M. (2016). *E governance practices and models: Options for Pakistan. Institute for strategic studies, research and analysis*. ISSRA.

Indergaard, M. (2019). A developmental network city? Double embeddedness in New York. *Cambridge Journal of Regions, Economy and Society*, rsz013. doi:10.1093/cjres/rsz013

Information Supplement: Best Practices for Implementing a Security Awareness Program. (2014). *Security Awareness Program.* Special Interest Group PCI Security Standards Council. Available at: https://www.pcisecuritystandards.org/documents/PCI_DSS_V1.0_Best_Practices_for_Implementing_Security_Awareness_Program.pdf

Information.gov.bn. (2019). *Information Department's Mission and Objectives*. Author.

Ingrams, A., Melitski, J., Holzer, M. & Manoharan, A. P. (2018). Big data and open government. *E-Government and Information Technology Management: Concepts and Best Practices*, 183.

Interactive Advertising Bureau (IAB). (2009). *Social Media Ad Metrics Definitions*. Retrieved from http://www.iab.net/media/file/Social-Media-Metrics-Definitions-0509.pdf

International Monetary Fund. (2019). *FinTech in sub-Saharan African countries: a game changer? Author.*

Internet World Stats. (2019). *Global Internet Users Statistics*. Author.

Ismagilova, E., Hughes, L., Dwivedi, Y. K., & Raman, K. R. (2019). Smart cities: Advances in research—An information systems perspective. *International Journal of Information Management*, *47*, 88–100. doi:10.1016/j.ijinfomgt.2019.01.004

Ivars-Baidal, J. A., Celdrán-Bernabeu, M. A., Mazón, J.-N., & Perles-Ivars, Á. F. (2019). Smart destinations and the evolution of ICTs: A new scenario for destination management? *Current Issues in Tourism*, *22*(13), 1581–1600. doi:10.1080/13683500.2017.1388771

Jaaskelainen, L. (n.d.). E-learning and digital education - Statistics & Facts. *Statista.*

Jaber, M., Imran, M. A., Tafazolli, R., & Tukmanov, A. (2016). 5G Backhaul Challenges and Emerging Research Directions: A Survey. *IEEE Access: Practical Innovations, Open Solutions*, *4*, 1743–1766. doi:10.1109/ACCESS.2016.2556011

Jain, A., Tailang, H., Goswami, H., Dutta, S., Sankhla, M. S., & Kumar, R. (2016). Social Engineering: Hacking a Human Being through Technology. *IOSR Journal of Computer Engineering*, *18*(5), 94–100. PMID:28540284

Jeffrey, A. (2013). *Social media measurement: A step-by-step approach using the AMEC valid metrics framework.* Institute for Public Relations. Retrieved from http://www.instituteforpr.org/wp-content/uploads/Social-Media-Measurement-Paper-Jeffrey-6-4-13.pdf

Jenks, M., & Burgess, R. (2000). *Compact Cities: Sustainable Urban Forms for Developing Countries.* London, UK: Spon Press.

Jiang, J., & Han, G. (2018). Routing Protocols for Unmanned Aerial Vehicles. *IEEE Communications Magazine*, *56*(1), 58–63. doi:10.1109/MCOM.2017.1700326

Jimenez, C. E., Solanas, A., & Falcone, F. (2014). E-government interoperability: Linking open and smart government. *Computer*, *47*(10), 22–24. doi:10.1109/MC.2014.281

Johnston, E., & Hansen, D. (2011). Design lessons for smart governance infrastructures. In D. Ink, A. Balutis, & T. Buss (Eds.), *American Governance 3.0: Rebooting the Public Square?* National Academy of Public Administration.

Jorba, A. R., Roca, J. C., & Scytl Online World Security, S. A. (2007). *Secure remote electronic voting system and cryptographic protocols and computer programs employed.* U.S. Patent 7,260,552.

Jose, D. V., & Vijyalakshmi, A. (2018). An Overview of Security in Internet of Things. *Procedia Computer Science*, *143*, 744–748. doi:10.1016/j.procs.2018.10.439

Joss, S., Sengers, F., Schraven, D., Caprotti, F., & Dayot, Y. (2019). The smart city as global discourse: Storylines and critical junctures across 27 cities. *Journal of Urban Technology*, *26*(1), 3–34. doi:10.1080/10630732.2018.1558387

JPKE-ILO. (2017). *Labour Force Survey 2017.* Department of Economic Planning and Development.

Jung, I., & Zawacki-Richter, O. (2019). Open and Distance Education in Asia, Africa and the Middle East National Perspectives in a Digital Age? *Springer Briefs in Open and Distance Education.*

Justification, S. (2018). The Effects of Digital Curriculum in the Classroom: Rigorous Evidence from Three Digital Math Programs. *Conference Symposium Justification.*

Kaika, M. (2017). 'Don't call me resilient again!': The New Urban Agenda as immunology… or… what happens when communities refuse to be vaccinated with 'smart cities' and indicators. *Environment and Urbanization*, *29*(1), 89–102. doi:10.1177/0956247816684763

Kajimoto, T. H. (2017). Vibration feedback latency affects material perception during rod tapping interactions. *IEEE Transactions on Haptics*, *10*(2), 288–295. doi:10.1109/TOH.2016.2628900 PMID:28113957

Kalathil, S., & Boas, T. C. (2010). *Open networks, closed regimes: The impact of the Internet on authoritarian rule.* Carnegie Endowment.

Kalia, P., Arora, R., & Law, P. (2016). Information Technology Act in India: e-Commerce value chain analysis. *NTUT Journal of Intellectual Property Law and Management*, *5*(2), 55–97.

Kalia, P., Arora, R., & Law, P. (2017). *Information technology act in India: E-commerce value chain analysis*. Academic Press.

Kallender, P., & Hughes, C. W. (2017). Japan's Emerging Trajectory as a 'Cyber Power': From Securitization to Militarization of Cyberspace. *The Journal of Strategic Studies*, *40*(1-2), 118–145. doi:10.1080/01402390.2016.1233493

Kamal, Z., Mohammed, A., Sayed, E., & Ahmed, A. (2017). *Internet of Things Applications, Challenges and Related Future Technologies*. World Scientific News. doi:10.1016/j.respol.2006.05.004

Kammoun, S., Loukil, S., & Ben Romdhane, Y. (2019). The Impact of FinTech on Economic Performance and Financial Stability in MENA Zone. In Impact of Financial Technology (FinTech) on Islamic Finance and Financial Stability. IGI Global. Doi:10.4018/978-1-7998-0039-2.ch013

Kaplan, A. M., & Haenlein, M. (2010). Users of the world, unite! The challenges and opportunities of Social Media. *Business Horizons*, *53*(1), 59–68. doi:10.1016/j.bushor.2009.09.003

Karlsson, F., Åström, J., & Karlsson, M. (2014). *Information security culture: State-of-the-art review between 2000 and 2013*. Information and Computer Security.

Karpinska-Krakowiak, M. (2014). Conceptualising and measuring consumer engagement in social media - implications for personal involvement. *International Journal of Contemporary Management*, 49-65.

Karunasena, K., & Deng, H. (2012). Critical factors for evaluating the public value of e-government in Sri Lanka. *Government Information Quarterly*, *29*(1), 76–84. doi:10.1016/j.giq.2011.04.005

Karyda, M. (2017). Fostering Information Security Culture In Organizations: A Research Agenda. *MCIS 2017 Proceedings*, 28.

Kasekende, L. (2014). What Role Does Financial Inclusion Play in the Policy Agenda for Inclusive Growth in Sub-Saharan Africa? *Development*, *57*(3-4), 481–487. doi:10.1057/dev.2015.32

Katiyar, S., Meka, K. R., Barbhuiya, F. A., & Nandi, S. (2011). Online Voting System Powered By Biometric Security Using Steganography. *Second International Conference on Emerging Applications of Information Technology*, *1*(11). 10.1109/EAIT.2011.70

Katz, J. E., & Haddon, L. (2019). Domestication and Mobile Telephony. In Machines that become Us. doi:10.4324/9780203786826-4

Kaur, S., & Singh, I. (2016). A Survey Report on Internet of Things Applications. *International Journal of Computer Science Trends and Technology*, *4*(2).

Kaushik, A. (2011). *Best Social Media Metrics: Conversation, Amplification, Applause, Economic Value*. Retrieved from https://www.kaushik.net/avinash/best-social-media-metrics-conversation-amplification-applause-economic-value/

Kayani, M. B., Haq, M. E., Perwez, M. R., & Humayun, H. (2011). *Analyzing Barriers in e-Government Implementation in Pakistan*. Academic Press.

Kearns, I. (2004). *Public Value and E-Government*. Institute for Public Policy Research.

Keegan, B. J., & Rowley, J. (2017). Evaluation and Decision Making in Social Media Marketing. *Management Decision*, *55*(1), 15–31. doi:10.1108/MD-10-2015-0450

Kelly, G., Mulgan, G., & Muers, S. (2002). *Creating public value: An analytical framework for public service reform*. Retrieved from http://www.allamreform.hu/letoltheto/kozfeladatok/kulfoldi/public_value2.pdf

Kem, O., Balbo, F., & Zimmermann, A. (2016). A Distributed Approach to Constructing Travel Solutions by Exploiting Web Resources. *2016 IEEE/WIC/ACM International Conference on Web Intelligence (WI)*, 713-716. 10.1109/WI.2016.0128

Kessier, J. (2017). *Defending Public Safety from Social Engineering & Boxing*. Academic Press.

Khan, F., Khan, S., & Zhang, B. (2010). *E-Government Challenges in Developing Countries : A Case Study of Pakistan*. doi:10.1109/ICMeCG.2010.49

Khan, N., Sakib, N., Jerin, I., Quader, S., & Chakrabarty, A. (2017). *Performance analysis of security algorithms for IoT devices*. Paper presented at the Humanitarian Technology Conference (R10-HTC), 2017 IEEE Region 10. 10.1109/R10-HTC.2017.8288923

Khan, R., Khan, S. U., Zaheer, R., & Khan, S. (2012). *Future internet: the internet of things architecture, possible applications and key challenges*. Paper presented at the 2012 10th international conference on frontiers of information technology. 10.1109/FIT.2012.53

Khawaja, W., Guvenc, I., Matolak, D. W., Feibig, U. C., & Schneckenberger, N. (2019). A Survey of Air-to-Ground Propagation Channel Modeling for Unmanned Aerial Vehicles. *IEEE Communications Surveys and Tutorials*, 1–1.

Khodashenas, P. S., Aznar, J., Legarrea, A., Ruiz, C., Siddiqui, M. S., Escalona, E., & Figuerola, S. (2016). 5G network challenges and realization insights. *International Conference on Transparent Optical Networks*. 10.1109/ICTON.2016.7550539

Kietzmann, J. H., Hermkens, K., McCarthy, I. P., & Silvestre, B. S. (2011). Social media? Get serious! Understanding the functional building blocks of social media. *Business Horizons*, *54*(3), 241–251. doi:10.1016/j.bushor.2011.01.005

Kim, T., Ramos, C., & Mohammed, S. (2017). Smart City and IoT. *Future Generation Computer Systems*, *76*, 159–162. doi:10.1016/j.future.2017.03.034

Kirushanth, S., & Kabaso, B. (2018). Telematics and Road Safety. *2018 2nd International Conference on Telematics and Future Generation Networks, TAFGEN 2018*. 10.1109/TAFGEN.2018.8580482

Kitcat, J. (n.d.). *Free E-Democracy Project*. Retrieved from http://www.gnu.org/software/free/

Kitchin, R. (2015). Making sense of smart cities: Addressing present shortcomings. *Cambridge Journal of Regions, Economy and Society*, *8*(1), 131–136. doi:10.1093/cjres/rsu027

Kizekova, A. (2012). *The Shanghai cooperation organisation: challenges in cyberspace*. Academic Press.

Klimkowska, A., Lee, I., & Choi, K. (2016). Possibilities of UAS for maritime monitoring. *The International Archives of the Photogrammetry, Remote Sensing and Spatial Information Sciences*, XLI-*B1*.

Kok, S., Abdullah, A., Jhanjhi, N. Z., & Supramaniam, M. (2019). Prevention of Crypto-Ransomware Using a Pre-Encryption Detection Algorithm. *Computers*, *8*(4), 79. doi:10.3390/computers8040079

Konstantinova, J., Jiang, A., Althoefer, K., Dasgupta, P., & Nanayakkara, T. (2014). Implementation of tactile sensing for palpation in robot-assisted minimally invasive surgery: A review. *IEEE Sensors Journal*, *14*(8), 2490–2501. doi:10.1109/JSEN.2014.2325794

Kooiman, J. (2003). *Governing as governance*. SAGE Publishing.

Kouicem, D. E., Bouabdallah, A., & Lakhlef, H. (2018). Internet of things security: A top-down survey. *Computer Networks*, *141*, 199–221. doi:10.1016/j.comnet.2018.03.012

Koyun, A., & Al Janabi, E. (2017). *Social engineering attacks. Journal of Multidisciplinary Engineering Science and Technology*.

Koyuncu, E. (2018). Power-Efficient Deployment of UAVs as Relays. *IEEE Workshop on Signal Processing Advances in Wireless Communications, SPAWC*. 10.1109/SPAWC.2018.8445865

Krishna, C. L., & Murphy, R. R. (2017). A review on cybersecurity vulnerabilities for unmanned aerial vehicles. In *2017 IEEE International Symposium on Safety, Security and Rescue Robotics (SSRR)*. Shanghai: IEEE. 10.1109/SSRR.2017.8088163

Kshetri, N. (2015). India's cybersecurity Landscape: The roles of the private sector and public-private partnership. *IEEE Security and Privacy*, *13*(3), 16–23. doi:10.1109/MSP.2015.61

Kshetri, N. (2016). Cybercrime and cybersecurity in India: Causes, consequences and implications for the future. *Crime, Law, and Social Change*, *66*(3), 313–338. doi:10.100710611-016-9629-3

Kshetri, N., & Kshetri, N. (2017). Cybersecurity in India: Regulations, governance, institutional capacity and market mechanisms. *Asian Research Policy*, *8*(1), 64–76.

Kulmala, R. (2010). Ex-Ante Assessment of the Safety Effects of Intelligent Transport Systems. *Accident; Analysis and Prevention*, *42*(4), 1359–1369. doi:10.1016/j.aap.2010.03.001 PMID:20441853

Kumar, P. V. (2016). *Growing cyber crimes in India: A survey*. Paper presented at the 2016 International Conference on Data Mining and Advanced Computing (SAPIENCE). 10.1109/SAPIENCE.2016.7684146

Kumar, S. P. (2017). Internet of Things for sophisticated e-governance: A special focus on agricultural sector. *International Journal of Trend in Research and Development.*

Kumar, R., Sachan, A., & Mukherjee, A. (2017, March). E-Government Adoption in India: a Mixed Method Approach. In *Proceedings of the 10th International Conference on Theory and Practice of Electronic Governance* (pp. 617-620). ACM. 10.1145/3047273.3047357

Kundi, G. M., Nawaz, A., Akhtar, R., Phil, M., & Student, I. (2014). Digital revolution, cyber-crimes and cyber legislation: A challenge to governments in developing countries. *Journal of Information Engineering and Applications*, *4*(4), 61–71.

Kundishora, S. M. (2010). *The Role of Information and Communication Technology ICT) in Enhancing Local Economic Development and Poverty Reduction*. Harare: Zimbabwe Academic and Research Network.

Lallement, D., Terrado, E., & Zhang, Y. (2006). Empowering information and communication technologies in isolated areas: Learning from the solar-net villages program in honduras. *Renewable & Sustainable Energy Reviews*, *10*(1), 46–53. doi:10.1016/j.rser.2004.08.002

Landsbergen, D. Jr, & Wolken, G. Jr. (2001). Realizing the promise: Government information systems and the fourth generation of information technology. *Public Administration Review*, *61*(2), 206–220. doi:10.1111/0033-3352.00023

Layne, K., & Lee, J. (2001). *Developing fully functional E-government : A four stage model*. Academic Press.

Leath, B., Dunn, L. W., Alsobrook, A., & Darden, M. L. (2018). Enhancing Rural Population Health Care Access and Outcomes through the Telehealth EcoSystemTM Model. *Online Journal of Public Health Informatics*, *10*(2). doi:10.5210/ojphi.v10i2.9311 PMID:30349636

Lei, T., & Tang, Y. (2019). Digital Governance Model for Big Data Era----Based on Typical Practices in Singapore. *Humanities and Social Sciences*, *7*(2), 76. doi:10.11648/j.hss.20190702.15

Lenzhofer, A., Reber, C., Diemers, D., & Kramer, S. (2013). Taking Wealth Management Digital. Retrieved from: http://www.strategyand.pwc.com/reports/taking-wealth-management digital.

Letaifa, S. B. (2015). How to strategize smart cities: Revealing the SMART model. *Journal of Business Research*, *68*(7), 1414–1419. doi:10.1016/j.jbusres.2015.01.024

Lever, C., Walls, R., Nadji, Y., Dagon, D., McDaniel, P., & Antonakakis, M. (2016). *Domain-Z: 28 registrations later measuring the exploitation of residual trust in domains.* Paper presented at the 2016 IEEE Symposium on Security and Privacy (SP). 10.1109/SP.2016.47

Levinson, D., & Huo, H. (2003). Effectiveness of VMS Using Empirical Loop Detector Data. Transportation Research Board Conference.

Levinson, H., Zimmerman, S., Clinger, J., Rutherford, S., Smith, R. L., Cracknell, J., & Soberman, R. (2003). Bus rapid transit. *Transit Cooperative Research Program Report, 90.*

Liang, Z., Quan, L., Jie, Y., & Juan, S. (2008). *Education IPTV for E-learning in Rural Area.* IEEE.

Li, B., Jiang, Y., Sun, J., Cai, L., & Wen, C. Y. (2016). Development and testing of a two-UAV communication relay system. *Sensors (Basel)*, *16*(10), 1696. doi:10.339016101696 PMID:27754369

Li, C., & Hwang, M. (2012). Security Enhancement of Chang-Lee Anonymous E-Voting Scheme. *International Journal of Smart Home*, *6*(2), 45–52.

Light, J., & David, D. (2008). An efficient security algorithm in mobile computing for resource Constrained mobile devices. *Proceedings of the 4thACM symposium on QoS and security for wireless And mobile networks.* 10.1145/1454586.1454608

Li, M. Ch'ng, E., Chong, A. & See, S. (2016). The new eye of smart city: novel citizen sentiment analysis in twitter. *2016 International Conference on Audio, Language and Image Processing (ICALIP)*, 557-562. 10.1109/ICALIP.2016.7846617

Lim, C. P., Chai, C. S., & Churchill, D. (2011). A framework for developing pre-service teachers' competencies in using technologies to enhance teaching and learning. *Educational Media International*, *48*(2), 69–83. doi:10.1080/09523987.2011.576512

Linders, D. (2012). From e-government to we-government: Defining a typology for citizen coproduction in the age of social media. *Government Information Quarterly*, *29*(4), 446–454. doi:10.1016/j.giq.2012.06.003

Linders, D., Wilson, S. C., & Bertot, J. C. (2012). *Open government as a vehicle for government transformation. In Public Sector Transformation through E-Government.* Routledge.

Lindsay, J. R., Cheung, T. M., & Reveron, D. S. (2015). *China and cybersecurity: Espionage, strategy, and politics in the digital domain.* Oxford University Press. doi:10.1093/acprof:oso/9780190201265.001.0001

Liu, Y., Weng, X., Wan, J., Yue, X., Song, H., & Vasilakos, A. V. (2017). Exploring data validity in transportation systems for smart cities. *IEEE Communications Magazine*, *55*(5), 26–33. doi:10.1109/MCOM.2017.1600240

Li, Y., Cheng, X., Cao, Y., Wang, D., & Yang, L. (2018). *Smart choice for the smart grid: Narrowband internet of things (NB-IoT).* IEEE Internet of Things Journal. doi:10.1109/JIOT.2017.2781251

Lombardi, P., Giordano, S., Farouh, H., & Yousef, W. (2012). Modelling the smart city performance. *Innovation (Abingdon)*, *25*(2), 137–149. doi:10.1080/13511610.2012.660325

Long, M. (2002). Beyond traditional boundaries: Government in the information age. *Australian Journal of Public Administration*, *61*(1), 3–12. doi:10.1111/1467-8500.00253

Loukil, S., Ben Romdhane, Y., Kammoun, S., & Ibenrissoul, A. (2019). Empirical Determinants of Fintech: Panel Data Analysis in the MENA Zone. *International Journal of Science and Engineering Invention*, *05*(05), 90–97. doi:10.23958/ijsei/vol05-i05/159

Lovett, J. (2011). *Social media metrics secrets.* Indianapolis, IN: Wiley Publishing, Inc.

Luiijf, H., Besseling, K., Spoelstra, M., & De Graaf, P. (2011). *Ten national cyber security strategies: A comparison.* Paper presented at the International Workshop on Critical Information Infrastructures Security.

Lu, M., Bagheri, M., James, A. P., & Phung, T. (2018). Wireless Charging Techniques for UAVs: A Review, Reconceptualization, and Extension. *IEEE Access: Practical Innovations, Open Solutions, 6*, 29865–29884. doi:10.1109/ACCESS.2018.2841376

Madison, I. (2012). *Why Your Social Media Metrics Are a Waste of Time.* Retrieved from https://hbr.org/2012/12/why-your-social-media-metrics

Mahendra, A. (2008). Vehicle Restrictions in Four Latin American Cities: Is Congestion Pricing Possible? *Transport Reviews, 28*(1), 105–133. doi:10.1080/01441640701458265

Maher Omar Alshammari, A. A. A. N. Z. (2017). Internet of Things (IoT): Charity Automation. *(IJACSA). International Journal of Advanced Computer Science and Applications, 8*(2), 166–170.

Maier, M., Chowdhury, M., Rimal, B. P., & Van, D. P. (2016). The tactile internet: Vision, recent progress, and open challenges. *IEEE Communications Magazine, 54*(5), 138–145. doi:10.1109/MCOM.2016.7470948

Makinda, S. M., Okumu, F. W., & Mickler, D. (2015). *The African Union: Addressing the challenges of peace, security, and governance.* Routledge. doi:10.4324/9781315688152

Malak Alamri, N. J. M. H. (2019). Blockchain for Internet of Things (IoT) Research Issues Challenges & Future Directions: A Review. *IJCSNS International Journal of Computer Science and Network Security, 19*(5), 244–258.

Malche, T., & Maheshwary, P. (2017). Internet of Things (IoT) for building smart home system. *Proceedings of the International Conference on IoT in Social, Mobile, Analytics and Cloud, I-SMAC 2017.* 10.1109/I-SMAC.2017.8058258

Malecki, E., & Moriset, B. (2008). *The digital economy: business organization, production processes and regional developments.* Abingdon, UK: Routledge.

Malkawi, M., Khasawneh, M., & Al-Jarrah, O. (2009). Modeling and Simulation of a Robust E-voting System. *Communications of the IBIMA, 8.*

Mallard, T. (2003). *New Zealand E-Government Strategy - June 2003 - Update.* Wellington: Minister of State Services.

Mansotra, V., Sharma, V., & Sambyal, G. (2009). *IT Initiatives in Education 'A Case Study for J&K State.* IEEE.

Marcon, G. (2014). Public Value Theory in the Context of Public Sector Modernization. In Public Value Management, Measurement and Reporting. 323-351. doi:10.1108/S2051-663020140000003014

Marcus, J. S., & Elixmann, D. (2012, Sept.). Re-thinking the Digital Agenda for Europe (DAE): A richer choice of technologies. *Studie für Liberty Global.*

Margetts, H., & Dunleavy, P. (2013). The second wave of digital-era governance: a quasi-paradigm for Government on the Web. *Philosophical Transactions of the Royal Society A: Mathematical, Physical and Engineering Sciences, 371*(1987), 20120382.

Maria Almulhim, N. I. N. Z. (2019). A Lightweight and Secure Authentication Scheme for IoT Based E-Health Applications. *IJCSNS International Journal of Computer Science and Network Security, 19*(1), 107–120.

Marketing Science Institute. (2016). *2016-2018 Research Priorities.* Retrieved from http://www.msi.org/research/2016-2018-research-priorities/

Marzouk, W. G. (2016). Usage and Effectiveness of Social Media Marketing in Egypt: An Organization Perspective. *Jordan Journal of Business Administration, 12*(1), 209–238. doi:10.12816/0030062

Matheus, R., Janssen, M., & Maheshwari, D. (2018). Data science empowering the public: Data-driven dashboards for transparent and accountable decision-making in smart cities. *Government Information Quarterly*, 101284. doi:10.1016/j.giq.2018.01.006

Mathur, A., Newe, T., Elgenaidi, W., Rao, M., Dooly, G., & Toal, D. (2017). A secure end-to-end IoT solution. *Sensors and Actuators. A, Physical, 263*, 291–299. doi:10.1016/j.sna.2017.06.019

Mavropoulos, O., Mouratidis, H., Fish, A., & Panaousis, E. (2018). Apparatus: A framework for security analysis in internet of things systems. *Ad Hoc Networks*. doi:10.1016/j.adhoc.2018.08.013

McCann, M., & Barlow, A. (2015). Use and measurement of social media for SMEs. *Journal of Small Business and Enterprise Development, 22*(2), 273–287. doi:10.1108/JSBED-08-2012-0096

McDonagh, D. (2008). Satisfying Needs beyond the Functional: The Changing Needs of the Silver Market Consumer. Presented at *the International Symposium on the Silver Market.*

McGuire, M., & Dowling, S. (2013). *Cyber crime: A review of the evidence.* Summary of key findings and implications. Home Office Research Report, 75.

McKinsey Global Institute. (2011). *Internet Matters: The Net's Sweeping Impact on Growth, Jobs, and Prosperity.* San Francisco: McKinsey & Company.

McLeod, B. (2017). *How Much Does It Cost to Advertise on Facebook?* Retrieved from https://www.bluecorona.com/blog/how-much-facebook-advertising-costs

McLuhan, M., & Fiore, Q. (1967) The medium is the message. New York: Academic Press.

Mehmood, Y., Ahmad, F., Yaqoob, I., Adnane, A., Imran, M., & Guizani, S. (2017). Internet-of-things-based smart cities: Recent advances and challenges. *IEEE Communications Magazine, 55*(9), 16–24. doi:10.1109/MCOM.2017.1600514

Melin, U., & Axelsson, K. (2009). Managing e-service development - comparing two e-government case studies, Transforming Government. People. *Process and Policy, 3*(3), 248–270.

Mellouli, S., Luna-Reyes, L. F., & Zhang, J. (2014). Smart government, citizen participation and open data. *Information Polity, 19*(1,2), 1–4. doi:10.3233/IP-140334

Mendes, T. D. P., Godina, R., Rodrigues, E., Matias, J., & Catalão, J. (2015). Smart Home Communication Technologies and Applications: Wireless Protocol Assessment for Home Area Network Resources. *Energies, 8*(7), 7279–7311. doi:10.3390/en8077279

Menon, G., & Guttikunda, S. (2010). *Electronic Road Pricing: Experience & Lessons from Singapore.* SIM-air Working Paper Series: 33-2010.

Menouar, H., Guvenc, I., Akkaya, K., Uluagac, A. S., Kadri, A., & Tuncer, A. (2017). UAV-enabled intelligent transportation systems for the smart city: Applications and challenges. *IEEE Communications Magazine, 55*(3), 22–28. doi:10.1109/MCOM.2017.1600238CM

Mer, S. B. (2015). Smart Vehicle-to-Vehicle Communication with 5G Technology. International Journal on Recent Innovation Trends in Computing and Communication, 3(5), 3241–3244.

Milakovich, M. (2012). *Digital governance: New technologies for improving public service and participation.* Routledge. doi:10.4324/9780203815991

Miles, M. B., & Huberman, A. M. (2004). *Analyse des données qualitatives* (2nd ed.). De Boeck.

Ministry of Communication and Information Technology. (n.d.). *Saudi Arabia Affirms Leading Position In Digital Transformation In Effort To Drive Global Economy In Argentina's G20 Summit*. Retrieved October 19, 2019 from https://www.mcit.gov.sa/en/media-center/news/99864

Ministry of Communications-AITI. (2016). *National ICT White Paper for 2016-2020*. Author.

Ministry of Education. (2005). *Ministry of Education - Circulars*. Available at: http://www.moe.gov.lk/english/images/stories/circulars/2005-29s.pdf

Ministry of Education. (2006). *School Census - Preliminary Report*. Statistics Branch.

Ministry of Education. (2009). *Year of English & IT*. Academic Press.

Ministry of Education. (2012). *Sri Lanka Education Information*. Colombo: Ministry of Education.

Ministry of Finance. (n.d.). *Ministry of Finance Launches Etimad - the Digital Services Platform*. Retrieved October 19, 2019 from https://www.mof.gov.sa/en/MediaCenter/news/Pages/news_20012017.aspx

Ministry of Health. (n.d.). *MOH Apps for Smartphones*. Retrieved October 20, 2019 from https://www.moh.gov.sa/en/Support/Pages/MobileApp.aspx

Ministry of Justice. (n.d.). *E-services*. Retrieved October 4, 2019 from https://www.moj.gov.sa/English/eServices/Pages/default.aspx

Misuraca, G., & Lusoli, W. (Eds.). (2010). *Envisioning digital europe 2030: Scenarios for ICT in future governance and policy modelling*. EUR 24614 EN, 12/2010. Retrieved from http://ipts.jrc.ec. europa.eu/publications/pub.cfm?id=3879

Misuraca, G., & Viscusi, G. (2015). Shaping public sector innovation theory : An interpretative framework for ICT-enabled governance innovation. *Electronic Commerce Research*, *15*(3), 303–322. doi:10.100710660-015-9184-5

Mitrakas, A. (2006). Information security and law in Europe: Risks checked? *Information & Communications Technology Law*, *15*(01), 33–53. doi:10.1080/13600830600557984

Mocrii, D., Chen, Y., & Musilek, P. (2018). IoT-based smart homes: A review of system architecture, software, communications, privacy and security. *Internet of Things*, *1-2*, 81–98. doi:10.1016/j.iot.2018.08.009

Moghaddasi, H., Mohammadpour, A., Bouraghi, H., Azizi, A., & Mazaherilaghab, H. (2018). Hospital Information Systems: The status and approaches in selected countries of the Middle East. *Electronic Physician*, *10*(5), 6829–6835. doi:10.19082/6829 PMID:29997768

Mohammed, Z. K., & Ahmed, E. S. (2017). Internet of Things Applications, Challenges and Related Future Technologies. *World Scientific News*, *67*(2), 126–148.

Mohassel, R. R., Fung, A., Raahemifar, K., & Mohammadi, F. A. (2014). A survey on Advanced Metering Infrastructure. *International Journal of Electrical Power & Energy Systems*, *63*, 473–484. doi:10.1016/j.ijepes.2014.06.025

Monga, A. (2008). E-government in India: Opportunities and challenges. *JOAAG*, *3*(2), 56.

Moore, M. H. (1995). *Creating public value: Strategic management in government*. London: Harvard University Press.

Moor, M. J., & Norris, D. F. (2005). Does managerial orientation matter? The adoption of reinventing government and egovernmentat the municipal level. *Information Systems Journal*, 43–60.

Motlagh, N. H., Taleb, T., & Arouk, O. (2016). Low-Altitude Unmanned Aerial Vehicles-Based Internet of Things Services: Comprehensive Survey and Future Perspectives. *IEEE Internet of Things Journal*, 899 - 922.

Mouton, F., Leenen, L., Malan, M. M., & Venter, H. S. (2014, July). Towards an ontological model defining the social engineering domain. In *IFIP International Conference on Human Choice and Computers* (pp. 266-279). Springer. 10.1007/978-3-662-44208-1_22

Mujahid, Y. H. (2002). Digital opportunity initiative for Pakistan. *The Electronic Journal on Information Systems in Developing Countries*, *8*(1), 1–14. doi:10.1002/j.1681-4835.2002.tb00050.x

Mukhopadhyay, C. (2016). A nested framework for transparency in Public Private Partnerships: Case studies in highway development projects in India. *Progress in Planning*, *107*, 1–36. doi:10.1016/j.progress.2015.02.001

Mundy, G. R. (2011). *Internet Deployment of Domain Name System (DNS) Security*. Academic Press.

Murphy, J., & Carmody, P. (2015). *Africa's information revolution: Technical regimes and production networks in South Africa and Tanzania*. Chichester, UK: Wiley.

Mussington, D., Arnold, B. J., Dupont, B., Hilts, S., Grayson, T., Leuprecht, C., . . . Tupler, J. (2018). *Governing Cyber Security in Canada*. doi:10.21236/ADA549761

Mutegi, L. (2017). *ICT sector to contribute up to 8% of Kenya's GDP, CIO/East Afric*. Retrieved from: http://www.cio.co.ke/news/ main-stories/ict-sector-to-contribute-upto-8-of-kenya%27s-gdp-by-2017

Mutegi, M. (2016). *Vodafone Launches MPesa Ghana in Africa Expansion Bid*. IPSOS 7.

Naik, S., & Maral, V. (2017). *Cyber security—IoT*. Paper presented at the Recent Trends in Electronics, Information & Communication Technology (RTEICT), 2017 2nd IEEE International Conference on.

Nalbandian, J., O'Neill Jr, R., Michael Wilkes, J., & Kaufman, A. (2013). Contemporary challenges in local government: Evolving roles and responsibilities, structures, and processes. *Public Administration Review*, *73*(4), 567–574. doi:10.1111/puar.12059

Nam, T., & Pardo, T. A. (2011). Conceptualizing smart city with dimensions of technology, people, and institutions. *Proceedings of the 12th annual international digital government research conference: digital government innovation in challenging times*, 282-291. 10.1145/2037556.2037602

Nambisan, S., & Majchrzak, A. (2017). Digital Innovation Management: Reinventing Innovation Management Research in a Digital World. *Management Information Systems Quarterly*, *41*(1), 223–238. doi:10.25300/MISQ/2017/41:1.03

Nam, T. (2011). Suggesting frameworks of citizen-sourcing via Government 2.0. *Government Information Quarterly*, (29): 12–20.

National Contact Center (Amer). (n.d.). Retrieved September 30, 2019 from https://www.yesser.gov.sa/EN/Building-Blocks/Pages/national_e-Government_contact_center.aspx

National Portal of India. (n.d.). Retrieved from https://india.gov.in/egovernance/national-e-governance-plan

National Water Company (NWC). (n.d.). Retrieved 21 October 2019 from https://www.nwc.com.sa/English/OurCompany/MediaCenter/NewsandEvents/News/Pages/smart-meters-in-the-regions-of-the-Kingdom-by-the.aspx

Navaneet, S. S. (2012). A model for Performance Enhancement of Steganography through Dynamic Key Cryptography. *International Journal of Advanced Networking and Applications*, *3*(6), 1395–1401.

Naveed, K., Watanabe, C., & Neittaanmäki, P. (2018). The Transformative Direction of Innovation toward an IoT-based Society: Increasing Dependency on Uncaptured GDP in Global ICT Firms. *Technology in Society*, *53*, 23–46. doi:10.1016/j.techsoc.2017.11.003

Nayyar, A., Jain, R., Mahapatra, B., & Singh, A. (2019). *Cyber Security Challenges for Smart Cities. Driving the Development, Management, and Sustainability of Cognitive Cities*. IGI Global.

Ndemo, B., & Weiss, T. (2017). Digital Kenya, Une révolution entrepreneuriale en devenir, Palgrave Studies of Entrepreneurship in Africa.

Neirotti, P., De Marco, A., Cagliano, A. C., Mangano, G., & Scorrano, F. (2014). Current trends in Smart City initiatives: Some stylised facts. *Cities (London, England)*, *38*, 25–36. doi:10.1016/j.cities.2013.12.010

Nenasala. (n.d.). *Establishment of Nenasalas*. Available at: http://www.nanasala.lk/

Nguyen, T. T., & Armitage, G. (2008). A survey of techniques for internet traffic classification using machine learning. *IEEE Communications Surveys and Tutorials*, *10*(4), 56–76. doi:10.1109/SURV.2008.080406

Niewels, F., Knoop, S., Jordan, R. A., & Michalke, T. (2013). *In-Vehicle Sensors. Encyclopedia of Automotive Engineering*. John Wiley & Sons.

Nitcheu, P. (2016). *Apport de l'économie numérique sur le développement économique au Cameroun*. halshs-01970291f.

NO2-A43F Nitrogen Dioxide Sensor datasheet. (2018). Retrieved 17 July 2019, from http://www.alphasense.com/WEB1213/wp-content/uploads/2018/12/NO2A43F.pdf

NO-A4 Nitric Oxide Sensor data sheet. (2016). Retrieved 17 July 2019, from http://www.alphasense.com/WEB1213/wp-content/uploads/2016/03/NO-A4.pdf

Noh, Y. H., & Yoo, K. (2008). Internet, inequality, and growth. *Journal of Policy Modeling*, *30*(6), 1005–1016. doi:10.1016/j.jpolmod.2007.06.016

Ntulo, G., & Otike, J. (2013). *E–Government: Its Role, Importance and Challenges*. School of Information Sciences. MoiUniversity.

Nuaimi, E. A., Neyadi, H., Mohamed, N., & Al-Jaroodi, J. (2015). Applications of big data to smart cities. *Journal of Internet Services and Applications*.

O'Connor, E., Allan, S., & Audretsch, B. (2016). *Entrepreneurial Ecosystems*. ISEN.

O'Dwyer, E., Pan, I., Acha, S., & Shah, N. (2019). Smart energy systems for sustainable smart cities: Current developments, trends and future directions. *Applied Energy*, *237*, 581–597. doi:10.1016/j.apenergy.2019.01.024

O'Mahony, S. (2005). *20 Nonprofit Foundations and Their Role in Community-Firm Software Collaboration*. Academic Press.

OECD. (2003). *The e-government imperative. OECD e-government studies*. Retrieved from www1.worlbank.org/publicsector/egov/E-governmentImperative.pdf

OECD. (2014). *Recommendation of the Council on Digital Government Strategies*. Retrieved from http://www.oecd.org/gov/digital-government/Recommendation-digital-government-strategies.pdf

OECD. (2016). *Measuring GPD in a digitalized economy*. STD/CSSP.

OECD. (2016). *Tax Challenges in the Digital Economy*. Paris: OECD.

OECD. (2016). *The Digital Economy*. Paris: OECD.

Okediran, O. O., Olabiyisi, S. O., Omidiora, E. O., & Ganiyu, R. A. (2011). A Survey of Remote Internet Voting Vulnerabilities. *World of Computer Science and Information Technology Journal (WCSIT)*.

Okediran, O. O., Omidiora, E. O., Olabiyisi, S. O., & Ganiyu, R. A. (2011b). A Comparative Study of Generic Cryptographic models for Secure Electronic Voting. *British Journals of Science*, *1*(2), 135–142.

Okediran, O. O., Omidiora, E. O., Olabiyisi, S. O., Ganiyu, R. A., & Alo, O. O. (2011a). A framework for a Multifaceted Electronic Voting System. *International Journal of Applied Sciences*, *1*(4), 135–142.

Okong'o, K., & Kyobe, M. (2018). Empirical Examination of E-Government in Developing Countries and its Value in Kenya's Public Service. *The Electronic Journal Information Systems Evaluation*, *21*(1), 35–45.

Olaniyan, O. M., Mapayi, T., & Adejumo, S. A. (2011). A Proposed Multiple Scan Biometric-Based System for Electronic Voting. *African Journal of Computing and ICT*, *4*(2), 9–16.

Olaniyi, O.M., Tayo, A.O., Olusayo, O.E., & Olusola, O.O. (2013). A survey of cryptographic and stegano-cryptographic models for secure electronic voting system. *Covenant Journal of Informatics and Communication Technology*, *2*(1).

Olaniyi, Arulogun, Omidiora, & Adeoye. (2013). Design of Electronic Voting System Using Multifactor Authentication and Cryptographic Hash Functions. *International Journal of Computer and Information Technology*, *2*(6), 1122–1130.

Olaniyi, O. M., Adewumi, D. O., Oluwatosin, E. A., Arulogun, O. T., & Bashorun, M. A. (2011). Framework for Multilingual Mobile E-Voting Service Infrastructure for Democratic Governance. *African Journal of Computing and ICT*, *4*(3), 23–32.

Olaniyi, O. M., Arulogun, O. T., & Omidiora, E. O. (2012). Towards an Improved Stegano-Cryptographic Model for Secured Electronic Voting. *African Journal of Computing and ICT*, *5*(6), 10–16.

Omar, A. M. (2016). *Innovation for Economic Development*. Academic Press.

OPC-N3 Particle Monitor datasheet. (2019). Retrieved 17 July 2019, from http://www.alphasense.com/WEB1213/wp-content/uploads/2019/03/OPC-N3.pdf

Osborn, D., & Gaebler, T. (1992). *Reinventing government. How the entrepreneurial spirit is transforming the public sector*. Addison-Wesley.

Osseiran. (2013). The Foundation of the Mobile and Wireless Communications System for 2020 and Beyond Challenges, Enablers and Technology Solutions. *Proc. IEEE Vehic. Technol. Conf. (VTC Spring'15)*.

Owyang, J. (2010). *Social Media Analytics*. Retrieved from https://www.slideshare.net/jeremiah_owyang/altimeter-report-social-marketing-analytics

Oztamur, D., & Karakadilar, I. S. (2014). Exploring the role of social media for SMEs: As a new marketing strategy tool for the firm performance perspective. *Procedia: Social and Behavioral Sciences*, *150*, 511–520. doi:10.1016/j.sbspro.2014.09.067

Pan, Y., & Ding, X. (2006, December). Anomaly based web phishing page detection. In *2006 22nd Annual Computer Security Applications Conference (ACSAC'06)* (pp. 381-392). IEEE. 10.1109/ACSAC.2006.13

Pan, B., & You, Y. (2017). Conceptualizing and Measuring Online Behavior Through Social Media Metrics. In *Analytics in Smart Tourism Design* (pp. 203–211). Springer International Publishing. doi:10.1007/978-3-319-44263-1_12

Panda, B. P., & Swain, D. K. (2009). *Effective Communications through e-Governance and e-Learning*. IBS Bhubaneswar.

Paneque-Gálvez, J., McCall, M., Napoletano, B., Wich, S., & Koh, L. (2014). Small drones for community-based forest monitoring: An assessment of their feasibility and potential in tropical areas. *Forests*, *5*(6), 1481–1507. doi:10.3390/f5061481

Pardos, A.B.C., Eucins, A.H., White, S.H., Del Rey, A.M., & Sanchez, G.R. (2007). A Simple Protocol for Yes-No Electronic Voting. *International Journal of Computer Science and Network Security, 7*(7), 72-76.

Paskaleva, K. A. (2009). Enabling the smart city: The progress of city e-governance in Europe. *International Journal of Innovative Research and Development*, *1*(4), 405–422. doi:10.1504/IJIRD.2009.022730

Patel, M., & Wang, J. (2010). Applications, challenges, and prospective in emerging body area networking technologies. *IEEE Wireless Communications*, *17*(1), 80–88. doi:10.1109/MWC.2010.5416354

Patil, V. M. (2010). Secure Electronic Voting System by Using Blind Signature and Cryptography for voter"s privacy and Authentication. *Journal of Signal and Image Processing*, *1*(1), 1–6.

Paul, S. C., Tay, Y. W. D., Panda, B., & Tan, M. J. (2018). Fresh and hardened properties of 3D printable cementitious materials for building and construction. *Archives of Civil and Mechanical Engineering*, *18*(1), 311–319. doi:10.1016/j.acme.2017.02.008

PDST. (n.d.). *Technology Overview in Schools - ICT Infrastructure.* PDST Technology in Education.

Pejkovska, M. (2018). *Potential negative effects of Fintech on the financial services sector.* Examples from the European Union.

Peña-López, I. (2016). *World development report 2016: Digital dividends.* Academic Press.

People are the weakest links in cybersecurity, says CSM. (2018, Jul. 22). *The Star Online*. Available: https://www.thestar.com.my/news/ nation/2018/07/22/people-are-the-weakest-links-in-cybersecurity-says-csm/

Pereira, G. V., Parycek, P., Falco, E., & Kleinhans, R. (2018). Smart governance in the context of smart cities: A literature review. *Information Polity*, *23*(2), 143–162. doi:10.3233/IP-170067

Perez, S. X. (2015). Impact of delay on telesurgical performance: Study on the robotic simulator dv-trainer. *International Journal of Computer Assisted Radiology and Surgery*, *11*. PMID:26450105

Peters, K., Chen, Y., Kaplan, A. M., Ognibeni, B., & Pauwels, K. (2013). Social media metrics—A framework and guidelines for managing social media. *Journal of Interactive Marketing*, *27*(4), 281–298. doi:10.1016/j.intmar.2013.09.007

Pettigrew, A. (1997, Sept.). Le new public management conduit à un nouveau modèle hybride public-privé. *Revue Française de Gestion*, 113-120.

Philippon, T. (2016). *The FinTech opportunity.* NBER Working Paper No.22476.

Photo Ionisation Detector datasheet. (2016). Retrieved 17 July 2019, from http://www.alphasense.com/WEB1213/wp-content/uploads/2016/09/PID-AH2.pdf

Platform, M. (n.d.). Retrieved September 17, 2019 from https://meras.gov.sa/en/about/

PMO. (2018). *Prime Minister's Office. Vision, Mission, and Strategy.* Retrieved from http://www.pmo.gov.bn/SitePages/Vision,%20Mission%20and%20Strategy.aspx

Pojani, D., & Stead, D. (2015). Sustainable urban transport in the developing world: Beyond megacities. *Sustainability*, *7*(6), 7784–7805. doi:10.3390u7067784

Portal, S. (n.d.). Retrieved September 30, 2019 from https://www.yesser.gov.sa/EN/BuildingBlocks/Pages/saudi_portal.aspx

Positive Technologies. (2019). *Cybersecurity threatscape 2018: trends and forecasts. Cyber Security Threat Trends 2018 and Forecasts for 2019: Attack Statistics and Facts*. Available: https://www.ptsecurity.com/ww-en/analytics/cybersecurity-threatscape-2018/

Post, S. (n.d.a). *Jamaee*. Retrieved 25 October 2019 from https://sp.com.sa/en/government-services/jamaee/

Post, S. (n.d.b). *National Address*. Retrieved 25 October 2019 from https://sp.com.sa/en/national-address/

Post, S. (n.d.c). *Wasel International*. Retrieved 25 October 2019 from http://www.waselalami.com.sa/ar/overview

Pramanik, M. I., Lau, R. Y., Demirkan, H., & Azad, M. A. K. (2017). Smart health: Big data enabled health paradigm within smart cities. *Expert Systems with Applications*, *87*, 370–383. doi:10.1016/j.eswa.2017.06.027

Praude, V., & Skulme, R. (2015). Social Media Campaign Metrics in Latvia. *Procedia: Social and Behavioral Sciences*, *213*, 628–634. doi:10.1016/j.sbspro.2015.11.462

Price, M. (2017). The global politics of Internet governance. *Technology and World Politics: An Introduction.*

Price, R. (2018). *Artificial intelligence-powered malware is coming, and it's going to be terrifying.* Retrieved: September 19, 2018, from https://www.businessinsider.com/darktrace-dave-palmerartificial-intelligence-poweredmalware-hacks-interview-2016-10?r=UK&IR=T

PS. (2003). Retrieved from http://cns.miis.edu/research/space/pakistan/mil.htm

Pundir, Y., Sharma, N., & Singh, Y. (2016). Internet of things (IoT): Challenges and future directions. *International Journal of Advanced Research in Computer and Communication Engineering*, *5*(3), 960–964.

Purser, S. (2014). *Standards for Cyber Security*. Academic Press.

Purusthomata, B. R., & Alwyn, R. P. (2009). Design and Implementation of Secure Internet Based Voting System with User Anonymity using Identity Based Encryption System. *Proceedings of IEEE International Conference on Services Computing*, 474-481.

Puschmann, Th. (2017). Fintech. *Business & Information Systems Engineering: The International Journal of Wirtschaftsinformatik*, *59*(1), 69-76. Doi:10.100712599-017-0464-6

Qadah, G. Z., & Taha, R. (2007). Electronic Voting Systems: Requirements, Design, And Implementation. *Computer Standards & Interfaces, 29*(3), 376-386. Doi:10.1016/j.csi.2006.06.001

Qaisar, N., & Ghufran, H. (2010). *E-Government Challenges in Public Sector : A case study of Pakistan*. Academic Press.

Qu, Z., Zhang, G., Cao, H., & Xie, J. (2017). LEO Satellite Constellation for Internet of Things. *IEEE Access: Practical Innovations, Open Solutions*, *5*, 18391–18401. doi:10.1109/ACCESS.2017.2735988

Radhakrishna, R. B. (2007). Tips for developing and testing questionnaires/instruments. *Journal of Extension*. Retrieved from https://www.joe.org/joe/2007february/tt2.php

Raghavan, R. (2018). Cyber Insurance–A Risk Mitigation Tool for Cyber Risk in India. *Bimaquest, 18*(1).

Rahman, N. A., & Hj Hassan, M. S. (2015). Social Media Research Trends in Malaysia: An Analysis of Three Major Communication Journals in Malaysia. *Communication. Mass Media*, 87-96.

Raju, K. (2004). A case for harnessing information technology for rural development. *Elseiver, 36*, 233-240.

Rasetyono, A. P. (2017). *IPTEK: Solusi Komprehensif Atasi Kemacetan Lalu-Lintas*. Retrieved from Kementerian Riset, Teknologi dan Pendidikan Tinggi (Kemenristekdikti) Republik Indonesia: http://dikti.go.id/iptek-solusi-komprehensif-atasi-kemacetan-lalu-lintas

Raskin, V., Hempelmann, C. F., & Triezenberg, K. E. (2004, July). Semantic forensics: An application of ontological semantics to information assurance. In *Proceedings of the 2nd Workshop on Text Meaning and Interpretation* (pp. 105-112). Association for Computational Linguistics. 10.3115/1628275.1628289

Raulot, A. (2019, Feb.). Bypassing phishing protections with email authentication. *Master Security and Network Engineering*, 5.

Ray, P. P. (2016). A survey of IoT cloud platforms. *Future Computing and Informatics Journal*, *1*(1-2), 35–46. doi:10.1016/j.fcij.2017.02.001

Rehman, M. (2011). *Factors influencing the adoption of E-government in Pakistan*. Academic Press.

Rehman, M., Esichaikul, V., & Kamal, M. (2012). *Factors influencing e-government adoption in Pakistan*. doi:10.1108/17506161211251263

Reuter, C., Ludwig, T., Kaufhold, M.-A., & Spielhofer, T. (2016). Emergency services' attitudes towards social media: A quantitative and qualitative survey across Europe. *International Journal of Human-Computer Studies*, *95*, 96–111. doi:10.1016/j.ijhcs.2016.03.005

Rhodes, R. A. W. (1996). The new governance: Governing without government. *Political Studies*, *44*(4), 652–667. doi:10.1111/j.1467-9248.1996.tb01747.x

Ricci, F. (2010). Mobile Recommender Systems. *Information Technology & Tourism*, *3*(12), 205–231. doi:10.3727/109830511X12978702284390

Rimbar, H. A. (2017). the influence of spell-checkers on students ability to generate repars of spelling errors. *Journal of Nusantara Studies*, *2*(1).

Risteska Stojkoska, B. L., & Trivodaliev, K. V. (2017). A review of Internet of Things for smart home: Challenges and solutions. *Journal of Cleaner Production*, *140*, 1454–1464. doi:10.1016/j.jclepro.2016.10.006

Rizal, M., Maulina, E., Purnomo, M., & Febrian, A. (2017). Electronic Road Pricing (ERP): A Systematic Mapping Study. *International Research Conference on Economic and Business 2017.*

Rizvi, S., Kurtz, A., Pfeffer, J., & Rizvi, M. (2018). *Securing the Internet of Things (IoT): A Security Taxonomy for IoT.* Paper presented at the 2018 17th IEEE International Conference On Trust, Security And Privacy In Computing And Communications/12th IEEE International Conference On Big Data Science And Engineering (TrustCom/BigDataSE).

Romaniuk, J. (2012). Are you ready for the next big thing? New media is dead! Long live new media! *Journal of Advertising Research*, *52*(4), 397–399. doi:10.2501/JAR-52-4-397-399

Ronaghan, S. A. (2002). *Benchmarking e-government a global perspective: Assessing the progress of UN member states*. New York, NY: United Nations DPIDG.

Rosenzweig, P. (2012). The international governance framework for cybersecurity. *Canada–United States Law Journal*, *37*, 405.

Rossiter, A., Nortcliffe, A., Griffin, A., & Middleton, A. (2015). Using student generated audio to enhance learning. *Journal of the Higher Education Academy, 4*(2).

Roy, A., Banik, S., & Karforma, S. (2011). Object oriented modeling of RSA digital signature in E-governance. *International Journal of Computer Engineering and Information Technology*, *26*(1), 2433.

Roy, A., & Karforma, S. (2011). Risk and Remedies of E-governance Systems. *Oriental Journal of Computer Systems and Technology*, *4*(2), 329–339.

Roy, A., & Karforma, S. (2012). A survey of digital Signatures and Its Applications. *CIT. Journal of Computing and Information Technology*, *3*(1), 45–46.

Rughiniş, C., & Rughiniş, R. (2014). Nothing ventured, nothing gained. Profiles of online activity, cyber-crime exposure, and security measures of end-users in European Union. *Computers & Security*, *43*, 111–125. doi:10.1016/j.cose.2014.03.008

Ruhi, U. (2014). Social Media Analytics as a business intelligence practice: current landscape & future prospects. *Journal of Internet Social Networking & Virtual Communities*. doi:10.5171/2012.920553

Sabillon, R., Cavaller, V., & Cano, J. (2016). National cyber security strategies: Global trends in cyberspace. *International Journal of Computer Science and Software Engineering*, *5*(5), 67.

Sadique, K. M., Rahmani, R., & Johannesson, P. (2018). Towards Security on Internet of Things: Applications and Challenges in Technology. *Procedia Computer Science*, *141*, 199–206. doi:10.1016/j.procs.2018.10.168

Safa, N. S., Von Solms, R., & Furnell, S. (2016). Information security policy compliance model in organizations. *Computers & Security, 56*, 70-82.

Safety, P. (2006). *Emergency Preparedness Canada (PSEPC)*. Canadian Disaster Database.

Salahdine, F., & Kaabouch, N. (2019). Social Engineering Attacks: A Survey. *Future Internet*, *11*(4), 89. doi:10.3390/fi11040089

Saleh, K., El-Morr, C., Mourtada, A., & Morad, Y. (2003). *Specifications Of A Mobile Electronic Voting System And A Mobile Agent Platform*. IADIS e-Society 2003, Lisbon, Portugal.

Samadi, S., Rad, A., Kazemi, F., & Jafarian, H. (2012). Performance Evaluation of Intelligent Adaptive Traffic Control Systems: A Case Study. *Journal of Transportation Technologies*, *2*(3), 248–259. doi:10.4236/jtts.2012.23027

Sandhya Sarma, K. N., & Umamaheswari, S. (2012). A Powerful Protocol for Electronic - Voting using a HYBRID Crypto Realm. Journal of computing and information technology. *International Journal of Clothing Science and Technology*, *3*(3).

Sari, O. (2018, October 26). *Data Breach Record of 2018 in the First Half.* Retrieved from https://www.keepnetlabs.com/social-engineering-stats/

Sarker, A. E. (2006). New public management in developing countries: An analysis of success and failure with particular reference to Singapore and Bangladesh. *International Journal of Public Sector Management*, *19*(2), 180–203. doi:10.1108/09513550610650437

Sarker, M. N. I., Wu, M., & Hossin, M. A. (2018). Smart governance through bigdata: Digital transformation of public agencies. *2018 International Conference on Artificial Intelligence and Big Data (ICAIBD)*, 62-70. 10.1109/ICAIBD.2018.8396168

Sarker, N. I., Wu, M., Liu, R., & Ma, C. (2019). *Challenges and Opportunities for Information Resource Management for E-Governance in Bangladesh*. Springer International Publishing. doi:10.1007/978-3-319-93351-1_53

Satola, D., & Judy, H. L. (2010). Towards a dynamic approach to enhancing international cooperation and collaboration in cybersecurity legal frameworks: Reflections on the proceedings of the workshop on cybersecurity legal issues at the 2010 United Nations Internet Governance Forum. *Wm. Mitchell L. Rev.*, *37*, 1745.

Saudi Arabia Builds First House Using 3d Printing Technology in the Middle East. (2018, November 5). *Al Arabiya English*. Retrieved 21 October 2019 from http://english.alarabiya.net/en/business/technology/2018/11/05/Saudi-Arabia-builds-first-house-using-3D-printing-technology-in-the-Middle-East.html

Saudi Arabia: ready to lead the world in smart city development? (2018, August 12). Retrieved October 2, 2019 from https://www.alj.com/en/perspective/saudi-arabia-ready-lead-world-smart-city-development/

Saudi Arabian Monetary Authority. (2018). *Saudi Arabian monetary authority launches fintechsaudi with the objective to make the kingdom a pioneer in the financial technology sector.* (Retrieved October 24, 2019 from http://www.sama.gov.sa/en-US/News/Pages/news30042018.aspx

Saudi Human Resources Gateway. (n.d.). Retrieved 23 October 2019 from https://hrdf.org.sa/Program/406/Saudi_Human_Resources_Gateway?bc=264

Saudi justice ministry launches 15 new e-services to streamline procedures. (2019, September 20). *Arab News*. Retrieved October 2, 2019 from https://www.arabnews.com/node/1557051/saudi-arabia

Saudi National Portal, The SADAD Payment System. (n.d.). Retrieved 21 October 2019 from https://www.saudi.gov.sa/wps/portal/snp/pages/ePayment

Saudi National Portal. (n.d.). *e-Procurement*. Retrieved 17 October 2019 from https://www.saudi.gov.sa/wps/portal/snp/pages/e-Procurement

Saudi National Portal. (n.d.). *Wasel*. Retrieved 25 October 2019 from https://www.saudi.gov.sa/wps/portal/snp/government/servicedetails/9746

Saudi Vision 2030. (n.d.a). *FekraTech*. Retrieved October 20, 2019 from https://fekratech.gov.sa/faq?lang=en

Saudi Vision 2030. (n.d.b). Retrieved October 20, 2019 from https://vision2030.gov.sa/en/vision-progress

Saugata, B., & Masud, R. R. (2007). Implementing E-Governance Using OECD Model(Modified) and Gartner Model (Modified) Upon Agriculture of Bangladesh. IEEE.

Saxena, K. B. (2005). Towards excellence in e-governance. *International Journal of Public Sector Management*, *18*(6), 498–513. doi:10.1108/09513550510616733

Scarfone, K., Benigni, D., & Grance, T. (2008). Cyber security standards. Wiley Handbook of Science and Technology for Homeland Security, 1-10.

Schedler, K., & Sharf, M.C. (2001). Exploring the interrelation between electronic government and the new public management: a managerial framework for electronic government. *13ième conférence Zurich*.

Scholl, H. J. (2005). Interoperability in e-Government: More than just smart middleware. *Proceedings of the 38th Annual Hawaii International Conference on System Sciences*, 123-123. 10.1109/HICSS.2005.336

Schultz, D. E., & Peltier, J. (2013). Social media's slippery slope: Challenges, opportunities and future research directions. *Journal of Research in Interactive Marketing*, *7*(2), 86–99. doi:10.1108/JRIM-12-2012-0054

Schumpeter, J. (1942). Creative Destruction. *Capitalism, Socialism, and Democracy*, *825*, 82-85.

Scoop Media. (2018b). *More than 2,000 cybersecurity attacks in Brunei last year*. Author.

Scoop. (2018a). *Darussalam Hotline now open for social media woes*. Retrieved from https://thescoop.co/2018/09/25/darussalam-hotline-now-open-for-social-media-woes/

Sekaran, U. (2006). *Research Methods for Business: A Skill Building Approach* (4th ed.). New Delhi: John Wiley & Sons Ltd.

Sepehrdoust, H. (2018). Impact of information and communication technology and financial development on economic growth of OPEC developing economies. *Kasetsart Journal of Social Sciences*. doi:10.1016/j.kjss.2018.01.008

Sfar, A. R., Natalizio, E., Challal, Y., & Chtourou, Z. (2018). A roadmap for security challenges in the Internet of Things. *Digital Communications and Networks*, *4*(2), 118–137. doi:10.1016/j.dcan.2017.04.003

Shabiralyani, G., Hasan, K. S., Hamad, N., & Iqbal, N. (2015). Impact of Visual Aids in Enhancing the Learning Process Case Research: District Dera Ghazi Khan. *Journal of Education and Practice*, *6*(19).

Shackelford, S. J. (2017). The Law of Cyber Peace. *Chi. J. Int'l L.*, *18*, 1.

Shackelford, S. J., & Bohm, Z. (2016). Securing North American critical infrastructure: A comparative case study in cybersecurity regulation. *Canada–United States Law Journal*, *40*, 61.

Shackelford, S. J., & Craig, A. N. (2014). Beyond the new digital divide: Analyzing the evolving role of national governments in internet governance and enhancing cybersecurity. *Stan. J. Int'l L.*, *50*, 119.

Shackelford, S. J., Russell, S., & Haut, J. (2015). Bottoms up: A comparison of voluntary cybersecurity frameworks. *UC Davis Bus. LJ*, *16*, 217.

Shah, T., & Venkatesan, S. (2018). *Authentication of IoT Device and IoT Server Using Secure Vaults*. Paper presented at the 2018 17th IEEE International Conference On Trust, Security And Privacy In Computing And Communications/12th IEEE International Conference On Big Data Science And Engineering (TrustCom/BigDataSE). 10.1109/TrustCom/BigDataSE.2018.00117

Shah, A., & Ahn, B. (2006). Intelligent Transportation Systems: Is it a Compatible Tool for Developing Countries? *Journal of Advanced Transportation*, *40*(3), 289–294. doi:10.1002/atr.5670400304

Shahghasemi, E., Tafazzoli, B., Akhavan, M., Mirani, G., & Khairkhah, T. (2013). *Electronic Government in Iran: A Case Study*. Online Journal of Social Sciences Research.

Shahgholian, M., & Gharavian, D. (2018). *Advanced Traffic Management Systems: An Overview and A Development Strategy*. Available: https://arxiv.org/abs/1810.02530

Shahid, N., & Aneja, S. (2017). Internet of Things: Vision, application areas and research challenges. *Proceedings of the International Conference on IoT in Social, Mobile, Analytics and Cloud, I-SMAC 2017*. 10.1109/I-SMAC.2017.8058246

Sha, K., Wei, W., Yang, T. A., Wang, Z., & Shi, W. (2018). On security challenges and open issues in Internet of Things. *Future Generation Computer Systems*, *83*, 326–337. doi:10.1016/j.future.2018.01.059

Shapiro, J. M. (2006). Smart cities: Quality of life, productivity, and the growth effects of human capital. *The Review of Economics and Statistics*, *88*(2), 324–335. doi:10.1162/rest.88.2.324

Shariatmadari, H., Ratasuk, R., Iraji, S., Laya, A., Taleb, T., Jäntti, R., & Ghosh, A. (2015). Machine-Type Communications: Current Status and Future Perspectives toward 5G Systems. *IEEE Comm.*, *53*(9), 10–17. doi:10.1109/MCOM.2015.7263367

Shi, Z. H. Z. (2009). Effects of Packet Loss and Latency on Temporal Discrimination of VisualHaptic Events. IEEE Trans. on Haptics, 1(1).

Silva, B. N., Khan, M., & Han, K. (2018). Towards sustainable smart cities: A review of trends, architectures, components, and open challenges in smart cities. *Sustainable Cities and Society*, *38*, 697–713. doi:10.1016/j.scs.2018.01.053

Sime Darby. (2016). *Innovating For The Future, Sime Darby Annual Report 2016*. Retrieved from http://www.simedarby.com/sites/default/files/annualreport-pdf/sime-darby-annual-report-2016.pdf

Simmons, R. (1998). Xavier: An autonomous mobile robot on the web. *Proc. IROS Workshop Robots on the Web.*

Simon, S. (2016). Mobile instant messaging support for teaching and learning in higher education. *The internet and higher education*, 31.

Simsek, A. A. (2016). The 5G-Enabled Tactile Internet: Applications, Requirements, and Architecture. *IEEE Wireless Communications and Networking Conference WS 6: IEEE WCNC.*

Simsek, M., Aijaz, A., Dohler, M., Sachs, J., & Fettweis, G. (2016). The 5G-Enabled Tactile Internet: Applications, Requirements, and Architecture. *IEEE Wireless Communications and Networking Conference WS 6: IEEE WCNC'2016 Workshop on 5G & Vertical Industry.*

Singh, A. S. (2017). Common procedures for development, validity and reliability of a questionnaire. *International Journal of Economics, Commerce and Management,* (5). Retrieved from http://ijecm.co.uk/wp-content/uploads/2017/05/5549.pdf

Skill International Pvt Ltd. (2010). *Outcome Evaluation Report of Nenasala Project ICTA Sri Lanka.* Available at: http://www.icta.lk/index.php?option=com_phocadownload&view=category&download=17:nenasala-outcome-evaluation-final-report-october-2010&id=2:reports&Itemid=270&lang=en

SME Corporation Malaysia. (2017). *2016/2017 SME Annual Report* (ISSN 1823-8734). Kuala Lumpur: National SME Development Council (NSDC).

SME Corporation Malaysia. (2018a). *SME Annual Report 2017/2018:SME Developments and Outlook.* Retrieved from http://www.smecorp.gov.my/index.php/en/laporan-tahunan/3342-laporan-tahunan-pks-2017-18

SME Corporation Malaysia. (2018b). *SME Masterplan (2012-2020).* Retrieved from www.smeinfo.com.my/policy-regulation/sme-masterplan

Smith, A., Papadaki, M., & Furnell, S. M. (2009, July). Improving awareness of social engineering attacks. In *IFIP World Conference on Information Security Education* (pp. 249-256). Springer.

Sodiya, A., Onashoga, S., & Adelani, D. I. (2011). Secure E-Voting Architecture. *Proceedings of Eighth International Conference on Information Technology: New Generations*, 342-347.

Solanas, A., Patsakis, C., Conti, M., Vlachos, I. S., Ramos, V., Falcone, F., ... Perrea, D. N. (2014). Smart health: A context-aware health paradigm within smart cities. *IEEE Communications Magazine*, *52*(8), 74–81. doi:10.1109/MCOM.2014.6871673

Song, H. J. (2004). *Building E-Governance through Reform* (Vol. 2). Ewha Womans University Press.

Sookhak, M., Tang, H., He, Y., & Yu, F. R. (2018). Security and privacy of smart cities: A survey, research issues and challenges. *IEEE Communications Surveys and Tutorials*, *21*(2), 1718–1743. doi:10.1109/COMST.2018.2867288

Sridhar, S., & Smys, S. (2017). *Intelligent security framework for iot devices cryptography based end-to-end security architecture.* Paper presented at the Inventive Systems and Control (ICISC), 2017 International Conference on. 10.1109/ICISC.2017.8068718

Srivastava, A. (2014). Basics in Education - Textbook for B. Ed. Course. National Council of Educational Research and Training.

Stanney, K. M., Graeber, D. A., & Kennedy, R. S. (2006). Virtual environment usage protocols. Handbook of standards and guidelines in ergonomics and human factors, 381-397.

Stefan, A., Stefan, A., Constantin, D., Mateescu, C., & Cartal, L. A. (2015). Aspects of kinematics and dynamics for payload UAVs. In *2015 7th International Conference on Electronics, Computers and Artificial Intelligence (ECAI)* (pp. WR-1). 10.1109/ECAI.2015.7301228

Steins, C. (2002, Sept.). E-government: the top ten technologies. *Planning.*

Stergiou, C., Psannis, K. E., Gupta, B. B., & Ishibashi, Y. (2018). Security, Privacy & Efficiency of Sustainable Cloud Computing for Big Data & IoT. *Sustainable Computing: Informatics and Systems.*

Stern, J. (2004). *Introduction to Online Teaching and Learning.* wlac.education.

Sterne, J. (2010). *Social media metrics: How to measure and optimize your marketing investment.* John Wiley & Sons.

Stockdale, R., Ahmed, A., & Scheepers, H. (2012). Identifying business value from the use of social media: An SME perspective. *PACIS 2012 Proceedings.* Retrieved from https://pdfs.semanticscholar.org/902b/4a16bb0f6364c77f1f7c06ed6e490cc1ed11.pdf

Stoddart, K. (2016). UK cyber security and critical national infrastructure protection. *International Affairs, 92*(5), 1079–1105. doi:10.1111/1468-2346.12706

Suciu, G., Vulpe, A., Halunga, S., Fratu, O., Todoran, G., & Suciu, V. (2013). Smart cities built on resilient cloud computing and secure internet of things. *2013 19th International Conference on Control Systems and Computer Science*, 513-518.

Sujata, M., & Banshidhar, M. (2010). A Secure Multi Authority Electronic Voting Protocol based on Blind Signature. *Proceedings of International Conference on Advances in Computer Engineering*, 271-273.

Sun, H., Wu, J., Wang, W., & Gao, Z. (2014). Reliability-based traffic network design with advanced traveller information systems. *Information Sciences, 287*, 121–130. doi:10.1016/j.ins.2014.07.036

Sussman, J. M. (2005). *Perspectives on Intelligent Transportation Systems.* Springer.

Sutherland, E. (2017). Governance of cybersecurity-The case of South Africa. *African Journal of Information and Communication, 20*(20), 83–112. doi:10.23962/10539/23574

Sutherland, E. (2018). *Digital Privacy in Africa: Cybersecurity.* Data Protection & Surveillance.

Tagliaferri, L. (2017). *An Introduction to Machine Learning.* DigitalOcean.

Talasila, M., Curtmola, R., & Borcea, C. (n.d.). *Mobile Crowd Sensing.* New Jersey Institute of Technology. Retrieved from https://web.njit.edu/~mt57/publications/Chapter4.pdf

Tambouris, E., Gorilas, S., & Boukis, G. (2001). *Investigation of Electronic Government.* E-gov Project. Retrieved from http://academic.research.microsoft.com/Publication/104353/investigationof- electronic-government/

Tan, T. C. (2015). *Building Information Modeling and Intelligent Green Buidings.* Commscope White Paper.

Tandon, G. (2018). *Piggy Backing & Tail Gating.* Security.

TAP. (2017). *Tunisie digitale 2020: seuls 5% des projets programmés ont été réalisés.* Retrieved from https://www.huffpostmaghreb.com/2017/02/16/tunisie-digitale_n_14796854.html

Taubman, G. (1998). A not-so World Wide Web: The Internet, China, and the challenges to nondemocratic rule. *Political Communication, 15*(2), 255–272. doi:10.1080/105846098093423 69

Tewari, A., & Gupta, B. (2018). Security, privacy and trust of different layers in Internet-of-Things (IoTs) framework. *Future Generation Computer Systems.* doi:10.1016/j.future.2018.04.027

Tewolde, G. S. (2012). Sensor and network technology for intelligent transportation systems. *2012 IEEE International Conference on Electro/Information Technology*, 1-7. 10.1109/EIT.2012.6220735

Thaler, R. (2016). Behavioral Economics: Past, Present, and Future. *American Economic Association*, *106*(7), 1577–1600.

The Borneo Post. (2016). *Brunei Times shuts down today after ten years*. Retrieved from https://www.theborneopost.com/2016/11/08/brunei-times-shuts-down-today-after-ten-years/

The Building Blocks of Digital Literacy. (2015). *Digital Literacy 21st Century Competences for Our Age*. Department of eLearning.

The World Bank. (2016). *Digital Dividends*. Washington, DC: The World Bank.

Thiem, D. B., Neupert, C., Bilz, J., Matich, S., Polzin, J., Werthschutzky, R., & Hatzfeld, C. (2017). User-interface for teleoperation with mixed-signal haptic feedback. *IEEE International Conference on Intelligent Robots and Systems*. 10.1109/IROS.2017.8202251

Thomas, K. (2014). *10 steps to salvation: Creating digital governance that works*. Confab Europe Barcelona.

Thomas, N. (2009). Cybersecurity in East Asia: Governing anarchy. *Asian Security*, *5*(1), 3–23. doi:10.1080/14799850802611446

Tian, J., Li, H. & Chen, R. (2018). Individual Behavior Change under Smart City Environment—A Proposal of Smart Citizen Concept with Four Dimensions. *iConference 2018 Proceedings*.

Tohari, A., Jainkun, H., & Song, H. (2009). An Efficient Mobile Voting System Security Scheme based on Elliptic Curve Cryptography. *Proceedings of Third International Conference on Network and System Security*, 474-479.

Tohidi, H. (2011). E-government and its different dimensions: Iran. *Procedia Computer Science*, *3*, 1101–1105. doi:10.1016/j.procs.2010.12.179

Tokuhama-Espinosa, T. (2017). *Elegant Complexity: The Theory of the Five Pillars of Neuoconstructivism in the Brain*. Harvard University Extension School, Latin American Faculty for Social Science Research.

Tribunnews. (2019). Retrieved from https://www.tribunnews.com/images/regional/view/1640382/layar-informasi-busway

Triki, T., & Faye, I. (2013). Financial inclusion in Africa: An overview. African Development Bank, 148.

Tsuchiya, M. (2016). Cyber security of financial sectors in Japan, South Korea and China. In *Managing Cyber Risk in the Financial Sector* (pp. 112–127). Routledge. doi:10.4324/9781315675930-5

Twizeyimana, J. D., & Andersson, A. (2019). The public value of E-Government - A literature review. *Government Information Quarterly*, *36*(2), 167–178. doi:10.1016/j.giq.2019.01.001

U.S. Energy Information Administration (EIA). (2019). Retrieved 18 July 2019, from https://www.eia.gov/

UN. (2017). Retrieved from https://www.un.org/development/desa/en/news/population/world-population-prospects-2017.html

UNDP. (2016). *Citizen Engagement in Public Service Delivery The Critical Role of Public Officials*. Author.

Un, E. (2008). *Government Survey 2008: From E-Government to Connected Governance/Dep. of economic and Social Affairs Div. for Publ. Administration and Development Management*. New York: United Nations.

United Nations E-Government Survey. (2008). *From e-Government to Connected. Governance*. Author.

United Nations E-Government Survey. (2012). *E-Government for the People*. Author.

United Nations E-government Survey. (2014). E-*government for the future we want*. Author.

United Nations E-government Survey. (2016). *E-government in support of sustainable development*. Author.

United Nations E-government Survey. (2018). *Gearing e-government to support Transformation towards Sustainable and resilient societies*. Author.

United Nations. (2002). *Benchmarking e-government: A Global perspective, assessing the progress of the UN member states*. Retrieved from www.golconferenceca/presentations/eGovernment_UN.pdf

United Nations. (2003). *World Public Sector Report 2003: E-Government at the Crossroads*. New York: UN.

United Nations. (2004). *Global e-government readiness report 2004: Towards access for Opportunity*. New York: UN. Retrieved from http://unpan1.un.org/intradoc/groups/public/documents/un/unpan019207.pdf

United Nations. (2018). *The 2018 United Nations E-Government Survey: Gearing e-government to support transformation towards sustainable and resilient societies*. Author.

Upadhyay, M. H. (2015). E-Courts in India and E-Judiciary in India. *International Multidisciplinary Research Journal, 7637*(7), 2–5.

Ursamajor. (2015). Retrieved from https://ursamajor.its-platform.eu/highlights/new-radar-traffic-sensor-system-along-a31-italy-ne

US News Deserts. (2018). *The Expanding News Desert*. Retrieved from https://www.usnewsdeserts.com/reports/expanding-news-desert/

Vaa, T., Penttinen, M., & Spyropoulou, I. (2007). Intelligent transport systems and effects on road traffic accidents: State of the art. *IET Intelligent Transport Systems*, *1*(2), 81–88. doi:10.1049/iet-its:20060081

Vaizey, E., & Boles, N. (2016). *Review of publicly funded digital skills qualifications*. Skills Funding Agency.

Valentine, J., & Holloway, S. (2001). A window on the wider world? Rural children's use of information and communication technologies. *Journal of Rural Studies*, *17*(4), 383–394. doi:10.1016/S0743-0167(01)00022-5

Valle-Cruz, D. (2019). Public value of e-government services through emerging technologies. *International Journal of Public Sector Management*, *32*(5), 530–545. doi:10.1108/IJPSM-03-2018-0072

Van Den Berg, D., Glans, R., De Koning, D., Kuipers, F. A., Lugtenburg, J., & Polachan, K. … Van Wijk, B. (2017). Challenges in haptic communications over the tactile internet. *IEEE Access: Practical Innovations, Open Solutions*. doi:10.1109/ACCESS.2017.2764181

Van Winden, W., & Van Den Buuse, D. (2017). Smart city pilot projects: Exploring the dimensions and conditions of scaling up. *Journal of Urban Technology*, *24*(4), 51–72. doi:10.1080/10630732.2017.1348884

Veeckman, C., Schuurman, D., Leminen, S., & Westerlund, M. (2013). *Linking Living Lab Characteristics and Their Outcomes: Towards a Conceptual Framework*. Technology Innovation Management Review.

Veillette, M., van der Stok, P., Pelov, A., & Bierman, A. (2018). CoAP Management Interface. Internet Engineering Task Force.

Ventre, D. (2014). *Chinese cybersecurity and defense*. John Wiley & Sons.

Vergelis, M., Shcherbakova, T., & Sidorina, T. (2019, March 12). *Spam and phishing in 2018*. Retrieved from https://securelist.com/spam-and-phishing-in-2018/89701/

Vertanews. (2019). Retrieved from http://www.vertanews.id/homepage/read/9014/Integritas+Transportasi+Antarmoda+Jakarta+Menuju+Kota+Global

Vianna, M. M. B., da Silva Portugal, L., & Balassiano, R. (2004). Intelligent transportation systems and parking management: Implementation potential in a Brazilian city. *Cities (London, England)*, *21*(2), 137–148. doi:10.1016/j.cities.2004.01.001

Viitanen, J., & Kingston, R. (2014). Smart cities and green growth: Outsourcing democratic and environmental resilience to the global technology sector. *Environment & Planning A*, *46*(4), 803–819. doi:10.1068/a46242

Vijayalakshmi, M. (2016). *Information and Communication Technology In Education.* Academic Press.

Visvizi, A., Lytras, M. D., Damiani, E., & Mathkour, H. (2018). Policy making for smart cities: Innovation and social inclusive economic growth for sustainability. *Journal of Science and Technology Policy Management*, *9*(2), 126–133. doi:10.1108/JSTPM-07-2018-079

Vivek, S. D., Beatty, S. E., Dalela, V., & Morgan, R. M. (2014). A generalized multidimensional scale for measuring customer engagement. *Journal of Marketing Theory and Practice*, *22*(4), 401–420. doi:10.2753/MTP1069-6679220404

von Heinegg, W. H. (2012). *Legal implications of territorial sovereignty in cyberspace.* Paper presented at the 2012 4th International Conference on Cyber Conflict (CYCON 2012).

Vorakulpipat, C., Siwamogsatham, S., Kamolsook, A., & Jamchudjai, P. (2010). *An Empirical Study of IT Adoption in Rural Areas of Thailand.* IEEE.

Wahlström, J., Skog, I., & Handel, P. (2017). Smartphone-based vehicle. *IEEE Transactions on Intelligent Transportation Systems*, *18*(10), 2802–2825. doi:10.1109/TITS.2017.2680468

Walker, L.E. (2016). *Deception of Phishing: Studying the Techniques of Social Engineering by Analyzing Modern-day Phishing Attacks on Universities*. Academic Press.

Wang, Z., & Kim, H. G. (2017). *Can Social Media Marketing Improve Customer Relationship Capabilities and Firm Performance*. Dynamic Capability Perspective. doi:10.1016/j.intmar.2017.02.004

Watanabe, C., Moriya, K., Tou, Y., & Neittaanmäki, P. (2018a). Consequences of the digital economy: Transformation of the growth concept. *International Journal of Managing Information Technology*, *10*(2), 21–39. doi:10.5121/ijmit.2018.10202

Watanabe, C., Moriya, K., Tou, Y., & Neittaanmäki, P. (2018b). Structural Sources of a Productivity Decline in the Digital Economy. *International Journal of Managing Information Technology*, *10*(1), 1–20. doi:10.5121/ijmit.2018.10101

Watanabe, C., Naveed, K., Tou, Y., & Neittaanmäki, P. (2018c). Measuring GDP in the digital economy: Increasing dependence on uncaptured GDP. *Technological Forecasting and Social Change*, *137*, 226–240. doi:10.1016/j.techfore.2018.07.053

Watts, A. C., Ambrosia, V. G., & Hinkley, E. A. (2012). Unmanned aircraft systems in remote sensing and scientific research: Classification and considerations of use. *Remote Sensing*, *4*(6), 1671–1692. doi:10.3390/rs4061671

We Are Social. (2018). *Global digital report 2018.* Retrieved from https://wearesocial. com/blog/2018/01/global-digital-report-2018

We are WWF Woking, Surrey. (2019). Retrieved 18 July 2019, from https://www.wwf.org.uk/

Weiner, M., Jorgovanovic, M., & Sahai, A. (2014). Design of a Low-Latency, High-Reliability Wireless Communication System for Control Applications. *2014 IEEE International Conference on Communications (ICC)*. 10.1109/ICC.2014.6883918

Welchman. (2017). Understanding Digital Governance Maturity. Academic Press.

What does Saudi Arabia's mega project 'NEOM' actually stand for? (2017, October 24). *Al Arbiya*. Retrieved October 2, 2019 from http://english.alarabiya.net/en/business/economy/2017/10/24/What-does-NEOM-mean-.html

Wiesel, T., Pauwels, K., & Arts, J. (2011). Practice prize paper—Marketing's profit impact: Quantifying online and off-line funnel progression. *Marketing Science*, *30*(4), 604–611. doi:10.1287/mksc.1100.0612

Williams, S. E. (2014). *Education and Economic Development: An Untapped Alliance*. The University of Southern Mississippi, The Aquila Digital Community.

Willner, A. (2018). The Industrial Internet of Things. Internet of Things A to Z. doi:10.1002/9781119456735.ch11

Windmill. (2019). Retrieved from https://www.windmill.co.uk/vehicle-sensing.html

Winters, J. V. (2011). Why are smart cities growing? Who moves and who stays. *Journal of Regional Science*, *51*(2), 253–270. doi:10.1111/j.1467-9787.2010.00693.x

Wojciech, B. (2017). Digital Era Governance–a new chapter of public management theory and practice. *MAZOWSZE Studia Regionalne*, (22), 117-129.

World Bank. (2002). *E-Governments*. Retrieved from: http://documents.worldbank.org/curated/en/527061468769894044/266390WP0E1Gov1gentina1Final1Report.doc#:~:targetText=E%2Dgovernment%20refers%20to%20the,and%20other%20arms%20of%20government

World Bank. (2011). *Sri Lanka - Education in Sri Lanka*. Available at: http://www.worldbank.lk/WBSITE/EXTERNAL/COUNTRIES/SOUTHASIAEXT/SRILANKAEXTN/0,contentMDK:21635561~pagePK:141137~piPK:141127~theSitePK:233047,00.html

World Economic Forum. (2014). *Network Readiness Index*. World Economic Forum.

Wright, L. (2001, May). Latin American busways: Moving people rather than cars. *Natural Resources Forum*, *25*(2), 121–134. doi:10.1111/j.1477-8947.2001.tb00754.x

Xiang, W., Gozalvez, J., Niu, N., Altintas, O., & Ekici, E. (2009). Wireless Access in Vehicular Environments. *EURASIP Journal on Wireless Communications and Networking*, *2009*(1), 576217. doi:10.1155/2009/576217

Xinhua Net. (2018). *Brunei introduces hotline to address fake news on social media*. Retrieved from http://www.xinhuanet.com/english/2018-09/25/c_137492233.htm

Xinhua Net. (2019). *Brunei sets up Digital Economy Council to explore economic potential*. Retrieved from http://www.xinhuanet.com/english/2019-07/15/c_138229056.htm

Xiong, Z., Sheng, H., Rong, W., & Cooper, D. E. (2012). Intelligent transportation systems for smart cities: A progress review. *Science China. Information Sciences*, *55*(12), 2908–2914. doi:10.100711432-012-4725-1

Xu, Y. (2010). E-government and Governance. In *China*. Zhejiang Gongshang University School of Public Hangzhou P.R. China.

Yang, K., & Callahan, K. (2007). Citizen involvement efforts and bureaucratic responsiveness: Participatory values, stakeholder pressures, and administrative practicality. *Public Administration Review*, *67*(2), 249–264. doi:10.1111/j.1540-6210.2007.00711.x

Yang, K., Li, Q., & Sun, L. (2018). Towards Automatic Fingerprinting of IoT Devices in the Cyberspace. *Computer Networks*.

Yaqoob, I., Hashem, I. A. T., Mehmood, Y., Gani, A., Mokhtar, S., & Guizani, S. (2017). Enabling communication technologies for smart cities. *IEEE Communications Magazine*, *55*(1), 112–120. doi:10.1109/MCOM.2017.1600232CM

Yuen, M. K. (2017, November 2). Don't fall prey to identity thieves. *The Star*, 12. Available https://www.thestar.com.my/news/nation /2017/11/02/dont-fall-prey-to-identity-thieves-rising -cases-of-criminals-stealing-personal-details-and-exploitin/

Yu, T., Wang, X., & Shami, A. (2018). UAV-Enabled Spatial Data Sampling in Large-Scale IoT Systems Using Denoising Autoencoder Neural Network. *IEEE Internet of Things Journal*, *6*(2), 1856–1865. doi:10.1109/JIOT.2018.2876695

Zamora, Barahona, & Palaco. (2016). Case: Digital Governance Office. *Journal of Business Research*, *69*(10).

Zander, S., Nguyen, T., & Armitage, G. (2005, November). Automated traffic classification and application identification using machine learning. In *The IEEE Conference on Local Computer Networks 30th Anniversary (LCN'05) l* (pp. 250-257). IEEE.

Zanella, A., Bui, N., Castellani, A., Vangelista, L. & Zorzi, M. (2014). Internet of things for smart cities. *IEEE Internet of Things Journal, 1*, 22-32.

Zhang, Q., Liu, J., & Zhao, G. (2018). *Towards 5G enabled tactile robotic telesurgery*. arXiv preprint arXiv:1803.03586

Zheng, L., Kwok, W.-M., Aquaro, V., & Qi, X. (2019). Digital Government, Smart Cities and Sustainable Development. *Proceedings of the 12th International Conference on Theory and Practice of Electronic Governance*, 291-301. 10.1145/3326365.3326403

Zhou, X. (2004). E-government in China: A content analysis of national and provincial web sites. *Journal of Computer-Mediated Communication, 9*(4).

Zhu, F., Li, Z., Chen, S., & Xiong, G. (2016). Parallel transportation management and control system and its applications in building smart cities. *IEEE Transactions on Intelligent Transportation Systems*, *17*(6), 1576–1585. doi:10.1109/TITS.2015.2506156

Zwass, V. (2006). The web-internet compound as the infrastructure of digital government. *Business Process Management Journal*, *12*(1), 7–12. doi:10.1108/14637150610643715

# About the Contributors

**Vasaki Ponnusamy** is an Assistant Professor at Universiti Tunku Abdul Rahman Malaysia. She obtained her Bachelor of Computer Science and MSc (Computer Science) from Science University of Malaysia and her PhD in IT from Universiti Teknologi PETRONAS (UTP), Malaysia (2013). She is currently working on cybersecurity, digital governance, biologically-inspired computing, wireless sensor network and energy harvesting.

**Khalid Rafique** is the Director General, Information Technology Board, Azad Jammu & Kashmir, Pakistan since April 2017. He has a lead role in the Digital Transformation campaign for public sector departments in Azad Jammu & Kashmir and unfolding of digital governance vision of the government through technology deployment. He has quite dynamically steered initiatives to materialize vision of public empowerment and facilitation through digitalization of public sector departments. Automation of Judicial system, Computerization of land record, Computerization of driving license, E-facilitation centers, Establishment of Tele-medicine centers, Online complaint addressal mechanism, dynamic interactive web interfaces for public sector departments, telepresence/video conferencing facilities etc are some of his major achievements in the last two and a half year. Dr. Khalid has also served as Senior Manager Business Development at Pakistan Telecommunication Limited- the largest hybrid (Fixed/ wireless) Telecom operator in Pakistan. He has basic Engineering degrees in Electrical & Telecommunications and a PhD degree in Emerging ICT ecosystem.

**Noor Zaman** received the Ph.D. degree in IT from UTP, Malaysia. He has great international exposure in academia, research, administration, and academic quality accreditation. He was with ILMA University, King Faisal University (KFU), Saudi Arabia for a decade, and currently with Taylor's University, Malaysia. He has 19 years of teaching & administrative experience. He has an intensive background of academic quality accreditation in higher education besides scientific research activities, he had worked a decade for academic accreditation and earned ABET accreditation twice for three programs at CCSIT, King Faisal University, Saudi Arabia. Dr. Noor Zaman has awarded as top reviewer 1% globally by WoS/ ISI (Publons) recently. He has edited/authored more than 11 research books with international reputed publishers, earned several research grants, and a great number of indexed research articles on his credit. He has supervised several postgraduate students including masters and Ph.D. Dr. Jhanjhi is an Associate Editor of IEEE ACCESS, Guest editor of several reputed journals, member of the editorial board of several research journals, Keynote Speaker for several IEEE international conferences around the globe, and active TPC member of reputed conferences around the globe.

* * *

**Muneer Ahmad** completed his PhD from Universiti Teknologi PETRONAS Malaysia in 2014 and currently serving as a faculty member in Department of Information Systems, Faculty of Computer Science & Information Technology, University of Malaya, Kuala Lumpur Malaysia. His research interests include signal processing, big data analysis, IoT and cloud computing.

**Malak Alamri** received her master's degree from King's College London, UK in 2014. She did her bachelor's degree in computer science from Taibah University in 2007. She has eight years of teaching and administrative experience. She is an active teacher and researcher. Currently, she is a PhD student at University of Southampton and her research interests include software engineering, Computer Vision and Mobile Applications.

**Robithoh Annur** is currently an Assistant professor in the Department of Computer and Communication Technology (DCCT), Faculty of Information and Communication Technology (FICT), UTAR. She received the B. Eng. and M. Eng. degrees from Gadjah Mada University, Yogyakarta, Indonesia and National University of Singapore, Singapore, respectively. She was a tutor in the Diploma of Electrical engineering in Gadjah Mada University before eventually completing her doctoral degree in Chulalongkorn University, Thailand. She has already published a series of papers both in international conferences and journals that some of the acknowledged as the best paper award, with a book chapter in a book entitled "Advanced Trends in Wireless Communications, 2011. Currently, she is focusing on the algorithms for tag identification in RFID system, which is simple to implement and performs better than existing known algorithms.

**Tahniyat Aslam** holds M.E Degree in "Telecommunication Engineering" from NED University of Engineering and Technology. She has done B.S in "Telecommunication Engineering" from Sir Syed University of Engineering and Technology. She has worked as a Visiting Teacher in NED University of Engineering and Technology. Currently, she is working as a Junior Lecturer in Electrical Engineering Department at Institute of Business Management. Her area of interest is Antenna and Microwave Systems and Network Security.

**Fatma Bouaziz** is assistant professor in Management at the Faculty of Economics and Management of Sfax - Tunisia. She got her Ph.D. in management in 2011 and her HDR in 2013 from the same faculty. Dr. Bouaziz is a member of the research laboratory "Perspectives and Research in Innovation, Strategy, and Business Management (LR18ES24). Her doctoral thesis deals with the risk management of e-government projects in the case of the Tunisian government. Her research interests include the use of ICTs by government, their organizational impacts and abilities to improve public services delivery, management of human resources and organizational resilience in the context of transition.

**Kennedy D. Gunawardena** is Professor of Accounting Information Systems at the department of Accounting, University of Sri Jayewardenepura, Sri Lanka. In addition to being the professor of Accounting in his department, He is the chairman of the Board of studies and Coordinator of the PhD in management program of Management, faculty of Management studies and commerce in the same university. He teaches Cost and Management Accounting, Accounting Information Systems, Artificial Neural Networks for Accounting and Research Methodology in Accounting and finance for the undergraduates and Postgraduates students.

**Bushra Hamid** is doing PhD in Computing from Riphah International University, Islamabad Pakistan. She has 10 years of teaching experience. Her research interest include Software Quality Engineering, Software Requirements Engineering, Software Engineering Education and Software Testing.

**Mamoona Humayun** has completed her PhD. in Computer Architecture from Harbin Institute of Technology, China. She has 12 years of teaching and administrative experience internationally. She is an active reviewer for a series of journals. She has supervised Various Masters and PhD thesis. Her research interests include Global software development, requirement engineering, knowledge management, Cyber security and wireless sensor networks.

**Souhaila Kammoun** is an assistant-Professor at IHEC of Sfax and researcher at CODECI, University of Sfax, Tunisia. She got her Phd in Economics at the University of Social Sciences of Toulouse 1, France. She has taught Economics in France and Tunisia. She was member of the PAQ-ES project committee, director of the Master's degree in Entrepreneurship and International Development (Tempus project funded by the European Union) and member of the scientific council at IHEC of Sfax. She has held the positions of Head of the Department of Economics, Director of Studies and Vice-Director. She was also Director of IHEC of Sfax, Tunisia.

**Azeem Khan** is working as a Lecturer in the American Degree Transfer Program, School of Liberal Arts and Sciences, Taylors University Sdn Bhd, Selangor, Malaysia. He has his Master's Degree from Kakatiya University, India and Ph.D. from the University of Malaya, Malaysia which is the top university of Malaysia and ranked among the best 100 universities in the world. Having 10+ years of teaching and 6 years of research experience, he has published more than 15 research articles in reputed ISI (Thompson Reuters) indexed peer-reviewed international journals and conferences in the fields of simulation and computational optimization.

**Muhammad Imran Majid** Muhammad Imran Majid is Head of Electrical Engineering Department, Institute of Business Management, a Chartered Engineer registered with Engineering Council UK and Commonwealth Scholarship Committee Advisory Council Member. He has a PhD in Electronics from University of Surrey, UK, MSc in Space Engineering from Umea University, Sweden and BSc in Electrical Engineering from UET Lahore. After his PhD, Imran consulted and later worked at University of Toronto Canada as Optics Team Lead for two of Canada's most advanced remote sensing satellites, NEMO-AM and NEMO-HD. He is also currently involved with Ontario Centre of Excellence as Technical Reviewer. Imran has about 10 years' experience in industrial consultancy, teaching and mentoring young professionals in high technology sector. He has worked for Telenor, SSBV, Surrey Satellite and various consultancies during the course of his career. He is Co-founder of UN NPO Space Generation Advisory Council where he coordinated activities in the Asia Pacific region from 1999 to 2008. He represented the Space Generation Forum Declaration during the 3rd UN Global Conference on Space, chaired sessions at 1st Ground Station Network Workshop in Tokyo Japan, UN/ESA Sympoias in Austria (2000-2003). Previously he was selected as a Buraq Space Cadet (1996), actively participated in 9th International

Aerospace School Uzbekistan 1997 and pioneered student exchanges at the world's oldest space port at Baikonur Kazakhstan in 1998. Imran also helped found the IET on Surrey Campus, UK for students at the University of Surrey, UK, and IET Karachi, mentored IEng candidates and is currently on the Information Communication and Technology Council IEP Advisory Panel based in Toronto Canada. His research interests are in resource optimization for future wireless systems, engineering goal driven project management and academic industrial partnerships to aid in economic growth. He is currently teaching Electronic Communication Systems, Opto-Electronics for BE and Project Management, Thesis I,2 courses for MS (EM) at IoBM.

**Noraniza Binti Md Jani** is a PhD student in the Faculty of Information and Communication Technology, Universiti Teknikal Malaysia Melaka, Malaysia. Her research interests are focusing on social media measurement and metrics development in online communities.

**Saira Muzafar** acquired her First Masters degree in Applied Physics with specialization in Electronics from Pakistan and Second Masters Degree in Telecommunications from Pakistan. Her areas of interest include Wireless Sensor Network (WSN), Internet of things, Smart Cities, Cloud Computing, Big Data, and Mobile Computing She is currently working as a Faculty member in the College of Computer Science and Information Technology, King Faisal University, in Saudi Arabia. She takes care of versatile operations including teaching and research activities. Contact Information College of Computer Science & Information Technology, ABET Accredited King Faisal University, Al Ahsa, Al Hofuf, Saudi Arabia.

**Abdul Malik Omar** is a Public Policy PhD candidate from the University of Brunei Darussalam. His research will cover Industry 4.0, Smart Cities, Digital Politics, and E-Governance. He received his MSc in Local Economic Development from the London School of Economics in 2016 and a BSc in International Enterprise and Business Development in 2014. He is also the author of Brunei's best-selling study guidebook for students, The Art of Learning (2019, Omar). His writings have appeared in The Diplomat, Investvine, Brunei Times and Borneo Bulletin. He has presented in numerous conferences in top universities, such as Cambridge University and Imperial College London.

**Mahfooz Osama** has earned an MBA degree in Telecommunication Management from Institute of Business Management (IoBM), Karachi and BE in Electronics from PAF-Karachi Institute of Economics & Technology. He has completed a CCNA Exploration certification from Aptech.

**Leslie Mark Pannir Selvam** is an undergraduate student at Universiti Tunku Abdul Rahman, Malaysia. He is specializing in Communication and Networking and his special interest is in Cybersecurity. He is working in Secured Electoral Systems for his undergraduate project.

**Mohd Hafiz Zakaria** is a Senior Lecturer in the Department of Interactive Media, Universiti Teknikal Malaysia Melaka. His primary research interests include social media, assistive tools for people with special needs and e-learning. Throughout his career he has been interested in different aspects of how technology can break barriers to social change and interactions. Hafiz also serves local and international companies as Media Consultant, besides tutoring IT related subjects for local communities. He continues to give talk about student development, digital natives and inclusive education.

# Index

## F



## H



## I



## K



## L



## M



## N



## O



## P



## Q



## R

## S



## T



## U



## V

www.ingramcontent.com/pod-product-compliance
Lightning Source LLC
LaVergne TN
LVHW060405300926
849162LV00006B/1345
* 9 7 8 1 7 9 9 8 1 8 5 1 9 *